AS/A Level course structure

This book has been written to support students studying for AQA
A Level Biology. The sections covered are shown in the contents list,
which also shows you the page numbers for the main topics within
each section. There is also an index at the back to help you find what
you are looking for. If you are studying for AS Biology, you will only
need to know the content in the blue box for the AS exams.

AS exam

Year 1 content

1 Biological molecules
2 Cells
3 Organisms exchange
   substances with their
   environment
4 Genetic information, variation,
   and relationships between
   organisms

Year 2 content

5 Energy transfers in and
   between organisms
6 Organisms respond to changes
   in their internal and external
   environment
7 Genetics, populations, evolution,
   and ecosystems
8 The control of gene expression

A level exam

A Level exams will cover content from Year 1 and Year 2 and will be at a
higher demand. You will also carry out practical activities throughout
your course. There are **twelve** required practicals: six from the AS and
six A-Level.

Contents

How to use t

Learning objective

→ At the beginning o
   topic, there is ~o the
   learning ob

→ These a...and allow you
   spe...tor your progress.

→ ...specification reference is
   also included.

Specification reference: 3.1.1

...ook contains many different features. Each feature is designed to ...ster and stimulate your interest in Biology, as well as supporting and developing the skills you will need for your examinations.

Terms that you will need to be able to define and understand are shown in **bold type** within the text.

Where terms are not explained within the same topic, they are highlighted in **bold orange text**. You can look these words up in the glossary.

Synoptic link

These highlight how the sections relate to each other. Linking different areas of biology together becomes increasingly important, and you will need to be able to do this.

There are also links to the mathematical skills on the specification. More detail can be found in the maths section.

Application features

These features contain important and interesting applications of biology in order to emphasise how scientists and engineers have used their scientific knowledge and understanding to develop new applications and technologies. There are also application features to develop your maths skills, with the icon $\sqrt{x}$, and to develop your practical skills, with the icon 🧪.

Extension features

These features contain material that is beyond the specification designed to stretch and provide you with a broader knowledge and understanding and lead the way into the types of thinking and areas you might study in further education. As such, neither the detail nor the depth of questioning will be required for the examinations. But this book is about more than getting through the examinations.

1   Extension and application features have questions that link the material with concepts that are covered in the specification. Answers can be found in the answers section at the back of the book.

Study tips

Study tips contain prompts to help you with your revision.

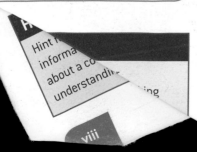
Summary questions

1   These are short questions that test your understanding of the topic and allow you to apply the knowledge and skills you have aquired. The questions are ramped in order of difficulty.

2   $\sqrt{x}$ Questions that will test and develop your mathematical and practical skills are labelled with the mathematical symbol ($\sqrt{x}$) and the practical symbol (🧪).

Section 8

Introduction at the opening of each section summarises what you need to know.

...pression

...technology

...ion

...el, control of metabolic activities is achieved by regulating
...the genome are transcribed and translated, and when
... Although the cells within an organism carry the same
...slate only part of them. In multicellular organisms, this
control of translation enables cells to have specialised functions and
to form specific tissues and organs. Cells formed from the zygote are
initially able to differentiate into any type of cell – they are totipotent. As
these cells become specialised they lose the ability to become a different
type of cell. In mature mammals, only a few cells retain the ability to
differentiate into other cells. These are called stem cells.

It has long been known that many factors control the expression of
genes and, thus, the phenotype of organisms. Some of these factors
are external, environmental factors, others are internal factors. What
was not generally disputed was the idea that these environmental
factors are never inherited by the following generation. Only those
processes such as mutations, which caused changes to the nucleotide
base sequence in a DNA molecule could be inherited. This view has
now been challenged by the discovery that environmental factors
can cause heritable changes in gene function without any change
to the base sequence of DNA. This so called epigenetic regulation of
transcription is being recognised as important.

We are increasingly able to control the expression of genes by altering
the epigenetics. This allows us to alter an organism's genomes and
the proteins they produce (proteomes). Along with our ability to
manipulate the transcription and translation of genes, this has opened
up many medical and technological applications. The use of DNA
technology allows us to clone genes for use in medical techniques such
as gene therapy. Other aspects include the use of DNA probes and
DNA hybridisation in the diagnosis and treatment of human diseases,
as well as the use of genetic fingerprinting for medical, forensic, and
breeding purposes.

Working scientifically

In studying this unit there will be opportunities to perform practical
exercises and so develop practical skills.

In performing these exercises you will have the chance to develop
practical skills such as:

- separating biological compounds using electrophoresis
- using microbiological aseptic techniques.

What you already know

The material in this unit is intended to be self-explanatory, but there is certain
information from GCSE that will be useful to your appreciation of this section.
This information includes:

- ○ Different genes control the development of different characteristics of an organism.
- ○ Differences in the characteristics of different individuals of the same kind may
be due to differences in:
  - the genes they have inherited (genetic causes)
  - the conditions in which they have developed (environmental causes)
  - a combination of both of the above.
- ○ The differences between Darwin's theory of evolution and conflicting theories,
such as that of Lamark.
- ○ In genetic engineering, genes from the chromosomes of humans and other
organisms can be 'cut out' using enzymes and transferred to cells of other
organisms.
- ○ Each person (apart from identical twins) has unique DNA. This can be used to
identify individuals in a process known as DNA fingerprinting.
- ○ Embryos can be screened for the alleles that cause genetic disorders.
- ○ Genes can also be transferred to the cells of animals, plants or microorganisms at an
early stage in their development so that they develop with desired characteristics.
- ○ New genes can be transferred to crop plants and crops that have had their genes
modified in this way are called genetically modified crops (GM crops). Examples
of genetically modified crops include ones that are resistant to insect attack or to
herbicides
- ○ Genetically modified crops generally show increased yields.
- ○ Concerns about GM crops include the effect on populations of wild flowers and
insects, and uncertainty about the effects of eating GM crops on human health.
- ○ Interpreting information about cloning techniques and genetic engineering
techniques.
- ○ Making informed judgements about the economic, social and ethical issues
concerning cloning and genetic engineering, including genetically modified
(GM) crops.

498

A checklist to help you assess your knowledge from KS4, before starting work on the section.

Visual summaries of each section show how some of the key concepts of that section interlink with other sections.

Section 7 Genetics, populations, evolution, and ecosystems

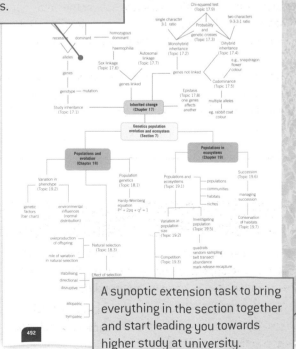

Practical skills

In this section you have met the following
practical skills:

- How to plot growth curves using a
logarithmic scale.
- How to carry out random sampling.
- Investigate the distribution of
organisms in a habitat using randomly
placed framed quadrats or a belt
transect.
- Use the mark-release-recapture
method to investigate the abundance
of a motile species.

Maths skills

In this section you have met the
following maths skills:

- Calculating ratios and percentages
of the offspring of genetic crosses.
- Understanding and calculating the
probability associated with genetic
inheritance.
- Using the chi-squared test to test
the significance of the difference
between observed and expected
results of genetic crosses.
- Solving, and changing the subject
in, algebraic equations such as the
Hardy-Weinberg equation.
- Using a logarithmic scale in relation
to quantities that range over several
orders of magnitude.
- Using the logarithmic function on a
calculator.
- Plotting two variables from
experimental data provided.
- Finding arithmetical means.

Extension task

Using your local newspaper, regional
television news or local community
websites in your area, identify a scheme in
your region designed to conserve a habitat.
Find out the purpose of this scheme and
the organisations involved in it.

Research the various sources of funds
that are available to support conservation
projects like the one you have identified.

Draft a letter to one source of funds
applying for a specified sum of money
to support the aims of your project.
Include in your letter a justification for the
conservation project and the benefits it
will bring to the community. Explain how
the money will be used, how it will further
the aims of the project and how you will
evaluate whether it has been well spent.

A synoptic extension task to bring everything in the section together and start leading you towards higher study at university.

Summaries of the key practical and math skills of the section.

492

ix

Non-SI units

Although examination boards use SI units, you may also encounter non-SI units elsewhere, for example:

- litre (cubic decimetre) (l, L, dm^3) for volume;
- Minute (min) for time;
- hour (h) for time;
- svedberg (S) (for sedimentation rate), used for ribosome particle size.

Unit prefixes

To accommodate the huge range of dimensions in our measurements, they may be further modified using appropriate prefixes. For example, one thousandth of a second is a millisecond (ms). This is illustrated in the Table 1.

▼ Table 1

Division	Factor	Prefix	Length		Mass		Volume		Time	
one thousand millionth	10^{-9}	nano	nanometre	nm	nanogram	ng	nanolitre	nl	nanosecond	ns
one millionth	10^{-6}	micro	micrometre	µm	microgram	µg	microlitre	µl	microsecond	µs
one thousandth	10^{-3}	milli	millimetre	mm	milligram	mg	millilitre	ml/cm^3	millisecond	ms
one hundredth	10^{-2}	centi	centimetre	cm						
whole unit			metre	m	gram	g	litre	$l/L/dm^3$	second	s
one thousand times	10^3	kilo	kilometre	km	kilogram	kg				

Converting between units

You may need to convert between units in order to be able to scale and express numbers in sensible forms. For example, rather than refer to the width of a cell in metres you would use micrometres (µm). This allows your measurements to be understood within the relevant scale of the observation.

Divide by 1000 for each step to convert in this direction →

nano-	micro-	milli-	whole unit	kilo-
e.g. nm	e.g. µm	e.g. mm	e.g. m	e.g. km

← Multiply by 1000 for each step to convert in this direction

▲ Figure 1

Examples:

Convert 1 m to mm: $1 \times 1000 = 1000\,mm$

Convert 1 m to µm: $1 \times 1000 = 1000\,mm$, then $1000 \times 1000 = 1\,000\,000\,µm$

Convert 1 l to cm^3: $1 \times 1000 = 1000\,cm^3$

Convert 20 000 µm to mm: $20\,000 \div 1000 = 20\,mm$

565

Chapter 23 Practical skills

Practical skills are at the heart of Biology. A good foundation [...] take your skills to a higher level. Biology is a dynamic su[...] constantly changing, largely as a result of developments in [...] specification there is a separate practical endorsement w[...] 12 practicals across the two years of the A level course. T[...] as part of your A level qualification. It is assessed by your [...] will be reported separately on your A level certificate. Yo[...] understanding and knowledge of practicals to the written [...] account for 15% of the total assessment – the majority of [...]

By undertaking the set practical activities in this course, [...] manipulative skills with specific apparatus and technique[...] deeper understanding into the processes of scientific inve[...] planning, implementing by making and processing measurements safely, analysing, and evaluating results will be reinforced and enhanced.

It is advantageous for you to answer practical questions when you have completed the practical – any questions on practical skills will have been written with the expectation that you will have carried out the practical activities. Having undertaken the practical, this helps with the teaching and learning of concepts in the specification. A richer practical experience will be gained if you do more practicals than the following twelve set practical activities in Table 1. Table 1 shows the 12 practicals which will be assessed in exams. For the practical endorsement the 12 practicals can consist of the required practicals or teacher devised practicals. For each activity, Table 1 references the relevant topic(s) in this book, the first six you will have already covered in your AS year of study.

▼ Table 1 A level required practical activities

	Practical	Topic
1	Investigation into the effect of a named variable on the rate of an enzyme-controlled reaction	1.8 Factors affecting enzyme action
2	Preparation of prepared squashes of cells from plant root tips; set-up and use of an optical microscope to identify the stages of mitosis in these stained squashes and calculation of a mitotic index	3.1 Methods of studying cells 3.7 Mitosis
3	Production of a dilution series of a solute to produce a calibration curve with which to identify the water potential of plant tissue	4.3 Osmosis
4	Investigation into the effect of a named variable on the permeability of cell-surface membranes	4.1 Structure of the cell-surface membrane
5	Dissection of animal or plant gas exchange systems, a mass transport system or of an organ within such a system	6.2 Gas exchange in insects 6.3 Gas exchange in fish 6.4 Gas exchange in the leaf of a plant 6.6 Mammalian lungs
6	Use of aseptic technique to investigate the effect of antimicrobial substances on microbial growth	9 Genetic diversity and adaptation
7	Use of chromatography to investigate the pigments isolated from leaves of different plants, e.g., leaves from shade-tolerant and shade intolerant plants or leaves of different colours	3.5.1 Photosynthesis
8	Investigation into the effect of a named factor on the rate of dehydrogenase activity in extracts of chloroplasts	3.5.1 Photosynthesis

581

Practice questions: Chapter 15

376

377

Callout boxes:

Mathematical section to support and develop your mathematical skills required for your course. Remember, at least 10% of your exam will involve mathematical skills.

Practical skills section with questions for each suggested practical on the specification. Remember, at least 15% of your exam will be based on practical skills.

Practice questions at the end of each chapter and each section, including questions that cover practical and maths skills. There are also additional practice questions at the end of the book.

Kerboodle

This book is supported by next generation Kerboodle, offering unrivalled digital support for independent study, differentiation, assessment, and the new practical endorsement.

If your school subscribes to Kerboodle, you will also find a wealth of additional resources to help you with your studies and with revision:

- Study guides
- Maths skills boosters and calculation worksheets
- On your marks activities to help you achieve your best
- Practicals and follow up activities to support the practical endorsement
- Interactive objective tests that give question-by-question feedback
- Animations and revision podcasts
- Self-assessment checklists.

Revise with ease using the study guides to guide you through each chapter and direct you towards the resources you need.

If you are a teacher reading this, Kerboodle also has plenty of further assessment resources, answers to the questions in the book, and a digital markbook along with full teacher support for practicals and the worksheets, which include suggestions on how to support and stretch your students. All of the resources are pulled together into teacher guides that suggest a route through each chapter.

Section 1
Biological molecules

Introduction

Biology covers a wide field of information over a considerable size range. On the one hand it involves the movement of electrons in photosynthesis and on the other, the migrations of populations around the Earth. In the same way, living organisms have an extremely diverse range of form and function.

This section explores the fundamental building blocks of these organisms – the molecules of which their cells are composed. Their cells are made up of only a few groups of molecules that react chemically with each other in very similar ways. More importantly, these molecules are all based on carbon.

Examples of biologically important carbon-based molecules that will be explored in this section:

- **Carbohydrates** that are a respiratory substrate from which cells release the energy required to carry out their functions. They also have structural roles in cell walls, and form part of glycoproteins and glycolipids which act as recognition sites in plasma membranes.
- **Lipids** form a major component of plasma membranes. They also make up certain hormones and act as respiratory substrates.
- **Proteins** display a very diverse range of structure and therefore function also. They too are found in plasma membranes but perhaps their most important role is as enzymes. In addition they are chemical messengers within and between cells as well as being important components of the blood, for example antibodies.
- **Nucleic acids** such as deoxyribonucleic acid (DNA) carry genetic information that determines the structure of proteins. Others, like ribonucleic acid (RNA) have a role in the synthesis of these proteins.

A review of biologically important molecules would not be complete without a mention of **water**. It is not a carbon-based molecule but, despite its simplicity, serves a wide range of roles in living organisms. It is the most common component of cells and all life as we know it relies on this simple molecule.

Working scientifically

The study of biological molecules provides many opportunities to carry out practical work and to develop practical skills. A required practical activity is an investigation into the effect of a named variable on the rate of an enzyme-controlled reaction. In carrying out this activity you should look to develop practical skills such as:

- using appropriate instrumentation to record quantitative measurements
- using laboratory glassware apparatus and qualitative reagents to identify biological molecules
- identifying variables that must be controlled and calculating the uncertainty of the measurements you make
- considering margins of error, accuracy and precision of data.

You will require a range of mathematical skills, including the ability to use a calculator's logarithmic functions, to plot two variables from experimental data and draw and use the slope of a tangent to a curve as a measure of rate of change.

What you already know

While the material in this unit is intended to be self-explanatory, there is certain information from GCSE that will prove helpful to the understanding of this section. This information includes:

- ☐ The glucose produced by plants during photosynthesis may be converted into insoluble starch for storage
- ☐ During aerobic respiration chemical reactions occur that use glucose and oxygen to release energy
- ☐ Some glucose in plants and algae is used to produce fat for storage, and cellulose which strengthens the cell wall and proteins
- ☐ Protein molecules are made up of long chains of amino acids. These long chains are folded to produce a specific shape that enables other molecules to fit into the protein
- ☐ Proteins act as structural components of tissues such as muscles, hormones, antibodies and catalysts
- ☐ Catalysts increase the rate of chemical reactions – biological catalysts are proteins called enzymes
- ☐ The shape of an enzyme is vital for its function. High temperatures change the shape of the enzyme
- ☐ Different enzymes work best at different pH values
- ☐ Some enzymes work outside the body's cells

Biological molecules
1.1 Introduction to biological molecules

Biological molecules are particular groups of chemicals that are found in living organisms. Their study is known as **molecular biology**. All molecules, whether biological or not, are made up of units called atoms.

Bonding and the formation of molecules

Atoms may combine with each other in a number of ways:

- **Covalent bonding** – atoms share a pair of electrons in their outer shells. As a result the outer shell of both atoms is filled and a more stable compound, called a molecule, is formed.

- **Ionic bonding** – ions with opposite charges attract one another. This electrostatic attraction is known as an ionic bond. For example, the positively charged sodium ion Na^+ and negatively charged chloride ion Cl^- form an ionic bond to make sodium chloride. Ionic bonds are weaker than covalent bonds.

- **Hydrogen bonding** – the electrons within a molecule are not evenly distributed but tend to spend more time at one position. This region is more negatively charged than the rest of the molecule. A molecule with an uneven distribution of charge is said to be **polarised**, in other words it is a **polar molecule**. The negative region of one polarised molecule and the positively charged region of another attract each other. A weak electrostatic bond is formed between the two. Although each bond is individually weak, they can collectively form important forces that alter the physical properties of molecules. This is especially true for water.

Polymerisation and the formation of macromolecules

Certain molecules, known as **monomers**, can be linked together to form long chains. These long chains of monomer sub-units are called **polymers** and the process by which they are formed is therefore called **polymerisation**. The monomers of a polymer are usually based on carbon. Many, such as polythene and polyesters, are industrially produced. Others, like polysaccharides, polypeptides and polynucleotides, are made naturally by living organisms. The basic sub-unit of a polysaccharide is a monosaccharide or single sugar (Topic 1.3), for example glucose. Polynucleotides are formed from mononucleotide sub-units. Polypeptides are formed by linking together peptides that have amino acids as their basic sub-unit (Topic 1.6).

Condensation and hydrolysis reactions

In the formation of polymers by polymerisation in organisms, each time a new sub-unit is attached a molecule of water is formed. Reactions that produce water in this way are termed **condensation reactions**. Therefore the formation of a polypeptide from amino acids and that of the polysaccharide starch from the monosaccharide glucose are both condensation reactions.

Polymers can be broken down through the addition of water. Water molecules are used when breaking the bonds that link the sub-units of a polymer, thereby splitting the molecule into its constituent parts. This type of reaction is called **hydrolysis** ('hydro' = water; 'lysis' = splitting). Thus polypeptides can be hydrolysed into amino acids, and starch can be hydrolysed into glucose. Figure 1 summarises atomic and molecular organisation.

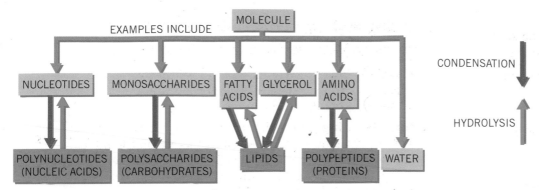

▲ **Figure 1** *Summary of atomic and molecular organisation*

Metabolism

All the chemical processes that take place in living organisms are collectively called metabolism.

The mole and molar solution

The mole is the SI unit for measuring the amount of a substance and is abbreviated to mol

One mole contains the same number of particles as there are in 12 g of carbon-12 atoms (^{12}C). 12 g of carbon-12 atoms contain 6.022×10^{23} carbon atoms. 6.022×10^{23} is called the Avogadro number or Avogadro constant.

A **molar solution** (M) is a solution that contains one mole of solute in each litre of solution. A mole is the molecular mass (molecular weight) expressed as grams (= one gram molecular mass).

As an example, to make a molar solution of sodium chloride we must first find its molecular mass.

The chemical formula for sodium chloride is NaCl, which means that a molecule of sodium chloride contains one sodium atom and one chlorine atom. The atomic weight of sodium (Na) is 23 while that of chlorine (Cl) is 35.5. The molecular mass of NaCl is therefore Na (23) + Cl (35.5) = NaCl (58.5).

Therefore, a 1 M solution of sodium chloride contains 58.5 grams of sodium chloride in 1 litre of solution.

Atoms, Isotopes and the formation of ions

Atoms

Atoms are the smallest units of a chemical element that can exist independently. An atom comprises a nucleus that contains particles called protons and neutrons (the hydrogen atom is the only exception as it has no neutrons). Tiny particles called electrons orbit the nucleus of the atom. The main features of these sub-atomic particles are:

- **Neutrons** – occur in the nucleus of an atom and have the same mass as protons but no electrical charge.

- **Protons** – occur in the nucleus of an atom and have the same mass as neutrons but do have a positive charge.

- **Electrons** – orbit in shells around the nucleus but a long way from it. They have such a small mass that their contribution to the overall mass of the atom is negligible. They are, however, negatively charged and their number determines the chemical properties of an atom.

In an atom the number of protons and electrons is the same and therefore there is no overall charge.

Two important terms are:

- the **atomic number** – the number of protons in an atom
- the **mass number** – the total number of protons and neutrons in an atom.

The atomic structure of three biological elements is given in Figure 2.

Isotopes

While the number of protons in an element always remains the same, the number of neutrons can vary. The different types of the atom so produced are called **isotopes**. Isotopes of any one element have the same chemical properties but differ in mass. Each type is therefore recognised by its different mass number. Isotopes, especially radioactive ones, are very useful in biology for tracing the route of certain elements in biological processes and for dating fossils.

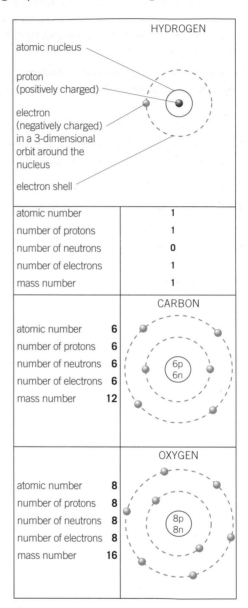

▲ **Figure 2** *Atomic structure of three commonly occurring biological elements*

The formation of ions

If an atom loses or receives an electron it becomes an **ion**.

- The loss of an electron leads to the formation of a positive ion, for example, the loss of an electron from a hydrogen atom produces a positively charged hydrogen ion, written as H^+.

- The receiving of an electron leads to the formation of a negative ion, for example, if a chlorine atom receives an additional electron it becomes a negatively charged chloride ion, written as Cl^-.

More than one electron may be lost or received, for example the loss of two electrons from a calcium atom forms the calcium ion, Ca^{2+}. Ions may be made up of more than one type of atom, for example a sulfate ion is formed when one sulfur atom and four oxygen atoms receive two electrons and form the sulfate ion, SO_4^{2-}.

1 a Name the element that contains one proton and one electron.

  b If an atomic particle with no overall charge is added to this element, state which general term can be used to describe the new form of the element.

  c Determine by what percentage the element's mass number is altered by the addition of this new particle.

  d Determine how the atomic number is affected by the addition of this new particle.

2 a State what is formed if a negatively charged particle is removed from a hydrogen atom.

  b State how the mass number is changed by the removal of this negatively charged particle.

1.2 Carbohydrates – monosaccharides

Learning objectives

→ Describe how carbohydrates are constructed.

→ Describe the structure of monosaccharides.

→ Describe how to carry out the Benedict's test for reducing and non-reducing sugars.

Specification reference: 3.1.2

Hint

In biology certain prefixes are commonly used to indicate numbers. There are two systems, one based on Latin and the other on Greek. The Greek terms which are used when referring to chemicals are:

- mono – one
- di – two
- tri – three
- tetra – four
- penta – five
- hexa – six
- poly – many

α-glucose

β-glucose

▲ **Figure 1** *Molecular arrangements of α-glucose and β-glucose (five carbon atoms at the intersection of the lines and one at the end of the vertical line at the top have been omitted for simplicity. Each line represents a covalent bond)*

As the word suggests, carbohydrates are carbon molecules (carbo) combined with water (hydrate). Some carbohydrate molecules are small while others are large.

Life based on carbon

Carbon atoms have an unusual feature. They very readily form bonds with other carbon atoms. This allows a sequence of carbon atoms of various lengths to be built up. These form a 'backbone' along which other atoms can be attached. This permits a large number of different types and sizes of molecule, all based on carbon. The variety of life that exists on Earth is a consequence of living organisms being based on the versatile carbon atom. Carbon-containing molecules are known as organic molecules. In living organisms, there are relatively few other atoms that attach to carbon. Life is therefore based on a small number of chemical elements.

The making of large molecules

Many organic molecules, including carbohydrates, are made up of a chain of individual molecules. Each of the individual molecules that make up these chains is given the general name **monomer**. Examples of monomers include monosaccharides, amino acids and nucleotides. Monomers can join together to form long chains called **polymers**. How this happens is explained in Topic 1.3. Biological molecules like carbohydrates and proteins are often polymers. These polymers are based on a surprisingly small number of chemical elements. Most are made up of just four elements: carbon, hydrogen, oxygen and nitrogen.

In carbohydrates, the basic monomer unit is a sugar, otherwise known as a saccharide. A single monomer is therefore called a **monosaccharide**. A pair of monosaccharides can be combined to form a **disaccharide**. Monosaccharides can also be combined in much larger numbers to form **polysaccharides**.

Monosaccharides

Monosaccharides are sweet-tasting, soluble substances that have the general formula $(CH_2O)_n$, where n can be any number from three to seven.

Examples of monosaccharides include glucose, galactose and fructose. Glucose is a hexose (6-carbon) sugar and has the formula $C_6H_{12}O_6$. However, the atoms of carbon, hydrogen and oxygen can be arranged in many different ways. For example, glucose has two isomers – α-glucose and β-glucose. Their structures are shown in Figure 1.

Test for reducing sugars 🧪

All monosaccharides and some disaccharides (e.g., maltose) are reducing sugars. Reduction is a chemical reaction involving the gain of electrons or hydrogen. A reducing sugar is therefore a sugar that can donate electrons to (or reduce) another chemical, in this case Benedict's reagent. The test for a reducing sugar is therefore known as the Benedict's test.

Benedict's reagent is an alkaline solution of copper(II) sulfate. When a reducing sugar is heated with Benedict's reagent it forms an insoluble red precipitate of copper(I) oxide. The test is carried out as follows:

- Add 2 cm^3 of the food sample to be tested to a test tube. If the sample is not already in liquid form, first grind it up in water.
- Add an equal volume of Benedict's reagent.
- Heat the mixture in a gently boiling water bath for five minutes.

Study tip

The Benedict's test may be a practical exercise but be certain to learn the *details* of the procedure. Add Benedict's and look for a red colour' is not enough.

Food sample dissolved in water.

Equal volume of Benedict's reagent added.

Heated in water bath. If reducing sugar present solution turns orange–brown.

▲ **Figure 2** *The Benedict's test*

▲ **Figure 3** *If a reducing sugar is present an orange-brown colour is formed*

Summary questions

1 Large molecules often contain carbon. Explain why this is.

2 State the general name for a molecule that is made up of many similar repeating units.

3 Explain why Benedict's reagent turns red when heated with a reducing sugar.

Semi-quantitative use of the Benedict's test

Table 1 shows the relationship between the concentration of reducing sugar and the colour of the solution and precipitate formed during the Benedict's test. The differences in colour mean that the Benedict's test is semi-quantitative, that is it can be used to estimate the approximate amount of reducing sugar in a sample.

none   very low   low   medium   high

▲ **Figure 4** *Results of Benedict's test according to the concentration of reducing sugar present*

The Benedict's test was carried out on five food samples. The results are shown in Table 1.

▼ **Table 1**

Sample	Colour of solution	
A	yellowish brown	m
B	green	vL
C	red	h
D	dark brown	m
E	yellowish green	l

1 List the letters in sequence of the increasing amount of reducing sugar in each sample.

2 Suggest a way, other than comparing colour changes, in which different concentrations of reducing sugar could be estimated.

3 Explain why it is not possible to distinguish between very concentrated samples, even when their concentrations are different.

Learning objectives

→ Explain how monosaccharides are linked together to form disaccharides.

→ Describe how α-glucose molecules are linked to form starch.

→ Describe the test for non-reducing sugars.

→ Describe the test for starch.

Specification reference: 3.1.2

In Topic 1.2 we saw that in carbohydrates, the monomer unit is called a monosaccharide. Pairs of monosaccharides can be combined to form a **disaccharide**. Monosaccharides can also be combined in much larger numbers to form **polysaccharides**.

Disaccharides

When combined in pairs, monosaccharides form a disaccharide. For example:

- Glucose joined to glucose forms maltose.
- Glucose joined to fructose forms sucrose.
- Glucose joined to galactose forms lactose.

When the monosaccharides join, a molecule of water is removed and the reaction is therefore called a **condensation reaction**. The bond that is formed is called a **glycosidic bond**.

When water is added to a disaccharide under suitable conditions, it breaks the glycosidic bond releasing the constituent monosaccharides. This is called **hydrolysis** (addition of water that causes breakdown).

Figure 1a) illustrates the formation of a glycosidic bond by the removal of water (condensation reaction). Figure 1b) shows the breaking of the glycosidic bond by the addition of water (hydrolysis reaction).

a *Formation of glycosidic bond by removal of water (condensation reaction)*

b *Breaking of glycosidic bond by addition of water (hydrolysis reaction)*

▲ Figure 1 *Formation and breaking of a glycosidic bond by condensation and hydrolysis*

Test for non-reducing sugars

Some disaccharides (e.g. maltose) are reducing sugars. To detect these we use the Benedict's test, as described in Topic 1.2, Carbohydrates – monosaccharides. Other disaccharides, such as sucrose, are known as non-reducing sugars because they do not change the colour of Benedict's reagent when they are heated with it. In order to detect a non-reducing sugar it must first be hydrolysed into its monosaccharide components by hydrolysis. The process is carried out as follows:

- If the sample is not already in liquid form, it must first be ground up in water.
- Add $2\,cm^3$ of the food sample being tested to $2\,cm^3$ of Benedict's reagent in a test tube and filter.
- Place the test tube in a gently boiling water bath for 5 minutes. If the Benedict's reagent does not change colour (the solution remains blue), then a reducing sugar is *not* present.
- Add another $2\,cm^3$ of the food sample to $2\,cm^3$ of dilute hydrochloric acid in a test tube and place the test tube in a gently boiling water bath for five minutes. The dilute hydrochloric acid will hydrolyse any disaccharide present into its constituent monosaccharides.
- Slowly add some sodium hydrogencarbonate solution to the test tube in order to neutralise the hydrochloric acid. (Benedict's reagent will not work in acidic conditions.) Test with pH paper to check that the solution is alkaline.
- Re-test the resulting solution by heating it with $2\,cm^3$ of Benedict's reagent in a gently boiling water bath for five minutes.
- If a non-reducing sugar was present in the original sample, the Benedict's reagent will now turn orange-brown. This is due to the reducing sugars that were produced from the hydrolysis of the non-reducing sugar.

Polysaccharides

Polysaccharides are polymers, formed by combining together many monosaccharide molecules. The monosaccharides are joined by glycosidic bonds that were formed by **condensation reactions**. As polysaccharides are very large molecules, they are insoluble. This feature makes them suitable for storage. When they are hydrolysed, polysaccharides break down into disaccharides or monosaccharides (Figure 2). Some polysaccharides, such as cellulose (see Topic 1.4), are not used for storage but give structural support to plant cells.

> **Hint**
>
> Polysaccharides illustrate an important principle: that a few basic monomer units can be combined in a number of different ways to give a large range of different biological molecules.

▲ **Figure 2** *The hydrolysis of a polysaccharide into disaccharides and monosaccharides*

Starch is a polysaccharide that is found in many parts of plants in the form of small granules or grains, for example starch grains in chloroplasts. It is formed by the joining of between 200 and 100 000 α-glucose molecules by glycosidic bonds in a series of condensation reactions. More details of starch and its functions are given in Topic 1.4.

1 Two drops of iodine solution added to test solution

2 If starch is present it turns the iodine a blue–black colour

▲ **Figure 3** *Test for starch*

Test for starch

Starch is easily detected by its ability to change the colour of the iodine in potassium iodide solution from yellow to blue–black (Figure 3). The test is carried out at room temperature. The test is carried out as follows:

- Place 2 cm³ of the sample being tested into a test tube (or add two drops of the sample into a depression on a spotting tile).
- Add two drops of iodine solution and shake or stir.
- The presence of starch is indicated by a blue–black coloration.

Summary questions

1 Identify which one, or more, monomer units make up each of the following carbohydrates.

  a lactose

  b sucrose

  c starch

2 Glucose $(C_6H_{12}O_6)$ combines with fructose $(C_6H_{12}O_6)$ to form the disaccharide sucrose. From your knowledge of how disaccharides are formed, deduce the formula of sucrose.

3 ⚠ To hydrolyse a disaccharide it can be boiled with hydrochloric acid but if hydrolysis is carried out by an enzyme a much lower temperature (40 °C) is required. Explain why.

1.4 Starch, glycogen and cellulose

In organisms, a wide range of different molecules with very different properties can be made from a limited range of smaller molecules. What makes the larger molecules different is the various ways in which the smaller molecules are combined to form them and small differences in the monomers used. You will look at some of these larger molecules by considering three important polysaccharides.

Starch

Starch is a polysaccharide that is found in many parts of a plant in the form of small grains. Especially large amounts occur in seeds and storage organs, such as potato tubers. It forms an important component of food and is the major energy source in most diets. Starch is made up of chains of α-glucose monosaccharides linked by glycosidic bonds that are formed by **condensation reactions**. The chains may be branched or unbranched. The unbranched chain is wound into a tight coil that makes the molecule very compact. The structure of a starch molecule is shown in Figure 1.

Learning objectives

→ Explain how α-glucose monomers are arranged to form the polymers of starch and glycogen.

→ Explain how β-glucose monomers are arranged to form the polymer cellulose.

→ Explain how the molecular structures of starch, glycogen and cellulose relate to their functions.

Specification reference: 3.1.2

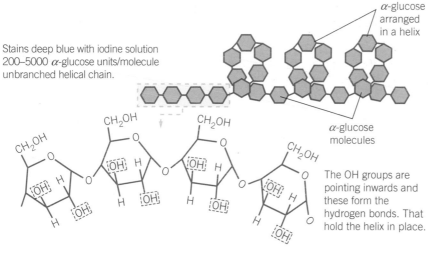

Stains deep blue with iodine solution
200–5000 α-glucose units/molecule
unbranched helical chain.

α-glucose arranged in a helix

α-glucose molecules

The OH groups are pointing inwards and these form the hydrogen bonds. That hold the helix in place.

▲ **Figure 1** *Structure of a starch molecule*

The main role of starch is energy storage, something its structure is especially suited for because:

- it is insoluble and therefore doesn't affect water potential, so water is not drawn into the cells by **osmosis**
- being large and insoluble, it does not diffuse out of cells
- it is compact, so a lot of it can be stored in a small space
- when hydrolysed it forms α-glucose, which is both easily transported and readily used in respiration
- the branched form has many ends, each of which can be acted on by enzymes simultaneously meaning that glucose monomers are released very rapidly.

Starch is never found in animal cells. Instead a similar polysaccharide, called glycogen, serves the same role.

▲ **Figure 2** *False colour scanning electron micrograph (SEM) of starch grains (blue) in the cells of a potato. Starch is a compact storage material.*

Glycogen

Glycogen is found in animals and bacteria but never in plant cells. Glycogen is very similar in structure to starch but has shorter chains and is more highly branched. It is sometimes called 'animal starch' because it is the major carbohydrate storage product of animals. In animals it is stored as small granules mainly in the muscles and the liver. The mass of carbohydrate that is stored is relatively small because fat is the main storage molecule in animals. Its structure suits it for storage because:

- it is insoluble and therefore does not tend to draw water into the cells by osmosis
- being insoluble, it does not diffuse out of cells
- it is compact, so a lot of it can be stored in a small space
- It is more highly branched than starch and so has more ends that can be acted on simultaneously by enzymes. It is therefore more rapidly broken down to form glucose monomers, which are used in respiration. This is important to animals which have a higher metabolic rate and therefore respiratory rate than plants because they are more active.

Cellulose

Cellulose differs from starch and glycogen in one major respect: it is made of monomers of β-glucose rather than α-glucose. This seemingly small variation produces fundamental differences in the structure and function of this polysaccharide.

Rather than forming a coiled chain like starch, cellulose has straight, unbranched chains. These run parallel to one another, allowing hydrogen bonds (Topic 1.6, Proteins) to form cross-linkages between adjacent chains. While each individual hydrogen bond adds very little to the strength of the molecule, the sheer overall number of them makes a considerable contribution to strengthening cellulose, making it the valuable structural material that it is. The arrangement of β-glucose chains in a cellulose molecule is shown in Figure 3.

Simplified representation of the arrangement of glucose chains

hydrogen bonds forming cross bridges

β-glucose molecules

The OH groups occur on both sides of the molecule and so it can form hydrogen bonds with other chains on both sides.

The cellulose chain, unlike that of starch, has adjacent glucose molecules rotated by 180°. This allows hydrogen bonds to be formed between the hydroxyl (–OH) groups on adjacent parallel chains that help to give cellulose its structural stability.

◀ **Figure 3** *Structure of a cellulose molecule*

The cellulose molecules are grouped together to form microfibrils (Figure 4) which, in turn, are arranged in parallel groups called fibres.

Cellulose is a major component of plant cell walls and provides rigidity to the plant cell. The cellulose cell wall also prevents the cell from bursting as water enters it by osmosis. It does this by exerting an inward pressure that stops any further influx of water. As a result, living plant cells are turgid and push against one another, making non-woody parts of the plant semi-rigid. This is especially important in maintaining stems and leaves in a turgid state so that they can provide the maximum surface area for photosynthesis.

In summary, the structure of cellulose is suited to its function of providing support and rigidity because:

- cellulose molecules are made up of β-glucose and so form long straight, unbranched chains
- these cellulose molecular chains run parallel to each other and are crossed linked by hydrogen bonds which add collective strength
- these molecules are grouped to form microfibrils which in turn are grouped to form fibres all of which provides yet more strength.

microfibril made up of cellulose molecules

a cellulose molecule composed of β-glucose molecules

▲ Figure 4 Structure of a cellulose microfibril

Synoptic link

More detail of the cell wall in plants is given in Topic 3.4 and its importance in supporting non-woody plant tissues is discussed in Topic 4.3.

Summary questions

From the following list of carbohydrates choose **one or more** that most closely fit each of the statements below. Each carbohydrate may be used once, more than once, or not at all.

α-glucose   starch   cellulose   β-glucose   glycogen

1 Stains deep blue with iodine solution.
2 Is known as 'animal starch'.
3 Found in plants.
4 Are polysaccharides.
5 Monosaccharide found in starch.
6 Has a structural function.
7 Can be hydrolysed.
8 Easily moves in and out of cells by facilitated diffusion.

Learning objectives

→ Describe the structure of triglycerides and how this relates to their function.

→ Describe the roles of lipids.

→ Describe the structure of a phospholipids and how this relates to their function.

→ Describe the test for a lipid.

Specification reference: 3.1.3

Hint

Fats are generally made of saturated fatty acids, while oils are made of unsaturated ones.

Fats are solid at room temperature (10–20 °C), whereas oils are liquid.

Lipids are a varied group of substances that share the following characteristics:

- They contain carbon, hydrogen and oxygen.
- The proportion of oxygen to carbon and hydrogen is smaller than in carbohydrates.
- They are insoluble in water.
- They are soluble in organic solvents such as alcohols and acetone.

The main groups of lipids are **triglycerides** (fats and oils) and phospholipids.

Roles of lipids

Lipids have many roles, one role of lipids is in the **cell membranes** (cell-surface membranes and membranes around organelles). Phospholipids contribute to the flexibility of membranes and the transfer of lipid-soluble substances across them. Other roles of lipids include:

- **source of energy**. When oxidised, lipids provide more than twice the energy as the same mass of carbohydrate and release valuable water.
- **waterproofing**. Lipids are insoluble in water and therefore useful as a waterproofing. Both plants and insects have waxy, lipid cuticles that conserve water, while mammals produce an oily secretion from the sebaceous glands in the skin.
- **insulation**. Fats are slow conductors of heat and when stored beneath the body surface help to retain body heat. They also act as electrical insulators in the myelin sheath around nerve cells.
- **protection**. Fat is often stored around delicate organs, such as the kidney.

Fats are solid at room temperature (10–20 °C) whereas oils are liquid.

Triglycerides

Triglycerides are so called because they have three (tri) fatty acids combined with glycerol (glyceride). Each fatty acid forms an ester bond with glycerol in a **condensation reaction** (Figure 1). **Hydrolysis** of a triglyceride therefore produces glycerol and three fatty acids.

▲ **Figure 1** *The formation of a triglyceride. The three fatty acids may all be the same, thereby forming a simple triglyceride, or they may be different, in which case a mixed triglyceride is produced. In either case it is a condensation reaction*

As the glycerol molecule in all triglycerides is the same, the differences in the properties of different fats and oils come from variations in the fatty acids. There are over 70 different fatty acids and all have a carboxyl (—COOH) group with a hydrocarbon chain attached. If this chain has no carbon–carbon double bonds, the fatty acid is then described as **saturated**, because all the carbon atoms are linked to the maximum possible number of hydrogen atoms, in other words they are saturated with hydrogen atoms. If there is a single double bond, it is **mono-unsaturated** – if more than one double bond is present, it is **polyunsaturated**. These differences are illustrated in Figure 2.

The structure of triglycerides related to their properties

- Triglycerides have a high ratio of energy-storing carbon–hydrogen bonds to carbon atoms and are therefore an excellent source of energy.

- Triglycerides have low mass to energy ratio, making them good storage molecules because much energy can be stored in a small volume. This is especially beneficial to animals as it reduces the mass they have to carry as they move around.

- Being large, non-polar molecules, triglycerides are insoluble in water. As a result their storage does not affect osmosis in cells or the **water potential** of them.

- As they have a high ratio of hydrogen to oxygen atoms, triglycerides release water when oxidised and therefore provide an important source of water, especially for organisms living in dry deserts.

Phospholipids

Phospholipids are similar to lipids except that one of the fatty acid molecules is replaced by a phosphate molecule (Figure 3). Whereas fatty acid molecules repel water (are hydrophobic), phosphate molecules attract water (are hydrophilic). A phospholipid is therefore made up of two parts:

- a **hydrophilic 'head'**, which interacts with water (is attracted to it) but not with fat

- a **hydrophobic 'tail'**, which orients itself away from water but mixes readily with fat.

Molecules that have two ends (poles) that behave differently in this way are said to be **polar**. This means that when these polar phospholipid molecules are placed in water they position themselves so that the hydrophilic heads are as close to the water as possible and the hydrophobic tails are as far away from the water as possible (Figure 4).

saturated
(no double bonds between carbon atoms)

mono-unsaturated
(one double bond between carbon atoms)

polyunsaturated
(more than one double bond between carbon atoms)

The double bonds cause the molecule to bend. They cannot therefore pack together so closely making them liquid at room temperature, i.e. they are oils.

▲ **Figure 2** *Saturated and unsaturated fatty acids*

Study tip

Do not use terms like 'water-loving' and 'water-hating'. Use the correct scientific terms '**hydrophilic**' and '**hydrophobic**'.

▲ **Figure 3** *Structure of a phospholipid*

Synoptic link

You will learn more about cell recognition in Topic 5.1, Defence mechanisms and their role in cell-surface membranes in Topic 4.1 Structure of cell surface membranes..

The structure of phospholipids related to their properties

- Phospholipids are polar molecules, having a hydrophilic phosphate head and a hydrophobic tail of two fatty acids. This means that in an aqueous environment, phospholipid molecules form a bilayer within cell-surface membranes. As a result, a hydrophobic barrier is formed between the inside and outside of a cell.
- The hydrophilic phosphate 'heads' of phospholipid molecules help to hold at the surface of the cell-surface membrane.
- The phospholipid structure allows them to form glycolipids by combining with carbohydrates within the cell-surface membrane. These glycolipids are important in cell recognition.

Test for lipids 🧪

The test for lipids is known as the emulsion test and is carried out as follows:

1 Take a completely dry and grease-free test tube.
2 To $2\,cm^3$ of the sample being tested, add $5\,cm^3$ of ethanol.
3 Shake the tube thoroughly to dissolve any lipid in the sample.
4 Add $5\,cm^3$ of water and shake gently.
5 A cloudy-white colour indicates the presence of a lipid.
6 As a control, repeat the procedures using water instead of the sample; the final solution should remain clear.

The cloudy colour is due to any lipid in the sample being finely dispersed in the water to form an emulsion. Light passing through this emulsion is refracted as it passes from oil droplets to water droplets, making it appear cloudy.

Summary questions

1 In the following passage state the most suitable word for each of the letters **a** to **e**.

Fats and oils make up a group of lipids called **a** which, when hydrolysed, form **b** and fatty acids. A fatty acid with more than one carbon–carbon double bond is described as **c**. In a phospholipid the number of fatty acids is **d**; these are described as **e** because they repel water.

2 List **two** differences between a triglyceride molecule and a phospholipid molecule.

3 Organisms that move, e.g. animals, and parts of organisms that move, e.g., some plant seeds, use lipids rather than carbohydrates as an energy store. Suggest **one** reason why this is so.

1.6 Proteins

Proteins are usually very large molecules. The types of carbohydrates and lipids in all organisms are relatively few and they are very similar. However, each organism has numerous proteins that differ from species to species. The shape of any one type of protein molecule differs from that of all other types of proteins. Proteins are very important molecules in living organisms. Indeed the word 'protein' is a Greek word meaning 'of first importance'. One group of proteins, enzymes, is involved in almost every living process. There is a vast range of different enzymes that between them perform a very diverse number of functions.

Structure of an amino acid

Amino acids are the basic **monomer** units which combine to make up a **polymer** called a polypeptide. Polypeptides can be combined to form proteins. About 100 amino acids have been identified, of which 20 occur naturally in proteins. The fact that the same 20 amino acids occur in all living organisms provides indirect evidence for evolution.

Every amino acid has a central carbon atom to which are attached four different chemical groups:

- amino group ($-NH_2$) – a basic group from which the amino part of the name amino acid is derived
- carboxyl group ($-COOH$) – an acidic group which gives the amino acid the acid part of its name
- hydrogen atom ($-H$)
- R (side) group – a variety of different chemical groups. Each amino acid has a different R group. These 20 naturally occurring amino acids differ only in their R (side) group.

The general structure of an amino acid is shown in Figure 1.

The formation of a peptide bond

In a similar way that monosaccharide monomers combine to form disaccharides (see Topic 1.3), so amino acid monomers can combine to form a dipeptide. The process is essentially the same: namely the removal of a water molecule in a **condensation** reaction. The water is made by combining an $-OH$ from the carboxyl group of one amino acid with an $-H$ from the amino group of another amino acid. The two amino acids then become linked by a new **peptide bond** between the carbon atom of one amino acid and the nitrogen atom of the other. The formation of a peptide bond is illustrated in Figure 3. In a similar way as a glycosidic bond of a disaccharide can be broken by the addition of water (hydrolysis), so the peptide bond of a dipeptide can also be broken by hydrolysis to give its two constituent amino acids.

The primary structure of proteins – polypeptides

Through a series of condensation reactions, many amino acid monomers can be joined together in a process called **polymerisation**. The resulting chain of many hundreds of amino acids is called a **polypeptide**. The sequence of amino acids in a polypeptide chain forms the primary structure of any protein. As we shall see in Topic 8.1, this sequence is

Learning objectives

→ Explain how amino acids are linked to form polypeptides – the primary structure of proteins.

→ Explain how polypeptides are arranged to form the secondary structure and then the tertiary structure of a protein.

→ Explain how the quaternary structure of a protein is formed.

→ Describe the test for proteins.

Specification reference: 3.1.4.1

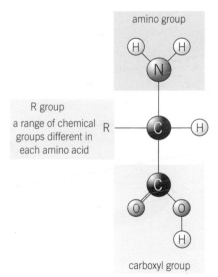

▲ **Figure 1** *The general structure of an amino acid*

$$H_2N \overset{\overset{\displaystyle R}{|}}{\underset{\underset{\displaystyle H}{|}}{-C-}} COOH$$

▲ **Figure 2** *Simplified structural formula of an amino acid*

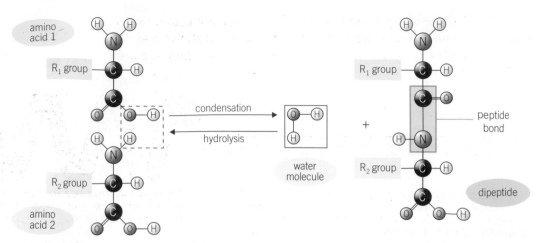

▲ **Figure 3** *The formation of a peptide bond*

determined by DNA. As polypeptides have many (usually hundreds) of the 20 naturally occurring amino acids joined in different sequences, it follows that there is an almost limitless number of possible combinations, and therefore types, of primary protein structure.

It is the primary structure of a protein that determines its ultimate shape and hence its function. A change in just a single amino acid in this primary sequence can lead to a change in the shape of the protein and may stop it carrying out its function. In other words, a protein's shape is very specific to its function. Change its shape and it will function less well, or differently.

A simple protein may consist of a single polypeptide chain. More commonly, however, a protein is made up of a number of polypeptide chains.

The secondary structure of proteins

The linked amino acids that make up a polypeptide possess both —NH and —C=O groups on either side of every peptide bond. The hydrogen of the —NH group has an overall positive charge while the O of the —C=O group has an overall negative charge. These two groups therefore readily form weak bonds, called **hydrogen bonds**. This causes the long polypeptide chain to be twisted into a 3-D shape, such as the coil known as an α-helix. Figure 4 illustrates the structure of an α-helix.

Tertiary structure of proteins

The α-helices of the secondary protein structure can be twisted and folded even more to give the complex, and often specific, 3-D structure of each protein (Figure 5). This is known as the tertiary structure. This structure is maintained by a number of different bonds. Where the bonds occur depends on the primary structure of the protein. These bonds include:

- **disulfide bridges** – which are fairly strong and therefore not easily broken.
- **ionic bonds** – which are formed between any carboxyl and amino groups that are not involved in forming peptide bonds. They are weaker than disulfide bonds and are easily broken by changes in pH.
- **hydrogen bonds** – which are numerous but easily broken.

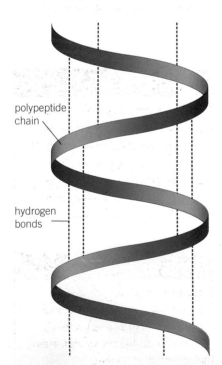

▲ **Figure 4** *Structure of the α-helix*

It is the 3-D shape of a protein that is important when it comes to how it functions. It makes each protein distinctive and allows it to recognise, and be recognised by, other molecules. It can then interact with them in a very specific way.

a The primary structure of a protein is the sequence of amino acids found in its polypeptide chains. This sequence determines its properties and shape. Following the elucidation of the amino acid sequence of the hormone insulin by Frederick Sanger in 1954, the primary structure of many other proteins is now known.

b The secondary structure is the shape which the polypeptide chain forms as a result of hydrogen bonding. This is most often a spiral known as the α-helix, although other configurations occur.

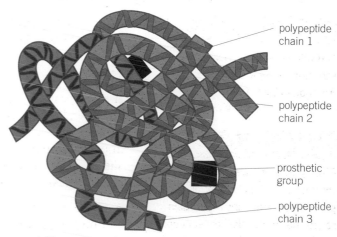

polypeptide chain 1

polypeptide chain 2

prosthetic group

polypeptide chain 3

c The tertiary structure is due to the bending and twisting of the polypeptide helix into a compact structure. All three types of bond, disulfide, ionic and hydrogen, contribute to the maintenance of the tertiary structure.

d The quaternary structure arises from the combination of a number of different polypeptide chains and associated non-protein (prosthetic) groups into a large, complex protein molecule, e.g., haemoglobin.

▲ **Figure 5** *Structure of proteins*

Quaternary structure of proteins

Large proteins often form complex molecules containing a number of individual polypeptide chains that are linked in various ways. There may also be non-protein (prosthetic) groups associated with the molecules (Figure 5d), such as the iron-containing haem group in haemoglobin. Remember that, although the 3-D structure is important to how a protein functions, it is the sequence of amino acids (primary structure) that determines the 3-D shape in the first place.

Test for proteins 🧪

The most reliable protein test is the Biuret test, which detects peptide bonds. It is performed as follows:

- Place a sample of the solution to be tested in a test tube and add an equal volume of sodium hydroxide solution at room temperature.
- Add a few drops of very dilute (0.05%) copper(II) sulfate solution and mix gently.
- A purple coloration indicates the presence of peptide bonds and hence a protein. If no protein is present, the solution remains blue.

> **Hint**
>
> Think of the polypeptide chain as a piece of string. In a fibrous protein many pieces of the string are twisted together into a rope, while in a globular protein the pieces of string, usually fewer, are rolled into a ball.

> **Study tip**
>
> You can simply refer to adding **Biuret** reagent to test for protein. A purple colour shows protein is present – a blue colour indicates that protein is absent.

Summary questions

1 Name the type of bond that joins amino acids together.

2 State the type of reaction involved in joining amino acids together.

3 List *four* different components that make up an amino acid.

Protein shape and function

Proteins perform many different roles in living organisms. Their roles depend on their molecular shape, which can be of two basic types.

- Fibrous proteins, such as collagen, have structural functions.
- Globular proteins, such as enzymes and haemoglobin, carry out metabolic functions.

It is the very different structure and shape of each of these types of proteins that enables them to carry out their functions.

Fibrous proteins

Fibrous proteins form long chains which run parallel to one another. These chains are linked by cross-bridges and so form very stable molecules. One example is **collagen**. Its molecular structure is as follows:

- The primary structure is an unbranched polypeptide chain.
- In the secondary structure the polypeptide chain is very tightly wound.
- Lots of the amino acid, glycine helps close packing.
- In the tertiary structure the chain is twisted into a second helix.
- Its quaternary structure is made up of three such polypeptide chains wound together in the same way as individual fibres are wound together in a rope.

Collagen is found in tendons. Tendons join muscles to bones. When a muscle contracts the bone is pulled in the direction of the contraction.

1 Explain why the quaternary structure of collagen makes it a suitable molecule for a tendon.

The individual collagen polypeptide chains in the fibres are held together by bonds between amino acids of adjacent chains.

2 Suggest how the cross-linkages between the amino acids of polypeptide chains increase the strength and stability of a collagen fibre.

The points where one collagen molecule ends and the next begins are spread throughout the fibre rather than all being in the same position along it.

3 Explain why this arrangement of collagen molecules is necessary for the efficient functioning of a tendon.

◄ **Figure 6** *Fine structure of the fibrous protein collagen*

each polypeptide forms a long, unfolded strand

the individual polypeptide strands are tightly wound around one another

long fibrous molecular shape

fibrous protein, e.g., collagen

◄ **Figure 7** *Structure of fibrous proteins*

Enzymes are globular proteins that act as catalysts. Catalysts alter the rate of a chemical reaction without undergoing permanent changes themselves. They can be reused repeatedly and are therefore effective in small amounts. Enzymes do not make reactions happen; they speed up reactions that already occur, sometimes by a factor of many millions.

Enzymes as catalysts lowering activation energy

Let us consider a typical chemical reaction:

$$\text{sucrose} + \text{water} \longrightarrow \text{glucose} + \text{fructose}$$
$$\text{(substrates)} \qquad\qquad \text{(products)}$$

For reactions like this to take place naturally a number of conditions must be satisfied:

* The sucrose and water molecules must collide with sufficient energy to alter the arrangement of their atoms to form glucose and fructose.
* The free energy of the products (glucose and fructose) must be less than that of the substrates (sucrose and water).
* Many reactions require an initial amount of energy to start. The minimum amount of energy needed to activate the reaction in this way is called the **activation energy**.

There is an activation energy level, like an energy hill or barrier, which must initially be overcome before the reaction can proceed. Enzymes work by lowering this activation energy level (Figure 1). In this way enzymes allow reactions to take place at a lower temperature than normal. This enables some metabolic processes to occur rapidly at the human body temperature of 37 °C, which is relatively low in terms of chemical reactions. Without enzymes these reactions would proceed too slowly to sustain life as we know it.

▲ **Figure 1** *How enzymes lower activation energy*

Learning objectives

→ Explain how enzymes speed up chemical reactions.

→ Describe how the structure of enzyme molecules relates to their function.

→ Explain the lock and key model of enzyme action.

→ Explain the induced-fit model of enzyme action.

Specification reference: 3.1.4.2

Hint

Free energy is the energy of a system that is available to perform work.

Hint

To help you understand the importance of enzymes, it is necessary to appreciate that they catalyse a wide range of reactions both inside the cell (intracellular) and outside the cell (extracellular). In doing so, enzymes determine the structures and functions of all parts of living matter from cells to complete organisms.

Enzyme structure

From Topic 1.6 you will be aware that enzymes, being globular proteins, have a specific 3-D shape that is the result of their sequence of amino acids (primary protein stucture). A specific region of the enzyme is functional, this is known as the **active site**. The active site is made up of a relatively small number of amino acids. The active site forms a small depression within the much larger enzyme molecule.

The molecule on which the enzyme acts is called the **substrate**. This fits neatly into this depression and forms an **enzyme–substrate complex** (Figure 2). The substrate molecule is held within the active site by bonds that temporarily form between certain amino acids of the active site and groups on the substrate molecule.

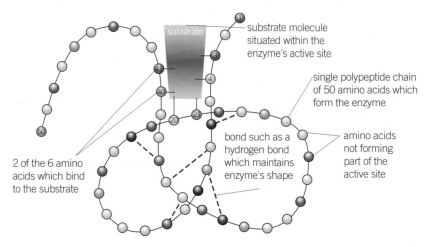

▲ **Figure 2** *Example of an enzyme–substrate complex showing the six out of the 50 amino acids that form the active site*

Induced fit model of enzyme action

Scientists often try to explain their observations by producing a representation of how something works. This is known as a scientific model. Examples include the physical models used to explain enzyme action. The induced fit model of enzyme action proposes that the active site forms as the enzyme and substrate interact. The proximity of the substrate (a change in the environment of the enzyme) leads to a change in the enzyme that forms the functional active site (Figure 3). In other words, the enzyme is flexible and can mould itself around the substrate in the way that a glove moulds itself to the shape of the hand. The enzyme has a certain general shape, just as a glove has, but this alters in the presence of the substrate. As it changes its shape, the enzyme puts a strain on the substrate molecule. This strain distorts a particular bond or bonds in the substrate and consequently lowers the activation energy needed to break the bond.

Any change in an enzyme's environment is likely to change its shape. The very act of colliding with its substrate is a change in its environment and so its shape changes – induced fit.

▲ **Figure 3** *Mechanism of enzyme action*

Summary questions

1 Define a catalyst.

2 Explain why enzymes are effective in tiny quantities.

3 Outline why changing one of the amino acids that make up the active site could prevent the enzyme from functioning.

4 Explain why changing certain amino acids that are not part of the active site also prevents the enzyme from functioning.

Lock and key model of enzyme action

One earlier model of enzyme action proposed that enzymes work in the same way as a key operates a lock – each key has a specific shape that fits and operates only a single lock. In a similar way, a substrate will only fit the active site of one particular enzyme. This model was supported by the observation that enzymes are specific in the reactions that they catalyse. The shape of the substrate (key) exactly fits the active site of the enzyme (lock). This is known as the **lock and key model**.

One limitation of this model is that the enzyme, like a lock, is considered to be a rigid structure. However, scientists had observed that other molecules could bind to enzymes at sites other than the active site. In doing so, they altered the activity of the enzyme. This suggested that the enzyme's shape was being altered by the binding molecule. In other words, its structure was not rigid but flexible. In true scientific fashion this led to an alternative model being proposed, one that better fitted the current observations. This was the called the induced fit model as described above. The induced fit model is therefore a modified version of the lock and key model.

1 Explain why the induced fit model is a better explanation of enzyme action than the lock and key model.

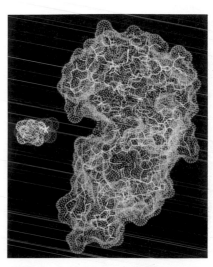

▲ **Figure 4** *Molecular computer graphics image of the enzyme ribonuclease A (right) and its substrate (left) approaching the enzyme's active site*

Hint

Enzymes have an active site but not all proteins are enzymes. Many proteins have binding sites or receptor sites that are not active sites. Some hormones are proteins and these have receptor sites but they are *not* active sites.

Learning objectives

→ Describe how the rate of an enzyme-controlled reaction is measured.

→ Explain how temperature affects the rate of an enzyme-controlled reaction.

→ Explain how pH affects the rate of an enzyme-controlled reaction.

→ Explain how substrate and enzyme concentration affect the rate of reaction.

Specification reference: 3.1.4.2

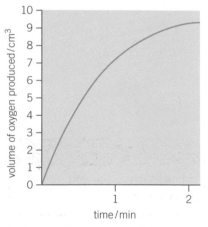

▲ **Figure 1** *Measurement of the formation of oxygen due to the action of catalase on hydrogen peroxide*

▲ **Figure 2** *Measurement of the disappearance of starch due to the action of amylase*

Before considering how pH and temperature affect enzymes, it is worth bearing in mind that, for an enzyme to work, it must:

* come into physical contact with its **substrate**,
* have an **active site** which fits the substrate.

Almost all factors that influence the rate at which an enzyme works do so by affecting one or both of the above. In order to investigate how enzymes are affected by various factors we need to be able to measure the rate of the reactions they catalyse.

Measuring enzyme-catalysed reactions

To measure the progress of an enzyme-catalysed reaction we usually measure its time-course, that is how long it takes for a particular event to run its course. The two changes most frequently measured are:

* the formation of the products of the reaction, for example the volume of oxygen produced when the enzyme catalase acts on hydrogen peroxide (Figure 1)
* the disappearance of the substrate, for example the reduction in concentration of starch when it is acted upon by amylase (Figure 2).

Although the graphs in Figures 1 and 2 differ, the explanation for their shapes is the same:

* At first there is a lot of substrate (hydrogen peroxide or starch) but no product (water and oxygen, or maltose).
* It is very easy for substrate molecules to come into contact with the empty active sites on the enzyme molecules.
* All enzyme active sites are filled at any given moment and the substrate is rapidly broken down into its products.
* The amount of substrate decreases as it is broken down, resulting in an increase in the amount of product.
* As the reaction proceeds, there is less and less substrate and more and more product.
* It becomes more difficult for the substrate molecules to come into contact with the enzyme molecules because there are fewer substrate molecules and also the product molecules may 'get in the way' of substrate molecules and prevent them reaching an active site.
* It therefore takes longer for the substrate molecules to be broken down by the enzyme and so its rate of disappearance slows, and consequently the rate of formation of product also slows. Both graphs 'tail off'.
* The rate of reaction continues to slow until there is so little substrate that any further decrease in its concentration cannot be measured.
* The graphs flatten out because all the substrate has been used up and so no new product can be produced.

Measuring rate of change $\sqrt{x}$

We can measure the change in the rate of a reaction at any point on the curve of a graph such as those in Figures 1 and 2. We do so by measuring the gradient at our chosen point. The gradient is equal to the gradient of the tangent to the curve at that point. This tangent is the point at which a straight line touches the curve but without cutting across it (see Maths skills chapter).

Accurately drawing the tangent to a curve is not easy but can be achieved by making use of the normal line. The normal line is a line that passes through a point at a 90° angle.

Let us look at an example. In Figure 3, you see the curve showing the formation of oxygen due to the action of catalase on hydrogen peroxide. Suppose you want to measure the rate of change in this reaction at point X as shown on Figure 3. You draw the tangent to the curve at this point as shown on Figure 3. Using this line you can find the gradient, in your case $= \frac{a}{b}$.

This technique is useful in a variety of practical situations, including ones involving measuring the rates of enzyme reactions.

Before we look at the effects of different factors on the rate of enzyme action, it is important to stress the fundamental experimental technique of changing only a single variable in each experiment. When investigating the effect of a named variable on the rate of an enzyme reaction all the other variables must be kept constant. For example, if measuring the effect of temperature, then pH, enzyme concentration and substrate concentration must be kept constant and all possible inhibitors should be absent. Another thing to remember is that the active site and the substrate are not 'the same', any more than a key and a lock are the same – in some senses they are more like opposites. The correct term is **complementary**.

Effect of temperature on enzyme action

A rise in temperature increases the **kinetic energy** of molecules. As a result, the molecules move around more rapidly and collide with each other more often. In an enzyme-catalysed reaction, this means that the enzyme and substrate molecules come together more often in a given time. There are more effective collisions resulting in more enzyme-substrate complexes being formed and so the rate of reaction increases.

Shown on a graph, this gives a rising curve. However, the temperature rise also begins to cause the hydrogen and other bonds in the enzyme molecule to break. This results in the enzyme, including its active site, changing shape. At first, the substrate fits less easily into this changed active site, slowing the rate of reaction. For many human enzymes this may begin at temperatures of around 45 °C.

At some point, usually around 60 °C, the enzyme is so disrupted that it stops working altogether. It is said to be denatured. **Denaturation** is a permanent change and, once it has occurred, the enzyme does not function again. Shown on a graph, the rate of this reaction follows

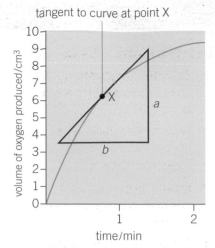

▲ **Figure 3** *Measuring the gradient at a point on a curve*

▲ **Figure 4** *Effect of temperature on the rate of an enzyme-controlled reaction*

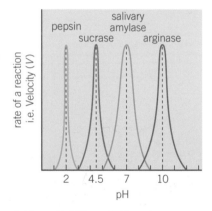

▲ **Figure 5** *Effect of pH on the rate of an enzyme-controlled reaction*

Maths link $\sqrt{x}$

MS 0.5, see Chapter 22.

Hint

When considering how factors affect enzyme action, think 'shape change'.

Study tip

Enzymes are not alive and so cannot be 'killed'. Use the correct term: *denatured*.

a falling curve. The actual effect of temperature on the rate of an enzyme reaction is a combination of these two factors (Figure 4). The optimum working temperature differs from enzyme to enzyme. Some work fastest at around 10 °C, while others continue to work rapidly at 80 °C. For example, enzymes used in biological washing powders and in the polymerase chain reaction (Topic 21.3). Many enzymes in the human body have an optimum temperature of about 40 °C. Our body temperatures have, however, evolved to be 37 °C. This may be related to the following:

- Although higher body temperatures would increase the metabolic rate slightly, the advantages are offset by the additional energy (food) that would be needed to maintain the higher temperature.

- Other proteins, apart from enzymes, may be denatured at higher temperatures.

- At higher temperatures, any further rise in temperature, for example, during illness, might denature the enzymes.

Different species of mammals and birds have different body temperatures. many birds, for example, have a normal body temperature of around 40 °C because they have a high metabolic rate for the high energy requirement of flight.

Effect of pH on enzyme action

The pH of a solution is a measure of its hydrogen ion concentration. Each enzyme has an optimum pH, that is a pH at which it works fastest (Figure 5). The pH of a solution is calculated using the formula: $pH = -\log_{10}[H^+]$. A hydrogen ion $[H^+]$ concentration of 1×10^{-9} therefore has a pH of 9. In a similar way to a change in temperature affecting the rate of enzyme action, a change in pH away from the optimum affects the rate of enzyme action. An increase or decrease in pH reduces the rate of enzyme action. If the change in pH is more extreme then, beyond a certain pH, the enzyme becomes denatured.

The pH affects how an enzyme works in the following ways:

- A change in pH alters the charges on the amino acids that make up the active site of the enzyme. As a result, the substrate can no longer become attached to the active site and so the enzyme–substrate complex cannot be formed.

- Depending on how significant the change in pH is, it may cause the bonds maintaining the enzyme's tertiary structure to break. The active site therefore changes shape.

The arrangement of the active site is partly determined by the hydrogen and ionic bonds between —NH$_2$ and —COOH groups of the polypeptides that make up the enzyme. The change in H$^+$ ions affects this bonding, causing the active site to change shape.

It is important to note that pH fluctuations inside organisms are usually small, this means they are far more likely to reduce an enzyme's activity than to denature it.

Effect of enzyme concentration on the rate of reaction

Once an **active site** on an enzyme has acted on its substrate, it is free to repeat the procedure on another substrate molecule. This means that enzymes, being catalysts, are not used up in the reaction and therefore work efficiently at very low concentrations. In some cases, a single enzyme molecule can act on millions of substrate molecules in one minute.

As long as there is an excess of substrate, an increase in the amount of enzyme leads to a proportionate increase in the rate of reaction. A graph of the rate of reaction against enzyme concentration will initially show a proportionate increase. This is because there is more substrate than the enzyme's active sites can cope with. If you therefore increase the enzyme concentration, some of the excess substrate can now also be acted upon and the rate of reaction will increase. If, however, the substrate is limiting, in other words there is not sufficient to supply all the enzyme's active sites at one time, then any increase in enzyme concentration will have no effect on the rate of reaction. The rate of reaction will therefore stabilise at a constant level, meaning the graph will level off. This is because the available substrate is already being used as rapidly as it can be by the existing enzyme molecules. These events are summarised in Figure 6.

> **Practical link** 🧪
>
> Required practical 1. Investigation into the effect of a named variable on the rate of an enzyme controlled reaction.

▲ Figure 6 *Effect of enzyme concentration on the rate of enzyme action*

▲ **Figure 7** *Enzymes in the algae in this hot spring remain functional at temperatures of 80 °C whereas in most organisms they are denatured at temperatures of 40 °C*

Effects of substrate concentration on the rate of enzyme action

If the concentration of enzyme is fixed and substrate concentration is slowly increased, the rate of reaction increases in proportion to the concentration of substrate. This is because, at low substrate concentrations, the enzyme molecules have only a limited number of substrate molecules to collide with, and therefore the active sites of the enzymes are not working to full capacity. As more substrate is added, the active sites gradually become filled, until the point where all of them are working as fast as they can. The rate of reaction is at its maximum (V_{max}). After that, the addition of more substrate will have no effect on the rate of reaction. In other words, when there is an excess of substrate, the rate of reaction levels off. A summary of the effect of substrate concentration on the rate of enzyme action is given in Figure 8.

▲ **Figure 8** *Effect of substrate concentration on the rate of an enzyme-controlled reaction*

Enzyme action

Different enzymes can function at a wide range of temperatures. Shrimps that live in Arctic waters have enzymes that function fastest at around 4 °C and are denatured at around 15 °C. By contrast, bacteria that live in hot springs have enzymes that function fastest at 95 °C and continue to operate effectively above 100 °C. These bacteria are called thermophilic (heat-loving) bacteria.

Enzyme X is produced by thermophilic bacteria and hydrolyses many proteins including haemoglobin and egg albumin.

Enzyme Y is found in the stomach of young mammals where it acts on a single soluble protein found in milk, causing it to coagulate (clot).

1   **a**   From the descriptions, comment on the differences in the specificity of the two enzymes.

     b   Enzymes X and Y are each used for different commercial purposes. Suggest what this might be in each case.

     c   Suggest a possible purpose of enzyme Y in the mammalian stomach.

     d   Use the information about the two enzymes to suggest a possible difference in the type of bonding found in the tertiary structure of each. Explain your reasoning.

An experiment was carried out with enzyme X in which the time taken for it to fully hydrolyse 5 g of its protein substrate was measured at different temperatures. It is important when investigating the effect of a named variable on the rate of an enzyme reaction that the other variables are kept constant. For example, if measuring the effect of temperature, then pH, enzyme concentration, substrate concentration must be kept constant and all possible inhibitors should be absent. The following data were obtained:

Temperature / °C	Time / min for hydrolysis of protein	Rate of reaction μ 1/time
15	5.8	
25	3.4	
35	1.7	
45	0.7	
55	0.6	
65	0.9	
75	7.1	

2   **a**   Calculate the relative rate of reaction for each temperature.

     b   Plot a graph that shows the effect of temperature on the rate of reaction of enzyme X.

     c   Measure the optimum temperature for the action of enzyme X.

     d   Suggest how you might determine this optimum temperature more precisely.

Maths link

MS 3.2 and 3.4, see Chapter 22.

Summary questions

1   Explain why enzymes function less well at lower temperatures.

2   Explain how high temperatures may completely prevent enzymes from functioning.

3   Enzymes produced by microorganisms are responsible for spoiling food. Using this fact and your knowledge of enzymes, deduce why each of the following procedures are carried out.

     a   Food is heated to a high temperature before being canned.

     b   Some foods, such as onions, are preserved in vinegar.

4   Calculate the pH of a solution that has a hydrogen ion concentration of 0.0001 M

1.9 Enzyme inhibition

Learning objectives

→ Describe the nature of enzyme inhibition.

→ Explain how competitive inhibitors and non-competitive inhibitors affect the active site.

Specification reference: 3.1.4.2

Enzyme inhibitors are substances that directly or indirectly interfere with the functioning of the active site of an enzyme and so reduce its activity. There are a number of types of enzyme inhibitor, two of which are:

- competitive inhibitors – which bind to the active site of the enzyme
- non-competitive inhibitors – which bind to the enzyme at a position other than the active site.

Competitive inhibitors

Competitive inhibitors have a molecular shape similar to that of the substrate. This allows them to occupy the active site of an enzyme. They therefore compete with the substrate for the available active sites (Figure 1). It is the difference between the concentration of the inhibitor and the concentration of the substrate that determines the effect that this has on enzyme activity. If the substrate concentration is increased, the effect of the inhibitor is reduced. The inhibitor is not permanently bound to the active site and so, when it leaves, another molecule can take its place. This could be a substrate or inhibitor molecule, depending on how much of each type is present. Sooner or later, all the substrate molecules will occupy an active site, but the greater the concentration of inhibitor, the longer this will take. An example of competitive inhibition occurs with an important respiratory enzyme that acts on succinate. Another compound, called malonate, can inhibit the enzyme because it has a very similar molecular shape to succinate. It therefore easily combines with the enzyme and blocks succinate from combining with the enzyme's active site. Another example is the inhibition of the enzyme transpeptidase by penicillin.

substrate molecule occupying the active site of the enzyme

substrate molecule unable to occupy the active site

enzyme molecule

inhibitor molecule occupying the active site of the enzyme

▲ **Figure 1** *Competitive inhibition*

1 *Inhibitor absent – the substrate attaches to the active site of the enzyme in the normal way. Reaction takes place as normal.*

substrate molecule occupying the active site of the enzyme

substrate molecule still fits the active site but not in a way that allows the reaction to proceed

enzyme molecule shape is changed due to presence of the inhibitor molecule

enzyme molecule

inhibitor molecule attached to enzyme molecule

2 *Inhibitor present – the inhibitor prevents the normal enzyme–substrate complex being formed. The reaction rate is reduced.*

▲ **Figure 2** *Non-competitive inhibition*

Non-competitive inhibitors

Non-competitive inhibitors attach themselves to the enzyme at a binding site which is not the active site. Upon attaching to the enzyme, the inhibitor alters the shape of the enzyme and thus its active site in such a way that substrate molecules can no longer occupy it, and so the enzyme cannot function (Figure 2). As the substrate and the inhibitor are not competing for the same site, an increase in substrate concentration does not decrease the effect of the inhibitor (Figure 3).

Control of metabolic pathways

A metabolic pathway is a series of reactions in which each step is catalysed by an enzyme. In the tiny space inside a single cell, there are many hundreds of different metabolic pathways. The pathways are not at all haphazard, but highly structured. The enzymes that control a pathway are often attached to the membrane of a cell organelle in a very precise sequence. Inside each organelle optimum conditions for the functioning of particular enzymes may be provided. To keep a steady concentration of a particular chemical in a cell, the same chemical often acts as an inhibitor of an enzyme at the start of a reaction.

Let us look at the example illustrated in Figure 4. The end product inhibits enzyme A. If for some reason the concentration of end product increases above normal, then there will be greater inhibition of enzyme A. As a result, less end product will be produced and its concentration will return to normal. If the concentration of the end product falls below normal there will be less of it to inhibit enzyme A. Consequently, more end product will be produced and, again, its concentration will return to normal. In this way, the concentration of any chemical can be maintained relatively constant. This is known as **end-product inhibition**. This type of inhibition is usually non-competitive.

1 Different conditions affect how enzymes work. Name one that might vary between one organelle and another.
2 Suggest why enzymes are attached to the inner membrane of an organelle 'in a very precise sequence'.
3 If an end product inhibits enzyme B rather than enzyme A, predict what would be:
   a the initial effect on the concentration of intermediate 1
   b the overall longer term effect on the concentration of the end product.
4 Suggest one advantage of end-product inhibition being non-competitive rather than competitive. Relate your answer to how the two types of inhibition take place.

▲ **Figure 4** *Inhibition*

Summary questions

1 Distinguish between a competitive and a non-competitve inhibitor
2 🧪 An enzyme-controlled reaction is inhibited by substance X. Suggest a simple way in which you could tell whether substance X is acting as a competitive or a non-competitive inhibitor.

▶ **Figure 3** *Comparison of competitive and non-competitive inhibition on the rate of an enzyme-controlled reaction at different substrate concentrations*

1 (a) The table shows some substances found in cells. Complete the table to show the properties of these substances. Put a tick in the box if the statement is correct.

Statement	Substance			
	Starch	Glycogen	Deoxyribose	DNA helicase
Substance contains only the elements carbon, hydrogen and oxygen				
Substance is made from amino acid monomers				
Substance is found in both animal cells and plant cells				

(4 marks)

(b) The diagram shows two molecules of β-glucose.

On the diagram, draw a box around the atoms that are removed when the two β-glucose molecules are joined by condensation. *(2 marks)*

(c) (i) Hydrogen bonds are important in cellulose molecules. Explain why. *(2 marks)*

(ii) A starch molecule has a spiral shape. Explain why this shape is important to its function in cells. *(1 mark)*

AQA Jan 2011

2 (a) Omega-3 fatty acids are unsaturated. What is an *unsaturated* fatty acid? *(2 marks)*

(b) Scientists investigated the relationship between the amount of omega-3 fatty acids eaten per day and the risk of coronary heart disease. The graph shows their results. Do the data show that eating omega-3 fatty acids prevents coronary heart disease? Explain your answer. *(3 marks)*

(c) Olestra is an artificial lipid. It is made by attaching fatty acids, by condensation, to a sucrose molecule. The diagram shows the structure of olestra. The letter R shows where a fatty acid molecule has attached.

(i)   Name bond **X**. *(1 mark)*

(ii) A triglyceride does **not** contain sucrose or bond **X**. Give **one** other way
      in which the structure of a triglyceride is different to olestra. *(1 mark)*

(iii) Starting with separate molecules of glucose, fructose and fatty acids,
      how many molecules of water would be produced when one molecule
      of olestra is formed? *(1 mark)*

AQA Jan 2011

3   A technician investigated the effect of temperature on the rate of an enzyme-controlled
        reaction. At each temperature, he started the reaction using the same volume of substrate
        solution and the same volume of enzyme solution.

▲ **Figure 2** *shows his results*

(a) Give **one** other factor the technician would have controlled. *(1 mark)*

(b) Calculate the rate of reaction at 25°C. *(2 marks)*

(c) Describe and explain the differences between the two curves. *(5 marks)*

AQA SAMS PAPER 1

Nucleic acids are a group of the most important molecules of which the best known are **ribonucleic acid** (**RNA**) and **deoxyribonucleic acid** (**DNA**). The double helix structure of deoxyribonucleic acid (DNA) makes it immediately recognisable. DNA carries genetic information. The identification of this extraordinary molecule as the material that passes on the features of organisms from one generation to the next is one of the most remarkable feats of experimental biology. The discovery of the precise molecular arrangement of DNA was no less remarkable. Despite its complex structure, DNA is made up of nucleotides that have just three basic components.

Nucleotide structure

Individual nucleotides are made up of three components:

- a pentose sugar (so called because it has five carbon atoms)
- a phosphate group
- a nitrogen-containing organic base. These are: cytosine **C**, thymine **T**, Uracil **U**, adenine **A** and guanine **G**.

The pentose sugar, phosphate group and organic base are joined, as a result of **condensation reactions**, to form a single nucleotide (**mononucleotide**) as shown in Figure 1. Two mononucleotides may, in turn, be joined as a result of a condensation reaction between the deoxyribose sugar of one mononucleotide and the phosphate group of another. The bond formed between them is called a **phosphodiester bond** (Figure 3). The new structure is called a **dinucleotide**. The continued linking of mononucleotides in this way forms a long chain known as a **polynucleotide**. In addition to DNA and RNA, some other biologically important molecules contain nucleotides. For simplicity the various components of nucleotides are represented by symbols, as shown in Table 1.

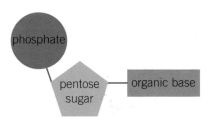

▲ **Figure 1** *Simplified structure of a nucleotide*

▼ **Table 1** *Components of nucleotides.*

Name of molecule	Symbol
phosphate	
pentose sugar	
adenine	adenine
guanine	guanine
cytosine	cytosine
thymine	thymine
uracil	uracil

Ribonucleic acid (RNA) structure

Ribonucleic acid is a polymer made up of nucleotides. It is a single, relatively short, polynucleotide chain in which the pentose sugar is always **ribose** and the organic bases are adenine, guanine, cytosine and **uracil** (Figure 2). One type of RNA transfers genetic information from DNA to the ribosomes. The ribosomes themselves are made up of proteins and another type of RNA. A third type of RNA is involved in protein synthesis.

DNA structure

In 1953, James Watson and Francis Crick worked out the structure of DNA, following pioneering work by Rosalind Franklin on the X-ray diffraction patterns of DNA. This opened the door for many of the major developments in biology over the next half-century.

In DNA the pentose sugar is deoxyribose and the organic bases are adenine, thymine, guanine and cytosine. DNA is made up of two strands of nucleotides (polynucleotides). Each of the two strands is extremely long, and they are joined together by **hydrogen bonds** formed between certain bases. In its simplified form, DNA can be thought of as a ladder in which the phosphate and deoxyribose molecules alternate to form the uprights and the organic bases pair together to form the rungs (Figure 4).

▲ **Figure 2** *Section of an RNA molecule*

Base pairing

The bases on the two strands of DNA attach to each other by hydrogen bonds. It is these hydrogen bonds that hold the two strands together. The base pairing is specific:

* Adenine always pairs with thymine
* Guanine always pairs with cytosine

As a result of these pairings, adenine is said to be **complementary** to thymine and guanine is said to be complementary to cytosine.

It follows that the quantities of adenine and thymine in DNA are always the same, and so are the quantities of guanine and cytosine. However, the ratio of adenine and thymine to guanine and cytosine varies from species to species.

The double helix

In order to appreciate the structure of DNA, you need to imagine the ladder-like arrangement of the two polynucleotide chains being twisted. In this way, the uprights of phosphate and deoxyribose wind around one another to form a double helix. They form the structural backbone of the DNA molecule.

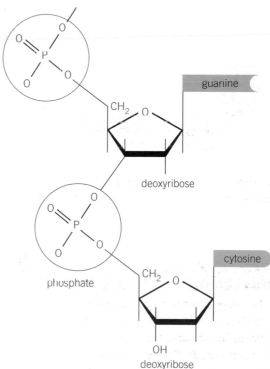

▲ **Figure 3** *The structure of a phosphodiester bond between a guanine nucleotide and a cytosine nucleotide*

The uprights are composed of deoxyribose–phosphate molecules, and the rungs are pairs of bases.

10 base pairs

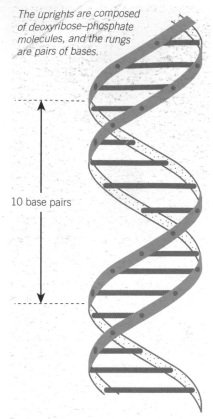

▲ **Figure 5** *The double helix structure of DNA*

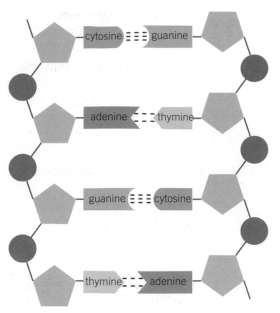

▲ **Figure 4** *Basic structure of DNA*
DNA structure may be likened to a ladder in which alternating phosphate and deoxyribose molecules make up the 'uprights' and pairs of organic bases comprise the 'rungs'. Note the base pairings are always cytosine–guanine and adenine–thymine. This ensures a standard 'rung' length. Note also that the 'uprights' run in the opposite direction to each other (i.e. are antiparallel).

Hint

In every molecule of DNA, the phosphate group, the deoxyribose and the four bases are always the same. What differs between one DNA molecule and another are the proportions, and more importantly the sequence, of each of the four bases.

Synoptic link

More detail on mutations is given in Topic 9.1, Gene mutations, and about protein synthesis in Topic 8.4, Polypeptide synthesis – transcription and splicing.

The stability of DNA

DNA is a stable molecule because:

- The phosphodiester backbone protects the more chemically reactive organic bases inside the double helix.
- Hydrogen bonds link the organic base pairs forming bridges (rungs) between the phosphodiester uprights. As there are three hydrogen bonds between cystine and guanine, the higher the proportion of C—G pairings, the more stable the DNA molecule.

There are other interactive forces between the base pairs that hold the molecule together (= base stacking).

Function of DNA

DNA is the hereditary material responsible for passing genetic information from cell to cell and generation to generation. In total, there are around 3.2 billion base pairs in the DNA of a typical mammalian cell. This vast number means that there is an almost infinite variety of sequences of bases along the length of a DNA molecule. It is this variety that provides the genetic diversity within living organisms.

The DNA molecule is adapted to carry out its functions in a number of ways:

- It is a very stable structure which normally passes from generation to generation without change. Only rarely does it mutate.

- Its two separate strands are joined only with hydrogen bonds, which allow them to separate during DNA replication (Topic 2.2) and protein synthesis.
- It is an extremely large molecule and therefore carries an immense amount of genetic information.
- By having the base pairs within the helical cylinder of the deoxyribose–phosphate backbone, the genetic information is to some extent protected from being corrupted by outside chemical and physical forces.
- Base pairing leads to DNA being able to replicate and to transfer information as mRNA.

The function of the remarkable molecule that is DNA depends on the sequence of base pairs that it possesses. This sequence is important to everything it does and, indeed, to life itself.

Summary questions

1 List the three basic components of a nucleotide.

2 Suggest why the base pairings of adenine with cytosine and guanine with thymine do not occur.

3 If the bases on one strand of DNA are TGGAGACT, determine the base sequence on the other strand.

4   If 19.9% of the base pairs in human DNA are guanine, calculate what percentage of human DNA is thymine. Show your reasoning.

 ## Unravelling the role of DNA

We now take for granted that DNA is the hereditary material that passes genetic information from cell to cell and generation to generation. This was not always the case because there were other contenders for this role, in particular proteins.

With the knowledge available at the time, scientists thought that proteins were the more likely candidate because of their considerable chemical diversity. DNA was considered to have too few components and to be chemically too simple to fulfil the role. However, not all scientists were convinced and so they set about finding experimental evidence to determine the true nature of hereditary material.

1 Assess the advantages of scientists questioning the validity of a current theory rather than automatically accepting it.

Scientists work by using **observations** and current knowledge to form a **hypothesis**. From this, they make **predictions** about the outcome of a particular **investigation**. By carrying out this investigation a number of times, they collect the experimental evidence that allows them to accept or reject their hypothesis.

2 Explain what is meant by the term 'hypothesis' in the scientific sense.

Investigations were needed to test the hypothesis that DNA was the hereditary material.

One investigation to test the hypothesis that DNA was the hereditary material involved experiments using mice and a bacterium that can cause pneumonia. The bacterium exists in two forms:

- a safe form that does not cause pneumonia, known as the R-strain,

- a harmful form that causes pneumonia, known as the S-strain.

Mice were separately injected with living bacteria of the safe form and dead bacteria from the harmful form.

The group of mice injected with the living safe form of bacteria remained healthy, as did the group injected with the dead harmful form of bacteria.

So, when mice were injected with both types together, it would not have been surprising to get a similar result. These mice, however, developed pneumonia. The experiment and the results are summarised in Figure 7.

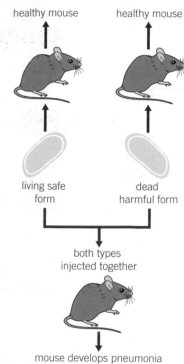

▲ **Figure 7** *Summary of an experiment to determine the nature of hereditary material in an organism*

Living bacteria of the harmful form were isolated from the mice with pneumonia. There are three possible explanations for this:

- Experimental error, for example, the harmful forms in the mixture were not all killed.

- The living safe form had mutated into the harmful form. This is possible but extremely unlikely, especially given that the experiment was repeated many times with the same result.

- Pneumonia is caused by a toxin. The harmful form of the bacterium has the information on how to make the toxin but, being dead, cannot do so. The safe form has the means of making the toxin but lacks the information on how to do so. The information on how to make the toxin may have been transferred from the harmful form to the safe form, which then produced it.

3 🧪 State what simple procedure could be carried out to discount the first explanation.

4 Mutations happen very rarely. Explain why this helps to discount the second explanation.

The third explanation was considered worthy of further investigation and so a series of experiments was designed and carried out as follows:

- The living harmful bacteria that were found in the mice with pneumonia, were collected.
- Various substances were isolated from these bacteria and purified.
- Each substance was added to suspensions of living safe bacteria to see whether it would transform them into the harmful form.
- The only substance that produced this transformation was purified DNA.
- When an enzyme that breaks down DNA was added, the ability to carry out the transformation ceased.

Other experiments provided further proof that DNA was the hereditary material and also suggested a mechanism by which it could be transferred from one bacterial cell to another.

- It had been observed that viruses infect bacteria, causing the bacteria to make more viruses.
- As the virus is made up of just protein and DNA, one or the other must possess the instructions that the bacteria use to make new viruses.
- The protein and DNA in the viruses were each labelled with a different radioactive element.
- One sample of bacteria was infected by viruses with radioactive protein while another sample was infected by viruses with radioactive DNA.
- In a later stage, the viruses and bacteria in both samples were separated from one another.
- Only the sample with bacteria that had been infected by viruses labelled with radioactive DNA showed signs of radioactivity.

This was evidence that DNA was the material that had provided the bacteria with the genetic information needed to make the viruses. It also showed how DNA can be passed from one bacterium to another, for example, by viruses.

5 A new scientific discovery often presents moral, economic and ethical issues. Justify why it is necessary for society to analyse the risks and benefits of these discoveries before they are developed.

A prime location

In order to understand how nucleotides are arranged in nucleic acids, it is necessary to know how the carbon atoms in the pentose molecule are numbered. Of particular importance is the numbering of the 3′ (3-prime) and 5′ (5-prime) carbon atoms. The 5′ carbon has an attached phosphate group, while the 3′ has a hydroxyl group.

Figure 8 shows a nucleotide with the 3′ and 5′ carbon atoms marked on its pentose sugar.

When nucleotides are organised into the double strands of a DNA molecule, one strand runs in the 5′ to 3′ direction while the other runs the opposite way – in the 3′ to 5′ direction. The two strands are therefore said to be antiparallel.

▲ **Figure 8** *Nucleotide showing positions of the 3-prime (3′) and 5-prime (5′) carbon atoms on the pentose sugar*

Nucleic acids can only be synthesised 'in vivo' in the 5′-to-3′ direction. This is because the enzyme DNA polymerase that assembles nucleotides into a DNA molecule can only attach nucleotides to the hydroxyl (OH) group on the 3′ carbon molecule.

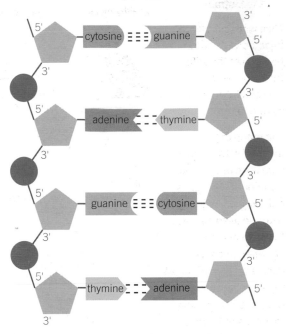

▲ **Figure 9** *DNA molecule showing the 3-prime and 5-prime carbon atoms labelled. Notice that one strand runs 5′ to 3′ while the other runs 3′ to 5′. They are* ***antiparallel***

1 Suggest what the term 'in vivo' means in the context of synthesising DNA.
2 From your knowledge of the way enzymes work, explain why DNA polymersae can only attach nucleotides to the hydroxyl (OH) group on the 3′ carbon molecule.

The cells that make up organisms are always derived from existing cells by the process of division. Cell division occurs in two main stages:

- **Nuclear division** is the process by which the nucleus divides. There are two types of nuclear division, mitosis and **meiosis**.
- **Cytokinesis** follows nuclear division and is the process by which the whole cell divides.

Before a nucleus divides its DNA must be replicated (copied). This is to ensure that all the daughter cells have the genetic information to produce the enzymes and other proteins that they need.

The process of DNA replication is clearly very precise because all the new cells are more or less genetically identical to the original one. How then does DNA replication take place? It is the semi-conservative model that is universally accepted.

Semi-conservative replication

For semi-conservative replication to take place there are four requirements:

- The four types of nucleotide, each with their bases of adenine, guanine, cytosine or thymine, must be present.
- Both strands of the DNA molecule act as a template for the attachment of these nucleotides.
- The enzyme DNA polymerase.
- A source of chemical energy is required to drive the process.

The process of semi-conservative replication is illustrated in Figure 1. It takes place as follows:

- The enzyme **DNA helicase** breaks the hydrogen bonds linking the base pairs of DNA.
- As a result the double helix separates into its two strands and unwinds.
- Each exposed polynucleotide strand then acts as a template to which complementary free nucleotides bind by specific base pairing
- Nucleotides are joined together in a condensation reaction by the enzyme **DNA polymerase** to form the 'missing' polynucleotide strand on each of the two original polynucleotide strands of DNA.
- Each of the new DNA molecules contains one of the original DNA strands, that is, half the original DNA has been saved and built into each of the new DNA molecules (Figure 2). The process is termed 'semi-conservative replication'.

a A representative portion of DNA, which is about to undergo replication.

b An enzyme, DNA helicase, causes the two strands of the DNA to separate by breaking the hydrogen bonds that join the complementary bases together.

c DNA helicase completes the splitting of the strand. Meanwhile, free nucleotides that have been activated bind specifically to their complementary bases.

free nucleotides

d Once the activated nucleotides are bound, they are joined together by DNA polymerase which makes phosphodiester bonds (bottom three nucleotides). The remaining unpaired bases continue to attract their complementary nucleotides.

e Finally, all the nucleotides are joined to form a complete polynucleotide chain using DNA polymerase. In this way, two identical strands of DNA are formed. As each strand retains half of the original DNA material, this method of replication is called the semi-conservative method.

original   new   new   original
strand   strand   strand   strand

▲ **Figure 1** *The semi-conservative replication of DNA*

Summary questions

1 If the bases on a portion of the original strand of DNA are ATGCTACG, determine the equivalent sequence of bases on the newly formed strand.

2 Explain why the process of DNA replication is described as semi-conservative.

3 If an inhibitor of DNA polymerase were introduced into a cell, explain what the effect would be on DNA replication.

original DNA molecule

DNA polymerase

replication   replication

replication fork

DNA polymerase   new DNA molecules

▲ **Figure 2** *Role of DNA polymerase in the semi-conservative replication of DNA*

Evidence for semi-conservative replication

This account illustrates how scientists use theories and models to attempt to explain observations. Scientific progress is made when experimental evidence is produced that supports a new theory or model.

When James Watson and Francis Crick worked out the structure of DNA in 1953, with the help of Rosalind Franklin's X-ray diffraction studies, they remarked in their paper:

> It has not escaped our notice that the specific pairing we have postulated immediately suggests a possible copying mechanism for the genetic material.

Their idea, namely the semi-conservative method, was, however, only one of two possible mechanisms. Both needed to be scientifically tested before a definite conclusion could be drawn. The two hypotheses were:

- **The conservative model** suggested that the original DNA molecule remained intact and that a separate daughter DNA copy was built up from new molecules of deoxyribose, phosphate and organic bases. Of the two molecules produced, one would be made of entirely new material while the other would be entirely original material (Figure 3).

- **The semi-conservative model** proposed that the original DNA molecule split into two separate strands, each of which then replicated its mirror image (i.e. the missing half). Each of the two new molecules would therefore have one strand of new material and one strand of original material (Figure 3).

If we look at Figure 3, we can see that the distribution of the strands from the original DNA molecule after replication is different in each model. To find out which mechanism was correct was therefore easy, at least in theory – simply label the original DNA in some way and then look at how it was distributed after replication. The next stage was to design an experiment to test which hypothesis was correct. Two scientists, Meselsohn and Stahl, achieved this in a neat and elegant experiment.

They based their work on three facts:

- All the bases in DNA contain nitrogen.
- Nitrogen has two forms: the lighter nitrogen ^{14}N and the **isotope** ^{15}N, which is heavier.
- Bacteria will incorporate nitrogen from their growing medium into any new DNA that they make.

= parental DNA   = daughter DNA

▲ **Figure 3** *Different models of DNA replication*

They reasoned that bacteria grown on a medium containing ^{14}N would have DNA that was lighter than bacteria grown on a medium containing ^{15}N. They labelled the original DNA of bacteria by growing them on a medium of ^{15}N. They then transferred the bacteria to a medium of ^{14}N for a single generation to allow it to replicate once. The mass of each 'new' DNA molecule would depend upon which method of replication had taken place (Figure 3). To separate out the different DNA types, they centrifuged the extracted DNA in a special solution. The lighter the DNA, the nearer the top of the centrifuge tube it collected. The heavier the DNA, the nearer the bottom of the tube it collected (see Topic 3.1). They also analysed the DNA after two, then three, generations. By interpreting the results they could determine which hypothesis was correct. Their work is summarised in Figure 4.

1 Name the part of the DNA molecule that contains nitrogen.
2 Explain why, after one generation, all the DNA is made up of an equal mixture of ^{14}N and ^{15}N.

3 Suppose DNA were replicated by the conservative model. Sketch a tube showing the position of DNA after one generation.

4 From Figure 4, copy the chart for tube 4. Draw bars on the chart to show the percentage of each of the three possible types of DNA.

5 After three generations (tube 5), calculate what percentage of the DNA will be made up of ^{14}N only.

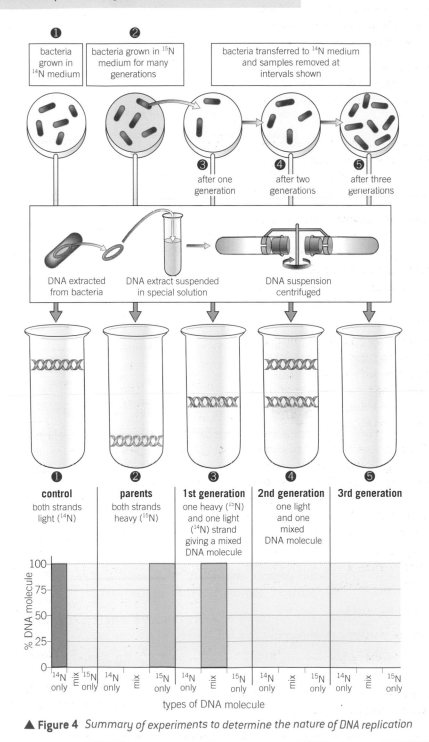

▲ **Figure 4** *Summary of experiments to determine the nature of DNA replication*

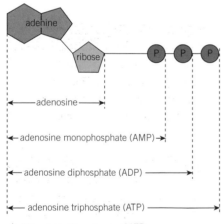

▲ **Figure 1** *Structure of ATP*

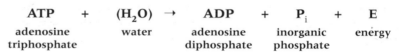

▲ **Figure 2** *Interconversion of ATP and ADP*

All living organisms require energy in order to remain alive. This energy comes initially from the Sun. Plants use solar energy to combine water and carbon dioxide into complex organic molecules by the process of photosynthesis. Both plants and animals then oxidise these organic molecules to make adenosine triphosphate (ATP), which is used as the main energy source to carry out processes within cells.

Structure of ATP

The ATP molecule (Figure 1) is a phosphorylated macromolecule. It has three parts:

- **adenine** – a nitrogen-containing organic base
- **ribose** – a sugar molecule with a 5-carbon ring structure (pentose sugar) that acts as the backbone to which the other parts are attached
- **phosphates** – a chain of three phosphate groups.

How ATP stores energy

Adenosine triphosphate (ATP) is a nucleotide and as the name suggests, has three phosphate groups. These are the key to how ATP stores energy. The bonds between these phosphate groups are unstable and so have a low **activation energy**, which means they are easily broken. When they do break they release a considerable amount of energy. Usually in living cells it is only the terminal phosphate that is removed, according to the equation:

$$\textbf{ATP} + \textbf{(H}_2\textbf{O)} \rightarrow \textbf{ADP} + \textbf{P}_i + \textbf{E}$$

| adenosine triphosphate | water | adenosine diphosphate | inorganic phosphate | energy |

As water is used to convert ATP to ADP, this is known as a **hydrolysis** reaction. The reaction is catalysed by the enzyme **ATP hydrolase** (ATPase).

Synthesis of ATP

The conversion of ATP to ADP is a reversible reaction and therefore energy can be used to add an inorganic phosphate to ADP to re-form ATP according to the reverse of the equation above. This reaction is catalysed by the enzyme **ATP synthase**. As water is removed in this process, the reaction is known as a **condensation reaction**. Figure 2 summarises the interconversion of ATP and ADP.

The synthesis of ATP from ADP involves the addition of a phosphate molecule to ADP. It occurs in three ways:

- in chlorophyll-containing plant cells during photosynthesis (photophosphorylation)
- in plant and animal cells during respiration (oxidative phosphorylation)
- in plant and animal cells when phosphate groups are transferred from donor molecules to ADP (substrate-level phosphorylation).

Roles of ATP

The same feature that makes ATP a good energy donor, namely the instability of its phosphate bonds, is also a reason why it is not a good long-term energy store. Fats, and carbohydrates such as glycogen, serve this purpose far better. ATP is therefore the **immediate energy source** of a cell. As a result, cells do not store large quantities of ATP, but rather just maintain a few seconds' supply. This is not a problem, as ATP is rapidly re-formed from ADP and inorganic phosphate (P_i) and so a little goes a long way. ATP is a better immediate energy source than glucose for the following reasons:

- Each ATP molecule releases less energy than each glucose molecule. The energy for reactions is therefore released in smaller, more manageable quantities rather than the much greater, and therefore less manageable, release of energy from a glucose molecule.

- The **hydrolysis** of ATP to ADP is a single reaction that releases immediate energy. The breakdown of glucose is a long series of reactions and therefore the energy release takes longer.

ATP cannot be stored and so has to be continuously made within the mitochondria of cells that need it. Cells, such as muscle fibres and the epithelium of the small intestine, which require energy for movement and active transport respectively, possess many large mitochondria.

ATP is used in energy-requiring processes in cells including:

- **metabolic processes**. ATP provides the energy needed to build up macromolecules from their basic units. For example, making starch from glucose or polypeptides from amino acids.

- **movement**. ATP provides the energy for muscle contraction. In muscle contraction, ATP provides the energy for the filaments of muscle to slide past one another and therefore shorten the overall length of a muscle fibre.

- **active transport**. ATP provides the energy to change the shape of carrier proteins in plasma membranes. This allows molecules or ions to be moved against a concentration gradient.

- **secretion**. ATP is needed to form the lysosomes necessary for the secretion of cell products.

- **activation of molecules**. The inorganic phosphate released during the hydrolysis of ATP can be used to phosphorylate other compounds in order to make them more reactive, thus lowering the activation energy in enzyme-catalysed reactions. For example – the addition of phosphate to glucose molecules at the start of **glycolysis**.

Hint

Think of the unstable bonds that link the phosphates in ATP as coiled springs. Due to these spring-like bonds the end phosphate is straining to break away from its nearest partner. Any small addition of energy and the end phosphate springs away, releasing all the energy that is stored in the 'spring', that is, stored in the bond.

Hint

ATP is synthesised during reactions that *release* energy and it is hydrolysed to provide energy for reactions that *require* it.

Study tip

Don't think about ATP as a 'high-energy' substance. ATP is an 'intermediate energy' substance that is used to transfer energy.

Synoptic link

Concentration gradients are looked at in more detail in Topic 4.4, while the role of lysosomes in secretion is covered in Topic 3.4.

Summary questions

1   ATP is sometimes referred to as 'an immediate energy source'. Explain why.

2   Explain how ATP can make an enzyme-catalysed reaction take place more readily.

3   State **three** roles of ATP in plant cells.

Learning objectives

→ Describe the structure of the water molecule.

→ State the properties of the water molecule.

→ Explain the importance of the water molecule to living organisms.

→ Describe inorganic ions and their roles.

Specification reference: 3.1.7 and 3.1.8

▲ **Figure 1** *Water molecules showing hydrogen bonding*

Water is a major component of cells. Although water is the most abundant liquid on Earth, it is certainly no ordinary molecule. Its unusual properties are due to its dipolar nature and the subsequent hydrogen bonding that this allows.

The dipolar water molecule

A water molecule is made up of two atoms of hydrogen and one of oxygen as shown in Figure 1. Although the molecule has no overall charge, the oxygen atom has a slight negative charge, while the hydrogen atoms have a slight positive one. In other words, the water molecule has both positive and negative poles and is therefore described as **dipolar**.

Water and hydrogen bonding

Different poles attract, and therefore the positive pole of one water molecule will be attracted to the negative pole of another water molecule. The attractive force between these opposite charges is called a hydrogen bond (Figure 2). Although each bond is fairly weak (about one-tenth as strong as a **covalent bond**), together they form important forces that cause the water molecules to stick together, giving water its unusual properties.

Specific heat capacity of water

Because water molecules stick together, it takes more energy (heat) to separate them than would be needed if they did not bond to one another. For this reason the boiling point of water is higher than expected. Without its hydrogen bonding, water would be a gas (water vapour.) at the temperatvures commonly found on Earth and life as we know it would not exist. For the same reason, it takes more energy to heat a given mass of water, that is water has a high specific heat capacity. Water therefore acts as a buffer against sudden temperature variations, making the aquatic environment a temperature-stable one. As organisms are mostly water, it also buffers them against sudden temperature changes especially in terrestrial environments.

Latent heat of vaporisation of water

Hydrogen bonding between water molecules means that it requires a lot of energy to evaporate 1 gram of water. This energy is called the **latent heat of vaporisation**. Evaporation of water such as sweat in mammals is therefore a very effective means of cooling because body heat is used to evaporate the water.

▲ **Figure 2** *Due to surface tension, pond skaters walk on water*

Cohesion and surface tension in water

The tendency of molecules to stick together is known as cohesion. With its hydrogen bonding, water has large cohesive forces and these allow it to be pulled up through a tube, such as a **xylem vessel** in plants. In the same way, where water molecules meet air they tend to be pulled back into the body of water rather than escaping from it. This force is called surface tension and means that the water surface acts like a skin and is strong enough to support small organisms such as pond skaters (Figure 3).

The importance of water to living organisms

Water is the main constituent of all organisms – up to 98% of a jellyfish is water and mammals are typically 65% water. Water is also where life on Earth arose and it is the environment in which many species still live. It is important for other reasons too.

Water in metabolism

- Water is used to break down many complex molecules by **hydrolysis**, for example, proteins to amino acids. Water is also produced in **condensation** reactions.
- Chemical reactions take place in an aqueous medium.
- Water is a major raw material in photosynthesis.

Water as a solvent

Water readily dissolves other substances:

- gases such as oxygen and carbon dioxide
- wastes such as ammonia and urea
- inorganic ions and small hydrophilic molecules such as amino acids, monosaccharides and ATP
- enzymes, whose reactions take place in solution.

Other important features of water

- Its evaporation cools organisms and allows them to control their temperature.
- It is not easily compressed and therefore provides support, for example the hydrostatic skeleton of animals such as the earthworm and turgor pressure in herbaceous plants.
- It is transparent and therefore aquatic plants can photosynthesise and also light rays can penetrate the jelly-like fluid that fills the eye and so reach the retina.

Inorganic ions

Inorganic ions are found in organisms where they occur in solution in the cytoplasm of cells and in body fluids and as well as part of larger molecules. They may be in concentrations that range from very high to very low.

Inorganic ions perform a range of functions. The specific function a particular ion per forms is related to its properties. For example, as we saw in Topic 1.6, iron ions are found in haemoglobin where they play a role in the transport of oxygen. Other examples we have looked at include the phosphate ions that form a structural role in DNA molecules (Topic 2.1) and a role in storing energy in ATP molecules (Topic 2.3). Hydrogen ions are important in determining the pH of solutions and therefore the functioning of enzymes (Topic 1.8). We shall see in Topic 4.5 that sodium ions are important in the transport of glucose and amino acids across plasma membranes.

▲ **Figure 3** *Evaporation of water during sweating helps to maintain body temperature*

Synoptic link

Xylem vessels will be covered in Topic 7.7 and condensation and hydrolysis were covered in Topic 1.3.

Summary questions

In the following passage, state the missing word indicated by the letters **a–f**.

A water molecule is said to be **a** because it has a positive and a negative pole as a result of the uneven distribution of **b** within it. This creates attractive forces called **c** between water molecules, causing them to stick together. This stickiness of water means that its molecules are pulled inwards at its surface. This force is called **d**. Water is able to split large molecules into smaller ones by a process known as **e**. Water is the raw material for the process of **f** in green plants.

1 (a) **Figure 1** shows one base pair of a DNA molecule.

▲ **Figure 1**

   (i)   Name part **F** of each nucleotide.   (*1 mark*)
   (ii) Scientists determined that a sample of DNA contained 18% adenine.

   What were the percentages of thymine and guanine in this sample of DNA?   (*2 marks*)

AQA SAMS PAPER 1

   (b) Organisms vary widely in the number of genes they have. Figure 2 shows the total
        length of DNA in six organisms plotted against the number of functional genes. The
        length of DNA is measured in numbers of base pairs.

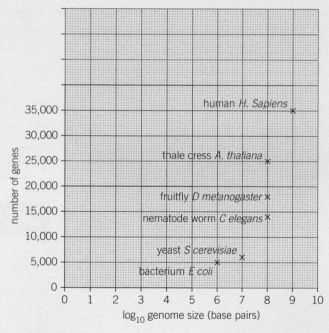

   (i)   A double-stranded DNA molecule is 2 μm long for every thousand base
         pairs. Use Figure 2 to calculate the total length of DNA in a human cell
         in metres. Show your working.   (*2 marks*)
   (ii) Calculate the ratio of DNA length to number of genes in humans and
         in *Escherichia coli* and suggest why there is this difference.   (*2 marks*)

2 **Figure 3** shows a short section of a DNA molecule.

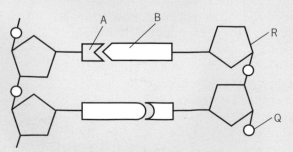

(a) Name parts **R** and **Q**. *(2 marks)*

(b) Name the type of bonds that join **A** and **B**. *(1 mark)*

(c) Ribonuclease is an enzyme. It is 127 amino acids long. What is the minimum number of DNA bases needed to code for ribonuclease? *(1 mark)*

(d) **Figure 2** shows the sequence of DNA bases coding for seven amino acids in the enzyme ribonuclease.

Figure 2

GTT TAC TAC TCT TCT TCT TTA

The number of each type of amino acid coded for by this sequence of DNA bases is shown in the table.

Amino acid	Number present
Arg	3
Met	2
Gln	1
Asn	1

Use the table and **Figure 2** to work out the sequence of amino acids in this part of the enzyme. Write your answer in the boxes below.

Gln						

(1 mark)

(e) Explain how a change in a sequence of DNA bases could result in a non-functional enzyme. *(3 marks)*

AQA Jan 2010

3 (a) Hydrogen bonds occur between water molecules. Explain how these affect the properties of water as a habitat for organisms. *(2 marks)*

(b) Give two inorganic ions within the human body and describe one function of each. *(4 marks)*

(c) Explain the importance of the hydrolysis reaction of ATP. *(2 marks)*

Section 1 Summary

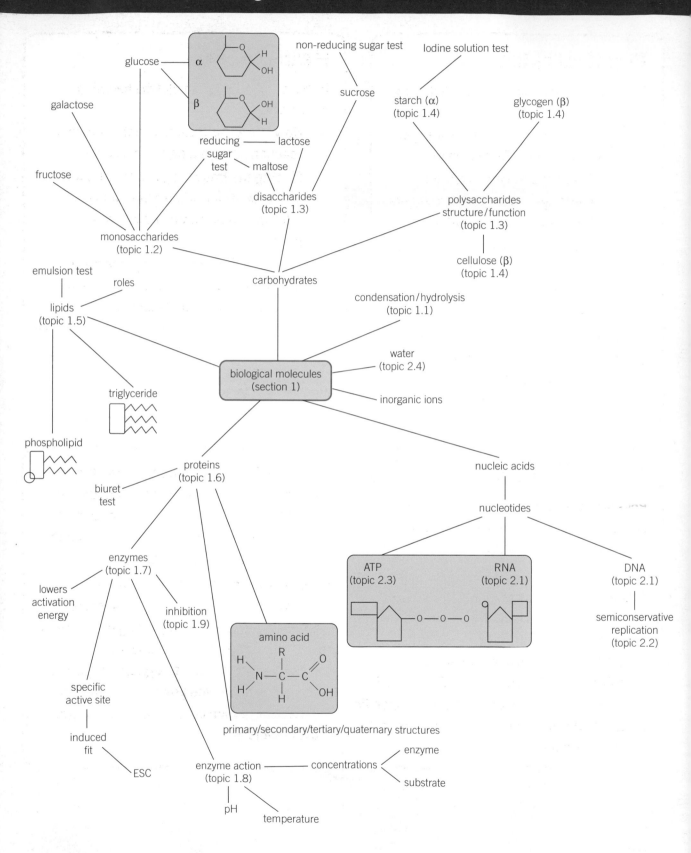

Practical skills

In this section you have met the following practical skills:

- How to carry out tests for a variety of food substances. For example:
  - Benedict's reagent to test for reducing and non-reducing sugars
  - iodine in potassium iodide solution to test for starch
  - the Biuret test for proteins
  - the emulsion test for lipids.
- The importance of keeping all other variables constant except the one being investigated, when carrying out investigations such as the effects of different factors on enzyme action.

Maths skills

In this section you have met the following maths skills:

- How to measure change in the rate of a reaction using a tangent to a curve.
- Finding the frequency of all the bases on a DNA strand when given only information on some of the bases present.
- Choosing the best way of presenting your results when investigating the rate of enzyme-controlled reactions.

Extension task

Research the following two topics using any source of information available to you, for example, textbooks, journals, the internet etc.

Proteins are very diverse in their structure and functions and so are involved in all processes of living organisms including: nutrition, respiration, transport, growth, excretion, support, movement, sensory perception, coordination and reproduction. For each of these, state the name of one different specific protein that is involved in each process and state its function.

A common riddle is: which came first, the chicken or the egg? A similar conundrum for scientists is: which came first, the enzyme needed to make the nucleic acid or the nucleic acid needed to make the enzyme?

It used to be thought that all enzymes were proteins but we now know that some reactions in cells are catalysed by non-protein molecules. Find out about these molecules and use the information to explain why these non-protein molecules provide an answer to the question, which came first — the enzyme or the nucleic acid?

Section 1 Practice questions

1 **(a)** Some seeds contain lipids. Describe how you could use the emulsion test to show that a seed contains lipids.   (*3 marks*)

   (b) A triglyceride is one type of lipid. The diagram shows the structure of a triglyceride molecule.

   (i) A triglyceride molecule is formed by condensation. From how many molecules is this triglyceride formed?   (*1 mark*)

   (ii) The structure of a phospholipid molecule is different from that of a triglyceride. Describe how a phospholipid is different.   (*2 marks*)

   (iii) Use the diagram to explain what is meant by an unsaturated fatty acid.   (*2 marks*)

                                                                                    AQA Jan 2012

2 Read the following passage.

   Aspirin is a very useful drug. One of its uses is to reduce fever
   and inflammation. Aspirin does this by preventing cells from producing
   substances called prostaglandins. Prostaglandins are produced by
   an enzyme-controlled pathway. Aspirin works by inhibiting one of the
   enzymes in this pathway. Aspirin attaches permanently to a   5
   chemical group on one of the monomers that make up the active site of this enzyme.

   The enzyme that is involved in the pathway leading to the production
   of prostaglandins is also involved in the pathway leading to the
   production of thromboxane. This is a substance that promotes blood
   clotting. A small daily dose of aspirin may reduce the risk of   10
   myocardial infarction (heart attack).
   Use information from the passage and your own knowledge to answer the following
   questions

   (a) Name the monomers that make up the active site of the enzyme (lines 6 – 7).

                                                                                    (*1 mark*)

   (b) The diagram shows the pathways by which prostaglandins and thromboxane
   are formed.

(i)   Aspirin only affects one of the enzymes in this pathway. Use information in lines 5 – 7 to explain why aspirin does **not** affect the other enzymes. (*2 marks*)

(ii) Which enzyme, **X**, **Y** or **Z**, is inhibited by aspirin? Explain the evidence from the passage that supports your answer. (*2 marks*)

(c) Aspirin is an enzyme inhibitor. Explain how aspirin prevents substrate molecules being converted to product molecules. (*2 marks*)

(d) Aspirin may reduce the risk of myocardial infarction (lines 8 – 12). Explain how. (*3 marks*)

AQA Jan 2012

3   The diagram shows part of a DNA molecule

(a) (i)   DNA is a polymer. What is the evidence from the diagram that DNA is a polymer? (*1 mark*)

(ii) Name the parts of the diagram labelled **C**, **D** and **E**. (*3 marks*)

(iii) In a piece of DNA, 34% of the bases were thymine.

Complete the table to show the names and percentages of the other bases.

Name of base	Percentage
Thymine	34
	34

(*2 marks*)

(b) A polypeptide has 51 amino acids in its primary structure.
(i)   What is the minimum number of DNA bases required to code for the amino acids in this polypeptide? (*1 mark*)

AQA Jan 2012

Section 2
Cells

Introduction

The cell is the fundamental unit of life. All organisms, whatever their type or size, are composed of cells. All new cells are derived from existing ones by one of the following the processes of binary fission (prokaryotic cells), mitosis and meiosis (eukaryotic cells). Cells contain the genetic material of an organism and metabolic processes take place within them.

Cells all share certain basic features and yet show remarkable diversity in both structure and function. Their differences are the result of additional features that have arisen over time. This provides indirect evidence for evolution. Eukaryotic cells have a nucleus, or nuclear region, at some stage of their existence. Cells are all surrounded by a cell-surface membrane and in eukaryotic cells there are also internal membranes.

The structure of the plasma membrane is basically the same whether it forms the cell-surface membrane or any of the internal membranes within the cell. It controls the passage of substances across it by passive and active transport. The cell-surface membrane acts as the boundary between the cell and its environment. It may exclude some substances while retaining others. Some substances may pass freely across it, while others are prevented from doing so at one moment, only to pass freely across on another occasion.

The cell-surface membrane is made up almost entirely of proteins and phospholipids. Certain proteins that are embedded in the membrane are involved in communication between cells (cell signalling) while others act as antigens, allowing the cell to be recognised by the immune system as either 'self' or 'non-self' (foreign). Antigens play an important role in defence against disease and immunity. Various cells such as T lymphocytes and B lymphocytes interact to combat infections and to prevent symptoms arising when there are future infections by the same pathogen. These immune responses can be artificially induced by the process of vaccination.

Working scientifically

The study of cells gives plenty of scope to carry out practical work and develop practical skills. Required practical activities are:

- The preparation of stained squashes of cells from plant root tips and the setting up and use of an optical microscope to identify the stages of mitosis in these stained squashes. You will also be required to use your observations to calculate a mitotic index.

- The production of a dilution series of a solute to produce a calibration curve with which to identify the water potential of a plant tissue.

- An investigation into the effect of a named variable on the permeability of cell-surface membranes.

You will require a range of mathematical skills – in particular the ability to use percentages, make order of magnitude calculations, plot two variables from experimental data and draw and determine the intercept of a graph.

What you already know

While the material in this unit should be understood without much prior knowledge, there is certain information from GCSE that will prove helpful. This information includes:

- ☐ Most human and animal cells have a nucleus, cytoplasm, cell membrane, mitochondria and ribosomes.

- ☐ Plant and algal cells also have a cell wall made of cellulose, which strengthens the cell. Plant cells often have chloroplasts and a permanent vacuole filled with cell sap.

- ☐ A bacterial cell consists of cytoplasm and a membrane surrounded by a cell wall – the genes are not in a distinct nucleus.

- ☐ In body cells the chromosomes are generally found in pairs. Body cells divide by mitosis. When a body cell divides by mitosis, copies of the genetic material are made then the cell divides once to form two genetically identical body cells. Mitosis occurs during asexual reproduction or growth or to produce replacement cells.

- ☐ Diffusion is the net movement of molecules from a region where they are of a higher concentration to a region with a lower concentration.

- ☐ Osmosis is the diffusion of water from a dilute to a more concentrated solution through a selectively permeable membrane that allows the passage of water molecules.

- ☐ Substances are sometimes absorbed against a concentration gradient. This involves the use of ATP from respiration. The process is called active transport. Active transport enables cells to absorb ions from very dilute solutions.

- ☐ Microorganisms that cause infectious disease are called pathogens. White blood cells help to defend against pathogens by ingesting them, producing antibodies and antitoxins.

- ☐ The immune system of the body produces specific antibodies that lead to the death of a particular pathogen. This leads to immunity from that pathogen.

- ☐ People can be immunised against a disease by introducing small quantities of dead or inactive forms of the pathogen into the body (vaccination).

3 Cell structure
3.1 Methods of studying cells

Learning objectives

→ Explain the principles of magnification and resolution.

→ Describe what cell fractionation is.

→ Explain how ultracentrifugation works.

Specification reference: 3.2.1.3

The cell is the basic unit of life. However, with a few exceptions, cells are not visible to the naked eye and their structure is only apparent when seen under a microscope.

Microscopy

Microscopes are instruments that produce a magnified image of an object. A simple convex glass lens can act as a magnifying glass but such lenses work more effectively if they are used in pairs in a compound light microscope. The relatively long wavelength of light rays means that a light microscope can only distinguish between two objects if they are 0.2 µm, or further, apart. This limitation can be overcome by using beams of **electrons** rather than beams of light. With their shorter wavelengths, the beam of electrons in the electron microscope can distinguish between two objects only 0.1 nm apart.

Magnification

The material that is put under a microscope is referred to as the **object**. The appearance of this material when viewed under the microscope is referred to as the **image**.

The magnification of an object is how many times bigger the image is when compared to the object.

$$\text{magnification} = \frac{\text{size of image}}{\text{size of real object}}$$

In practice, it is more likely that you will need to calculate the size of an object when you know the size of the image and the magnification. In this case:

$$\text{size of real object} = \frac{\text{size of image}}{\text{magnification}}$$

The important thing to remember when calculating the magnification is to ensure that the units of length (Table 1) are the same for both the object and the image.

Worked example

An object that measures 100 nm in length appears 10 mm long in a photograph. What is the magnification of the object?

$$\frac{\text{size of image}}{\text{size of real object}} = \frac{10\,\text{mm}}{100\,\text{nm}}$$

Now convert the measurements to the same units – normally the smallest – which in this case is nanometres. There are 10 000 000 nanometres in 10 millimetres and therefore the magnification is:

$$\frac{\text{size of image}}{\text{size of real object}} = \frac{10\,000\,000\,\text{nm}}{100\,\text{nm}} = \frac{100\,000}{1} = \times 100\,000 \text{ times}$$

These figures can also be expressed in standard form as follows:

$$\frac{\text{size of image}}{\text{size of real object}} = \frac{10^7}{10^2} = \frac{10^5}{1} = \times 10^5$$

Study tip

Make sure that you use scientific terms correctly. For example, light has a longer wavelength than a beam of electrons. It's not correct to say that optical microscopes have a longer wavelength than electron microscopes though.

Maths link

MS 1.8, 0.1, 0.2 and 2.2, see Chapter 22.

▼ **Table 1** *Units of length*

Unit	Symbol	Equivalent in metres
kilometre	km	10^3
metre	m	1
millimetre	mm	10^{-3}
micrometre	µm	10^{-6}
nanometre	nm	10^{-9}

▲ **Figure 1** *The equation triangle for calculating the size of image, magnification and size of object*

Resolution

The resolution, or resolving power, of a microscope is the minimum distance apart that two objects can be in order for them to appear as separate items. Whatever the type of microscope, the resolving power depends on the wavelength or form of radiation used. In a light microscope it is about 0.2 μm. This means that any two objects which are 0.2 μm or more apart will be seen separately, but any objects closer than 0.2 μm will appear as a single item. In other words, greater resolution means greater clarity, that is the image produced is clearer and more precise.

Increasing the magnification increases the size of an image, but does not always increase the resolution. Every microscope has a limit of resolution. Up to this point increasing the magnification will reveal more detail but beyond this point increasing the magnification will not do this. The object, while appearing larger, will just be more blurred.

Cell fractionation 🧪

In order to study the structure and function of the various organelles that make up cells, it is necessary to obtain large numbers of isolated organelles.

Cell fractionation is the process where cells are broken up and the different organelles they contain are separated out.

Before cell fractionation can begin, the tissue is placed in a cold, buffered solution of the same water potential as the tissue. The solution is:

- cold – to reduce enzyme activity that might break down the organelles
- is of the same water potential as the tissue – to prevent organelles bursting or shrinking as a result of osmotic gain or loss of water
- buffered – so that the pH does not fluctuate. Any change in pH could alter the structure of the organelles or affect the functioning of enzymes.

There are two stages to cell fractionation:

Homogenation

Cells are broken up by a homogeniser (blender).This releases the organelles from the cell. The resultant fluid, known as homogenate, is then filtered to remove any complete cells and large pieces of debris.

Ultracentrifugation

Ultracentrifugation is the process by which the fragments in the filtered homogenate are separated in a machine called a centrifuge. This spins tubes of homogenate at very high speed in order to create a centrifugal force. For animal cells, the process is as follows:

- The tube of filtrate is placed in the centrifuge and spun at a slow speed.
- The heaviest organelles, the nuclei, are forced to the bottom of the tube, where they form a thin sediment or pellet.
- The fluid at the top of the tube (supernatant) is removed, leaving just the sediment of nuclei.
- The supernatant is transferred to another tube and spun in the centrifuge at a faster speed than before.

▲ **Figure 2** *The effect of progressive magnification of a portion of human skin*

Summary questions

1 Distinguish between magnification and resolution.

2 $\sqrt{x}$ An organelle that is 5 μm in diameter appears under a microscope to have a diameter of 1 mm. Calculate how many times the organelle has been magnified.

3 $\sqrt{x}$ A cell organelle called a ribosome is typically 25 nm in diameter. Calculate its diameter when viewed under an electron microscope that magnifies it 400 000 times.

4 $\sqrt{x}$ At a magnification of ×12 000 a structure appears to be 6 mm long. Determine its actual length.

5 ⚗ Chloroplasts have a greater mass than mitochondria but a smaller mass than nuclei. Starting with a sample of plant cells, describe briefly how you would obtain a sample rich in chloroplasts. Use Figure 3 to help you.

6 $\sqrt{x}$ Using the magnifications given in Figure 2, calculate the actual size of the following organelles as measured along the line labelled X–Y. In your answer, use the most appropriate units from Table 1.

   a The organelle in box e

   b The organelle in box g

Maths link $\sqrt{x}$

MS 1.8 and 2.2, see Chapter 22.

- The next heaviest organelles, the mitochondria, are forced to the bottom of the tube.
- The process is continued in this way so that, at each increase in speed, the next heaviest organelle is sedimented and separated out.

A summary of cell fractionation is given in Figure 3.

1. Tissue is cut up and kept in a cold, buffered solution.

2. Cut-up tissue further broken up in a homogeniser.

3. Homogenised tissue is spun in an ultracentrifuge at a low speed for 10 minutes.

4. Result of being spun in ultracentrifuge at low speed (e.g. 1 000 × gravity).

supernatant 1

sediment 1

5. Spun in ultracentrifuge at medium speed (e.g. 3 500 × gravity).

supernatant 2

sediment 2

6. Spun in ultracentrifuge at high speed (e.g. 16 500 × gravity).

supernatant 3

sediment 3

Organelles to be separated out	Speed of centrifugation /revolutions min^{-1}
nuclei	1 000
mitochondria	3 500
lysosomes	16 500

▲ **Figure 3** *Summary of cell fractionation*

The techniques of cell fractionation and ultracentrifugation enabled considerable advances in biological knowledge. They allowed a detailed study of the structure and function of organelles, by showing what isolated components do.

◀ **Figure 4** *An ultracentrifuge used to separate the various components of cell homogenate*

Light microscopes have poor resolution as a result of the relatively long wavelength of light. In the 1930s, however, a microscope was developed that used a beam of electrons instead of light. This is called an electron microscope and it has two main advantages:

- The electron beam has a very short wavelength and the microscope can therefore resolve objects well – it has a high resolving power.
- As electrons are negatively charged the beam can be focused using electromagnets (Figure 2).

The best modern electron microscopes can resolve objects that are just 0.1 nm apart – 2000 times better than a light microscope. Because electrons are absorbed or deflected by the molecules in air, a near-vacuum has to be created within the chamber of an electron microscope in order for it to work effectively.

Learning objectives

→ Explain how electron microscopes work.

→ Explain the differences between a transmission electron microscope and a scanning electron microscope.

→ Describe the limitations of the transmission and the scanning electron microscopes.

Specification reference: 3.2.1.3

light source

electron source

condenser lens

magnetic condenser

object

object

objective lenses

magnetic objective

intermediate image

intermediate image

eyepiece lenses (projector)

magnetic projector

human eye

fluorescent screen

▲ **Figure 2** *Comparison of radiation pathways in light and electron microscopes*

▲ **Figure 1** *Scientist looking at a sample using a transmission electron microscope (TEM)*

There are two types of electron microscope:

- the transmission electron microscope (TEM)
- the scanning electron microscope (SEM).

The transmission electron microscope

The TEM consists of an electron gun that produces a beam of electrons that is focused onto the specimen by a condenser electromagnet. In a TEM, the beam passes through a thin section of the specimen. Parts of this specimen absorb electrons and therefore appear dark. Other parts of the specimen allow the electrons to pass through and so appear bright. An image is produced on a screen and this can be photographed to give a **photomicrograph**. The resolving power of the TEM is 0.1 nm although this cannot always be achieved in practice because:

- difficulties preparing the specimen limit the resolution that can be achieved
- a higher energy electron beam is required and this may destroy the specimen.

The main limitations of the TEM are:

- The whole system must be in a vacuum and therefore living specimens cannot be observed.
- A complex 'staining' process is required and even then the image is not in colour.
- The specimen must be extremely thin.
- The image may contain artefacts. Artefacts are things that result from the way the specimen is prepared. Artefacts may appear on the finished photomicrograph but are not part of the natural specimen. It is therefore not always easy to be sure that what we see on a photomicrograph really exists in that form.

In the TEM the specimens must be extremely thin to allow electrons to penetrate. The result is therefore a flat, 2-D image. We can partly get over this by taking a series of sections through a specimen. We can then build up a 3-D image of the specimen by looking at the series of

◀ **Figure 3** *Part of an animal cell seen under a TEM*

photomicrographs produced. However, this is a slow and complicated process. One way in which this problem has been overcome is the development of the SEM.

The scanning electron microscope

All the limitations of the TEM also apply to the SEM, except that specimens need not be extremely thin as electrons do not penetrate. Basically similar to a TEM, the SEM directs a beam of electrons on to the surface of the specimen from above, rather than penetrating it from below. The beam is then passed back and forth across a portion of the specimen in a regular pattern. The electrons are scattered by the specimen and the pattern of this scattering depends on the contours of the specimen surface. We can build up a 3-D image by computer analysis of the pattern of scattered electrons and secondary electrons produced. The basic SEM has a lower resolving power than a TEM, around 20 nm, but is still ten times better than a light microscope.

▲ **Figure 4** *False-colour (SEM) of a pollen grain from a marigold plant*

▲ **Figure 5** *False-colour SEM of human red blood cells*

Summary questions

1 Explain how the electron microscope is able to resolve objects better than the light microscope.

2 Explain why specimens have to be kept in a near-vacuum in order to be viewed effectively using an electron microscope.

3 State which of the biological structures in the following list can be resolved using each of the microscopes below:
   plant cell (100 μm)   DNA molecule (2 nm)   virus (100 nm)
   actin molecule (3.5 nm)   a bacterium (1 μm)

   a a light microscope

   b a transmission electron microscope

   c a scanning electron microscope.

4 In practice, the theoretical resolving power of an electron microscope cannot always be achieved. Explain why not.

5 ⓥ In a photomicrograph, an organelle measures 25 mm when its actual size is 5 μm. Calculate the magnification of this photomicrograph.

Maths link ⓥ

MS 1.8, see Chapter 22.

Maths link ⓥ

MS 2.2 and 1.8, see Chapter 22.

Learning outcomes

→ Explain how to calibrate an eyepiece graticule.

→ Explain how to measure cell size using an eyepiece graticule.

→ Learn how to calculate the size of a specimen and/or magnifications from drawings and photographs.

Specification reference: 3.2.1.3

Measuring cells

When using a light microscope, we can measure the size of objects using an **eyepiece graticule**. The graticule is a glass disc that is placed in the eyepiece of a microscope. A scale is etched on the glass disc. This scale is typically 10 mm long and is divided into 100 sub-divisions as shown in Figure 1. The scale is visible when looking down the eyepiece of the microscope.

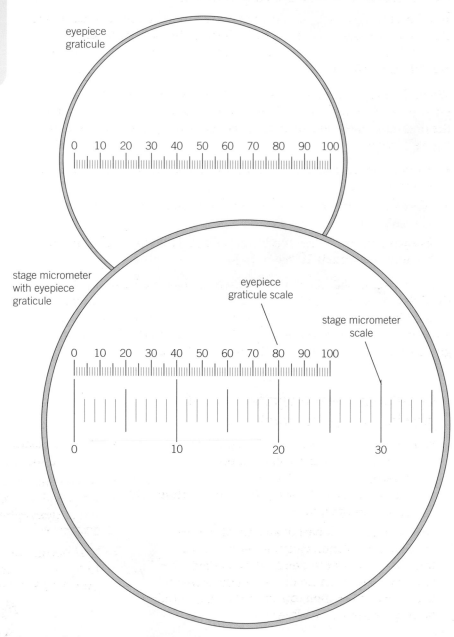

▲ **Figure 1** *An eyepiece graticule and how it is calibrated*

The scale on the eyepiece graticule cannot be used directly to measure the size of objects under a microscope's objective lens because each objective lens will magnify to a different degree. The graticule must first be calibrated for a particular objective lens. Once calibrated in this way, the graticule can remain in position for future use, provided the same objective lens is used.

It is therefore sensible to record the results of the calibration for a particular objective lens and to leave this attached to the microscope. This will save you having to recalibrate each time you want to measure the size of the object being viewed under the microscope.

Calibrating the eyepiece graticule

To calibrate an eyepiece graticule you need to use a special microscope slide called a **stage micrometer**. This slide also has a scale etched onto it. Usually the scale is 2 mm long and its smallest sub-divisions are 0.01 mm (10 μm).

When the eyepiece graticule scale and the stage micrometer scales are lined-up as shown in Figure 1, it is possible to calculate the length of the divisions on the eyepiece graticule. For example, you can see in Figure 1, that:

* 10 units on the micrometer scale are equivalent to 40 units on the graticule scale
* therefore 1 unit on the micrometer scale equals 4 units on the graticule scale
* as each unit on the micrometer scale equals 10 μm, each unit on the graticule equals 10 ÷ 4 = 2.5 μm.

It is easy to calculate the scale for different objective lenses by dividing the differences in magnification. For example, if an objective lens magnifying ×40 gives a calibration of 25 μm per graticule unit, then an objective lens magnifying ×400 (10 times greater) will mean a graticule unit is equivalent to 25 μm ÷ 10 = 2.5 μm.

Calculating linear magnifications of drawings and photographs

You may need to calculate the magnification of a drawing or photograph of an object under a microscope. You have looked at how the calculation is made. Now try an example using an actual photograph.

Look at Figure 2 of part of an animal cell seen under a transmission electron microscope. On this photograph a red line X—Y is marked. This line represents a length of 5 μm of the actual cell. Using this information, calculate the magnification of the photograph as follows:

* If you measure the length of X—Y as drawn on the photograph you find it is 23 mm long.
* As the line represents 5 μm in the cell, you also need to convert your measurement in the photograph to microns (μm). 23 μm = 23 000 μm.
* If 23 000 μm on the photograph is equivalent to 5 μm in the cell, then the magnification must be 23 000 ÷ 5 = 4600 times.

▲ **Figure 2** *Part of an animal cell seen under a TEM.*

Calculating actual sizes of specimens from drawings and photographs

You have seen how the size of an object can be calculated when you know the magnification and the image size. You can see how this is done again using a photograph. In Figure 2, the magnification is ×4600. Calculate the actual size of a mitochondrion shown in the photo.

This mitochondrion is labelled A. To calculate its actual size:

- Measure the diameter of mitochondrion A. As it is not truly spherical you need to calculate the mean of a number of different diameters. The mean is 20 mm.

- The size of the image is 20 mm (20 000 µm).

- The actual size of the mitochondrion equals the size of the image ÷ magnification or 20 000 µm ÷ 4600 = 4.3 µm.

Maths link √x̄

MS 1.2 and 1.8, see Chapter 22.

Maths link √x̄

MS 1.8, see Chapter 22.

Summary questions

In the following passage, state the missing word indicated by each letter **a–h**.

To measure the size of an object under a **a** microscope you can use an **b** graticule and a **c** micrometer. Before you can use the graticule to measure the size of objects it must first be **d**. To do this you line up the scale on the eyepiece with that on the micrometer using an objective lens that magnifies 400 times. Suppose this shows that 50 graticule units are equivalent to 10 micrometer units. If each micrometer unit is 10 µm, then each graticule unit equals **e** µm. If an objective lens magnifying 100 times is used, each graticule unit would be equivalent to **f** µm. A photograph of a cell under an electron microscope is magnified 5000 times. On the photograph the nucleus measures 100 mm in diameter. The actual size of the nucleus is therefore **g** µm. A chloroplast that is 5 µm in diameter measures 15 mm in a drawing made of a plant cell as seen under a microscope. The magnification of this drawing is therefore **h** times.

Each cell can be regarded as a metabolic compartment, a separate place where the chemical processes of that cell occur. Cells are often adapted to perform a particular function. Depending on that function, each cell type has an internal structure that suits it for its job. This is known as the **ultrastructure** of the cell. **Eukaryotic** cells have a distinct nucleus and possess membrane-bounded organelles. They differ from **prokaryotic** cells, such as bacteria. More details of these differences are given in Topic 3.6. Using an electron microscope, we can see the structure of organelles within cells, details of which are described below. The most important of these organelles are described below, with the exception of the cell-surface membrane.

The nucleus

The nucleus (Figure 1) is the most prominent feature of a eukaryotic cell, such as an epithelial cell. The nucleus contains the organism's hereditary material and controls the cell's activities. Usually spherical and between 10 and 20 μm in diameter, the nucleus has a number of parts.

- The **nuclear envelope** is a double membrane that surrounds the nucleus. Its outer membrane is continuous with the endoplasmic reticulum of the cell and often has ribosomes on its surface. It controls the entry and exit of materials in and out of the nucleus and contains the reactions taking place within it.
- **Nuclear pores** allow the passage of large molecules, such as messenger RNA, out of the nucleus. There are typically around 3000 pores in each nucleus, each 40–100 nm in diameter.
- **Nucleoplasm** is the granular, jelly-like material that makes up the bulk of the nucleus.
- **Chromosomes** consist of protein-bound, linear DNA.
- The **nucleolus** is a small spherical region within the nucleoplasm. It manufactures ribosomal RNA and assembles the ribosomes. There may be more than one nucleolus in a nucleus.

The functions of the nucleus are to:

- act as the control centre of the cell through the production of mRNA and tRNA and hence protein synthesis (see Topic 8.4)
- retain the genetic material of the cell in the form of DNA and chromosomes
- manufacture ribosomal RNA and ribosomes.

The mitochondrion

Mitochondria (Figures 2 and 3) are usually rod-shaped and 1–10 μm in length. They are made up of the following structures:

- Around the organelle is a **double membrane** that controls the entry and exit of material. The inner of the two membranes is folded to form extensions known as cristae.

Learning objectives

→ Describe the structure and functions of the nucleus, mitochondria, chloroplasts, rough and smooth endoplasmic reticulum, Golgi apparatus, Golgi vesicles and lysosomes.

→ Describe the structure and function of the cell wall in plants, algae and fungi.

→ Describe the structure and function of the cell vacuole in plants.

Specification reference: 3.2.1.1

Hint

When you look at a group of animal cells, such as epithelial cells, under a light microscope you cannot see the cell-surface membrane because it is too thin to be observed. What you actually see is the boundary between cells.

Synoptic link

The cell-surface membrane is covered in Topic 4.1, and DNA is covered in Topics 2.1, 2.2 and 8.2.

In places the nuclear membrane is continuous with the endoplasmic reticulum.

▲ **Figure 1** *The nucleus*

▲ **Figure 2** *The basic structure of a mitochondrion (top); false-colour TEM of a mitochondrion (bottom)*

- **Cristae** are extensions of the inner membrane, which in some species extend across the whole width of the mitochondrion. These provide a large surface area for the attachment of enzymes and other proteins involved in respiration.
- The **matrix** makes up the remainder of the mitochondrion. It contains protein, lipids, ribosomes and DNA that allows the mitochondria to control the production of some their own proteins. Many enzymes involved in respiration are found in the matrix.

Mitochondria are the sites of the aerobic stages of respiration (the Krebs cycle and the oxidative phosphorylation pathway. They are therefore responsible for the production of the energy-carrier molecule, **ATP**, from respiratory substrates such as glucose. Because of this, the number and size of the mitochondria, and the number of their cristae, are high in cells that have a high level of metabolic activity and therefore require a plentiful supply of ATP. Examples of metabolically active cells include muscle and epithelial cells. Epithelial cells in the intestines require a lot of ATP in the process of absorbing substances from the intestines by **active transport**.

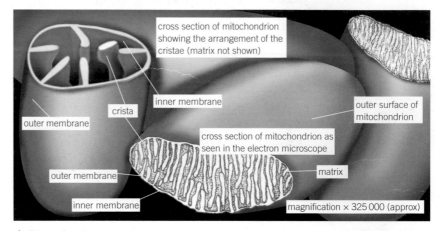

cross section of mitochondrion showing the arrangement of the cristae (matrix not shown)

inner membrane

outer surface of mitochondrion

crista

outer membrane

cross section of mitochondrion as seen in the electron microscope

outer membrane

matrix

inner membrane

magnification × 325 000 (approx)

▲ **Figure 3** *Mitochondria*

Link

A level students will learn about the function of chloroplasts function in photosynthesis in Chapter 11 Photosynthesis

Hint

Chloroplasts have DNA and may have evolved from free-living prokaryotic cells, but they are organelles, not cells.

Chloroplasts

Chloroplasts (Figure 4) are the organelles that carry out photosynthesis (see Topic 11.2). They vary in shape and size but are typically disc-shaped, 2–10 µm long and 1 µm in diameter. The following are their main features:

- **The chloroplast envelope** is a double plasma membrane that surrounds the organelle. It is highly selective in what it allows to enter and leave the chloroplast.
- **The grana** are stacks of up to 100 disc-like structures called **thylakoids**. Within the thylakoids is the photosynthetic pigment called **chlorophyll**. Some thylakoids have tubular extensions that join up with thylakoids in adjacent grana. The grana are where the first stage of photosynthesis (light absorption) takes place.
- **The stroma** is a fluid-filled matrix where the second stage of photosynthesis (synthesis of sugars) takes place. Within the stroma are a number of other structures, such as starch grains.

Chloroplasts are adapted to their function of harvesting sunlight and carrying out photosynthesis in the following ways:

- The granal membranes provide a large surface area for the attachment of chlorophyll, electron carriers and enzymes that carry out the first stage of photosynthesis. These chemicals are attached to the membrane in a highly ordered fashion.

- The fluid of the stroma possesses all the enzymes needed to make sugars in the second stage of photosynthesis.

- Chloroplasts contain both DNA and ribosomes so they can quickly and easily manufacture some of the proteins needed for photosynthesis.

Study tip

Not all plant cells have chloroplasts. Think about root cells. These are below the soil surface where light rarely penetrates and so no photosynthesis is possible.

a *Chloroplast structure*

c *False colour TEM of chloroplast*

b *Grana and thylakoids*

▲ **Figure 4** *Chloroplast structure*

Endoplasmic reticulum

The endoplasmic reticulum (ER) is an elaborate, three-dimensional system of sheet-like membranes, spreading through the cytoplasm of the cells. It is continuous with the outer nuclear membrane. The membranes enclose a network of tubules and flattened sacs called cisternae (see Figure 5). There are two types of ER:

- **Rough endoplasmic reticulum (RER)** has ribosomes present on the outer surfaces of the membranes. Its functions are to:

  a provide a large surface area for the synthesis of proteins and glycoproteins

  b provide a pathway for the transport of materials, especially proteins, throughout the cell.

- **Smooth endoplasmic reticulum (SER)** lacks ribosomes on its surface and is often more tubular in appearance. Its functions are to:

  a synthesise, store and transport lipids

  b synthesise, store and transport carbohydrates.

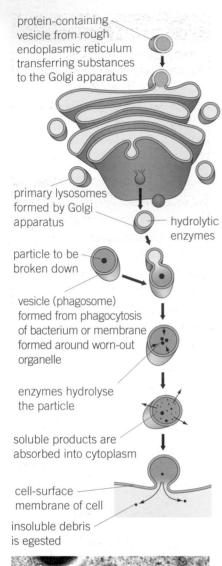

protein-containing vesicle from rough endoplasmic reticulum transferring substances to the Golgi apparatus

primary lysosomes formed by Golgi apparatus

hydrolytic enzymes

particle to be broken down

vesicle (phagosome) formed from phagocytosis of bacterium or membrane formed around worn-out organelle

enzymes hydrolyse the particle

soluble products are absorbed into cytoplasm

cell-surface membrane of cell

insoluble debris is egested

▲ **Figure 6** *The Golgi apparatus and the formation and functioning of a lysosome (top); false-colour TEM of a Golgi apparatus (orange) (bottom)*

It follows that cells that manufacture and store large quantities of carbohydrates, proteins and lipids have a very extensive ER. Such cells include liver and secretory cells, for example the epithelial cells that line the intestines.

ribosomes

cisterna

lamellae of endoplasmic reticulum made up of double membrane

▲ **Figure 5** *Structure of RER (above); false-colour TEM of a section through RER (RER; red) (right)*

Golgi apparatus

The Golgi apparatus occurs in almost all eukaryotic cells and is similar to the SER in structure except that it is more compact. It consists of a stack of membranes that make up flattened sacs, or **cisternae**, with small rounded hollow structures called vesicles. The proteins and lipids produced by the ER are passed through the Golgi apparatus in strict sequence. The Golgi modifies these proteins often adding non-protein components, such as carbohydrate, to them. It also 'labels' them, allowing them to be accurately sorted and sent to their correct destinations. Once sorted, the modified proteins and lipids are transported in Golgi vesicles which are regularly pinched off from the ends of the Golgi cisternae (Figure 6). These vesicles may move to the cell surface, where they fuse with the membrane and release their contents to the outside.

The functions of the Golgi apparatus are to:

* add carbohydrate to proteins to form glycoproteins
* produce secretory enzymes, such as those secreted by the pancreas
* secrete carbohydrates, such as those used in making cell walls in plants
* transport, modify and store lipids
* form lysosomes.

The Golgi apparatus is especially well developed in secretory cells, such as the epithelial cells that line the intestines.

- The granal membranes provide a large surface area for the attachment of chlorophyll, electron carriers and enzymes that carry out the first stage of photosynthesis. These chemicals are attached to the membrane in a highly ordered fashion.
- The fluid of the stroma possesses all the enzymes needed to make sugars in the second stage of photosynthesis.
- Chloroplasts contain both DNA and ribosomes so they can quickly and easily manufacture some of the proteins needed for photosynthesis.

a *Chloroplast structure*

c *False colour TEM of chloroplast*

b *Grana and thylakoids*

▲ **Figure 4** *Chloroplast structure*

Study tip

Not all plant cells have chloroplasts. Think about root cells. These are below the soil surface where light rarely penetrates and so no photosynthesis is possible.

Endoplasmic reticulum

The endoplasmic reticulum (ER) is an elaborate, three-dimensional system of sheet-like membranes, spreading through the cytoplasm of the cells. It is continuous with the outer nuclear membrane. The membranes enclose a network of tubules and flattened sacs called cisternae (see Figure 5). There are two types of ER:

- **Rough endoplasmic reticulum (RER)** has ribosomes present on the outer surfaces of the membranes. Its functions are to:
    - a provide a large surface area for the synthesis of proteins and glycoproteins
    - b provide a pathway for the transport of materials, especially proteins, throughout the cell.
- **Smooth endoplasmic reticulum (SER)** lacks ribosomes on its surface and is often more tubular in appearance. Its functions are to:
    - a synthesise, store and transport lipids
    - b synthesise, store and transport carbohydrates.

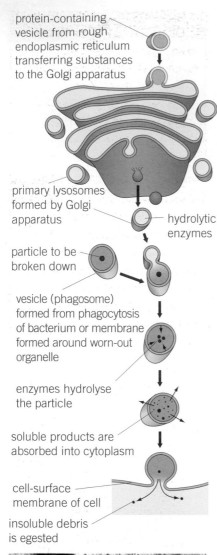

protein-containing vesicle from rough endoplasmic reticulum transferring substances to the Golgi apparatus

primary lysosomes formed by Golgi apparatus

hydrolytic enzymes

particle to be broken down

vesicle (phagosome) formed from phagocytosis of bacterium or membrane formed around worn-out organelle

enzymes hydrolyse the particle

soluble products are absorbed into cytoplasm

cell-surface membrane of cell

insoluble debris is egested

▲ **Figure 6** *The Golgi apparatus and the formation and functioning of a lysosome (top); false-colour TEM of a Golgi apparatus (orange) (bottom)*

It follows that cells that manufacture and store large quantities of carbohydrates, proteins and lipids have a very extensive ER. Such cells include liver and secretory cells, for example the epithelial cells that line the intestines.

ribosomes

cisterna

lamellae of endoplasmic reticulum made up of double membrane

▲ **Figure 5** *Structure of RER (above); false-colour TEM of a section through RER (RER; red) (right)*

Golgi apparatus

The Golgi apparatus occurs in almost all eukaryotic cells and is similar to the SER in structure except that it is more compact. It consists of a stack of membranes that make up flattened sacs, or **cisternae**, with small rounded hollow structures called vesicles. The proteins and lipids produced by the ER are passed through the Golgi apparatus in strict sequence. The Golgi modifies these proteins often adding non-protein components, such as carbohydrate, to them. It also 'labels' them, allowing them to be accurately sorted and sent to their correct destinations. Once sorted, the modified proteins and lipids are transported in Golgi vesicles which are regularly pinched off from the ends of the Golgi cisternae (Figure 6). These vesicles may move to the cell surface, where they fuse with the membrane and release their contents to the outside.

The functions of the Golgi apparatus are to:

- add carbohydrate to proteins to form glycoproteins
- produce secretory enzymes, such as those secreted by the pancreas
- secrete carbohydrates, such as those used in making cell walls in plants
- transport, modify and store lipids
- form lysosomes.

The Golgi apparatus is especially well developed in secretory cells, such as the epithelial cells that line the intestines.

Lysosomes

Lysosomes are formed when the vesicles produced by the Golgi apparatus contain enzymes such as proteases and lipases. They also contain lysozymes, enzymes that hydrolyse the cell walls of certain bacteria. As many as 50 such enzymes may be contained in a single lysosome. Up to 1.0 μm in diameter, lysosomes isolate these enzymes from the rest of the cell before releasing them, either to the outside or into a **phagocytic** vesicle within the cell (Figure 6).

The functions of lysosomes are to:

- hydrolyse material ingested by phagocytic cells, such as white blood cells and bacteria
- release enzymes to the outside of the cell (exocytosis) in order to destroy material around the cell
- digest worn out organelles so that the useful chemicals they are made of can be re-used
- completely break down cells after they have died (autolysis).

Given the roles that lysosomes perform, it is not surprising that they are especially abundant in secretory cells, such as epithelial cells, and in **phagocytic** cells.

Ribosomes

Ribosomes are small cytoplasmic granules found in all cells. They may occur in the cytoplasm or be associated with the RER. There are two types, depending on the cells in which they are found:

- **80S** – found in eukaryotic cells, is around 25 nm in diameter.
- **70S** – found in prokaryotic cells, mitochondria and chloroplasts, is slightly smaller.

Ribosomes have two subunits – one large and one small (Figure 7) – each of which contains ribosomal RNA and protein. Despite their small size, they occur in such vast numbers that they can account for up to 25 % of the dry mass of a cell. Ribosomes are the site of in protein synthesis.

Cell wall

Characteristic of all plant cells, the cell wall consists of microfibrils of the polysaccharide cellulose, embedded in a matrix. Cellulose microfibrils have considerable strength and so contribute to the overall strength of the cell wall. Cell walls have the following features:

- They consist of a number of polysaccharides, such as cellulose.
- There is a thin layer, called the **middle lamella**, which marks the boundary between adjacent cell walls and cements adjacent cells together.

The functions of the cellulose cell wall are:

- to provide mechanical strength in order to prevent the cell bursting under the pressure created by the osmotic entry of water

Hint

To help you understand the functions of the Golgi apparatus, think of it as the cell's post office, but receiving, sorting and delivering proteins and lipids, rather than letters.

Hint

Lysosomes can be thought of as refuse disposal operatives. They remove useless and potentially dangerous material (e.g., bacteria) and reuse the useful parts, disposing of only that which cannot be recycled.

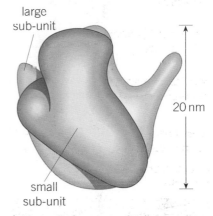

large sub-unit

20 nm

small sub-unit

▲ **Figure 7** *Structure of a ribosome*

Synoptic link

Look back to Topic 1.4, to refresh your knowledge of cellulose. Osmosis will be covered in Topic 4.3.

Study tip

Plant cells have a cell-surface membrane *and* a cell wall, not just a cell wall.

- to give mechanical strength to the plant as a whole
- to allow water to pass along it and so contribute to the movement of water through the plant.

The cell walls of algae are made up of either cellulose or glycoproteins, or a mixture of both.

The cell walls of fungi do not contain cellulose but comprise a mixture of a nitrogen-containing polysaccharide called **chitin**, a polysaccharide called glycan and glycoproteins.

Vacuoles

A fluid-filled sac bounded by a single membrane may be termed a vacuole. Within mature plant cells there is usually one large central vacuole. The single membrane around it is called the **tonoplast**. A plant vacuole contains a solution of mineral salts, sugars, amino acids, wastes and sometimes pigments such as anthocyanins.

Plant vacuoles serve a variety of functions:

- They support herbaceous plants, and herbaceous parts of woody plants, by making cells turgid.
- The sugars and amino acids may act as a temporary food store.
- The pigments may colour petals to attract pollinating insects.

Relating cell ultrastructure to function

As each organelle has its own function, it is possible to deduce, with reasonable accuracy, the role of a cell by looking at the number and size of the organelles it contains. For example, as mitochondria produce ATP that is used as a temporary energy store, it follows that cells with many mitochondria are likely to require a lot of ATP and therefore have a high rate of metabolism. Even within each mitochondrion, the more dense and numerous the cristae, the greater the metabolic rate of the cell possessing these mitochondria.

Summary questions

1 State in which process ribosomes are important.

2 List **three** carbohydrates that are absorbed by an epithelial cell of the small intestine.

3 State the organelle that is being referred to in each of the following descriptions:

   a It possesses structures called cristae.

   b It contains chromatin.

   c It synthesises glycoproteins.

   d It digests worn out organelles.

4 The following list gives a type of cell and a brief description of its role. Suggest *two* organelles that might be numerous and/or well developed in each of the cells.

   a A sperm cell swims a considerable distance carrying the male chromosomes.

   b One type of white blood cell engulfs and digests foreign material.

   c Liver cells manufacture proteins and lipids at a rapid rate.

In multicellular organisms, cells are specialised to perform specific functions. Similar cells are then grouped together into tissues, tissues into organs and organs into systems for increased efficiency.

Cell specialisation

To stay alive, all cells of a multicellular organism perform certain basic functions. However, no one cell can provide the best conditions for all functions. Therefore the cells of multicellular organisms are each specialised in different ways to perform a particular role. Each specialised cell has evolved more or fewer of certain organelles and structures to suit the role it carries out.

The first group of cells in an embryo are all initially identical. As it matures, each cell takes on its own individual characteristics that suit it to the function that it will perform when it is mature. In other words, each cell becomes specialised in structure to suit the role that it will carry out.

All the cells in an organism, such as a human, are produced by mitotic divisions from the fertilised egg. It follows that they all contain exactly the same **genes**. How then does the cell become specialised? Every cell contains the genes needed for it to develop into any one of the many different cells in an organism. But only some of these genes are switched on (expressed) in any one cell, at any one time. Different genes are switched on in each type of specialised cell. The rest of the genes are switched off.

It is not just the shape of different cells that varies, but also the numbers of each of their organelles. For example, a muscle or sperm cell will have many mitochondria, while a bone cell has very few. White blood cells have many lysosomes while a muscle cell has very few.

The cells of a multicellular organism have therefore evolved to become more and more suited to one specialised function. These cells are adapted to their own particular function and perform it more effectively. As a result, the whole organism functions efficiently.

Tissues

For working efficiency, cells are normally aggregated together. Such a collection of similar cells that perform a specific function is known as a **tissue**. Examples of tissues include:

- **epithelial tissues** (see Topic 4.5), which are found in animals and consist of sheets of cells. They line the surfaces of organs and often have a protective or secretory function. There are many similar types, including those made up of thin, flat cells that line organs where diffusion takes place, for example the alveoli of the lungs (see Topic 6.8), and ciliated epithelium that lines a duct such as the trachea (see Topic 6.6). The cilia are used to move mucus over the epithelial surface.

Learning objectives

→ Discuss the advantages of cellular differentiation.

→ Describe how cells are arranged into tissues.

→ Describe how tissues are arranged into organs.

→ Describe how organs are arranged into organ systems.

Specification reference: 3.2.1.1

Link

A level students will also learn about the roles of unspecialised cells in Topic 20.2 Stem cells and totipotency

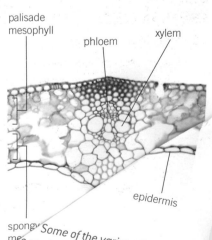

palisade mesophyll

phloem

xylem

epidermis

spongy mesophyll

Some of the various tissues make up the organ called a leaf

circulatory
system
organ system

heart

organ

muscle
tissue

muscle cell
cell

▲ **Figure 2** *The circulatory system as an example of an organ system*

- **xylem** (see Topic 7.8), which occurs in plants and is made up of a number of similar cell types. It is used to transport water and mineral ions throughout the plant and also gives mechanical support.

Organs

Just as cells are aggregated into tissues, so tissues are aggregated into organs. An organ is a combination of tissues that are coordinated to perform a variety of functions, although they often have one predominant major function. In animals, for example, the stomach is an organ that is involved in the digestion of certain types of food. It is made up of tissues such as:

- muscle to churn and mix the stomach contents
- epithelium to protect the stomach wall and produce secretions
- connective tissue to hold together the other tissues.

In plants, a leaf (Figure 1) is an organ made up of the following tissues:

- palisade mesophyll made up of leaf palisade cells that carry out photosynthesis
- spongy mesophyll adapted for gaseous diffusion
- epidermis to protect the leaf and allow gaseous diffusion
- phloem to transport organic materials away from the leaf
- xylem to transport water and ions into the leaf.

It is not always easy to determine which structures are organs. Blood capillaries, for example, are not organs whereas arteries and veins are both organs. All three structures have the same major function, namely the transport of blood. However, capillaries are made up of just one tissue – epithelium – whereas arteries and veins are made up of many tissues, for example, epithelial, muscle and other tissues.

Organ systems

Organs work together as a single unit known as an organ system. These systems may be grouped together to perform particular functions more efficiently. There are a number of organ systems in humans.

- The **digestive system** digests and processes food. It is made up of organs that include the salivary glands, oesophagus, stomach, duodenum, ileum, pancreas and liver.
- The **respiratory system** is used for breathing and gas exchange. It is made up of organs that include the trachea, bronchi and lungs.
- The **circulatory system** (Figure 2) pumps and circulates blood. It is made up of organs that include the heart, arteries and veins.

Summary questions

1 Explain what is meant by a tissue.

2 Explain why an artery is described as an organ whereas a blood capillary is not.

3 State whether each of the following is a tissue or an organ.

   a heart   **b** xylem   **c** lungs   **d** epithelium.

3.6 Prokaryotic cells and viruses

Although cells come in a diverse variety of size, shape and function, they are of two main types:

- **Eukaryotic cells** are larger and have a nucleus bounded by nuclear membranes (nuclear envelope).
- **Prokaryotic** cells are smaller and have no nucleus or nuclear envelope.

The structure of a generalised prokaryotic cell is shown in Figure 1. The differences between prokaryotic and eukaryotic cells are listed in Table 1.

Learning objectives

→ Describe the structure of prokaryotic cells.

→ Distinguish prokaryotic cells from eukaryotic ones.

Specification reference: 3.2.1.2

Structure of a bacterial cell

Bacteria occur in every habitat in the world – they are versatile, adaptable and very successful. Much of their success is a result of their small size, normally ranging from 0.1 to 10 μm in length. Their cellular structure is relatively simple (Figure 1). All bacteria possess a **cell wall**, which is made up of murein. This is a polymer of polysaccharides and peptides. Many bacteria further protect themselves by secreting a **capsule** of mucilaginous slime around this wall.

Inside the cell wall is the **cell-surface membrane**, within which is the cytoplasm that contains 70S ribosomes. These ribosomes are smaller than those in the cytoplasm of eukaryotic cells (80S), but nevertheless still synthesise proteins. Bacteria store food reserves as glycogen granules and oil droplets. The genetic material in bacteria is in the form of a **circular strand of DNA**. Separate from this are smaller circular pieces of DNA, called **plasmids**. These can reproduce themselves independently and may give the bacterium resistance to harmful chemicals, such as antibiotics. Plasmids are used extensively as vectors (carriers of genetic information) in genetic engineering. The roles of the main structures in a bacterial cell are summarised in Table 2.

▼ **Table 1** *Comparison of prokaryotic and eukaryotic cells*

Prokaryotic cells	Eukaryotic cells
no true nucleus, only an area where DNA is found	distinct nucleus, with a nuclear envelope
(Pro) DNA is not associated with proteins	DNA is associated with proteins called histones.
some DNA may be in the form of circular strands called plasmids	There are no plasmids and DNA is linear.
no membrane-bounded organelles	membrane-bounded organelles, such as mitochondria, are present
no chloroplasts, only bacterial chlorophyll associated with the cell-surface membrane in some bacteria	chloroplasts present in plants and algae
ribosomes are smaller (70S)	ribosomes are larger (80S)
cell wall made of **murein** (peptidoglycan)	where present, cell wall is made mostly of cellulose (or chitin in fungi)
may have an outer mucilaginous layer called a capsule	no capsule

▼ **Table 2** *Roles of structures found in a bacterial cell*

Cell structure	Role
cell wall	physical barrier that excludes certain substances and protects against mechanical damage and osmotic lysis
capsule	protects bacterium from other cells and helps groups of bacteria to stick together for further protection
cell-surface membrane	acts as a differentially permeable layer, which controls the entry and exit of chemicals
circular DNA	possesses the genetic information for the replication of bacterial cells
plasmid	possesses genes that may aid the survival of bacteria in adverse conditions, e.g. produces enzymes that break down antibiotics

▲ **Figure 2** *False-colour TEM of the cholera bacterium,* Vibrio cholerae

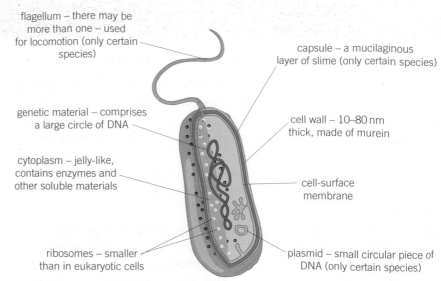

flagellum – there may be more than one – used for locomotion (only certain species)

capsule – a mucilaginous layer of slime (only certain species)

genetic material – comprises a large circle of DNA

cell wall – 10–80 nm thick, made of murein

cytoplasm – jelly-like, contains enzymes and other soluble materials

cell-surface membrane

ribosomes – smaller than in eukaryotic cells

plasmid – small circular piece of DNA (only certain species)

▲ **Figure 1** *Structure of a generalised bacterial cell*

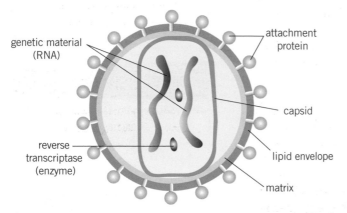

genetic material (RNA)

attachment protein

capsid

reverse transcriptase (enzyme)

lipid envelope

matrix

▲ **Figure 3** *Structure of the human immunodeficiency virus (HIV)*

Viruses

Viruses are acellular, non-living particles. They are smaller than bacteria, ranging in size from 20–300 nm. They contain **nucleic acids** such as DNA or RNA as genetic material but can only multiply inside living host cells. The nucleic acid is enclosed within a protein coat called the **capsid**. Some viruses, like the human immunodeficiency virus, are further surrounded by a lipid envelope. The lipid envelope, or if this is not present, the capsid, have **attachment proteins** which are essential to allow the virus to identify and attach to a host cell.

Summary questions

1 Table 3 lists some of the features of cells. For the letter in each box, write down **one** of the following:
'present' if the feature always occurs
'absent' if it never occurs
'sometimes' if it occurs in some cells but not others.

▼ **Table 3** *Features of prokaryotic and eukaryotic cells*

Feature	Prokaryotic cell	Eukaryotic cell
nuclear envelope	A	B
cell wall	C	D
flagellum	E	F
ribosomes	G	H
plasmid	I	J
cell-surface membrane	K	L
mitochondria	M	N

2 If a bacterium is 6 µm long and a virus is 150 nm long, calculate how many times larger the bacterium is than the virus.

Maths link √x̄

MS 0.1, see Chapter 22.

Cell division can take place by either mitosis or meiosis:

- **Mitosis** produces two daughter cells that have the same number of **chromosomes** as the parent cell and each other.

- **Meiosis** produces four daughter cells, each with half the number of chromosomes of the parent cell. We shall consider meiosis in Topic 9.2.

  The structure of a chromosome is shown in Figure 1.

Mitosis

Mitosis is division of a cell that results in each of the daughter cells having an exact copy of the DNA of the parent cell. Except in the rare event of a **mutation**, the genetic make-up of the two daughter nuclei is also identical to that of the parent nucleus. Mitosis is always preceded by a period during which the cell is not dividing. This period is called **interphase**. It is a period of considerable cellular activity that includes a very important event, the replication of DNA. The two copies of DNA after replication remain joined at a place called the centromere. Although mitosis is a continuous process, it can be divided into four stages for convenience:

Prophase

In prophase, the chromosomes first become visible, initially as long thin threads, which later shorten and thicken. Animal cells contain two cylindrical organelles called centrioles, each of which moves to opposite ends (called poles) of the cell. From each of the centrioles, **spindle fibres** develop, which span the cell from pole to pole. Collectively, these spindle fibres are called the **spindle apparatus**. As plant cells lack centrioles but do develop a spindle apparatus, centrioles are clearly not essential to spindle fibre formation. The nucleolus disappears and the nuclear envelope breaks down, leaving the chromosomes free in the cytoplasm of the cell. These chromosomes are drawn towards the equator of the cell by the spindle fibres attached to the centromere.

Metaphase

By metaphase the chromosomes are seen to be made up of two chromatids. Each chromatid is an identical copy of DNA from the parent cell. The chromatids are joined by the centromere (Topic 8.2). It is to this centromere that some microtubules from the poles are attached, and the chromosomes are pulled along the spindle apparatus and arrange themselves across the equator of the cell.

Anaphase

In anaphase, the centromeres divide into two and the spindle fibres pull the individual chromatids making up the chromosome apart. The chromatids move rapidly to their respective, opposite poles of the cell and we now refer to them as chromosomes. The energy for the process is provided by mitochondria, which gather around the spindle fibres. If cells are treated with chemicals that destroy the spindle, the chromosomes remain at the equator, unable to reach the poles.

Learning objectives

→ Describe what mitosis is.

→ State when DNA replication takes place.

→ Explain the importance of mitosis.

Specification reference: 3.2.2

Study tip

It is important to remember that the replication of DNA takes place during interphase before the nucleus and the cell divide.

Synoptic link

The replication of DNA was covered in Topic 2.2, DNA replication.

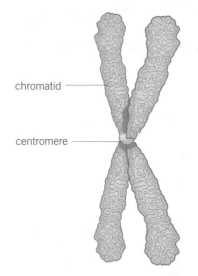

chromatid

centromere

▲ **Figure 1** *Structure of a chromosome*

Synoptic link

We will learn about the centromere in Topic 8.2, DNA and chromosomes.

Telophase and cytokinesis

In this stage, the chromosomes reach their respective poles and become longer and thinner, finally disappearing altogether, leaving only widely spread **chromatin**. The spindle fibres disintegrate and the nuclear envelope and nucleolus re-form. Finally the cytoplasm divides in a process called **cytokinesis**. The process is illustrated and explained in Figure 1.

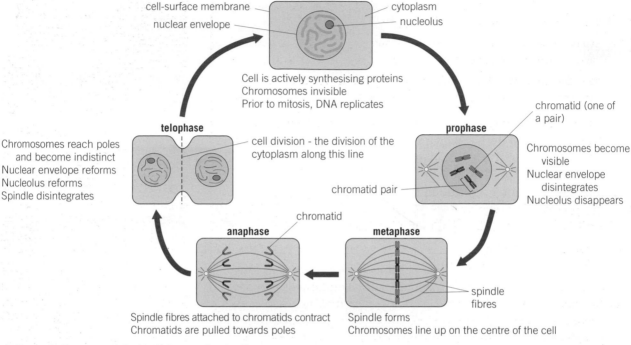

▲ **Figure 1** *The stages of mitosis in an animal cell*

Practical link

Required practical 2. Preparation of stained squashes of cells from plant root tips; set-up and use of an optical microscope to identify the stages of mitosis in these stained squashes and calculation of a mitotic index. Also measuring the apparent size of cells in the root tip and calculating their actual size.

Cell division in prokaryotic cells

Cell division in prokaryotic cells takes place by a process called **binary fission** as follows:

- The circular DNA molecule replicates and both copies attach to the cell membrane.
- The plasmids also replicate.
- The cell membrane begins to grow between the two DNA molecules and begins to pinch inward, dividing the cytoplasm into two.
- A new cell wall forms between the two molecules of DNA, dividing the original cell into two identical daughter cells, each with a single copy of the circular DNA and a variable number of copies of the plasmids.

Replication of viruses

As viruses are non-living, they cannot undergo cell division. Instead they replicate by attaching to their host cell with the attachment proteins on their surface. They then inject their nucleic acid into the host cell. The genetic information on the injected viral nucleic acid then provides the 'instructions' for the host cell's metabolic processes to start producing the viral components, nucleic acid, enzymes and structural proteins, which are then assembled into new viruses.

The importance of mitosis

Mitosis is important in organisms as it produces daughter cells that are genetically identical to the parent cells. Why is it essential to make exact copies of existing cells? There are three reasons:

- **growth**. When two **haploid** cells (e.g., a sperm and an ovum) fuse together to form a **diploid** cell, it has all the genetic information needed to form the new organism. If the new organism is to resemble its parents, all the cells that grow from this original cell must be genetically identical. Mitosis ensures that this happens.

- **repair**. If cells are damaged or die it is important that the new cells produced have an identical structure and function to the ones that have been lost.

- **reproduction** single-celled organisms divide by mitosis to give two new organisms. Each new organism is genetically identical to the parent organism.

> 1 Suggest an advantage and a disadvantage of having offspring that are genetically identical to their parents.

▲ **Figure 2** *Stages of mitosis*

Summary question

1 In the following passage about mitosis, state the most appropriate word that is represented by each of the letters.

The period when a cell is not dividing is called **a**. The stage of mitosis when the chromosomes are first visible as distinct structures is called **b**. During this stage thin threads develop that span the cell from end to end and together form a structure called the **c**. Towards the end of this stage, the **d** breaks down and the **e** disappears. The stage when the chromosomes arrange themselves across the centre of the cell is called **f**. During the stage called **g** the chromatids move to opposite ends of the cell.

Recognising the stages of mitosis $\sqrt{x}$

The photographs in Figure 2 show cells at various stages of mitosis.

Mitosis is a continuous process. When mitosis is viewed under a microscope, the observer only gets a snapshot of the process at one moment in time. The ratio of the number of cells undergoing mitosis to the total number of cells is called the mitotic index. The number of cells at each stage of mitosis is proportional to the time each cell spends undergoing that stage. Table 1 shows the number of cells at each stage of mitosis during one observation.

▼ **Table 1**

Stage	Number of cells
interphase	890
prophase	73
metaphase	20
anaphase	9
telophase	8

Maths link $\sqrt{x}$

MS 0.3, see Chapter 22.

1 State the names of the five different stages represented by the letters A–E in Figure 2. In each case give a reason for choosing your answer.

2 $\sqrt{x}$ From Table 1, if one complete cycle takes 20 hours, calculate how many minutes were spent in metaphase. Show your working.

3 $\sqrt{x}$ Calculate in what percentage of the cells the chromosomes would have been visible. Show your working.

Learning objectives

→ Describe the three stages of the cell cycle.

→ Describe what happens during interphase.

→ Explain how mitosis is controlled.

→ Describe how cancer and its treatment relate to the cell cycle.

Specification reference: 3.2.2

Only some cells in multicellular organisms retain the ability to divide. Those that do not divide continuously, but undergo a regular cycle of division separated by periods of cell growth. This is known as the **cell cycle** and has three stages:

1 **interphase**, which occupies most of the cell cycle, and is sometimes known as the resting phase because no division takes place

2 **nuclear division**, when the nucleus divides either into two (mitosis) or four (meiosis)

3 **division of the cytoplasm (cytokinesis)**, which follows nuclear division and is the process by which the cytoplasm divides to produce two new cells (mitosis) or four new cells (meiosis) (Topic 9.2).

The length of a complete cell cycle varies greatly amongst organisms. Typically, a mammalian cell takes about 24 hours to complete a cell cycle, of which about 90% is interphase.

The various stages of the cell cycle are shown in Figure 1.

Figure 2 shows the variations in mass of a diploid cell and the DNA within it during the cell cycle.

▲ **Figure 2** *Variation in the mass of a diploid cell and the DNA within it during the cell cycle*

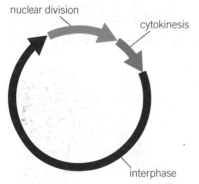

▲ **Figure 1** *The cell cycle*

Cancer and the control of mitosis

Cancer is a group of diseases (around 200 in total) caused by a growth disorder of cells. It is the result of damage to the genes that regulate mitosis and the cell cycle. This leads to uncontrolled growth and division of cells. As a consequence, a group of abnormal cells, called a tumour, develops and constantly expands in size. Tumours can develop in any organ of the body, but are most commonly found in the lungs, prostate gland (male), breast and ovaries (female), large intestine, stomach, oesophagus and pancreas. A tumour becomes cancerous if it changes from benign to malignant.

Most cells divide by mitosis, either to increase the size of a tissue during development (growth) or to replace dead and worn out cells (repair). The rate of mitosis can be affected by the environment of the cell and by growth factors. It is also controlled by two types of gene. A mutation to one of these genes results in uncontrolled mitosis. The mutant cells so formed are usually structurally and functionally different from normal cells. Most mutated cells die. However, any that survive are capable of dividing to form clones of themselves and forming tumours. Malignant tumours grow rapidly, are less compact and are more likely to be life-threatening, while benign ones grow more slowly, are more compact and are less likely to be life-threatening.

Treatment of cancer

The treatment of cancer often involves killing dividing cells by blocking a part of the cell cycle. In this way the cell cycle is disrupted and cell division, and hence cancer growth, ceases. Drugs used to treat cancer (chemotherapy) usually disrupt the cell cycle by:

- preventing DNA from replicating
- inhibiting the metaphase stage of mitosis by interfering with spindle formation.

The problem with such drugs is that they also disrupt the cell cycle of normal cells. However, the drugs are more effective against rapidly dividing cells. As cancer cells have a particularly fast rate of division, they are damaged to a greater degree than normal cells. Those normal body cells, such as hair-producing cells, that divide rapidly are also vulnerable to damage. This explains the hair loss frequently seen in patients undergoing cancer treatment.

Link
A level students will learn more about cancer in Topic 20.5 Gene expression and cancer.

Summary questions

1 List the three main stages of the cell cycle.

2 $\sqrt{x}$ Using Figure 2, state at what time(s), or during which period, each of the following occur:
   a cell division
   b replication of DNA.

Treating cancer $\sqrt{x}$

The graph in Figure 3 shows the effect of a chemotherapy drug that kills dividing cells. It was given to a cancer patient once every three weeks starting at time zero. The graph plots the changes in the number of healthy cells and cancer cells in a tissue over the treatment period of 12 weeks.

▲ **Figure 3** *Changes in the number of healthy cells and cancer cells in a tissue during a chemotherapy treatment of 12 weeks*

5 It would be possible to kill more cancer cells if the same dose of the drug was given more frequently or the frequency was kept the same but a larger dose of the drug was used each time. Suggest why
   a the drug was not given more frequently
   b the dose of the drug was not increased.

▲ **Figure 4** *Patient undergoing chemotherapy treatment*

1 An amoeba is a single-celled, eukaryotic organism. Scientists used a transmission electron microscope to study an amoeba. The diagram shows its structure.

(a) (i) Name organelle **Y**. *(1 mark)*

(ii) Name **two** other structures in the diagram which show that the amoeba is a eukaryotic cell. *(2 marks)*

(b) What is the function of organelle **Z**? *(1 mark)*

(c) The scientists used a transmission electron microscope to study the structure of the amoeba. Explain why. *(2 marks)*

AQA June 2012

2 The photograph shows part of the cytoplasm of a cell.

2 (a) (i) Organelle X is a mitochondrion. What is the function of this organelle? *(1 mark)*

(ii) Name organelle Y. *(1 mark)*

(b) This photograph was taken using a transmission electron microscope. The structure of the organelles visible in the photograph could not have been seen using an optical (light) microscope. Explain why. *(2 marks)*

AQA Jan 2013

3 The diagram shows a chloroplast as seen with an electron microscope.

magnification × 30 000

(a) Name **X** and **Y**. *(2 marks)*
(b) Describe the function of a chloroplast. *(2 marks)*
(c) Calculate the maximum length of this chloroplast in micrometres (μm).
Show your working. *(2 marks)*

AQA Jan 2012

4 **(a)** The table shows some features of cells. Complete the table by putting a tick in the box
if the feature is present in the cell.

Feature	Cell		
	Cholera bacterium	Epithelial cell from intestine	Epithelial cell from alveolus of lung
Cell-surface membrane			
Flagellum			
Nucleus			

(3 marks)

(b) The diagram shows part of an epithelial cell from an insect's gut.

lumen of gut
microvillus
golgi apparatus
ribosomes on endoplasmic reticulum
mitochondrion
blood

This cell is adapted for the three functions listed below. Use the diagram to explain
how this cell is adapted for each of these functions.

Use a **different** feature in the diagram for each of your answers.
(i) the active transport of substances from the cell into the blood *(2 marks)*
(ii) the synthesis of enzymes *(2 marks)*
(iii) rapid diffusion of substances from the lumen of the gut into the
cytoplasm *(1 mark)*

AQA Jan 2012

Learning objectives

→ Describe the structure of the cell-surface membrane.

→ Describe the functions of the various components of the cell-surface membrane.

→ Explain the fluid-mosaic model of cell membrane structure.

Specification reference: 3.2.3

All membranes around and within all cells (including those around and within cell organelles) have the same basic structure and are known as **plasma membranes**.

The cell-surface membrane is the name specifically given to the plasma membrane that surrounds cells and forms the boundary between the cell cytoplasm and the environment. It allows different conditions to be established inside and outside a cell. It controls the movement of substances in and out of the cell. Before we look at how the cell-surface membrane achieves this, we need first to look in more detail at the molecules that form its structure.

Phospholipids

We looked at the molecular structure of a phospholipid in Topic 1.5. Phospholipids form a bilayer (see Figure 1). They are important components of cell-surface membranes for the following reasons:

- The hydrophilic heads of both phospholipid layers point to the outside of the cell-surface membrane attracted by water on both sides.

- The hydrophobic tails of both phospholipid layers point into the centre of the cell membrane, repelled by the water on both sides.

Lipid-soluble material moves through the membrane via the phospholipid portion. The functions of phospholipids in the membrane are to:

- allow lipid-soluble substances to enter and leave the cell

- prevent water-soluble substances entering and leaving the cell

- make the membrane flexible and self-sealing.

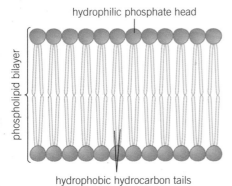

hydrophilic phosphate head

phospholipid bilayer

hydrophobic hydrocarbon tails

▲ **Figure 1** *A simplified diagram of a phospholipid bilayer.*

Proteins

Proteins are interspersed throughout the cell surface membrane. They are embedded in the phospholipid bilayer in two main ways:

- Some proteins occur in the surface of the bilayer and never extend completely across it. They act either to give mechanical support to the membrane or, in conjunction with glycolipids, as cell receptors for molecules such as hormones.

- Other proteins completely span the phospholipid bilayer from one side to the other. Some are **protein channels**, which form water-filled tubes to allow water-soluble ions to diffuse across the membrane. Others are **carrier proteins** that bind to ions or molecules like glucose and amino acids, then change shape in order to move these molecules across the membrane.

The functions of the proteins in the membrane are to:

- provide structural support
- act as channels transporting water-soluble substances across the membrane
- allow active transport across the membrane through carrier proteins
- form cell-surface receptors for identifying cells
- help cells adhere together
- act as receptors, for example for hormones.

Cholesterol

Cholesterol molecules occur within the phospholipid bilayer of the cell-surface membrane. They add strength to the membranes. Cholesterol molecules are very hydrophobic and therefore play an important role in preventing loss of water and dissolved ions from the cell. They also pull together the fatty acid tails of the phospholipid molecules, limiting their movement and that of other molecules but without making the membrane as a whole too rigid.

The functions of cholesterol in the membrane are to:

- reduce lateral movement of other molecules including phospholipids
- make the membrane less fluid at high temperatures
- prevent leakage of water and dissolved ions from the cell.

Glycolipids

Glycolipids are made up of a carbohydrate covalently bonded with a lipid. The carbohydrate portion extends from the phospholipid bilayer into the watery environment outside the cell where it acts as a cell-surface receptor for specific chemicals, for example the human ABO blood system operates as a result of glycolipids on the cell-surface membrane.

The functions of glycolipids in the membrane are to:

- act as recognition sites
- help maintain the stability of the membrane
- help cells to attach to one another and so form tissues.

Glycoproteins

Carbohydrate chains are attached to many extrinsic proteins on the outer surface of the cell membrane. These glycoproteins also act as cell-surface receptors, more specifically for hormones and neurotransmitters.

The functions of glycoproteins in the membrane are to:

- act as recognition sites
- help cells to attach to one another and so form tissues
- allows cells to recognise one another, for example **lymphocytes** can recognise an organism's own cells.

Study tip

When representing a phospholipid it is important to be accurate. It has a *single* phosphate head and *two* fatty acid tails. All too often students show too many heads and/or too many tails.

Hint

All plasma membranes found around and inside cells have the same phospholipid bilayer structure. What gives plasma membranes their different properties are the different substances they contain – especially proteins.

Link

A level students will find out more about the role of glycoproteins as cell surface receptors for hormones and neurotransmitters in Topic 15.6 Transmission across a synapse and Topic 16.3 Hormones and the regulation of blood glucose concentration.

Functions of membranes within cells
control the entry and exit of materials in discrete organelles such as mitochondria and chloroplasts
separate organelles from cytoplasm so that specific metabolic reactions can take place within them
provide an internal transport system, e.g., endoplasmic reticulum
isolate enzymes that might damage the cell, e.g., lysosomes
provide surfaces on which reactions can occur, e.g., protein synthesis using ribosomes on rough endoplasmic reticulum

Practical link 🧪

Required practical 4. Investigation into the effect of a named variable on the permeability of cell-surface membranes.

Summary questions

1 State the overall function of the cell-surface membrane.

2 State which end of the phospholipid molecule lies towards the inside of the cell-surface membrane.

3 State through which molecule in the cell-surface membrane each of the following are likely to pass in order to get in or out of a cell.

   a a molecule that is soluble in lipids

   b a mineral ion

4 From your knowledge of the cell-surface membrane, suggest two properties that a drug should possess if it is to enter a cell rapidly.

Permeability of the cell-surface membrane

The cell-surface membrane controls the movement of substances into and out of the cell. In general most molecules do not freely diffuse across it because many are:

• not soluble in lipids and therefore cannot pass through the phospholipid layer

• too large to pass through the channels in the membrane

• of the same charge as the charge on the protein channels and so, even if they are small enough to pass through, they are repelled

• electrically charged (in other words are polar) and therefore have difficulty passing through the non-polar hydrophobic tails in the phospholipid bilayer.

Fluid-mosaic model of the cell-surface membrane

The way in which all the various molecules are combined into the structure of the cell-surface membrane is shown in Figure 2. This arrangement is known as the **fluid-mosaic model** for the following reasons:

• **fluid** because the individual phospholipid molecules can move relative to one another. This gives the membrane a flexible structure that is constantly changing in shape

• **mosaic** because the proteins that are embedded in the phospholipid bilayer vary in shape, size and pattern in the same way as the stones or tiles of a mosaic.

glycoprotein – branching carbohydrate portion of a protein which acts as a recognition site for chemicals, e.g., hormones

glycolipid – acts as a recognition site, e.g., for cholera toxins

cholesterol – adds strength

hydrophilic heads of phospholipid molecules – point outwards

7 nm

channel

protein molecule lying on the surface

protein molecule partly embedded

hydrophobic tails of phospholipid molecules – point inwards

channel protein molecule spanning the phospholipid layer

▲ **Figure 2** *The fluid-mosaic model of the cell-surface membrane*

The exchange of substances between cells and the environment occurs in ways that require metabolic energy (active transport) and in ways that do not (passive transport). Diffusion is an example of passive transport.

Explanation of simple diffusion

As all movement involves energy, it is possibly confusing to describe diffusion as passive transport. In this sense, 'passive' means that the energy comes from the natural, inbuilt motion of particles, rather than from some external source such as ATP. To help understand diffusion and other passive forms of transport it is necessary to understand that:

- all particles are constantly in motion due to the kinetic energy that they possess
- this motion is random, with no set pattern to the way the particles move around
- particles are constantly bouncing off one another as well as off other objects, for example, the sides of the vessel in which they are contained.

Given these facts, particles that are concentrated together in part of a closed vessel will, of their own accord, distribute themselves evenly throughout the vessel as a result of diffusion.

Diffusion is therefore defined as:

the net movement of molecules or ions from a region where they are more highly concentrated to one where their concentration is lower until evenly distributed.

We saw in Topic 4.1 that most molecules do not easily pass across the cell-surface membrane. Amongst the few molecules that can diffuse across membranes are small, non-polar molecules such as oxygen and carbon dioxide.

Facilitated diffusion

We saw in Topic 4.1 that plasma membranes are not readily permeable to molecules. Only small, non-polar molecules like oxygen can diffuse across them easily. Charged ions and polar molecules do not diffuse easily because of the hydrophobic nature of the fatty-acid tails of the phospholipids in the membrane. The movement of these molecules is made easier (facilitated) by transmembrane channels and carriers that span the membrane. The process is therefore called facilitated diffusion.

Facilitated diffusion is a passive process. It relies only on the inbuilt motion (kinetic energy) of the diffusing molecules. There is no external input of ATP from respiration. Like diffusion, it occurs down a concentration gradient, but it differs in that it occurs at specific points on the plasma membrane where there are special protein molecules. Two types of protein are involved – **protein channels** and **carrier proteins**. Each has a different mechanism.

Learning objectives

- → Explain what diffusion is and how it occurs.
- → Explain what affects the rate of diffusion.
- → Distinguish between facilitated diffusion and diffusion.

Specification reference: 3.2.3

Hint

Remember that diffusion is the **net** movement of particles. All particles move at random in diffusion; it is just that more move in one direction than in the other. This is due to concentration differences.

Hint

Diffusion only occurs between different concentrations of the *same* substance. For example, it may occur between different concentrations of oxygen or between different concentrations of carbon dioxide. It *never* occurs between different concentrations of oxygen and carbon dioxide.

plasma membrane

low concentration

high concentration

carrier protein

Particles are taken up by carrier proteins which as a result change shape.

particle

The change in shape causes the particles to be released on the other side of the membrane.

▲ **Figure 2** *Facilitated diffusion involving carrier proteins*

Protein channels

These proteins form water-filled hydrophilic channels across the membrane. They allow specific water-soluble ions to pass through. The channels are selective, each opening in the presence of a specific ion. If the particular ion is not present, the channel remains closed. In this way, there is control over the entry and exit of ions. The ions bind with the protein causing it to change shape in a way that closes it to one side of the membrane and opens it to the other side.

Carrier proteins

An alternative form of facilitated diffusion involves carrier proteins that span the plasma membrane. When a molecule such as glucose that is specific to the protein is present, it binds with the protein. This causes it to change shape in such a way that the molecule is released to the inside of the membrane (Figure 2). No external energy is needed for this. The molecules move from a region where they are highly concentrated to one of lower concentration, using only the kinetic energy of the molecules themselves.

Summary questions

1 State *three* factors that affect the rate of diffusion.

2 Contrast facilitated diffusion and diffusion.

3 Explain why facilitated diffusion is a passive process.

4 Glucose molecules are transported into cells through the pores in the proteins that span the phospholipid bilayer. Explain why they do not pass easily through the phospholipid bilayer.

5 List *two* changes to the structure of cell-surface membranes that would increase the rate at which glucose is transported into a cell.

6 √x̄ Oxygen is required by cells for respiration. This diffuses into the blood through the epithelial layers of the alveoli and blood capillaries. Calculate by how much each of the following changes would increase or decrease the rate of diffusion of oxygen.

   a The surface area of the alveoli is doubled.

   b The surface area of the alveoli is halved and the oxygen concentration gradient is doubled.

   c The oxygen concentration gradient is halved and the total thickness of the epithelial layers is doubled.

   d The oxygen concentration of the blood is halved and the carbon dioxide concentration of the alveoli is doubled.

In the last topic we learned about diffusion. We now turn our attention to a special case of diffusion, known as osmosis. Osmosis only involves the movement of water molecules.

What is osmosis?

Osmosis is defined as:

the passage of water from a region where it has a higher water potential to a region where it has a lower water potential through a selectively permeable membrane.

Cell-surface membranes and other plasma membranes such as those around organelles are selectively permeable, that is, they are permeable to water molecules and a few other small molecules, but not to larger molecules.

Solutions and water potential

A solute is any substance that is dissolved in a solvent, for example, water. The solute and the solvent together form a solution.

Water potential is represented by the Greek letter psi (Ψ), and is measured in units of pressure, usually kiloPascals (kPa). Water potential is the pressure created by water molecules. Under standard conditions of temperature and pressure (25 °C and 100 kPa), pure water is said to have a water potential of zero.

It follows that:

- the addition of a solute to pure water will lower its water potential
- the water potential of a solution (water + solute) must always be less than zero, that is, a negative value
- the more solute that is added (i.e., the more concentrated a solution), the lower (more negative) its water potential
- water will move by osmosis from a region of higher (less negative) water potential (e.g., −20 kPa) to one of lower (more negative) water potential (e.g., −30 kPa).

One way of finding the water potential of cells or tissues is to place them in a series of solutions of different water potentials. Where there is no net gain or loss of water from the cells or tissues, the water potential inside the cells or tissues must be the same as that of the external solution.

Explanation of osmosis

Consider the hypothetical situation in Figure 1 overleaf, in which a selectively permeable plasma membrane separates two solutions.

- The solution on the left has a low concentration of solute molecules while the solution on the right has a high concentration of solute molecules.
- Both the solute and water molecules are in random motion due to their **kinetic energy**.

Learning objectives

→ Describe the nature of osmosis.

→ State the water potential of pure water.

→ Describe the effect of solutes on water potential.

→ Explain how water potential affects water movement.

→ Explain what happens when animal cells and plant cells are placed into pure water.

Specification reference: 3.2.3

Practical link

Required practical 3. Production of a dilution series of a solute to produce a calibration curve with which to identify the water potential of plant tissue.

Link

A level students will discover how the water potential of the blood is controlled in Topic 16.5 Control of blood water potential and Topic 16.6 Role of the nephron in osmoregulation

Hint

Remember that, while diffusion can be the movement of *any* molecule, osmosis is the movement of water molecules *only*.

Study tip

Remember that *all* water potential values are negative. The highest water potential is *zero*. Therefore the lower the water potential, the more negative it becomes.

- The selectively permeable plasma membrane, however, only allows water molecules across it and not solute molecules.
- The water molecules diffuse from the left-hand side, which has the higher water potential, to the right-hand side, which has the lower water potential, that is, down a water potential gradient (Figure 2).
- At the point where the water potentials on either side of the plasma membrane are equal, a dynamic equilibrium is established and there is no net movement of water.

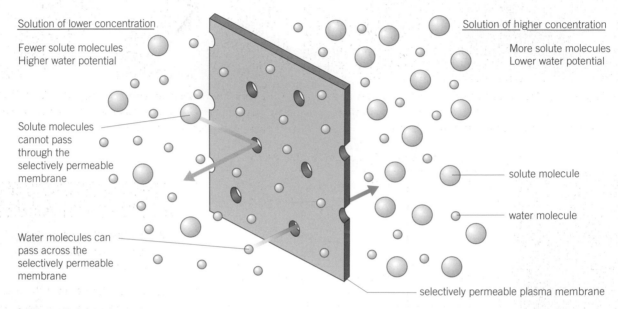

Solution of lower concentration

Fewer solute molecules
Higher water potential

Solute molecules cannot pass through the selectively permeable membrane

Water molecules can pass across the selectively permeable membrane

Solution of higher concentration

More solute molecules
Lower water potential

solute molecule

water molecule

selectively permeable plasma membrane

▲ **Figure 1** *Osmosis*

Understanding water potential

The highest value of water potential, that of pure water, is zero, and so all other values are negative. The more negative the value, the lower the water potential.

Osmosis and animal cells

Animal cells, such as red blood cells, contain a variety of solutes dissolved in their watery cytoplasm. If a red blood cell is placed in pure water it will absorb water by osmosis because it has a lower water potential. Cell surface membranes are very thin (7 nm) and, although they are flexible, they cannot stretch to any great extent. The cell-surface membrane will therefore break, bursting the cell and releasing its contents (in red blood cells this is called haemolysis). To prevent this happening, animal cells normally live in a liquid which has the same water potential as the cells. In our example, the liquid is the blood plasma. This and red blood cells have the same water potential. If a red blood cell is placed in a solution with a water potential lower than its own, water leaves by osmosis and the cell shrinks and becomes shrivelled (see Table 1).

key

x kPa   water potential of cell
➡   direction of water movement

water moves from higher water potential to lower water potential. The highest water potential is zero

▲ **Figure 2** *Movement of water between cells along a water potential gradient*

▲ Figure 3 *SEM of red blood cells that have been placed in a solution of lower water potential. Water has left by osmosis and the cells have become shrunken and shrivelled*

▲ Figure 4 *Onion epidermal cells showing plasmolysis. The protoplasts, with their vacuoles containing red liquid, have shrunk and pulled away from the cell walls*

▼ Table 1 *Summary of osmosis in an animal cell, for example, a red blood cell*

Water potential (ψ) of external solution compared to cell solution	higher (less negative)	equal	lower (more negative)
Net movement of water	enters cell	neither enters nor leaves	leaves cell
State of cell	swells and bursts	no change	shrinks
	contents, including haemoglobin, are released; remains of cell-surface membrane	normal red blood cell	haemoglobin is more concentrated, giving cell a darker appearance; cell shrunken and shrivelled

Summary questions

1 Explain what is meant by a selectively permeable membrane.

2 Under standard conditions of pressure and temperature, what is the water potential of pure water?

3 √x̄ Four cells have the following water potentials:

Cell A = −200 kPa
Cell B = −250 kPa
Cell C = −100 kPa
Cell D = −150 kPa.

Determine the order in which the cells have to be placed for water to pass from one cell to the next if they are arranged in a line.

Osmosis and plant cells

▼ **Table 2** *Summary of osmosis in a plant cell*

Water potential (ψ) of external solution compared to cell solution	higher (less negative)	equal	lower (more negative)
Net movement of water	enters cell	neither enters nor leaves	leaves cell
Protoplast	swells	no change	shrinks
Condition of cell	turgid	incipient plasmolysis	plasmolysed
	protoplast pushed against cell wall nucleus cellulose cell wall protoplast	protoplast beginning to pull away from the cell wall	protoplast completely pulled away from the cell wall

For the purposes of the following explanations, the plant cell can be divided into three parts:

- the **central vacuole**, which contains a solution of salts, sugars and organic acids in water
- the **protoplast**, consisting of the outer cell-surface membrane, nucleus, cytoplasm and the inner vacuole membrane
- the **cellulose cell wall**, a tough, inelastic covering that is permeable to even large molecules.

Like animal cells, plant cells also contain a variety of solutes, mainly dissolved in the water of the large cell vacuole that each possesses. When placed in pure water they also absorb water by osmosis because of their lower (more negative) water potential. Unlike animal cells, however, they are unable to control the composition of the fluid around their cells. Indeed, plant cells are normally permanently bathed in almost pure water, which is constantly absorbed from the plant's roots. Water entering a plant cell by osmosis causes the protoplast to swell and press on the cell wall. Because the cell wall is capable of only very limited expansion, a pressure builds up on it that resists the entry of further water. In this situation, the protoplast of the cell is kept pushed against the cell wall and the cell is said to be **turgid**.

If the same plant cell is placed in a solution with a lower water potential than its own, water leaves by osmosis. The volume of the cell decreases. A stage is reached where the protoplast no longer presses on the cellulose cell wall. At this point the cell is said to be at **incipient plasmolysis**. Further loss of water will cause the cell contents to shrink further and the protoplast to pull away from the cell wall. In this condition the cell is said to be **plasmolysed**. These events are summarised in Table 2.

1   Explain why an animal cell placed in pure water bursts while a plant cell placed in pure water does not.

2   $\sqrt{x}$ Plant cells that have a water potential of −600 kPa are placed in solutions of different water potentials. Determine in each of the following cases whether, after 10 minutes, the cells would be turgid, plasmolysed or at incipient plasmolysis.
   a   Solution A = −400 kPa
   b   Solution B = −600 kPa
   c   Solution C = −900 kPa
   d   Solution D = pure water

3   $\sqrt{x}$ An animal cell with a water potential of −700 kPa was placed in each of the solutions. Deduce in which solutions the cell is likely to burst.

Learning objectives

→ Explain the process of active transport.

→ Describe the conditions required for active transport.

Specification reference: 3.2.3

We have looked at diffusion and osmosis, both of which are passive processes, that is they occur without the use of metabolic energy. The transport of some molecules in and out of cells involves a process that uses metabolic energy. This process is active transport.

What is active transport?

Active transport is:

the movement of molecules or ions into or out of a cell from a region of lower concentration to a region of higher concentration using ATP and carrier proteins.

In active transport ATP is used to:

- directly move molecules
- individually move molecules using a concentration gradient which has already been set up by (direct) active transport. This is known as **co-transport** and is further explained in Topic 4.5.

It differs from passive forms of transport in the following ways:

- Metabolic energy in the form of **ATP** is needed.
- Substances are moved against a concentration gradient, that is from a lower to a higher concentration.
- Carrier protein molecules which act as 'pumps' are involved.
- The process is very selective, with specific substances being transported.

Direct active transport of a single molecule or ion is described below.

- The carrier proteins span the plasma membrane and bind to the molecule or ion to be transported on one side of it.
- The molecule or ion binds to receptor sites on the carrier protein.
- On the inside of the cell/organelle, ATP binds to the protein, causing it to split into ADP and a phosphate molecule. As a result, the protein molecule changes shape and opens to the opposite side of the membrane.
- The molecule or ion is then released to the other side of the membrane.
- The phosphate molecule is released from the protein which causes the protein to revert to its original shape, ready for the process to be repeated. The phosphate molecule then recombines with the ADP to form ATP during respiration.

These events are illustrated in Figure 1. It is important to distinguish between active transport and facilitated diffusion. Both use carrier proteins but facilitated diffusion occurs *down* a concentration gradient, while active transport occurs *against* a concentration gradient. This means that facilitated diffusion does not require metabolic energy, while active transport does. The metabolic energy is provided in the form of ATP.

Study tip

Carrier proteins have a specific tertiary structure and will only transport particular substances across a membrane. They have binding sites – these are different to the active sites of enzymes.

Hint

Different carrier proteins are involved in facilitated diffusion and active transport *but* any given protein carrier is very specific about what it carries and by which method.

Sometimes more than one molecule or ion may be moved in the same direction at the same time by active transport. Occasionally, the molecule or ion is moved into a cell/organelle at the same time as a different one is being removed from it. One example of this is the **sodium–potassium pump**.

In the sodium–potassium pump, sodium ions are actively removed from the cell/organelle while potassium ions are actively taken in from the surroundings. This process is essential to a number of important processes in the organism, including the creation of a nerve impulse.

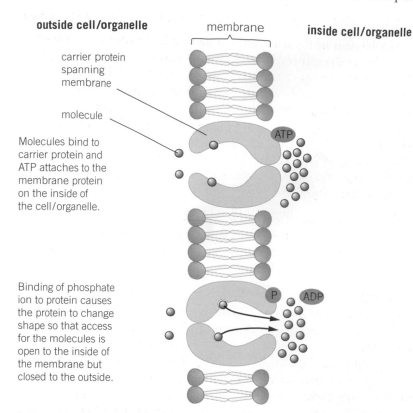

outside cell/organelle   membrane   inside cell/organelle

carrier protein spanning membrane

molecule

Molecules bind to carrier protein and ATP attaches to the membrane protein on the inside of the cell/organelle.

Binding of phosphate ion to protein causes the protein to change shape so that access for the molecules is open to the inside of the membrane but closed to the outside.

▲ **Figure 1** *Active transport*

Summary questions

1 State *one* similarity and *one* difference between active transport and facilitated diffusion.

2 The presence of many mitochondria is typical of cells that carry out active transport. Explain why this is so.

3 In the production of urine, glucose is initially lost from the blood but is then reabsorbed into the blood by cells in the kidneys. Explain why it is important that this reabsorption occurs by active transport rather than by diffusion.

Co-transport and absorption of glucose in the ileum

To illustrate how the various forms of movement across membranes occur in a particular situation, you can look at how products of digestion like glucose and amino acids are absorbed in the ileum (small intestine). Firstly let us look at how the rate of transport across membranes and into cells may be increased.

Increasing the rate of movement across membranes

The epithelial cells lining the ileum possess **microvilli** (see Figure 1). These are finger-like projections of the cell-surface membrane about 0.6 µm in length. They are collectively termed a 'brush border' because, when viewing them under a light microscope, they look like the bristles on a brush. The microvilli provide more surface area for the insertion of carrier proteins through which diffusion, facilitated diffusion and active transport can take place. Another mechanism to increase transport across membranes is to increase the number of protein channels and carrier proteins in any given area of membrane (i.e., increase their density).

The role of diffusion in absorption

Diffusion (Topic 4.2) is the net movement of molecules or ions from a region where they are highly concentrated to a region where their concentration is lower.

As carbohydrates and proteins are being digested continuously, there is normally a greater concentration of glucose and amino acids within the ileum than in the blood. There is therefore a concentration gradient down which glucose moves by facilitated diffusion from inside the ileum into the blood. Given that the blood is constantly being circulated by the heart, the glucose absorbed into it is continuously being removed by the cells as they use it up during respiration. This helps to maintain the concentration gradient between the inside of the ileum and the blood (Figure 2). This means the rate of movement by facilitated diffusion across epithelial cell-surface membranes is increased.

Role of active transport in absorption

At best, diffusion only results in the concentrations either side of the intestinal epithelium becoming equal. This means that not all the available glucose and amino acids can be absorbed in this way and some may pass out of the body. The reason why this does not happen is because glucose and amino acids are also being absorbed by active transport (see Topic 4.4). This means that all the glucose and amino acids should be absorbed into the blood.

The actual mechanism by which they are absorbed from the small intestine is an example of **co-transport**. This term is used because either glucose or amino acids are drawn into the cells along with sodium ions that have been actively transported out by the sodium–potassium pump (see Topic 4.4). It takes place in the following manner (see Figure 3):

Learning objectives

→ Describe the part villi and microvilli play in absorption.

→ Explain how the products of carbohydrate digestion are absorbed in the ileum.

→ Explain the roles of diffusion, active transport and co-transport in the process.

Specification reference: 3.2.3

▲ **Figure 1** *Microvilli on an epithelial cell from the small intestine*

Study tip

Do not confuse villi and microvilli. Villi (Topic 6.10) are 1 mm projections of the wall of the ileum while microvilli are 0.6 µm projections of the cell-surface membrane of the epithelial cells that line this wall. Microvilli are therefore more than one thousand times smaller than villi.

▲ **Figure 2** *Absorption of monosaccharides (e.g., glucose) by diffusion in the ileum*

cavity (lumen) of ileum

▲ **Figure 3** *Co-transport of a glucose molecule*

1 Sodium ions are actively transported out of epithelial cells, by the sodium–potassium pump, into the blood. This takes place in one type of protein-**carrier molecule** found in the cell-surface membrane of the epithelial cells.

2 This maintains a much higher concentration of sodium ions in the lumen of the intestine than inside the epithelial cells.

3 Sodium ions diffuse into the epithelial cells down this concentration gradient through a different type of protein carrier (co-transport protein) in the cell-surface membrane. As the sodium ions diffuse in through this second carrier protein, they carry either amino acid molecules or glucose molecules into the cell with them.

4 The glucose/amino acids pass into the blood plasma by facilitated diffusion using another type of carrier.

Both sodium ions and glucose/amino acid molecules move into the cell, but while the sodium ions move *down* their concentration gradient, the glucose molecules move *against* their concentration gradient. It is the sodium ion concentration gradient, rather than **ATP** directly, that powers the movement of glucose and amino acids into the cells. This makes it an indirect rather than a direct form of active transport.

Summary questions

1 State **three** ways in which the rate of movement across membranes can be increased.

2 Explain why the term 'co-transport' is used to describe the transport of glucose into cells.

3 In each of the following events in the glucose co-transport system, state whether the movements are active or passive.

  a Sodium ions move out of the epithelial cell.

  b Sodium ions move into the epithelial cell.

  c Glucose molecules move into the epithelial cell.

Synoptic link

The absorption of amino acids and lipids in the ileum will be covered in Topic 6.10, the structure of glucose was described in Topic 1.2 and the structure of amino acids in Topic 1.6.

Oral rehydration therapy 🧪

There are a number of diarrhoeal diseases that infect the intestines. Diarrhoea kills many people, especially the very young, and yet a treatment to prevent death is relatively simple. This treatment is **oral rehydration therapy**.

Diarrhoea is an intestinal disorder where watery faeces are produced frequently. The causes include:

- damage to the epithelial cells lining the intestine
- loss of microvilli due to toxins
- excessive secretion of water due to toxins, for example cholera toxin

As a result of diarrhoea, insufficient fluid is taken into, and/or excessive fluid is lost from, the body. Either way, dehydration results and may be fatal.

To treat diarrhoeal diseases it is vital to rehydrate the patient. Just drinking water is ineffective for two reasons:

- Water is not being absorbed from the intestine – indeed, as in the case of cholera, water is actually being lost from cells.
- Drinking water does not replace the electrolytes (ions) that are being lost from the intestinal cells.

It is possible to replace the water and electrolytes intravenously by a drip, but this requires trained personnel and means the patient is confined to bed for much of the time. What is required is a suitable mixture of substances that can safely be taken by mouth and which will be absorbed by the intestine.

But how can the patient be rehydrated if the intestine is not absorbing water? As it happens, there is more than one type of carrier protein in the plasma membranes of epithelial cells that absorbs sodium ions. The trick is to develop a rehydration solution that uses these alternative pathways. As sodium ions are absorbed, so the water potential of the cells falls and water enters the cells by osmosis. Therefore, a rehydration solution needs to contain:

- **water** – to rehydrate the tissues
- **sodium** ions – to replace the sodium ions lost from the epithelium of the intestine and to optimise use of the alternative sodium-glucose carrier proteins
- **glucose** – to stimulate the uptake of sodium ions from the intestine and to provide energy
- **potassium** ions – to replace lost potassium ions and to stimulate appetite
- **other electrolyes** – such as chloride ions and citrate ions to help prevent electrolyte imbalance and a condition called metabolic acidosis.

The ingredients can be mixed and packaged as a powder and then the solution made up with boiled water as needed. It can then be administered by people with minimal training. The solution must be given regularly, and in large amounts, throughout the illness.

Oral rehydration solutions do not prevent or cure diarrhoea. They simply rehydrate and nourish the patient until the diarrhoea is cured by some other means.

When commercial products are not available it is possible to use an inexpensive, home-made rehydration solution. This can be made up of eight level teaspoons of sugar + 1 level teaspoon of table salt dissolved in 1 litre of boiled water.

1 List *two* reasons why glucose is included in the mixture.
2 Table salt is sodium chloride. List *three* reasons why it is included in the mixture.
3 Explain why is it essential that the water is boiled.
4 Bananas are rich in potassium. It is sometimes recommended that mashed banana is added to the mixture. List *two* reasons why this might help the patient recover.
5 Suggest another advantage of adding … se before drinking the mixture, e… of children.

6 Sports drinks contain a high proportion of glucose to help replace that used during strenuous exercise. Explain in terms of water potential why these drinks are therefore not suitable to rehydrate those suffering from diarrhoea.

The development of oral rehydration solutions resulted from a long process of scientific experimentation.

Early rehydration solutions led to side effects, especially in children. These were caused by excess sodium and so mixtures with lower sodium content, but more glucose, were tested. Unfortunately the additional glucose lowered the water potential in the lumen of the ileum so much that it started to draw out even more water from the epithelial cells. This made the dehydration even worse. Lowering the glucose content reduced this effect but, as glucose also acted as a respiratory substrate, it reduced the amount of energy being supplied to the patient. One answer was to use starch in place of some of the glucose.

7 Explain why using starch is better than using glucose.

Starch is broken down steadily by amylase and maltase in the ileum into its glucose **monomers**. By experimenting with different concentrations of starch, a rehydration solution was developed that released glucose at the optimum rate for it to be taken up as it was produced, without it adversely influencing the water potential. Further scientific research is being carried out to find the best source of starch.

Rice starch is a popular choice for two main reasons:

- It is readily available in many parts of the world, especially those where diarrhoeal diseases are common.

- It also provides other nutrients like amino acids. Not only are these nutrients nutritionally valuable but they also help the uptake of sodium ions from the ileum.

s rice flour produces a very viscous solution, it is not to swallow.

8 Suggest **one** possible method of reducing the viscosity of the rice flour solution and explain how it works.

We have seen how the development of an improved medicine takes place in a number of stages, each of which must be tested for its safety. While initial testing can be done on tissue cultures and animals, to be sure of a drug's effectiveness and safety, it must eventually be tested on humans.

9 Consider why drugs must ultimately be tested on humans.

Testing is normally carried out in four phases.

- A small number (20–80) of usually healthy people are given a tiny amount of the drug to test for side effects rather than to see if the drug is effective. The dose may be increased gradually in a series of such trials. This stage takes around six months.

- The drug is given to a slightly larger number of people (100–300) who have the condition the drug is designed to treat. This is to see that it works and to look at any safety issues. This stage takes up to two years.

- A large-scale trial of many thousands of patients takes place. Many are given a dummy drug called a **placebo**. Often, neither the scientists nor the patients know who has taken the real drug and who has taken the placebo until after the trial. This is known as a double-blind trial. These trials take many years.

- If the drug passes all these stages it may be granted a licence, but its use and effects are still monitored over many years to check on any long-term effects.

10 Suggest a reason why a placebo is necessary to ensure that the results of a drug trial can be relied upon.
11 Suggest why the results of a 'double-blind' trial might be more reliable than one in which the patients knew whether they were taking the real drug or a placebo.

1 A student investigated the effect of putting cylinders cut from a potato into sodium chloride solutions of different concentration. He cut cylinders from a potato and weighed each cylinder. He then placed each cylinder in a test tube. Each test tube contained a different concentration of sodium chloride solution. The tubes were left overnight. He then removed the cylinders from the solutions and reweighed them.

(a) Before reweighing, the student blotted dry the outside of each cylinder. Explain why. *(2 marks)*

The student repeated the experiment several times at each concentration of sodium chloride solution. His results are shown in the graph.

(b) The student made up all the sodium chloride solutions using a 1.0 mol dm⁻³ sodium chloride solution and distilled water.

Complete the table to show how he made 20 cm³ of a 0.2 mol dm⁻³ sodium chloride solution.

Volume of 1.0 mol dm⁻³ sodium chloride solution	Volume of distilled water

(1 mark)

(c) The student calculated the *percentage* change in mass rather than the change in mass. Explain the advantage of this. *(2 marks)*

(d) The student carried out several repeats at each concentration of sodium chloride solution. Explain why the repeats were important. *(2 marks)*

(e) Use the graph to find the concentration of sodium chloride solution that has the same water potential as the potato cylinders.

$mol\ dm^{-3}$

(1 mark)

AQA Jan 2011

2 Some substances can cross the cell-surface membrane of a cell by simple diffusion through the phospholipid bilayer. Describe other ways by which substances cross this membrane. *(5 marks)*

AQA Jan 2013

3 **(a)** Give **two** ways in which active transport is different from facilitated diffusion. *(2 marks)*

Scientists investigated the effect of a drug called a proton pump inhibitor. The drug is given as a tablet to people who produce too much acid in their stomach. It binds to a carrier protein in the surface membrane of cells lining the stomach. This carrier protein usually moves hydrogen ions into the stomach by active transport.

The scientists used two groups of people in their investigation. All the people produced too much acid in their stomach. People in group **P** were given the drug. Group **Q** was the control group.

The graph shows the results.

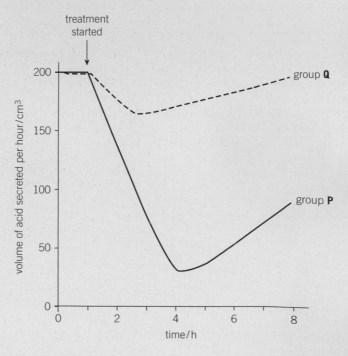

(b) **(i)**   The scientists used a control group in this trial. Explain why.   *(1 mark)*
     (ii) Suggest how the control group would have been treated.   *(2 marks)*
(c) Describe the effect of taking the drug on acid secretion.   *(1 mark)*

AQA June 2011

(d) Calculate the percentage decrease in acid secretion of group P compared to group Q after 8 hours.   *(2 marks)*

4 Scientists investigated the percentages of different types of lipid in plasma membranes from different types of cell. **Table 2** shows some of their results.

▼ Table 2

Type of lipid	Percentage of lipid in plasma membrane by mass		
	Cell lining ileum of mammal	Red blood cell of mammal	The bacterium *Escherichia coli*
Cholesterol	17	23	0
Glycolipid	7	3	0
Phospholipid	54	60	70
Others	22	14	30

(a) The scientists expressed their results as **Percentage of lipid in plasma membrane by mass**. Explain how they would find these values. *(2 marks)*

(b) Cholesterol increases the stability of plasma membranes. Cholesterol does this by making membranes less flexible.
Suggest **one** advantage of the different percentage of cholesterol in red blood cells compared with cells lining the ileum. *(1 mark)*

(c) *E. coli* has no cholesterol in its cell-surface membrane. Despite this, the cell maintains a constant shape. Explain why. *(2 marks)*

AQA SAMS AS PAPER 1

▲ **Figure 1** *Baby boy covered with measles rash. Measles is a highly infectious viral disease that mainly affects young children before they have acquired immunity to it*

Hint

The defensive mechanisms can be likened to the defences of a castle hundreds of years ago. The physical barrier is like the walls of the castle, the phagocytes are like the foot soldiers patrolling in the castle, who seek out and kill any intruders, and the lymphocytes are like specialised soldiers who respond to specific threats and use the intelligence gained from previous attacks to recognise, and quickly destroy, future intruders.

NB Don't use these descriptions in an examination!

Tens of millions of humans die each year from infectious diseases. Many more survive and others appear never to be affected in the first place. Why are there these differences?

Any infection is, in effect, an interaction between the **pathogen** and the body's various defence mechanisms. Sometimes the pathogen overwhelms the defences and the individual dies. Sometimes the body's defence mechanisms overwhelm the pathogen and the individual recovers from the disease. Having overwhelmed the pathogen, however, the body's defences seem to be better prepared for a second infection from the same pathogen and can kill it before it can cause any harm. This is known as **immunity** and is the main reason why some people are unaffected by certain pathogens.

There is a complete range of intermediates between the stages described above. Much depends on the overall state of health of an individual. A fit, healthy adult will rarely die of to an infection. Those in ill health, the young and the elderly are usually more vulnerable.

Defence mechanisms

The human body has a range of defences to protect itself from pathogens (Figure 2). Some are general and immediate defences like the skin forming a barrier to the entry of pathogens and phagocytosis (see Topic 5.2). Others are more specific, less rapid but longer-lasting. These responses involve a type of white blood cell called a lymphocyte and take two forms:

- cell-mediated responses involving T lymphocytes
- humoral responses involving B lymphocytes.

Before we look in detail at these defence mechanisms, let us first consider how the body distinguishes its own cells from foreign material.

Recognising your own cells

To defend the body from invasion by foreign material, lymphocytes must be able to distinguish the body's own cells and molecules (**self**) from those that are foreign (**non-self**). If they could not do this, the lymphocytes would destroy the organism's own tissues.

Each type of cell, self or non-self, has specific molecules on its surface that identify it. While these molecules can be of a variety of types, it is the proteins that are the most important. This is because proteins have enormous variety and a highly specific tertiary structure. It is this variety of specific 3-D structure that distinguishes one cell from another. It is these protein molecules which usually allow the immune system to identify:

- pathogens, for example the human immunodeficiency virus (see Topic 5.7).
- non-self material such as cells from other organisms of the same species.

- toxins including those produced by certain pathogens like the bacterium that causes cholera.
- abnormal body cells such as cancer cells.

All of the above are potentially harmful and their identification is the first stage in removing the threat they pose. Although this response is clearly advantageous to the organism, it has implications for humans who have had tissue or organ transplants. The immune system recognises these as non-self even though they have come from individuals of the same species. It therefore attempts to destroy the transplant. To minimise the effect of this tissue rejection, donor tissues for transplant are normally matched as closely as possible to those of the recipient. The best matches often come from relatives that are genetically close. In addition, immunosuppressant drugs are often administered to reduce the level of the immune response that still occurs.

▲ **Figure 3** *False-colour SEM of a single human lymphocyte (blue) and red blood cells (red)*

It is important to remember that specific lymphocytes are not produced in response to an infection, but that they already exist – all ten million different types. Given that there are so many different types of lymphocytes, there is a high probability that, when a pathogen gets into the body, one of these lymphocytes will have a protein on its surface that is complementary to one of the proteins of the pathogen. In other words, the lymphocyte will 'recognise' the pathogen. Not surprisingly with so many different lymphocytes, there are very few of each type. When an infection occurs, the one type already present that has the complementary proteins to those of the pathogen is stimulated to divide to build up its numbers to a level where it can be effective in destroying it. This is called clonal selection and you will learn more about it in Topic 5.4. This explains why there is a time lag between exposure to the pathogen and body's defences bringing it under control.

▲ **Figure 2** *Summary of defence mechanisms*

How lymphocytes recognise cells belonging to the body
- There are probably around ten million different lymphocytes present at any time, each capable of recognising a different chemical shape.
- In the fetus, these lymphocytes are constantly colliding with other cells.
- Infection in the fetus is rare because it is protected from the outside world by the mother and, in particular, the placenta.
- Lymphocytes will therefore collide almost exclusively with the body's own material (self).
- Some of the lymphocytes will have receptors that exactly fit those of the body's own cells.
- These lymphocytes either die or are suppressed.
- The only remaining lymphocytes are those that might fit foreign material (non-self), and therefore only respond to foreign material.
- In adults, lymphocytes produced in the bone marrow initially only encounter self-antigens.
- Any lymphocytes that show an immune response to these self-antigens undergo programmed cell death (apoptosis) before they can differentiate into mature lymphocytes.
- No clones of these anti-self lymphocytes will appear in the blood, leaving only those that might respond to non-self antigens.

Summary questions

1 State *two* differences between a specific and a non-specific defence mechanism.

2 After a pathogen gains entry to the body it is often a number of days before the body's immune system begins to control it. Suggest a possible reason why this is so.

3 In the above case, suggest why it would be inaccurate to say that the body takes days to 'respond' to the pathogen.

If a pathogen is to infect the body it must first gain entry. Clearly then, the body's first line of defence is to form a physical or chemical barrier to entry. Should this fail, the next line of defence is the white blood cells. There are two types of white blood cell: **phagocytes** and **lymphocytes**. Phagocytes ingest and destroy the pathogen by a process called phagocytosis before it can cause harm. Lymphocytes are involved in immune responses (Topics 5.3 and 5.5).

Despite various barriers pathogens still frequently gain entry and the next line of defence is then phagocytosis.

▲ **Figure 1** *Human skin forms a tough outer layer that forms a barrier to the entry of pathogens*

Phagocytosis

Large particles, such as some types of bacteria, can be engulfed by cells in the vesicles formed from the cell-surface membrane. This process is called phagocytosis. In the blood, the types of white blood cells that carry out phagocytosis are known as **phagocytes**. They provide an important defence against the pathogens that manage to enter the body. Some phagocytes travel in the blood but can move out of blood vessels into other tissues. Phagocytosis is illustrated in Figure 2 and is summarised below and in Figure 3.

- Chemical products of pathogens or dead, damaged and abnormal cells act as attractants, causing phagocytes to move towards the pathogen (e.g., a bacterium).

- Phagocytes have several receptors on their cell-surface membrane that recognise, and attach to, chemicals on the surface of the pathogen.

- They engulf the pathogen to form a vesicle, known as a **phagosome**.

- Lysosomes move towards the vesicle and fuse with it.

- Enzymes called **lysozymes** are present within the lysosome. These lysozymes destroy ingested bacteria by hydrolysis of their cell walls. The process is the same as that for the digestion of food in the intestines, namely the hydrolysis of larger, insoluble molecules into smaller, soluble ones.

- The soluble products from the breakdown of the pathogen are absorbed into the cytoplasm of the phagocyte.

▲ **Figure 2** *False-colour SEM of a phagocyte (red) engulfing tuberculosis bacteria (yellow), a process known as phagocytosis*

1 The phagocyte is attracted to the pathogen by chemical products of the pathogen. It moves towards the pathogen along a concentration gradient

pathogen (e.g., bacterium)

chemical products of pathogen

phagocyte

2 The phagocyte has several receptors on its cell-surface membrane that attach to chemicals on the surface of the pathogen

lysosome

nucleus

3 Lysosomes within the phagocyte migrate towards the phagosome formed by engulfing the bacterium

phagosome forming

4 The lysosomes release their lysozymes into the phagosome, where they hydrolyse the bacterium

phagosome

lysosomes release lysozymes into phagosome

5 The hydrolysis products of the bacterium are absorbed by the phagocyte

breakdown debris of pathogen

▲ **Figure 3** *Summary of phagocytosis*

Summary questions

1 In the following passage, state the missing word indicated by each letter **a**–**d**.

Pathogens that invade the body may be engulfed by cells which carry out **a**. The engulfed pathogen forms a vesicle known as a **b**. Once engulfed the pathogen is broken down by enzymes called **c** released from organelles called **d**.

2 Among other places, lysozymes are found in tears. Suggest a reason why this is so.

Learning objectives

→ State the definition of an antigen.

→ Describe the two main types of lymphocyte.

→ Explain the role of T cells (T lymphocytes) in cell-mediated immunity.

Specification reference: 3.2.4

The initial response of the body to infection is non-specific (see Topic 5.2). The next phase is the primary immune response that confers immunity. Immunity is the ability of organisms to resist infection by protecting against disease-causing microorganisms or their toxins that invade their bodies. It involves the recognition of foreign material (antigens).

Antigens

An antigen is any part of an organism or substance that is recognised as non-self (foreign) by the immune system and stimulates an immune response. Antigens are usually proteins that are part of the cell-surface membranes or cell walls of invading cells, such as microorganisms, or abnormal body cells, such as cancer cells. The presence of an antigen triggers the production of an antibody as part of the body's defence system (see Topic 5.4).

Lymphocytes

Immune responses such as phagocytosis are **non-specific** (see Topic 5.2) and occur whatever the infection. The body also has **specific** responses that react to specific antigens. These are slower in action at first, but they can provide long-term immunity. This specific immune response depends on a type of white blood cell called a **lymphocyte**. Lymphocytes are produced by stem cells in the bone marrow. There are two types of lymphocyte, each with its own role in the immune response:

- **B lymphocytes (B cells)** are so called because they mature in the bone marrow. They are associated with humoral immunity, that is, immunity involving antibodies that are present in body fluids, or 'humour' such as blood plasma. This is described in more detail in Topic 5.5.

- **T lymphocytes (T cells)** are so called because they mature in the thymus gland. They are associated with cell-mediated immunity, that is immunity involving body cells.

Cell-mediated immunity

Lymphocytes respond to an organism's own cells that have been infected by non-self material from a different species, for example a virus. They also respond to cells from other individuals of the same species because these are genetically different. These therefore have different antigens on their cell-surface membrane from the antigens on the organism's own cells. T lymphocytes can distinguish these invader cells from normal cells because:

- phagocytes that have engulfed and hydrolysed a pathogen present some of a pathogen's antigens on their own cell-surface membrane

- body cells invaded by a virus present some of the viral antigens on their own cell-surface membrane.

Study tip

It is always necessary to describe events in detail using the appropriate scientific terms. For example, in immunity questions vague references to 'cells fighting disease' should be avoided.

- transplanted cells from individuals of the same species have different antigens on their cell-surface membrane
- cancer cells are different from normal body cells and present antigens on their cell-surface membranes.

Cells that display foreign antigens on their surface are called **antigen-presenting cells** because they can present antigens of other cells on their own cell-surface membrane.

T lymphocytes will only respond to antigens that are presented on a body cell (rather than to antigens within the body fluids). This type of response is called **cell-mediated immunity** or the **cellular response**. The role of the receptors on T cells is important. The receptors on each T cell respond to a single antigen. It follows that there is a vast number of different types of T cell, each one responding to a different antigen (see Topic 5.1). The stages in the response of T lymphocytes to infection by a pathogen are summarised in Figure 1 and explained below.

1 Pathogens invade body cells or are taken in by phagocytes.

2 The phagocyte places antigens from the pathogen on its cell-surface membrane.

3 Receptors on a specific helper T cell (T_H cell) fit exactly onto these antigens.

4 This attachment activates the T cell to divide rapidly by mitosis and form a clone of genetically identical cells.

5 The cloned T cells:

  a develop into memory cells that enable a rapid response to future infections by the same pathogen

  b stimulate phagocytes to engulf pathogens by phagocytosis

  c stimulate B cells to divide and secrete their antibody

  d activate cytotoxic T cells (T_C cells).

> ### Hint
>
> Three terms that are frequently confused are *antigen*, *antibody* and *antibiotic*. When dealing with immunity put *antibiotic* out of your mind – it has nothing to do with immunity.

▲ **Figure 1** *Summary of the role of T cells in cell-mediated immunity*

How cytotoxic T cells kill infected cells

Cytotoxic T cells (T_C cells) kill abnormal cells and body cells that are infected by pathogens, by producing a protein called perforin that makes holes in the cell-surface membrane. These holes mean the cell membrane becomes freely permeable to all substances and the cell dies as a result. This illustrates the vital importance of cell-surface membranes in maintaining the integrity of cells and hence their survival. The action of T cells is most effective against viruses because viruses replicate inside cells. As viruses use living cells in which to replicate, this sacrifice of body cells prevents viruses multiplying and infecting more cells.

▲ **Figure 2** *False-colour SEM of two human cytotoxic T cells (yellow) attacking a cancer cell (red)*

Summary questions

1 Define an antigen
2 State *two* similarities between T cells and B cells.
3 State *two* differences between T cells and B cells.

Bird flu

Avian (bird) flu is caused by one of many strains of the influenza virus. Although it is adapted primarily to infect birds, the H5N1 strain of the virus can infect other species, including humans. Avian flu affects the lungs and can cause the immune system to go into overdrive. This results in a massive overproduction of T cells.

1 From your knowledge of cell-mediated immunity and lung structure suggest why humans infected with the H5N1 virus may sometimes die from suffocation.
2 Suggest a reason why any spread of bird flu across the world is likely to be very rapid.

We saw in Topic 5.3 that the first phase of the specific response to infection is the mitotic division of specific T cells to form a clone of the relevant T cells to build up their numbers. Some of these T cells produce factors that stimulate B cells to divide. It is these B cells that are involved in the next phase of the immune response: humoral immunity.

Humoral immunity

Humoral immunity is so called because it involves **antibodies** (see Topic 5.5), and antibodies are soluble in the blood and tissue fluid of the body. An old-fashioned word for body fluids is 'humour'. There are many different types of B cell, possibly as many as ten million, and each B cell starts to produce a specific antibody that responds to one specific antigen. When an **antigen**, for example, a protein on the surface of a pathogen, foreign cell, toxin, damaged or abnormal cell, enters the blood or tissue fluid, there will be one B cell that has an antibody on its surface whose shape exactly fits the antigen, that is, they are complementary. The antibody therefore attaches to this complementary antigen. The antigen enters the B cell by **endocytosis** and gets presented on its surface (processed). T_H cells bind to these processed antigens and stimulate this B cell to divide by mitosis (see Topic 3.7) to form a clone of identical B cells, all of which produce the antibody that is specific to the foreign antigen. This is called **clonal selection** and accounts for the body's ability to respond rapidly to any of a vast number of antigens.

In practice, a typical pathogen has many different proteins on its surface, all of which act as antigens. Some pathogens, such as the bacterium that causes cholera, also produce toxins. Each toxin molecule also acts as an antigen. Therefore many different B cells make clones, each of which produces its own type of antibody. As each clone produces one specific antibody these antibodies are referred to as **monoclonal antibodies** (see Topic 5.5). In each clone, the cells produced develop into one of two types of cell:

- **Plasma cells** secrete antibodies usually into blood plasma. These cells survive for only a few days, but each can make around 2000 antibodies every second during its brief lifespan. These antibodies lead to the destruction of the antigen. The plasma cells are therefore responsible for the immediate defence of the body against infection. The production of antibodies and memory cells (see below) is known as the **primary immune response**.
- **Memory cells** are responsible for the **secondary immune response**. Memory cells live considerably longer than plasma cells, often for decades. These cells do not produce antibodies directly, but circulate in the blood and tissue fluid. When they encounter the same antigen at a later date, they divide rapidly and develop into plasma cells and more memory cells. The plasma cells produce the antibodies needed to destroy the pathogen, while the new memory cells circulate in readiness for any future infection. In this way, memory cells provide long-term immunity against the original infection. An increased quantity of antibodies is secreted at a faster rate than in the primary immune response. It ensures that a new infection is destroyed before it can cause any harm – and individuals

Learning objectives

→ Explain the role of B cells (B lymphocytes) in humoral immunity.

→ Explain the roles of plasma cells and antibodies in the primary immune response.

→ Explain the role of memory cells in the secondary immune response.

→ Explain how antigenic variation affects the body's response to infection.

Specification reference: 3.2.4

Hint

Remember that B cells with the appropriate antibody to bind to antigens of a pathogen are not produced in response to the pathogen. They are present from birth. Being present, they simply *multiply* in response to the pathogen.

▲ **Figure 1** *Primary and secondary responses to an antigen*

are often totally unaware that they have ever been infected. Figure 1 illustrates the relative amounts of antibody produced in the primary and secondary immune responses.

The role of B cells in immunity is explained below and summarised in Figure 2.

▲ **Figure 2** *Summary of role of B cells in humoral immunity*

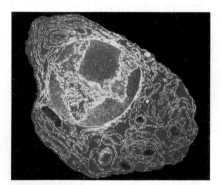

▲ **Figure 3** *False-colour TEM of a plasma cell. Plasma cells are mature B lymphocytes that secrete antibodies. Note the well-developed rough endoplasmic reticulum (yellow dotted lines) where the antibodies are synthesised.*

1 The surface antigens of an invading pathogen are taken up by a B cell.

2 The B cell processes the antigens and presents them on its surface.

3 Helper T cells (activated in the process described in Topic 5.3) attach to the processed antigens on the B cell thereby activating the B cell.

4 The B cell is now activated to divide by **mitosis** to give a clone of plasma cells.

5 The cloned plasma cells produce and secrete the specific antibody that exactly fits the antigen on the pathogen's surface.

6 The antibody attaches to antigens on the pathogen and destroys them (see Topic 5.5).

7 Some B cells develop into memory cells. These can respond to future infections by the same pathogen by dividing rapidly and developing into plasma cells that produce antibodies. This is the secondary immune response.

Summary questions

1 Explain why the secondary immune response is much more rapid than the primary one.

2 Contrast the cell-mediated and humoral responses to a pathogen.

3 Plasma cells can produce around 2000 protein antibodies each second. Suggest **three** cell organelles that you might expect to find in large quantities in a plasma cell, and explain why.

In Topic 5.4 we saw how B cells respond to **antigens** by producing antibodies. Let us now look at antibodies and how they work in more detail.

Antibodies

Antibodies are proteins with specific binding sites synthesised by B cells. When the body is infected by non-self material, a B cell produces a specific antibody. This specific antibody reacts with an antigen on the surface of the non-self material by binding to them. Each antibody has two identical binding sites. The antibody binding sites are complementary to a specific antigen. The massive variety of antibodies is possible because they are made of proteins – molecules that occur in an almost infinite number of forms.

Antibodies are made up of four polypeptide chains. The chains of one pair are long and are called **heavy chains**, while the chains of the other pair are shorter and are known as **light chains**. Each antibody has a specific binding site that fits very precisely onto a specific antigen to form what is known as an **antigen–antibody complex**. The binding site is different on different antibodies and is therefore called the **variable region**. Each binding site consists of a sequence of amino acids that form a specific 3-D shape that binds directly to a specific antigen. The rest of the antibody is known as the **constant region**. This binds to receptors on cells such as B cells. The structure of an antibody is illustrated in Figure 1.

How the antibody leads to the destruction of the antigen

It is important to understand that antibodies do not destroy antigens directly but rather prepare the antigen for destruction. Different antibodies lead to the destruction of an antigen in a range of ways. Take the example of when the antigen is a bacterial cell – antibodies assist in its destruction in two ways:

- They cause agglutination of the bacterial cells (Figure 2). In this way clumps of bacterial cells are formed, making it easier for the phagocytes to locate them as they are less spread-out within the body.
- They then serve as markers that stimulate phagocytes to engulf the bacterial cells to which they are attached.

Learning objectives

→ Describe the structure of an antibody.

→ Describe the functions of antibodies.

→ Describe the nature of a monoclonal antibody.

→ Explain how monoclonal antibodies are produced.

→ Explain how monoclonal antibodies are used to target specific substances and cells.

Specification reference: 3.2.4

Hint

One molecule fitting neatly with another is a recurring theme throughout biology. We met it with enzymes (see Topic 1.7) and with T cells (see Topic 5.3) and it features again here. While the 'lock and key' image is helpful, remember that, with the induced fit model of enzyme action, the molecules are flexible rather than rigid. This is the same for antibodies. The image of a hand fitting a glove is therefore perhaps a better one when it comes to understanding the process.

Study tip

Agglutination is possible because each antibody has two antigen binding sites.

◀ **Figure 1** *Structure of an antibody (left); molecular model of an antibody (right). This Y-shaped protein is produced by B lymphocytes as part of the immune response*

antigen-binding sites

light chain

heavy chain

receptor binding site

variable region (different in different antibodies)

constant region

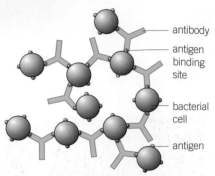

antibody

antigen binding site

bacterial cell

antigen

Each antibody attaches to two bacterial cell, causing them to clump together

▲ **Figure 2** *Agglutination*

Monoclonal antibodies

We have seen that a bacterium or other microorganism entering the body is likely to have many hundreds of different antigens on its surface. Each antigen will induce a different B cell to multiply and form a clone of itself. Each of these clones will produce a different antibody. It is of considerable medical value to be able to produce antibodies outside the body. It is even better if a single type of antibody can be isolated and cloned. Such antibodies are known as **monoclonal antibodies**.

Monoclonal antibodies have a number of useful functions in science and medicine.

Targeting medication to specific cell types by attaching a therapeutic drug to an antibody

As an antibody is very specific to particular antigen (protein), monoclonal antibodies can be used to target specific substances and specific cells. One type of cell they can target is cancer cells. Monoclonal antibodies can be used to treat cancer in a number of ways. By far the most successful so far is direct monoclonal antibody therapy.

- Monoclonal antibodies are produced that are specific to antigens on cancer cells.

- These antibodies are given to a patient and attach themselves to the receptors on their cancer cells.

- They attach to the surface of their cancer cells and block the chemical signals that stimulate their uncontrolled growth.

An example is herceptin, a monoclonal antibody used to treat breast cancer. The advantage of direct monoclonal antibody therapy is that since the antibodies are not toxic and are highly specific, they lead to fewer side effects than other forms of therapy.

Another method, called indirect monoclonal antibody therapy, involves attaching a radioactive or cytotoxic drug (a drug that kills cells) to the monoclonal antibody. When the antibody attaches to the cancer cells, it kills them.

For obvious reasons, monoclonal antibodies used in this way are referred to as 'magic bullets' and can be used in smaller doses, as they are targeted on specific sites. Using them in smaller doses is not only cheaper but also reduces any side effects the drug might have.

Medical diagnosis

Monoclonal antibodies are an invaluable tool in diagnosing disease with over a hundred different diagnostic products based on them. They are used for the diagnosis of influenza, hepatitis and chlamydia infections where they produce a much more rapid result than conventional methods of diagnosis. They are important in diagnosing certain cancers. For example, men with prostate cancer often produce more of a protein called prostate specific antigen (PSA) leading to unusually high levels of it in the blood. By using a monoclonal antibody that interacts with this antigen, it is possible to obtain a measure of the level of PSA in a sample of blood. While a higher than normal level of PSA is not itself diagnostic of the disease, it gives an early warning of its possibility and the need for further tests. The use of antibodies in the ELISA test is discussed in Topic 5.7.

Pregnancy testing

It is important that a mother knows as early as possible that she is pregnant, not least because there are certain actions she can take to ensure the welfare of herself and her unborn baby. The use of pregnancy testing kits that can easily be used at home has made possible the early detection of a pregnancy. These kits rely on the fact that the placenta produces a hormone called human chorionic gonadatrophin (hCG) and that this is found in the mother's urine. Monoclonal antibodies present on the test strip of a home pregnancy testing kit are linked to coloured particles. If hCG is present in the urine it binds to these antibodies. The hCG-antibody-colour complex moves along the strip until it is trapped by a different type of antibody creating a coloured line.

Ethical use of monoclonal antibodies

The development of monoclonal antibodies has provided society with the power and opportunity to treat diseases in hitherto unknown ways. However, with this power and opportunity comes responsibility. The use of monoclonal antibodies raises some ethical issues.

*   Production of monoclonal antibodies involves the use of mice. These mice are used to produce both antibodies and tumour cells. The production of tumour cells involves deliberately inducing cancer in mice. Despite the specific guidelines drawn up to minimise any suffering, some people still have reservations about using animals in this way.

*   Monoclonal antibodies have been used successfully to treat a number of diseases, including cancer and diabetes, saving many lives. There have also been some deaths associated with their use in the treatment of multiple sclerosis. It is important that patients have full knowledge of the risks and benefits of these drugs before giving permission for them to be used (=informed consent).

*   Testing for the safety of new drugs presents certain dangers. In March 2006, six healthy volunteers took part in the trial of a new monoclonal antibody (TGN1412) in London. Within minutes they suffered multiple organ failure, probably as a result of T cells overproducing chemicals that stimulate an immune response or attacking the body tissues. All the volunteers survived, but it raises issues about the conduct of drug trials.

Society must use the issues raised here, combined with current scientific knowledge about monoclonal antibodies, to make decisions about their use. We must balance the advantages that a new medicine provides with the dangers that its use might bring. Only then can we make informed decisions at individual, local, national and global levels about the ethical use of drugs such as monoclonal antibodies.

▲ **Figure 3** *Home pregnancy testing kit showing a positive result. These kits use monoclonal antibodies*

▲ **Figure 4** *A scientist adding monoclonal antibodies to human tissue samples in order to detect cancer*

Study tip

When discussing social and ethical issues such as these, do not resort to comments such as "Who are we to play God?". General arguments should be supported with sound biology.

Summary questions

1 Suggest why antibodies made of proteins, rather than carbohydrates or fats, are more likely to be effective against a wide range of diseases.

2 Distinguish between an antigen and an antibody.

3 Discuss whether drug trials should be limited to volunteers who are terminally ill with a condition that the monoclonal antibody is designed to treat.

Producing monoclonal antibodies

The production of a large quantities of a specific antibody has long been recognised as useful. The problem had always been that B cells are short-lived and only divide inside a living organism. Nowadays, large quantities of a single antibody can be produced outside the body.

There was much competition among scientific research teams to overcome the problem of getting B cells to grow indefinitely outside of the body and a variety of methods was investigated. Cesar Milstein and Georges Kohler evaluated these methods in relation to the behaviour of cancer cells. In 1975, they produced a solution to the problem by developing the following procedure:

- A mouse is exposed to the non-self material against which an antibody is required.

- The B cells in the mouse then produce a mixture of antibodies, which are extracted from the spleen of the mouse.

- To enable these B cells to divide outside the body, they are mixed with cells that divide readily outside the body, for example, cells from a cancer tumour.

- Detergent is added to the mixture to break down the cell-surface membranes of both types of cell and enable them to fuse together. The fused cells are called **hybridoma cells**.

- The hybridoma cells are separated under a microscope and each single cell is cultured to form a clone. Each clone is tested to see whether it is producing the required antibody.

- Any clone producing the required antibody is grown on a large scale and the antibodies are extracted from the growth medium.

- Because these antibodies come from a clone formed from a single B cell, they are called **monoclonal antibodies**.

▲ **Figure 5** *False-colour SEM of hybridoma cell used to produce monoclonal antibodies*

As these monoclonal antibodies come from mouse tissue, they have to be modified to make them like human cells before they can be used. This process is called humanisation.

1 From your knowledge of membrane structure, suggest a reason why detergent might causes B cells and tumour cells to fuse.

2 When the detergent is added to the cells, the mixture is gently agitated. Suggest a reason why.

3 Explain why cells from cancer tumours are used to fuse with the B cells?

4 Some B cells and tumour cells fuse together. Suggest which other cells might also fuse together.

5 Explain why it is necessary to carry out 'humanisation' of the monoclonal antibodies.

6 One way to eliminate the need for humanisation would be to inject humans with an antigen and then extract the antibodies produced in response to it. Suggest reasons why this is considered unethical.

Immunity is the ability of an organism to resist infection. This immunity takes two forms.

- **Passive immunity** is produced by the introduction of antibodies into individuals from an outside source. No direct contact with the pathogen or its antigen is necessary to induce immunity. Immunity is acquired immediately. As the antibodies are not being produced by the individuals themselves, the antibodies are not replaced when they are broken down, no memory cells are formed and so there is no lasting immunity. Examples of passive immunity include anti-venom given to the victims of snake bites and the immunity acquired by the fetus when antibodies pass across the placenta from the mother.

- **Active immunity** is produced by stimulating the production of antibodies by the individuals' own immune system. Direct contact with the pathogen or its antigen is necessary. Immunity takes time to develop. It is generally long-lasting and is of two types:

  - **Natural active immunity** results from an individual becoming infected with a disease under normal circumstances. The body produces its own antibodies and may continue to do so for many years.

  - **Artificial active immunity** forms the basis of vaccination (immunisation). It involves inducing an immune response in an individual, without them suffering the symptoms of the disease.

Vaccination is the introduction of the appropriate disease antigens into the body, either by injection or by mouth. The intention is to stimulate an immune response against a particular disease. The material introduced is called **vaccine** and, in whatever form (see below), it contains one or more types of antigen from the pathogen. These antigens stimulate the immune response as described in Topics 5.3 and 5.4. The response is slight because only a small amount of antigen has been introduced, However, the crucial factor is that memory cells (see Topic 5.4) are produced. These remain in the blood and allow a greater, and more immediate, response to a future infection with the pathogen. The result is that there is a rapid production of antibodies and the new infection is rapidly overcome before it can cause any harm and with few, if any, symptoms.

When carried out on a large scale, this provides protection against disease not only for individuals, but also for whole populations.

Features of a successful vaccination programme

It is important to understand that vaccination is used as a precautionary measure to prevent individuals contracting a disease. It is not a means of treating individuals who already have the disease. Some programmes of vaccination against diseases have had considerable success. Yet, in other instances, similar measures have been less successful. The success of a vaccination programme depends on a number of factors:

- A suitable vaccine must be economically available in sufficient quantities to immunise most of the vulnerable population.

Learning objectives

→ Describe the nature of vaccines.

→ Describe the features of an effective vaccination programme.

→ Explain why vaccination rarely eliminates a disease.

→ Discuss the ethical issues associated with vaccination programmes.

Specification reference: 3.2.4

▲ **Figure 1** *The development of new vaccines is a highly technological process requiring sterile conditions*

- There must be few side-effects, if any, from vaccination. Unpleasant side-effects may discourage individuals in the population from being vaccinated.

- Means of producing, storing and transporting the vaccine must be available. This usually involves technologically advanced equipment, hygienic conditions and refrigerated transport.

- There must be the means of administering the vaccine properly at the appropriate time. This involves training staff with appropriate skills at different centres throughout the population.

- It must be possible to vaccinate the vast majority of the vulnerable population to produce **herd immunity**.

Herd immunity

Herd immunity arises when a sufficiently large proportion of the population has been vaccinated to make it difficult for a pathogen to spread within that population. The concept is based on the idea that pathogens are passed from individual to individual when in close contact. Where the vast majority of the population is immune, it is highly improbable that a susceptible individual will come in contact with an infected person. In this way those individuals who are not immune to the disease are nevertheless protected.

Herd immunity is important because it is never possible to vaccinate everyone in a large population. For example, babies and very young children are not vaccinated because their immune system is not yet fully functional. It could also be dangerous to vaccinate those who are ill or have compromised immune systems. The percentage of the population that must be vaccinated in order to achieve herd immunity is different for each disease. To achieve herd immunity, vaccination is best carried out at one time. This means that, for a certain period, there are very few individuals in the population with the disease and the transmission of the pathogen is interrupted.

Why vaccination may not eliminate a disease

Even when these criteria for successful vaccination are met, it can still prove extremely difficult to eradicate a disease. The reasons are as follows:

- Vaccination fails to induce immunity in certain individuals, for example people with defective immune systems.

- Individuals may develop the disease immediately after vaccination but before their immunity levels are high enough to prevent it. These individuals may harbour the pathogen and reinfect others.

- The pathogen may mutate frequently, so that its antigens change suddenly rather than gradually. This means that vaccines suddenly become ineffective because the new antigens on the pathogen are no longer recognised by the immune system. As a result the immune system does not produce the antibodies to destroy the pathogen. This **antigenic variability** happens with the influenza virus, which changes its antigens frequently. Immunity is therefore short-lived and individuals may develop repeated bouts of influenza during their lifetime.

Link

A level students will learn more about mutations in Topic 20.1 Gene mutations

- There may be so many varieties of a particular pathogen that it is almost impossible to develop a vaccine that is effective against them all. For example, there are over 100 varieties of the common cold virus and new ones are constantly evolving.

- Certain pathogens 'hide' from the body's immune system, either by concealing themselves inside cells, or by living in places out of reach, such as within the intestines, for example, the cholera pathogen.

- Individuals may have objections to vaccination for religious, ethical or medical reasons. For example, unfounded concerns over the measles, mumps and rubella (MMR) triple vaccine has led a number of parents to opt for separate vaccinations for their children, or to avoid vaccination altogether.

▲ **Figure 2** *Vaccination programmes for children have considerably reduced deaths from infectious diseases*

The ethics of using vaccines

As vaccinations have saved millions of lives, it is easy to accept vaccination programmes without question. However, they do raise ethical issues that need to be addressed if such programmes are to command widespread support. The production and use of vaccines raises the following questions:

- The production of existing vaccines, and the development of new ones, often involves the use of animals. How acceptable is this?

- Vaccines have side-effects that may sometimes cause long-term harm. How can the risk of side-effects be balanced against the risk of developing a disease that causes even greater harm?

- On whom should vaccines be tested? How should such trials be carried out? To what extent should individuals be asked to accept risk in the interests of the public health?

- Is it acceptable to trial a new vaccine with unknown health risks only in a country where the targeted disease is common, on the basis that the population there has most to gain if it proves successful?

- To be fully effective the majority, and preferably all, of the population should be vaccinated. Is it right, in the interests of everyone's health, that vaccination should be compulsory? If so, should this be at any time, or just when there is a potential epidemic? Can people opt out? If so, on what grounds: religious belief, medical circumstances, personal belief?

- Should expensive vaccination programmes continue when a disease is almost eradicated, even though this might mean less money for the treatment of other diseases?

- How can any individual health risks from vaccination be balanced against the advantages of controlling a disease for the benefit of the population at large?

Summary questions

1 Distinguish between active immunity and passive immunity.

2 Explain why vaccinating against influenza is not always effective.

MMR vaccine

In 1988, a combined vaccine for measles, mumps and rubella (MMR) was introduced into the UK to replace three separate vaccines. All three diseases are potentially disabling. Mumps can lead to orchitis in men possibly causing sterility and measles is potentially lethal. Ten years later a study was published in a well-respected medical journal. This suggested that there was a higher incidence of autism among children who had received the triple MMR vaccine than those who had received separate vaccinations. Autism is a condition in which individuals have impaired social interaction and communication skills.

In the wake of the media furore that followed, many parents decided to have their children vaccinated separately for the three diseases, while others opted for no vaccination at all. Parents of autistic children recalled that symptoms of the disorder emerged at around 14 months of age – shortly after the children had been given the MMR vaccination, adding to public concern about the MMR vaccine. The incidence of measles, mumps and rubella rose.

The vast majority of scientists now think that the vaccine is safe. A number of facts have emerged since the first research linking the MMR vaccine to autism.

- The author of the research had a conflict of interests. He was also being paid by the Legal Aid Board to discover whether parents who claimed their children had been damaged by MMR had a case. Some children were included in both studies.

- Further studies, including one in Japan involving over 30 000 children, have found no link between the MMR vaccine and autism.

- The sample size of the initial research was very small relative to later studies.

- The journal that published the initial research has publicly declared that, had it known all the facts, it would not have published the work.

1 Autism experts point out that many of the symptoms of autism first occur around the age of 14 months. Explain why this information is relevant to the debate on whether MMR vaccine and autism are linked.

2 Discuss how an organisation funding research might influence the outcome of that research without dishonestly altering the findings.

▲ **Figure 3** *MMR vaccination phial*

Even without this additional evidence, care has to be exercised when looking at data, specially where there are correlations between two factors, In this example, almost all the population had been vaccinated with the MMR vaccine. There would therefore be a correlation between people who had been vaccinated and almost everything – what they ate, where they lived etc. For example, data would have shown that that the majority of children who died in road accidents had been given the MMR vaccine. It does not follow that MMR causes road accidents. It was clearly a difficult choice for parents. Some parents, understandably, opted for separate vaccinations. Others mistrusted vaccinations in general and left their children unprotected. As a result, some children have developed disabilities that could have been avoided. On the other hand, had the research proved valid, it would have been those who held faith with the MMR vaccine who would have been putting their children's health at risk. It was a real dilemma.

The public sometimes believe that all such evidence must be true and accept it uncritically. However, all scientific evidence should be initially treated with caution – after all it is fellow scientists who are often quickest to criticise. There are various reasons for this caution:

- To be universally accepted, a scientific theory must first be critically appraised and confirmed by other scientists in the field. The confirmation of a theory takes time.

- Some scientists may not be acting totally independently but may be funded by other people or organisations who are anticipating a particular outcome from the research.

- Scientists' personal beliefs, views and opinions may influence the way they approach or represent their research.

- The facts, as presented by media headline writers, companies, governments and other organisations, may have been biased or distorted to suit their own interests.

- New knowledge may challenge accepted scientific beliefs; theories are being modified all the time.

The human immunodeficiency virus causes the disease **acquired immune deficiency syndrome (AIDS)**. Among contagious diseases it is a relative newcomer, having been first diagnosed in 1981. In this topic we will look at the structure of HIV and how it leads to the symptoms of AIDS.

Structure of the human immunodeficiency virus

The structure of HIV is shown in Figure 1. On the outside is a **lipid envelope**, embedded in which are peg-like **attachment proteins**. Inside the envelope is a protein layer called the **capsid** that encloses two single strands of **RNA** and some enzymes. One of these enzymes is **reverse transcriptase**, so-called because it catalyses the production of DNA from RNA – the reverse reaction to that carried out by transcriptase. The presence of reverse transcriptase, and consequent ability to make DNA from RNA, means that HIV belongs to a group of viruses called **retroviruses**.

Learning objectives

→ Describe the structure of the human immunodeficiency virus.

→ Explain how the human immunodeficiency virus replicates.

→ Explain how the human immunodeficiency virus causes AIDS.

→ Describe the treatment and control of AIDS.

→ Explain how the ELISA test works.

→ Explain why antibiotics are ineffective against viruses.

Specification reference: 3.2.4

genetic material (RNA)

attachment protein

capsid

lipid envelope

reverse transcriptase (enzyme)

matrix

▲ **Figure 1** *Structure of HIV*

Replication of the human immunodeficiency virus

Being a virus, HIV cannot replicate itself. Instead it uses its genetic material to instruct the host cell's biochemical mechanisms to produce the components required to make new HIV. It does so as follows:

- Following infection HIV enters the bloodstream and circulates around the body.

- A protein on the HIV readily binds to a protein called CD4. While this protein occurs on a number of different cells, HIV most frequently attaches to helper T cells (see Topic 5.3).

- The protein capsid fuses with the cell-surface membrane. The RNA and enzymes of HIV enter the helper T cell.

- The HIV reverse transcriptase converts the virus's RNA into DNA.

Link

Reverse transcriptase is an important enzyme in recombinant DNA technology and A level students will discover more about this in Topic 21.1 Producing DNA fragments.

▲ **Figure 2** *Colourised TEM of HIV*

- The newly made DNA is moved into the helper T cell's nucleus where it is inserted into the cell's DNA.
- The HIV DNA in the nucleus creates **messenger RNA** (mRNA), using the cell's enzymes. This mRNA contains the instructions for making new viral proteins and the RNA to go into the new HIV.
- The mRNA passes out of the nucleus through a nuclear pore and uses the cell's protein synthesis mechanisms to make HIV particles.
- The HIV particles break away from the helper T cell with a piece of its cell-surface membrane surrounding them which forms their lipid envelope.

Once infected with HIV a person is said to be **HIV positive**. However, the replication of HIV often goes into dormancy and only recommences, leading to AIDS, many years later.

How HIV causes the symptoms of AIDS

The human immunodeficiency virus specifically attacks helper T cells. HIV causes AIDS by killing or interfering with the normal functioning of helper T cells. An uninfected person normally has between 800 and 1200 helper T cells in each mm^3 of blood. In a person suffering from AIDS this number can be as low as $200\,mm^{-3}$. We have seen (Topic 5.3) that helper T cells are important in cell-mediated immunity. Without a sufficient number of helper T cells, the immune system cannot stimulate B cells to produce antibodies or the cytotoxic T cells that kill cells infected by pathogens. Memory cells may also become infected and destroyed. As a result, the body is unable to produce an adequate immune response and becomes susceptible to other infections and cancers. Many AIDS sufferers develop infections of the lungs, intestines, brain and eyes, as well as experiencing weight loss and diarrhoea. It is these secondary, diseases that ultimately cause death.

HIV does not kill individuals directly. By infecting the immune system, HIV prevents it from functioning normally. As a result those infected by HIV are unable to respond effectively to other pathogens. It is these infections, rather than HIV, that ultimately cause ill health and eventual death.

The ELISA test

ELISA stands for **enzyme linked immunosorbant assay**. It uses antibodies to not only detect the presence of a protein in a sample but also the quantity. It is extremely sensitive and so can detect very small amounts of a molecule. To understand how the test works, imagine that we are trying to find whether a particular protein, in this case an antigen, is present in a sample. The procedure is as follows:

- Apply the sample to a surface, for example a slide, to which all the antigens in the sample will attach.
- Wash the surface several times to remove any unattached antigens.
- Add the antibody that is specific to the antigen we are trying to detect and leave the two to bind together.
- Wash the surface to remove excess antibody.

- Add a second antibody that binds with the first antibody. This second antibody has an enzyme attached to it.
- Add the colourless substrate of the enzyme. The enzyme acts on the substrate to change it into a coloured product.
- The amount of the antigen present is relative to the intensity of colour that develops.

This basic technique can be used to detect HIV and the pathogens of diseases including tuberculosis and hepatitis. ELISA is especially useful where the quantity of an antigen needs to be measured. In testing for particular drugs in the body for example. The mere presence of a drug is often less important than its quantity as many drugs are found naturally in low concentrations. ELISA is therefore very useful in both drug and allergen tests.

Why antibiotics are ineffective against viral diseases like AIDS

Antibiotics work in a number of different ways. One is by preventing bacteria from making normal cell walls.

In bacterial cells, as in plant cells, water constantly enters by osmosis. This entry of water would normally cause the cell to burst. That it doesn't burst is due to the wall that surrounds all bacterial cells. This wall is made of **murein** (peptidoglycan) a tough material that is not easily stretched. As water enters the cell by osmosis, the cell expands and pushes against the cell wall. Being relatively inelastic, the cell wall resists expansion and so halts further entry of water. Antibiotics like penicillin inhibit certain enzymes required for the synthesis and assembly of the peptide cross-linkages in bacterial cell walls. This weakens the walls, making them unable to withstand pressure. As water enters naturally by osmosis, the cell bursts and the bacterium dies.

Viruses rely on the host cells to carry out their metabolic activities and therefore lack their own metabolic pathways and cell structures. As a result antibiotics are ineffective because there are no metabolic mechanisms or cell structures for them to disrupt. Viruses also have a protein coat rather than a murein cell wall and so do not have sites where antibiotics can work. In any case, when viruses are within an organism's own cells, antibiotics cannot reach them.

Synoptic link

Osmosis was covered in Topic 4.3 and the structure of RNA in Topic 2.1.

Summary questions

1 Explain why is HIV called a retrovirus.
2 Distinguish between HIV and AIDS.
3 Tuberculosis (TB) is a lung disease spread through the air. Suggest a possible reason why the widespread use of condoms might help reduce the incidence of TB in a population.

1   **(a)**   What is a pathogen?   *(1 mark)*

      (b)   When a pathogen enters the body it may be destroyed by phagocytosis. Describe how.   *(4 marks)*

      (c)   When a pathogen causes an infection, plasma cells secrete antibodies which destroy this pathogen. Explain why these antibodies are only effective against a specific pathogen.   *(2 marks)*

                                                                  AQA June 2012

2   The diagram shows an antibody molecule.

      (a)   What is the evidence from the diagram that this antibody has a quaternary structure?   *(1 mark)*

      (b)   Scientists use this antibody to detect an antigen on the bacterium that causes stomach ulcers. Explain why the antibody will only detect this antigen.   *(3 marks)*

                                                                  AQA Jan 2012

3   The table shows the cumulative rise in cases of the infectious disease Ebola over a five week period in 2014.

Week	Number of cases	Number of deaths
1	70	20
2	112	40
3	168	95
4	200	119
5	230	134
6	250	148

      (a)   Plot a graph of the above information with the number of weeks on the X axis   *(1 mark)*

      (b)   Calculate the rate of increase in number of cases of Ebola in the time period shown on the graph.   *(2 marks)*

      (c)   A scientist suggests that the increase in the number of cases in the following six months will be exponential. Explain how plotting the next six months' data on a log scale would show whether the increase is exponential.   *(1 mark)*

      (d)   Ebola is a rare disease in the human population, but can be passed on to humans from wild animals. Suggest, using your knowledge of the immune system, why the disease spreads fast once it is present in one human in an urban area.   *(4 marks)*

4   Read the following passage.

      Microfold cells are found in the epithelium of the small intestine. Unlike other epithelial cells in the small intestine, microfold cells do not have adaptations for the absorption of food.

      Microfold cells help to protect against pathogens that enter the intestine. They have receptor proteins on their cell-surface membranes that bind to antigens on the   5
      surface of pathogens. The microfold cells take up the antigens and transport them to cells of the immune system. Antibodies are then produced which give protection against the pathogen.

      Scientists believe that it may be possible to develop vaccines that make use of microfold cells. These vaccines could be swallowed in tablet form.   10

   Use information from the passage and your own knowledge to answer the following questions.

      (a)   Microfold cells do not have adaptations for the absorption of food (lines 2–3). Give two adaptations that other epithelial cells have for the absorption of food. *(2 marks)*

(b) **(i)** Microfold cells have receptor proteins on their cell-surface membranes that bind to antigens (line 5). What is an antigen? *(1 mark)*

(ii) Microfold cells take up the antigens and transport them to cells of the immune system (lines 6–7). Antigens are not able to pass through the cell-surface membranes of other epithelial cells. Suggest **two** reasons why. *(2 marks)*

(c) Scientists believe that it may be possible to develop vaccines that make use of microfold cells (lines 9–10). Explain how this sort of vaccine would lead to a person developing immunity to a pathogen. *(5 marks)*

AQA June 2013

5 Read the passage below.

Most cases of cervical cancer are caused by infection with Human Papilloma Virus (HPV). This virus can be spread by sexual contact. There are many types of HPV, each identified by a number. Most of these types are harmless but types 16 and 18 are most likely to cause cervical cancer.

A vaccine made from HPV types 16 and 18 is offered to girls aged 12 to 13. Three   5
injections of the vaccine are given over six months. In clinical trials, the vaccine has proved very effective in protecting against HPV types 16 and 18. However, it will be many years before it can be shown that this vaccination programme has reduced cases of cervical cancer. Until then, smear tests will continue to be offered to women, even if they have been vaccinated. A smear test allows abnormal cells in   10
the cervix to be identified so that they can be removed before cervical cancer develops.

The Department of Health has estimated that 80% of girls aged 12 to 13 need to be vaccinated to achieve herd immunity to HPV types 16 and 18. Herd immunity is where enough people have been vaccinated to reduce significantly the spread of   15
HPV through the population.

Use information from this passage and your own knowledge to answer the following questions.

(a) HPV vaccine is offered to girls aged 12 to 13 (line 5). Suggest why it is offered to this age group. *(1 mark)*

(b) The vaccine is made from HPV types 16 and 18 (line 5). Explain why this vaccine may **not** protect against other types of this virus. *(2 marks)*

(c) Three injections of the vaccine are given (lines 5 to 6). Use your knowledge of immunity to suggest why. *(2 marks)*

(d) It will be many years before it can be shown that this vaccination programme has reduced cases of cervical cancer (lines 7 to 9). Suggest two reasons why. *(2 marks)*

(e) Smear tests will continue to be offered to women, even if they have been vaccinated (lines 9 to 10). Suggest why women who have been vaccinated still need to be offered smear tests. *(1 mark)*

AQA Jan 2011

The table shows the uptake of the vaccine in one year in four health authority areas.

health authority area	number of girls aged 12–13	number of girls vaccinated
A	14 053	11 151
B	12 789	10 743
C	11 892	8 662
D	8 054	6 524

(f) Analyse mathematically which of these areas would be most likely to show a reduction in the spread of HPV through the population (lines 14 to 16). Use calculations to support your answer. *(2 marks)*

Section 2 Summary

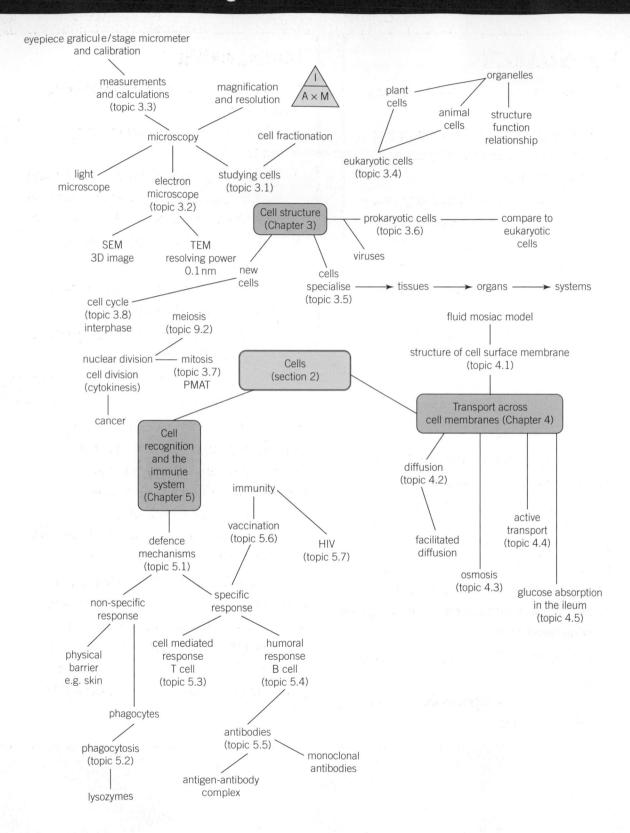

eyepiece graticule/stage micrometer
and calibration

measurements
and calculations
(topic 3.3)

magnification
and resolution

$\dfrac{I}{A \times M}$

microscopy

cell fractionation

plant
cells

organelles

animal
cells

structure
function
relationship

eukaryotic cells
(topic 3.4)

light
microscope

electron
microscope
(topic 3.2)

studying cells
(topic 3.1)

Cell structure
(Chapter 3)

prokaryotic cells
(topic 3.6)

compare to
eukaryotic
cells

SEM
3D image

TEM
resolving power
0.1 nm

new
cells

viruses

cells
specialise
(topic 3.5)

tissues → organs → systems

cell cycle
(topic 3.8)
interphase

meiosis
(topic 9.2)

fluid mosiac model

structure of cell surface membrane
(topic 4.1)

nuclear division
cell division
(cytokinesis)

mitosis
(topic 3.7)
PMAT

Cells
(section 2)

cancer

Cell
recognition
and the
immune
system
(Chapter 5)

immunity

Transport across
cell membranes (Chapter 4)

vaccination
(topic 5.6)

HIV
(topic 5.7)

diffusion
(topic 4.2)

active
transport
(topic 4.4)

defence
mechanisms
(topic 5.1)

facilitated
diffusion

osmosis
(topic 4.3)

glucose absorption
in the ileum
(topic 4.5)

non-specific
response

specific
response

physical
barrier
e.g. skin

cell mediated
response
T cell
(topic 5.3)

humoral
response
B cell
(topic 5.4)

phagocytes

antibodies
(topic 5.5)

monoclonal
antibodies

phagocytosis
(topic 5.2)

antigen-antibody
complex

lysozymes

Practical skills

In this section you have met the following practical skills:

- How to use an optical microscope to measure the size of an object using an eyepiece graticule
- How to calibrate an eyepiece graticule
- Understanding the processes of cell fractionation, homogenation and ultracentrifugation
- Using scientific knowledge to solve practical problems
- Considering ethical issues when carrying out experiments.

Maths skills

In this section you have met the following maths skills:

- Using percentages in calculating a mitotic index
- Calculating magnifications and altering the magnification formula to determine the size of the image and/or real object
- Making use of appropriate units in calculations
- Changing the subject of an equation
- Calculating arithmetical means
- Interpreting graphs and finding values using the intercept of a graph.

Extension task

You are a research scientist working for a major pharmaceutical company that is developing immunosuppressant drugs to prevent rejection of newly transplanted tissue. Your section leader has asked you to consider the features that any new immunosuppressant drug should have. She has asked you to produce a report under the following headings:

Possible method(s) of transporting the drug across cell membranes.
Cells of the immune system that the drug should be targeted at.

How the effects of the drug on the targeted cells will reduce the likelihood of the transplanted tissue being rejected.
Possible side effects on other cells.
Likely detrimental effects of the immunosuppressant drug on the recipients and how these might be overcome.

Using the information given in the section of the book you have just completed, and additional information from textbooks, journals and the internet, write a report in around 1000 words. Use clear scientific terminology and specific biological terms.

1 **Figure 6** represents part of a DNA molecule.

(i) Draw a box around a single nucleotide. (*1 mark*)

Table 4 shows the percentage of bases in each of the strands of a DNA molecule.

DNA strand	% of each base			
	A	C	G	T
strand 1	16			
strand 2		21	34	

(ii) Complete **Table 4** by adding the missing values. (*2 marks*)

(iii) During replication, the two DNA strands separate and each acts as a template for the production of a new strand. As new DNA strands are produced, nucleotides can only be added in the 5' to 3' direction.
Use **Figure 6** and your knowledge of enzyme action and DNA replication to explain why new nucleotides can only be added in a 5' to 3' direction. (*4 marks*)

AQA SAMS A LEVEL PAPER 1

2 The diagram shows an epithelial cell from the small intestine.

(a) (i) Name organelle **Y**. *(1 mark)*

(ii) There are large numbers of organelle Y in this cell. Explain how these organelles help the cell to absorb the products of digestion. *(2 marks)*

(b) This diagram shows the cell magnified 1000 times. Calculate the actual length of the cell between points **P** and **Q**. Give your answer in μm. Show your working.

(1 mark)

(c) Coeliac disease is a disease of the human digestive system. In coeliac disease, the structures labelled **Z** are damaged.

Although people with coeliac disease can digest proteins they have low concentrations of amino acids in their blood.

Explain why they have low concentrations of amino acids in their blood. *(2 marks)*

AQA Jan 2010

3 The human immunodeficiency virus (HIV) leads to the development of acquired immunodeficiency syndrome (AIDS). Eventually, people with AIDS die because they are unable to produce an immune response to pathogens.

Scientists are trying to develop an effective vaccine to protect people   5
against HIV. There are three main problems. HIV rapidly enters host cells. HIV causes the death of T cells that activate B cells. HIV shows a lot of antigenic variability.

Scientists have experimented with different types of vaccine for HIV. One type contains HIV in an inactivated form. A second type contains   10
attenuated HIV which replicates in the body but does not kill host cells. A third type uses a different, non-pathogenic virus to carry genetic information from HIV into the person's cells. This makes the person's cells produce HIV proteins. So far, these types of vaccine have not been considered safe to use in a mass vaccination programme.   15

Use the information in the passage and your own knowledge to answer the following questions.

(a) People with AIDS die because they are unable to produce an immune response to pathogens (lines 2–4). Explain why this leads to death. *(3 marks)*

(b) Explain why each of the following means that a vaccine might **not** be effective against HIV.

(i) HIV rapidly enters host cells (lines 6–7). *(2 marks)*

(ii) HIV shows a lot of antigenic variability (lines 7–8). *(2 marks)*

(c) So far, these types of vaccine have not been considered safe to use in a mass vaccination programme (lines 14–15). Suggest why they have not been considered safe. *(3 marks)*

AQA Jan 2013

Section 3
Organisms exchange substances with their environment

Introduction

All cells and organisms exchange material between themselves and their environment. To enter or leave an organism, substances must pass across a plasma membrane. Single-celled and small multicellular organisms can satisfactorily exchange materials over their body surfaces using diffusion alone, especially if their metabolic rate is low. As organisms evolved and became larger, their surface area to volume ratios decreased and specialised respiratory surfaces evolved to meet the increasing requirement to exchange ever larger quantities of materials.

Where large size is combined with a high metabolic rate there is a requirement for a mass transport system to move substances between the exchange surface and the cells of which the organism is composed. In animals these systems often involve circulating a specialised transport medium (blood) through vessels using a pump (heart).

Plants do not move from place to place and have a relatively low metabolic rate and consequently reduced demand for oxygen and glucose. Coupled with their large surface area, essential for obtaining light for photosynthesis, they have not evolved a pumped circulatory system. Plants do, however, transport water up from their roots to the leaves and distribute the products of photosynthesis. Their mass transport system comprises vessels too – xylem and phloem, but the movement of fluid within them is largely a passive process.

The internal environment of a cell or organism differs from the environment around it. The cells of large multicellular animals are surrounded by tissue fluid, the composition of which is kept within a suitable metabolic range. In both plants and animals, it is the mass transport system that maintains the final diffusion gradients which allows substances to be exchanged across cell-surface membranes.

Working scientifically

Studying exchange between organisms and the environment allows you to carry out practical work and to develop practical skills. A required practical activity is the dissection of an animal or plant gas exchange system or mass transport system or of an organ within such a system.

You will require a range of mathematical skills; in particular the ability to change the subject of an equation and calculate the surface areas and volumes of various shapes.

What you already know

The material in this unit is intended to be self-explanatory. However, there is some knowledge from GCSE that will aid your understanding of this section. This information includes:

- ⬭ The effectiveness of a gas-exchange surface is increased by having a large surface area, being thin, having an efficient blood supply and being ventilated.

- ⬭ In humans the surface area of the lungs is increased by alveoli and that of the small intestine by villi. The villi provide a large surface area with an extensive network of capillaries to absorb the products of digestion by diffusion and active transport.

- ⬭ Breathing in involves the ribcage moving out and up and the diaphragm becoming flatter. Breathing out involves these changes being reversed.

- ⬭ In plants, water and mineral ions are absorbed by roots, the surface area of which is increased by root hairs.

- ⬭ Plants have stomata in their leaves through which carbon dioxide and oxygen are exchanged with the atmosphere by diffusion. The size of stomata is controlled by guard cells that surround them and help control water loss.

- ⬭ In flowering plants, xylem tissue transports water and mineral ions from the roots to the stem and leaves and phloem tissue carries dissolved sugars from the leaves to the rest of the plant.

- ⬭ In animals a circulatory system transports substances using a heart, which is a muscular organ with four main chambers – left and right atria and ventricles.

- ⬭ Blood flows from the heart to the organs through arteries and returns through veins. Arteries have thick walls containing muscle and elastic fibres. Veins have thinner walls and often have valves to prevent back-flow of blood.

- ⬭ Blood is a tissue and consists of plasma in which red blood cells, white blood cells and platelets are suspended.

- ⬭ Red blood cells have no nucleus and are packed with haemoglobin. In the lungs haemoglobin combines with oxygen to form oxyhaemoglobin. In other organs oxyhaemoglobin splits up into haemoglobin and oxygen.

- ⬭ White blood cells have a nucleus and form part of the body's defence system against microorganisms.

6.1 Exchange between organisms and their environment

Learning objectives

→ Explain how the size of an organism and its structure relate to its surface area to volume ratio.

→ Describe how larger organisms increase their surface area to volume ratio.

→ Explain how surfaces are specially adapted to facilitate exchange.

Specification reference: 3.3.1

The external environment is different from the internal environment found within an organism and within its cells. To survive, organisms transfer materials between the two environments. This transfer takes place at exchange surfaces and always involves crossing cell plasma membranes. The environment around the cells of multicellular organisms is called **tissue fluid**. The majority of cells are too far from exchange surfaces for diffusion alone to supply or remove their tissue fluid with the various materials needed to keep its composition relatively constant. Therefore, once absorbed, materials are rapidly distributed to the tissue fluid and the waste products returned to the exchange surface for removal. This involves a mass transport system. It is this mass transport system that maintains the diffusion gradients that bring materials to and from the cell-surface membranes.

The size and metabolic rate of an organism will affect the amount of each material that is exchanged. For example, organisms with a high metabolic rate exchange more materials and so require a larger surface area to volume ratio. In turn this is reflected in the type of exchange surface and transport system that evolved to meet the requirements of each organism. In this chapter we will investigate the adaptations of exchange surfaces and transport systems in a variety of organisms.

Examples of things that need to be interchanged between an organism and its environment include: respiratory gases (oxygen and carbon dioxide); nutrients (glucose, fatty acids, amino acids, vitamins, minerals); excretory products (urea and carbon dioxide); and heat.

Except for heat, these exchanges can take place in two ways:

- passively (no metabolic energy is required), by **diffusion** and **osmosis**
- actively (metabolic energy is required), by **active transport**.

Surface area to volume ratio

Exchange takes place at the surface of an organism, but the materials absorbed are used by the cells that mostly make up its volume. For exchange to be effective, the exchange surface(s) of the organism must be large compared with its volume.

Small organisms have a surface area that is large enough, compared with their volume, to allow efficient exchange across their body surface. However, as organisms become larger, their volume increases at a faster rate than their surface area (Table 1). Because of this, simple diffusion of substances across the outer surface can only meet the needs of relatively inactive organisms. Even if the outer surface could supply enough of a substance, it would still take too long for it to reach the middle of the organism if diffusion alone was the method of transport. Organisms have evolved one or more of the following features:

Study tip

In a cell the lowest oxygen concentration is inside the mitochondria, where oxygen is used up in respiration. Mitochondria also contain the highest concentration of carbon dioxide. This maintains the diffusion gradient for these gases in and out of the cell.

Maths link √x

MS 4.1, see Chapter 22.

▼ **Table 1** *How the surface area to volume ratio gets smaller as an object becomes larger*

Length of edge of a cube / cm	Surface area of whole cube (area of one side × 6 sides) / cm²	Volume of cube (length × width × height) / cm³	Ratio of surface area to volume (surface area ÷ volume)
1	1 × 6 = 6	1 × 1 × 1 = 1	$\frac{6}{1} = 6.0 : 1$
2	4 × 6 = 24	2 × 2 × 2 = 8	$\frac{24}{8} = 3.0 : 1$
3	9 × 6 = 54	3 × 3 × 3 = 27	$\frac{54}{27} = 2.0 : 1$
4	16 × 6 = 96	4 × 4 × 4 = 64	$\frac{96}{64} = 1.5 : 1$
5	25 × 6 = 150	5 × 5 × 5 = 125	$\frac{150}{125} = 1.2 : 1$
6	36 × 6 = 216	6 × 6 × 6 = 216	$\frac{216}{216} = 1.0 : 1$

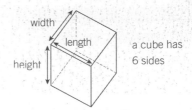

▲ **Figure 1** *Calculating volume*

- a flattened shape so that no cell is ever far from the surface (e.g. a flatworm or a leaf)
- specialised exchange surfaces with large areas to increase the surface area to volume ratio (e.g., lungs in mammals, gills in fish).

You may be asked to calculate the surface area to volume ratio of cells with different shapes. To make these calculations reasonably straightforward, cells or organisms may have to be assumed to have a uniform shape although in practice they almost never do.

Maths link √x̄

MS 0.3 and 4.1, see Chapter 22.

Calculating the surface area to volume ratio of cells with different shapes

For example, let us assume a cell has the shape of a sphere that is 10 μm in diameter. The surface area of a sphere is calculated using the formula: $4\pi r^2$

In our example: $r = 5\,\mu m$ (radius = half the diameter) and we will use the value of π as 3.14.

Therefore the surface area of the cell = 4 × 3.14 × (5 × 5) = 314 μm²

The volume of a sphere is calculated using the formula: $\frac{4}{3}\pi r^3$

Therefore the volume of the cell = $\frac{4}{3}$ × 3.14 × (5 × 5 × 5) = 523.33 μm³

The surface area to volume ratio is therefore 314 ÷ 523.33 = 0.6: 1

Features of specialised exchange surfaces

To allow effective transfer of materials across specialised exchange surfaces by diffusion or active transport, exchange surfaces show the following characteristics:

- a large surface area relative to the volume of the organism which increases the rate of exchange
- very thin so that the diffusion distance is short and therefore materials cross the exchange surface rapidly
- selectively permeable to allow selected materials to cross

Hint

Remember that substances not only have to move into cells through the cell-surface membrane but also into organelles like mitochondria through the plasma membrane that surrounds them. All plasma membranes are therefore thin not just cell-surface membranes.

Summary questions

1 Name **four** general things that need to be exchanged between organisms and their environment.

2   Calculate the surface area to volume ratio of a cube that has sides 10 mm long.

3 Name **three** factors that affect the rate of diffusion of substances into cells.

- movement of the environmental medium, for example, air, to maintain a diffusion gradient
- A transport system to ensure the movement of the internal medium, for example blood, in order to maintain a diffusion gradient.

We saw in Topic 4.2 that the relationship between certain of these factors can be expressed as:

$$\text{diffusion} \propto \frac{\text{surface area} \times \text{difference in concentration}}{\text{length of diffusion path}}$$

Being thin, specialised exchange surfaces are easily damaged and dehydrated. They are therefore often located inside an organism. Where an exchange surface is located inside the body, the organism needs to have a means of moving the external medium over the surface, e.g. a means of ventilating the lungs in a mammal.

Significance of the surface area to volume ratio in organisms

The graph in Figure 2 shows the surface area to volume ratios of different-sized cubes. The ratios are actually 1:1, 2:1, 3:1 etc. but are shown as single numbers for ease of plotting.

1 Microscopic organisms obtain their oxygen by diffusion in across their body surface. Using the graph, explain how they are able to obtain sufficient oxygen for their needs.

2 The blue whale (Figure 3) is the largest organism on the planet. It spends much of its life in cold waters with temperatures between 0 °C and 6 °C. Use the graph to explain one way in which large size is an advantage to blue whales.

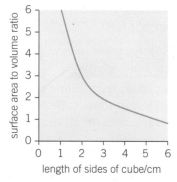

▲ **Figure 2** *Surface area to volume ratios*

Maths link

MS 0.3 and 4.1, see Chapter 22.

Calculating a surface area to volume ratio

Consider the shape shown in Figure 4, which has dimensions marked on it. Use the information below to calculate the ratio of surface area to volume of this shape (to two decimal places).

The area of a disc (like those at the ends of an enclosed cylinder) is calculated using the formula πr^2

The external surface area of an enclosed cylinder is calculated using the formula $2\pi rh + 2\pi r^2$.

▲ **Figure 4**

Gas exchange in single-celled organisms

Single-celled organisms are small and therefore have a large surface area to volume ratio. Oxygen is absorbed by diffusion across their body surface, which is covered only by a cell-surface membrane. In the same way, carbon dioxide from respiration diffuses out across their body surface. Where a living cell is surrounded by a cell wall, this is no additional barrier to the diffusion of gases.

Gas exchange in insects

As with all terrestrial organisms, insects have evolved mechanisms to conserve water. The increase in surface area required for gas exchange conflicts with conserving water because water will evaporate from it. How insects overcome water loss is discussed in Topic 6.5. For gas exchange, insects have evolved an internal network of tubes called **tracheae**. The tracheae are supported by strengthened rings to prevent them from collapsing. The tracheae divide into smaller dead-end tubes called **tracheoles**. The tracheoles extend throughout all the body tissues of the insect. In this way atmospheric air, with the oxygen it contains, is brought directly to the respiring tissues, as there is a short diffusion pathway from a tracheole to any body cell.

Respiratory gases move in and out of the tracheal system in three ways.

- **Along a diffusion gradient**. When cells are respiring, oxygen is used up and so its concentration towards the ends of the tracheoles falls. This creates a diffusion gradient that causes gaseous oxygen to diffuse from the atmosphere along the tracheae and tracheoles to the cells. Carbon dioxide is produced by cells during respiration. This creates a diffusion gradient in the opposite direction. This causes gaseous carbon dioxide to diffuse along the tracheoles and tracheae from the cells to the atmosphere. As diffusion in air is much more rapid than in water, respiratory gases are exchanged quickly by this method.
- **Mass transport**. The contraction of muscles in insects can squeeze the trachea enabling mass movements of air in and out. This further speeds up the exchange of respiratory gases.
- **The ends of the tracheoles are filled with water**. During periods of major activity, the muscle cells around the tracheoles respire carry out some anaerobic respiration. This produces lactate, which is soluble and lowers the water potential of the muscle cells. Water therefore moves into the cells from the tracheoles by osmosis. The water in the ends of the tracheoles decreases in volume and in doing so draws air further into them. This means the final diffusion pathway is in a gas rather than a liquid phase, and therefore diffusion is more rapid. This increases the rate at which air is moved in the tracheoles but leads to greater water evaporation.

Gases enter and leave tracheae through tiny pores, called **spiracles**, on the body surface. The spiracles may be opened and closed by a valve. When the spiracles are open, water vapour can evaporate from

Learning objectives

→ Describe how single-celled organisms exchange gases.

→ Explain how terrestrial insects balance the need to exchange gases with the need to conserve water

→ Explain how insects exchange gases.

Specification reference: 3.3.2

Synoptic link

As a starting point for this topic, it would be useful to revise Topics 4.2 and 4.3.

Hint

Every cell of an insect is only a very short distance from one of the tracheae or tracheoles and so the diffusion pathway is always short.

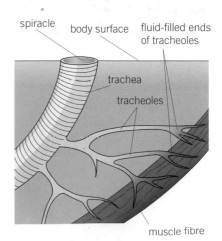

▲ **Figure 1** *Part of an insect tracheal system*

the insect. For much of the time insects keep their spiracles closed to prevent this water loss. Periodically they open the spiracles to allow gas exchange. Part of an insect tracheal system is illustrated in Figure 1.

The tracheal system is an efficient method of gas exchange. It does, however, have some limitations. It relies mostly on diffusion to exchange gases between the environment and the cells. For diffusion to be effective, the diffusion pathway needs to be short which is why insects are of a small size. As a result the length of the diffusion pathway limits the size that insects can attain. Not that being small has hindered insects. They are one of the most successful groups of organisms on Earth.

Practical link ⚗

Required practical 5. Dissection of animal or plant gas exchange system or mass transport system or of an organ within such a system.

Summary questions

1 Name the process by which carbon dioxide is removed from a single-celled organism.

2 Explain why there is a conflict in terrestrial insects between gas exchange and conserving water.

3 Explain how the tracheal system limits the size of insects.

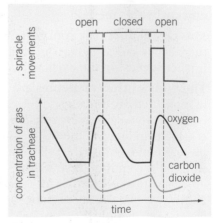

▲ **Figure 2** *Scanning electron micrograph (SEM) of a spiracle of an insect*

⚛ Spiracle movements $\sqrt{x}$ ⚗

An experiment was carried out to measure the concentration of oxygen and carbon dioxide in the tracheal system of an insect over a period of time. During the experiment the opening and closing of the insect's spiracles was observed and recorded. The results are shown in Figure 3.

1 Describe what happens to the concentration of oxygen in the tracheae when the spiracles are closed.

2 Suggest an explanation for this change in the concentration of oxygen when the spiracles are closed.

3 Use the information provided by the graph to suggest what causes the spiracles to open.

4 Suggest an advantage of these spiracle movements to a terrestrial insect.

5 Fossil insects have been discovered that are larger than insects that occur on Earth today. What does this suggest about the composition of the atmosphere at the time when these fossil insects lived.

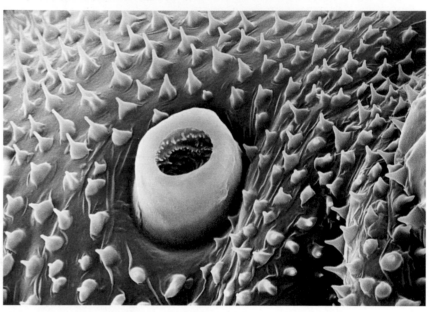

▲ **Figure 3**

Fish have a waterproof, and therefore a gas-tight, outer covering. Being relatively large they also have a small surface area to volume ratio. Their body surface is therefore not adequate to supply and remove their respiratory gases and so, like insects and humans, they have evolved a specialised internal gas exchange surface: the gills.

Structure of the gills

The gills are located within the body of the fish, behind the head. They are made up of **gill filaments**. The gill filaments are stacked up in a pile, rather like the pages in a book. At right angles to the filaments are **gill lamellae**, which increase the surface area of the gills. Water is taken in through the mouth and forced over the gills and out through an opening on each side of the body. The position and arrangement of the gill filaments and gill lamellae are shown in Figure 1. From this figure you will notice that the flow of water over the gill lamellae and the flow of blood within them are in opposite directions. This is known as a **countercurrent flow**.

Learning objectives

→ Describe the structure of fish gills.

→ Describe how water is passed along fish gills.

→ Explain the difference between parallel flow and countercurrent flow.

→ Explain how countercurrent flow increases the rate of gas exchange.

Specification reference: 3.3.2

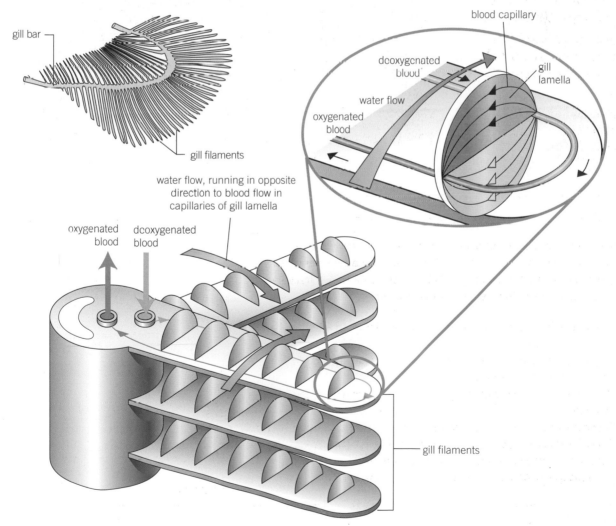

▲ **Figure 1** *Arrangement of gills in a fish and direction of water flow over them*

Summary questions

1 In relation to fish gills, describe what is meant by countercurrent flow.

2 Outline why countercurrent flow is an efficient means of exchanging gases across the gills of fish.

3 Mackerel are active, fast-swimming fish while plaice spend most of their lives moving slowly on the sea bed. There are differences in the gills of these two types of fish. Suggest what these differences might be

4 Water flow over fish gills is one-way whereas the flow of air in and out of the lungs is two-way. Suggest why one-way flow is an advantage to fish.

It is important for ensuring that the maximum possible gas exchange is achieved. If the water and blood flowed in the same direction, far less gas exchange would take place.

The countercurrent exchange principle

The essential feature of the countercurrent exchange system is that the blood and the water that flow over the gill lamellae do so in opposite directions. This arrangement means that:

- Blood that is already well loaded with oxygen meets water, which has its maximum concentration of oxygen. Therefore **diffusion** of oxygen from the water to the blood takes place.

- Blood with little oxygen in it meets water which has had most, but not all, of its oxygen removed. Again, diffusion of oxygen from the water to blood takes place.

As a result, a diffusion gradient for oxygen uptake is maintained across the entire width of the gill lamellae. In this way, about 80% of the oxygen available in the water is absorbed into the blood of the fish. If the flow of water and blood had been in the same direction (parallel flow), the diffusion gradient would only be maintained across part of the length of the gill lamellae and only 50% of the available oxygen would be absorbed by the blood.

countercurrent flow

distance along gill plate

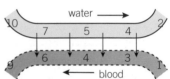

numbers represent relative oxygen concentrations

Diffusion of oxygen
A diffusion gradient is maintained all the way across the gill lamellae. Almost all the oxygen from the water diffuses into the blood.

parallel flow

distance along gill plate

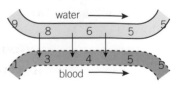

numbers represent relative oxygen concentrations

Diffusion of oxygen
A diffusion gradient is maintained for only half of the distance across the gill lamellae. Only 50% of the oxygen from the water diffuses into the blood.

▲ **Figure 2** *Parallel flow and countercurrent flow in the gills of a fish*

6.4 Gas exchange in the leaf of a plant

Like animal cells, all plant cells require oxygen and produce carbon dioxide during respiration. When it comes to gas exchange, however, plants show one important difference from animals. Some plant cells carry out photosynthesis. During photosynthesis, plant cells take in carbon dioxide and produce oxygen. At times the gases produced in one process can be used for the other. This reduces gas exchange with the external air. Overall, this means that the volumes and types of gases that are being exchanged by a plant leaf change. This depends on the balance between the rates of photosynthesis and respiration.

- When photosynthesis is taking place, although some carbon dioxide comes from respiration of cells, most of it is obtained from the external air. In the same way, some oxygen from photosynthesis is used in respiration but most of it **diffuses** out of the plant.

- When photosynthesis is not occurring, for example, in the dark, oxygen diffuses into the leaf because it is constantly being used by cells during respiration. In the same way, carbon dioxide produced during respiration diffuses out.

Structure of a plant leaf and gas exchange

In some ways, gas exchange in plants is similar to that of insects (see Topic 6.2).

- No living cell is far from the external air, and therefore a source of oxygen and carbon dioxide.

- Diffusion takes place in the gas phase (air), which makes it more rapid than if it were in water.

Overall, therefore, there is a short, fast diffusion pathway. In addition, the air spaces inside a leaf have a very large surface area compared with the volume of living tissue. There is no specific transport system for gases, which simply move in and through the plant by diffusion. Most gaseous exchange occurs in the leaves, which show the following adaptations for rapid diffusion:

- many small pores, called stomata, and so no cell is far from a stoma and therefore the diffusion pathway is short (Figure 1)

- numerous interconnecting air-spaces that occur throughout the mesophyll so that gases can readily come in contact with mesophyll cells

- large surface area of mesophyll cells for rapid diffusion.

The structure of a leaf is shown in Figure 2.

Stomata

Stomata are minute pores that occur mainly, but not exclusively, on the leaves, especially the underside. Each stoma (singular) is surrounded by a pair of special cells (guard cells). These cells can open and close the stomatal pore (Figure 3). In this way they can control the rate of gaseous exchange. This is important because terrestrial organisms lose water by evaporation. Plants have evolved to balance the conflicting needs of gas exchange and control of water loss. They do this by closing stomata at times when water loss would be excessive.

Study tip

The diffusion gradients in and out of the leaf are maintained by mitochondria carrying out respiration and chloroplasts carrying out photosynthesis.

Hint

Remember that plant cells respire all the time, but only plant cells with chloroplasts photosynthesise – and then only when the conditions are right.

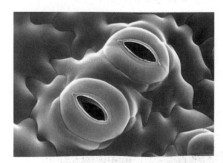

▲ **Figure 1** *False-colour SEM of open stomata on the surface a leaf*

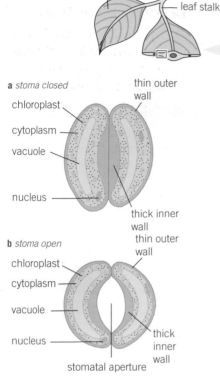

a *leaf structure*

leaf blade

leaf stalk

a *stoma closed*

thin outer wall

chloroplast

cytoplasm

vacuole

nucleus

thick inner wall

b *stoma open*

thin outer wall

chloroplast

cytoplasm

vacuole

nucleus

thick inner wall

stomatal aperture

▲ **Figure 3** *Surface view of a stoma closed and open*

Maths link √x̄

MS 0.3, 1.1 and 3.4, see Chapter 22.

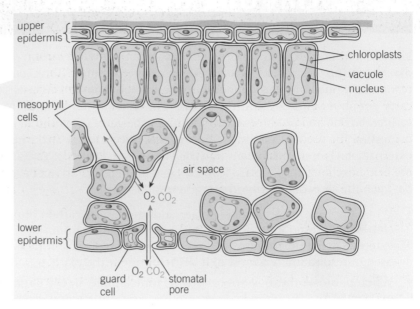

upper epidermis

chloroplasts

vacuole

nucleus

mesophyll cells

air space

O_2 CO_2

lower epidermis

guard cell

O_2 CO_2

stomatal pore

b *vertical section through a dicotyledonous leaf*

▲ **Figure 2** *Section through a leaf of a dicotyledonous plant showing gas exchange when photosynthesis is taking place*

Summary questions

1 State **two** similarities between gas exchange in a plant leaf and gas exchange in a terrestrial insect.

2 State **two** differences between gas exchange in a plant leaf and gas exchange in a terrestrial insect.

3 Explain the advantage to a plant of being able to control the opening and closing of stomata.

Exchange of carbon dioxide

The graph in Figure 4 shows the volume of carbon dioxide produced by a sample of tomato plants at different light intensities.

volume of carbon dioxide /cm³ hour⁻¹

CO_2 taken in

CO_2 given out

light intensity/ arbitrary units

▲ **Figure 4**

1 Name the process which produces carbon dioxide in the tomato plants.

2 Name the process which uses up carbon dioxide in the tomato plants.

3 Explain why, at point X, carbon dioxide is neither taken up nor given out by the tomato plants.

4 √x̄ A plant at a light intensity of 10000 lux produced 115 cm³hour⁻¹ of carbon dioxide. When the light intensity was increased to 15000 lux the amount of carbon dioxide produced was 160 cm³hour⁻¹. Calculate the percentage increase in carbon dioxide at 15000 lux to four significant figures.

5 Some herbicides cause the stomata of plants to close. Suggest how these herbicides might lead to the death of a plant.

6 Suggest what information is provided by the point at which the line of the graph meets the y-axis.

In terrestrial organisms like insects and plants problems arise from the opposing needs of an efficient gas-exchange system and the requirement to conserve water. The features that make a good gas-exchange system are the same features that increase water loss. In order to survive, terrestrial organisms must limit their water loss without compromising the efficiency of their gas-exchange systems. The gas exchange surfaces of terrestrial organisms are inside the body. The air at the exchange surface is more or less 100% saturated with water vapour. This means there is less evaporation of water from the exchange surface.

Limiting water loss in insects

Most insects are terrestrial (live on land). The problem for all terrestrial organisms is that water easily evaporates from the surface of their bodies and they can become dehydrated. They have evolved adaptations to conserve water.

However, efficient gas exchange requires a thin, permeable surface with a large area. These features conflict with the need to conserve water. Overall, as a terrestrial organism, the insect has to balance the opposing needs of exchanging respiratory gases with limiting water loss.

Insects have evolved the following adaptations that reduce water loss:

- **Small surface area to volume ratio** to minimise the area over which water is lost.
- **Waterproof coverings** over their body surfaces. In the case of insects this covering is a rigid outer skeleton of chitin that is covered with a waterproof cuticle.
- **Spiracles** are the openings of the tracheae at the body surface and these can be closed to reduce water loss. This conflicts with the need for oxygen and so occurs largely when the insect is at rest.

These features mean that insects cannot use their body surface to diffuse respiratory gases in the way a single-celled organism does. Instead they have an internal network of tubes called **tracheae** that carry air containing oxygen directly to the tissues (see Topic 6.2).

Limiting water loss in plants

While plants also have waterproof coverings, they cannot have a small surface area to volume ratio. This is because they photosynthesise, and photosynthesis requires a large leaf surface area for the capture of light and for the exchange of gases. So how do plants limit water loss?

To reduce water loss, terrestrial plants have a waterproof covering over parts of the leaves and the ability to close stomata when necessary. Certain plants with a restricted supply of water, have also evolved a range of other adaptations to limit water loss through **transpiration**. These plants are called **xerophytes**.

Xerophytes are plants that are adapted to living in areas where water is in short supply. Without these adaptations these plants would become desiccated and die.

Learning objectives

→ Explain how terrestrial plants and insects balance the need for gas-exchange and the need to conserve water.

Specification reference: 3.3.2

▲ **Figure 1** *Conifers have needle-like leaves to reduce water loss*

▲ **Figure 2** *Holly has leaves with a thick waxy cuticle that reduces water loss*

Climate change affects rainfall and rate of evaporation of water. As a result, the distribution of plant species changes. As regions become drier, so the number of xerophytic plants in them increases.

▲ **Figure 3** *This cactus stores water in its swollen stem. The leaves are needle-like to reduce their surface area and hence water loss*

The main way of surviving in habitats where there is a high rate of water loss and a limited water supply is to reduce the rate at which water can be lost through evaporation. As the vast majority of water loss occurs through the leaves, it is these organs that usually show most modifications. Examples of these modifications include:

- **a thick cuticle**. Although the waxy cuticle on leaves forms a waterproof barrier, up to 10% of water loss can still occur by this route. The thicker the cuticle, the less water can escape by this means, for example holly.

- **rolling up of leaves**. Most leaves have their stomata largely, or entirely, confined to the lower epidermis. The rolling of leaves in a way that protects the lower epidermis from the outside helps to trap a region of still air within the rolled leaf. This region becomes saturated with water vapour and so has a very high **water potential**. There is no water potential gradient between the inside and outside of the leaf and therefore no water loss. Marram grass rolls its leaves.

- **hairy leaves**. A thick layer of hairs on leaves, especially on the lower epidermis, traps still, moist air next to the leaf surface. The water potential gradient between the inside and the outside of the leaves is reduced and therefore less water is lost by evaporation. One type of heather plant has this modification.

- **stomata in pits or grooves**. These again trap still, moist air next to the leaf and reduce the water potential gradient. Examples of plants using this mechanism include pine trees.

- **a reduced surface area to volume ratio of the leaves**. We saw in Topic 6.1 that the smaller the surface area to volume ratio, the slower the rate of diffusion. By having leaves that are small and roughly circular in cross-section, as in pine needles, rather than leaves that are broad and flat, the rate of water loss can be considerably reduced. This reduction in surface area is balanced against the need for a sufficient area for photosynthesis to meet the requirements of the plant.

Summary questions

1 Insects and plants face the same problems when it comes to living on land. What is the main problem they share?

2 State **one** modification to reduce water loss that is shared by plants and insects.

3 Insects limit water loss by having a small surface area to volume ratio. Why is this not a feasible way of limiting water loss in plants?

4 Plants such as marram grass roll up their leaves, with the lower epidermis on the inside, to reduce water loss.

   a Explain how rolling up their leaves helps to reduce water loss.

   b Why would rolling the leaf the other way (with the upper epidermis on the inside) not be effective in reducing water loss?

When explaining adaptations of xerophytic plants to reduce water loss always relate these adaptations to reducing the water potential gradient and therefore slower diffusion, less water loss from air spaces and hence reduced evaporation of water.

Not only desert plants have problems obtaining water

Xerophytes are typically thought of as desert plants, which show a wide range of adaptations for coping with hot, dry conditions. However, similar adaptations may also be seen in plants found in sand dunes or other dry, windy places in temperate climates where rainfall is high and temperature relatively low. These adaptations are essential because the rain quickly drains away through the sand and out of the reach of the roots, making it difficult for these plants to obtain water. Plants living on salt marshes near the coast may have their roots drenched in water but find it difficult to absorb it. In addition, coastal regions are exposed to high wind speeds, which increase transpiration rates. Plants living in cold regions often have difficulty obtaining water for much of the year. Most plants living in these habitats show xerophytic modifications to enable them to reduce transpiration and so survive.

▲ **Figure 4** *Sand dunes*

1 List **two** reasons why plants growing on sand dunes (Figure 4) need to have xerophytic features even though there is plentiful rainfall.
2 Explain in terms of water potential why salt marsh plants have difficulty absorbing water, despite having plenty around their roots.
3 Explain why plants in cold regions 'have difficulty obtaining water from the soil for much of the year'.
4 Plants living in cold regions often reduce water loss by having leaves with a small surface area to volume ratio. This reduces the surface area available to capture light for photosynthesis. Photosynthesis is, in part, an enzyme-controlled process. Suggest a reason why having a smaller leaf area does not reduce the rate of photosynthesis in the same way as it would for plants in warmer climates.

Learning objectives

→ Describe how the human gas-exchange system is arranged.

→ Explain the functions of the human gas-exchange system.

Specification reference: 3.3.2

All aerobic organisms require a constant supply of oxygen to release energy in the form of **ATP** during respiration. The carbon dioxide produced in the process needs to be removed as its build-up could be harmful to the body.

The volume of oxygen that has to be absorbed and the volume of carbon dioxide that must be removed are large in mammals because:

• they are relatively large organisms with a large volume of living cells

• they maintain a high body temperature which is related to them having high metabolic and respiratory rates.

As a result mammals have evolved specialised surfaces, called **lungs**, to ensure efficient gas exchange between the air and their blood.

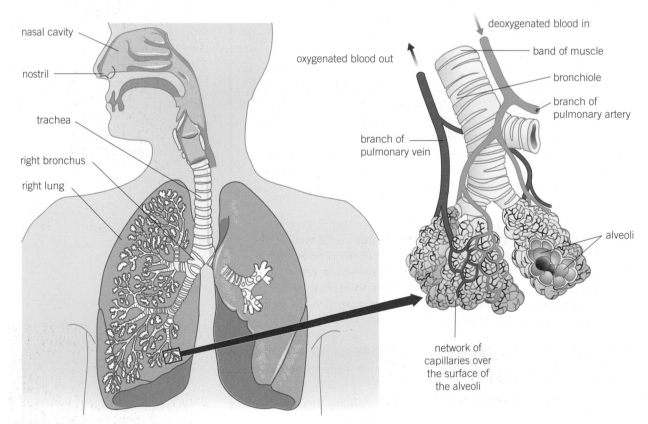

▲ **Figure 1** *The gross structure of the human gas-exchange system*

Mammalian lungs

The lungs are the site of gas exchange in mammals. They are located inside the body because:

- air is not dense enough to support and protect these delicate structures
- the body as a whole would otherwise lose a great deal of water and dry out.

The lungs are supported and protected by a bony box called the **ribcage**. The ribs can be moved by the muscles between them. The lungs are ventilated by a tidal stream of air, thereby ensuring that the air within them is constantly replenished. The main parts of the human gas-exchange system and their structure and functions are described below.

- The **lungs** are a pair of lobed structures made up of a series of highly branched tubules, called bronchioles, which end in tiny air sacs called alveoli.
- The **trachea** is a flexible airway that is supported by rings of cartilage. The cartilage prevents the trachea collapsing as the air pressure inside falls when breathing in. The tracheal walls are made up of muscle, lined with ciliated epithelium and goblet cells.
- The **bronchi** are two divisions of the trachea, each leading to one lung. They are similar in structure to the trachea and, like the trachea, they also produce mucus to trap dirt particles and have cilia that move the dirt-laden mucus towards the throat. The larger bronchi are supported by cartilage, although the amount of cartilage is reduced as the bronchi get smaller.
- The **bronchioles** are a series of branching subdivisions of the bronchi. Their walls are made of muscle lined with epithelial cells. This muscle allows them to constrict so that they can control the flow of air in and out of the alveoli.
- The **alveoli** are minute air-sacs, with a diameter of between $100\,\mu m$ and $300\,\mu m$, at the end of the bronchioles. Between the alveoli there are some **collagen** and elastic fibres. The alveoli are lined with epithelium. The elastic fibres allow the alveoli to stretch as they fill with air when breathing in. They then spring back during breathing out in order to expel the carbon dioxide-rich air. The alveolar membrane is the gas-exchange surface.

Hint

The ending '-ioles' is commonly used in biology to denote a smaller version of a structure. Hence 'bronchioles' are small bronchi, and 'arterioles' are small arteries.

▲ **Figure 2** False-colour X-ray of the bronchus and bronchioles of a healthy human lung

Summary questions

1 State **two** reasons why humans need to absorb large volumes of oxygen from the lungs.

2 List in the correct sequence all the structures that air passes through on its journey from the gas-exchange surface of the lungs to the nose.

3 Explain how the cells lining the trachea and bronchus protect the alveoli from damage.

▲ **Figure 3** False-colour SEM of a section of the epithelium of the trachea showing ciliated cells (green)

Hint

Do not write about 'respiration' when you mean 'breathing' and vice versa.

Hint

There are two basic physical laws that will help you to understand the movement of air during breathing:

Within a closed container, as the volume of a gas increases its pressure decreases. Similarly, as the volume of a gas decreases so the pressure increases.

Gases move from a region where their pressure is higher to a region where their pressure is lower.

To maintain diffusion of gases across the alveolar epithelium, air is constantly moved in and out of the lungs. This process is called breathing, or **ventilation**. When the air pressure of the atmosphere is greater than the air pressure inside the lungs, air is forced into the lungs. This is called **inspiration** (inhalation). When the air pressure in the lungs is greater than that of the atmosphere, air is forced out of the lungs. This is called **expiration** (exhalation). The pressure changes within the lungs are brought about by the movement of three sets of muscles:

- the diaphragm, which is a sheet of muscle that separates the thorax from the abdomen
- the intercostal muscles, which lie between the ribs. There are two sets of intercostal muscles:
  - the **internal intercostal muscles**, whose contraction leads to expiration
  - the **external intercostal muscles**, whose contraction leads to inspiration.

Figure 1 shows the arrangement of these various muscles.

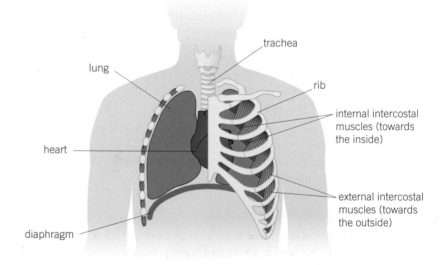

▲ **Figure 1** *The arrangement of the diaphragm and intercostal muscles*

Inspiration

Breathing in is an active process (it uses energy) and occurs as follows:

- The external intercostal muscles contract, while the internal intercostal muscles relax.
- The ribs are pulled upwards and outwards, increasing the volume of the thorax.
- The diaphragm muscles contract, causing it to flatten, which also increases the volume of the thorax.
- The increased volume of the thorax results in reduction of pressure in the lungs.
- Atmospheric pressure is now greater than pulmonary pressure, and so air is forced into the lungs.

Expiration

Breathing out is a largely passive process (it does not require much energy) and occurs as follows:

- The internal intercostal muscles contract, while the external intercostal muscles relax.
- The ribs move downwards and inwards, decreasing the volume of the thorax.
- The diaphragm muscles relax and so it is pushed up again by the contents of the abdomen that were compressed during inspiration. The volume of the thorax is therefore further decreased.
- The decreased volume of the thorax increases the pressure in the lungs.
- The pulmonary pressure is now greater than that of the atmosphere, and so air is forced out of the lungs.

During normal quiet breathing, the recoil of the elastic tissue in the lungs is the main cause of air being forced out (like air being expelled from a partly inflated balloon). Only under more strenuous conditions such as exercise do the various muscles play a major part.

Summary questions √x̄

1. From the graphs in Figure 3, calculate the rate of breathing of this person. Give your answer in breaths per minute. Show how you arrived at your answer.

2. If the volume of air in the lungs when the person inhaled was 3 000 cm³ calculate the volume of air in the lungs after the person had exhaled. Show your working.

3. Explain how muscles create the change of pressure in the alveoli over the period 0 to 0.5 s.

Pulmonary ventilation √x̄

It is sometimes useful to know how much air is taken in and out of the lungs in a given time. To do this we use a measure called pulmonary ventilation rate. Pulmonary ventilation rate is the total volume of air that is moved into the lungs during 1 minute. To calculate it we multiply together two factors:

- tidal volume, which is the volume of air normally taken in at each breath when the body is at rest. This is usually around 0.5 dm³.
- breathing (ventilation) rate, that is, the number of breaths taken in 1 minute. This is normally 12–20 breaths in a healthy adult.

Pulmonary ventilation rate is expressed as dm³ min⁻¹.

To summarise:

pulmonary ventilation rate = tidal volume × breathing rate
(dm³ min⁻¹)　(dm³)　(min⁻¹)

1. A person has a pulmonary ventilation rate of 10.2 dm³ min⁻¹ and a tidal volume of 0.6 dm³. Calculate the person's breathing rate.

BREATHING IN (inspiration)

air in
ribcage moves outwards
diaphragm moves downwards

BREATHING OUT (expiration)

air out
ribcage moves inwards
position of ribcage when breathing in
diaphragm moves upwards

▲ **Figure 2** Position of ribs and diaphragm during inspiration and expiration

▲ **Figure 3** The volume and pressure changes that occurred in the lungs of a person during breathing while at rest

Maths link √x̄

MS 0.1, 2.2, 2.4 and 3.1, see Chapter 22.

Learning objectives

→ Describe the essential features of exchange surfaces.

→ Explain how gases are exchanged in the alveoli of humans.

Specification reference: 3.3.2

The site of gas exchange in mammals is the epithelium of the alveoli. These alveoli are minute air sacs some 100–300 μm in diameter and situated in the lungs. To ensure a constant supply of oxygen to the body, a diffusion gradient must be maintained at the alveolar surface. We saw in Topic 6.1 that, to enable efficient transfer of materials across them, exchange surfaces are thin, partially permeable and have a large surface area. To maintain a diffusion gradient, there also has to be movement of both the environmental medium (for example, air) and the internal medium (for example, blood).

▲ **Figure 1** *Alveoli*

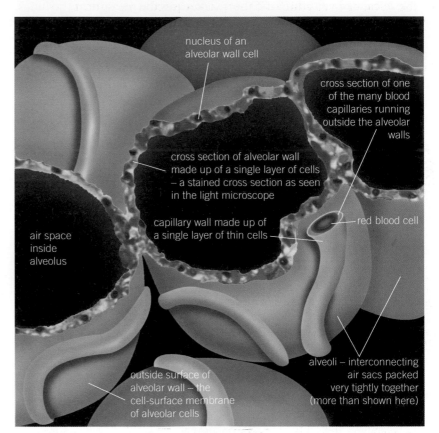

▲ **Figure 2** *External appearance of a group of alveoli*

Being thin, these specialised exchange surfaces are easily damaged and therefore are often located inside an organism for protection. Where an exchange surface, such as the lungs, is located inside the body, the organism has some means of moving the external medium over the surface, for example a means of ventilating the lungs in a mammal. This is because diffusion alone is not fast enough to maintain adequate transfer of oxygen and carbon dioxide along the trachea, bronchi and bronchioles. Breathing is basically a form of mass transport.

Role of the alveoli in gas exchange

There are about 300 million alveoli in each human lung. Their total surface area is around 70 m² – about half the area of a tennis court. Their structure is shown in Figures 1 and 2. Each alveolus is lined with

▲ **Figure 3** *False-colour SEM of a section of human lung tissue showing alveoli surrounded by blood capillaries*

epithelial cells only 0.05 μm to 0.3 μm thick. Around each alveolus is a network of pulmonary capillaries, so narrow (7–10 μm) that red blood cells are flattened against the thin capillary walls in order to squeeze through. These capillaries have walls that are only a single layer of cells thick (0.04–0.2 μm). Diffusion of gases between the alveoli and the blood will be very rapid because:

- red blood cells are slowed as they pass through pulmonary capillaries, allowing more time for diffusion
- the distance between the alveolar air and red blood cells is reduced as the red blood cells are flattened against the capillary walls
- the walls of both alveoli and capillaries are very thin and therefore the distance over which diffusion takes place is very short
- alveoli and pulmonary capillaries have a very large total surface area
- breathing movements constantly ventilate the lungs, and the action of the heart constantly circulates blood around the alveoli. Together, these ensure that a steep concentration gradient of the gases to be exchanged is maintained
- blood flow through the pulmonary capillaries maintains a concentration gradient.

The diffusion of gases in an alveolus is illustrated in Figure 4.

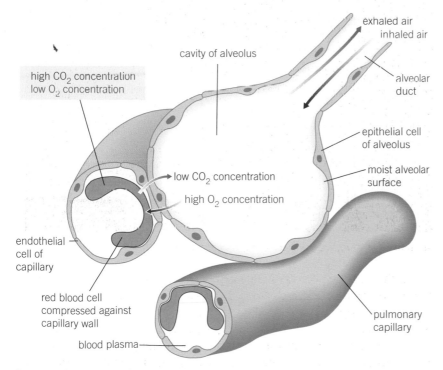

high CO_2 concentration
low O_2 concentration

cavity of alveolus

exhaled air
inhaled air

alveolar duct

low CO_2 concentration
high O_2 concentration

epithelial cell of alveolus

moist alveolar surface

endothelial cell of capillary

red blood cell compressed against capillary wall

blood plasma

pulmonary capillary

▲ **Figure 4** *Diffusion of gases in an alveolus*

Hint

The diffusion pathway is short because the alveoli have only a single layer of epithelial cells and the blood capillaries have only a single layer of endothelial cells. Don't say they have cells with thin membranes.

Summary questions

1 Explain how each of the following features contributes to the efficiency of gas exchange in alveoli.

   a The wall of each alveolus is not more than 0.3 μm thick.

   b There are 300 million alveoli in each lung.

   c Each alveolus is covered by a dense network of pulmonary blood capillaries.

   d Each pulmonary capillary is very narrow.

2 √x̄ If the number of alveoli in each lung was increased to 600 million and the pulmonary ventilation was doubled, calculate how many times greater the rate of diffusion would be.

Correlations and causal relationships

A **correlation** occurs when a change in one of two variables is reflected by a change in the other variable.

The interpretation of the data in Figure 5 shows that there is a correlation between drinking alcohol and breast cancer. What we *cannot* do from these data, however, is to conclude that drinking alcohol is the **cause** of breast cancer. The data seem to suggest this is the case but there is no actual evidence here to prove it. There needs to be a clear causal connection between drinking alcohol and breast cancer before you we can say that the case is proven. These data alone show only a correlation and not a cause. It could be that women who are stressed drink more alcohol and that it is the stress, rather than the alcohol, that causes breast cancer. To prove that drinking alcohol is the cause of breast cancer we would need experimental evidence to show that some component of the alcoholic drink led directly to women getting breast cancer. Recognising the distinction between a correlation and a causal relationship is a necessary and important skill.

Figure 6 shows how the incidence of lung cancer changes with the number of cigarettes smoked a day. What can we conclude from this data? Well, nothing really. We can see that the more cigarettes that are smoked, the greater are the number of deaths from lung cancer. In other words, there is a positive correlation between the two factors. However, we cannot conclude that it is the cigarette smoke that causes lung cancer. It may just be coincidence, or it could be that smokers are more stressed or drink more alcohol and these factors might be the cause of the cancer. Even though this graph does not itself establish a link, scientists have produced compelling experimental evidence to show that smoking tobacco definitely can cause lung cancer.

1 State a correlation shown in figure 6.
2 Explain why the information provided does not show a causal relationship with the correlation you have identified.

Study tip

It is important to be clear that a correlation does *not* mean that there is a causal link.

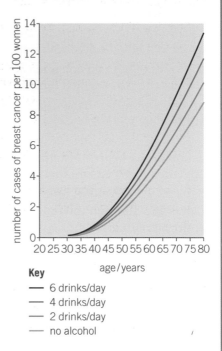

Key
— 6 drinks/day
— 4 drinks/day
— 2 drinks/day
— no alcohol

▲ **Figure 5**

▲ **Figure 6** *Annual incidence of lung cancer per 100 000 men in the USA correlated to the daily consumption of cigarettes*

Risk factors for lung disease √x̄

There are a number of specific risk factors that increase the probability of someone suffering from lung disease. In this context 'lung disease' refers to chronic obstructive pulmonary disease (COPD), which includes emphysema and chronic bronchitis. These risk factors include:

- **Smoking**. 90% of people suffering from COPD are, or have been, heavy smokers.

- **Air pollution**. Pollutant particles and gases (e.g., sulfur dioxide) increase the likelihood of COPD, especially in areas of heavy industry.

- **Genetic make-up**. Some people are genetically more likely to get lung disease, others less so; this explains why some lifelong smokers never get lung disease while others die early.

- **Infections**. People who frequently get other chest infections also show a higher incidence of COPD.

- **Occupation**. People working with harmful chemicals, gases and dusts that can be inhaled have an increased risk of lung disease.

Here is an analysis of some data relating to the most significant risk factor – smoking.

The world's longest-running survey of smoking began in the UK in 1951. This survey and other ones elsewhere in the world have revealed a number of general statistical facts about smokers. Look at Figure 7. What does it tell us

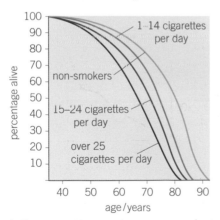

▲ **Figure 7** *Life expectancy related to the number of cigarettes smoked*

- All the lines start at 100%. This shows that the whole of this group of the population were alive at the start of the survey. What else does this tell us? As the scale of the independent variable (age in years) has its origin at 35 it suggests that everyone in this group must have been at least 35 years old at the start of the survey.

Maths link √x̄

MS 0.3, see Chapter 22.

- All the lines follow approximately the same pattern: they decline slowly at first and then at an increasing rate until at some point all lines cross the x-axis. This describes the shape but what does it actually show? Namely that only a few people die between the ages of 35 and 60 but that, after age 60, the death rate becomes increasingly rapid until, at some point everyone in the group has died.

- What about the differences between the four separate coloured lines? Each represents a different group distinguished by the number of cigarettes smoked each day. At every age beyond 35 years, the more cigarettes smoked, the fewer people remain alive. This difference is more marked the greater the age.

- At what age did the members of each group die? Well the line representing the group who smoked more than 25 cigarettes a day crosses the x-axis at 82 years, showing that no one in the group lived beyond that age. By contrast, some of the non-smokers lived beyond 90 years.

- What is the overall interpretation? Namely that the more cigarettes smoked per day, the earlier, on average, a smoker dies.

The interpretation of the data in Figure 7 shows there is a **correlation** between smoking and premature death. This does not, however, prove that smoking is the cause of an early death. The data seem to suggest this is the case but there is no evidence here to prove that it is so. There needs to be a clear causal connection between smoking and death before you can say that the case is proven. These data alone show only a correlation and not a cause. To prove that smoking is the cause of early death in smokers the correct scientific process needs to be followed. There are three main stages:

1. Establish a hypothesis to try to explain the correlation; this should be based on current knowledge.
2. Design and perform experiments to test the hypothesis.
3. Establish the causal link and formulate theories to explain it.

This is precisely what happened in establishing the causal link between smoking and lung cancer.

1 List **four** risk factors associated with lung disease.

2 Use Figure 7 to determine what percentage of non-smokers are likely to survive to age 80.

3 Calculate how many times greater is the likelihood of a non-smoker living to age 70 than someone who smokes over 25 cigarettes a day.

4 About 10 to 15 years after giving up smoking the risk of death approaches that of non-smokers. Use this information to explain to a 40-year-old who smokes 30 cigarettes a day the likely impact on her life expectancy of giving up smoking immediately.

5 Data showing a causal link between smoking and lung disease has led to statutory restrictions on the sources of risk factors. Suggest some restrictions that have been introduced and how these might reduce the incidence of lung disease.

6 Pulmonary fibrosis is a lung disease that causes the epithelium of the lungs to become irreversibly thickened. It also leads to reduced elasticity of the lungs.
One symptom of the disease is shortness of breath, especially when exercising. Suggest why this symptom arises.

7 One measure of lung function is Forced Expiratory Volume (FEV). This is the volume of air that can forcibly be blown out in one second, after full inspiration. Suggest how pulmonary fibrosis might effect FEV and explain why.

Smoking and lung cancer

Life insurance companies have calculated that, on average, smoking a single cigarette lowers an individual's life expectancy by 10.7 minutes – longer than it takes to smoke the cigarette! While this is a statistical deduction rather than a scientific one, there is now clear scientific evidence to support the view that smoking cigarettes damages your health and reduces life expectancy. One type of evidence comes from correlations between cigarette smoking and certain diseases.

Figure 8 shows deaths from lung cancer in the UK correlated to the number of cigarettes smoked per year during a period in the last century. Study it carefully and then answer the questions.

1 Determine in which decade smoking reached its peak for the following:
   a males   b females

2 Explain how the graphs show that there is a correlation between the number of cigarettes smoked and deaths from lung cancer in both sexes.

3 In both sexes, the number of deaths per year from lung cancer increased over the period 1910 to 1970. Suggest **three** possible reasons for this.

4 Suggest a reason why there is a time lag between the number of cigarettes smoked and a corresponding change in the number of deaths from lung cancer.

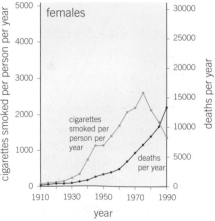

▲ **Figure 8** *Incidence of deaths from lung cancer in the UK correlated to cigarettes smoked per year (1910–90)*

The human digestive system is made up of a long muscular tube and its associated glands. The glands produce **enzymes** that hydrolyse large molecules into small ones ready for absorption. The digestive system (Figure 1) is therefore an exchange surface through which food substances are absorbed.

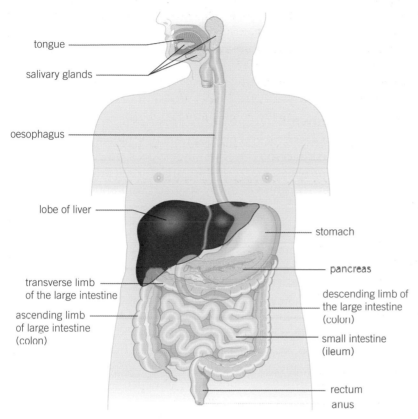

▲ **Figure 1** *Human digestive system*

Learning objectives

→ Describe the structure and function of the major parts of the digestive system.

→ Explain how the digestive system breaks down food both physically and chemically.

→ Explain the role of enzymes in digestion of carbohydrates, lipids and proteins.

Specification reference: 3.3.3

Study tip

Digestion is the process in which *large* molecules are hydrolysed by enzymes into *small* molecules, which can be absorbed and assimilated.

Major parts of the digestive system

- The **oesophagus** carries food from the mouth to the stomach.

- The **stomach** is a muscular sac with an inner layer that produces enzymes. Its role is to store and digest food, especially proteins. It has glands that produce enzymes which digest protein.

- The **ileum** is a long muscular tube. Food is further digested in the ileum by enzymes that are produced by its walls and by glands that pour their secretions into it. The inner walls of the ileum are folded into villi, which gives them a large surface area. The surface area of these villi is further increased by millions of tiny projections, called microvilli, on the epithelial cells of each villus. This adapts the ileum for its purpose of absorbing the products of digestion into the bloodstream.

- The **large intestine** absorbs water. Most of the water that is absorbed is water from the secretions of the many digestive glands.

- The **rectum** is the final section of the intestines. The faeces are stored here before periodically being removed via the anus in a process called **egestion**.

Hint

The contents of the intestines are *not* inside the body. Molecules and ions only truly enter the body when they cross the cells and cell-surface membranes of the epithelial lining of the intestines.

Hint

All organisms are made up of the same biological molecules and therefore your food consists almost entirely of other organisms, or parts of them. You must first hydrolyse them into molecules that are small enough to pass across cell-surface membranes.

Synoptic link

It will help you understand this Topic if you revisit Topics 1.3, 1.5, 1.6 and 1.7.

- The **salivary glands** are situated near the mouth. They pass their secretions via a duct into the mouth. These secretions contain the enzyme amylase, which **hydrolyses** starch into maltose.
- The **pancreas** is a large gland situated below the stomach. It produces a secretion called pancreatic juice. This secretion contains proteases to hydrolyse proteins, lipase to hydrolyse lipids and amylase to hydrolyse starch.

What is digestion?

In humans, as with many organisms, digestion takes place in two stages:

1 physical breakdown,

2 chemical digestion.

Physical breakdown

If the food is large, it is broken down into smaller pieces by means of structures such as the teeth. This not only makes it possible to ingest the food but also provides a large surface area for chemical digestion. Food is churned by the muscles in the stomach wall and this also physically breaks it up.

Chemical digestion

Chemical digestion hydrolyses large, insoluble molecules into smaller, soluble ones. It is carried out by enzymes. All digestive enzymes function by **hydrolysis**. Hydrolysis is the splitting up of molecules by adding water to the chemical bonds that hold them together. Enzymes are specific and so it follows that more than one enzyme is needed to hydrolyse a large molecule. Usually one enzyme hydrolyses a large molecule into sections and these sections are then hydrolysed into smaller molecules by one or more additional enzymes. There are different types of digestive enzymes, three of which are particularly important:

- **Carbohydrases** hydrolyse carbohydrates, ultimately to monosaccharides.
- **Lipases** hydrolyse lipids (fats and oils) into glycerol and fatty acids.
- **Proteases** hydrolyse proteins, ultimately to amino acids.

You can now look at these three groups of digestive enzymes in more detail.

Carbohydrate digestion

It usually takes more than one enzyme to completely hydrolyse a large molecule. Typically one enzyme hydrolyses the molecule into smaller sections and then other enzymes hydrolyse these sections further into their **monomers**. These enzymes are usually produced in different parts of the digestive system. It is obviously important that enzymes are added to the food in the correct sequence. This is true of starch digestion.

Firstly the enzyme **amylase** is produced in the mouth and the pancreas. Amylase hydrolyses the alternate glycosidic bonds of the starch molecule to produce the disaccharide maltose. The maltose is in turn hydrolysed into the monosaccharide α-glucose by a second enzyme, a disaccharidase called **maltase**. Maltase is produced by the lining of ileum.

In humans the process takes place as follows:

* Saliva enters the mouth from the salivary glands and is thoroughly mixed with the food during chewing.

* Saliva contains **salivary amylase**. This starts hydrolysing any starch in the food to maltose. It also contains mineral salts that help to maintain the pH at around neutral. This is the optimum pH for salivary amylase to work.

* The food is swallowed and enters the stomach, where the conditions are acidic. This acid **denatures** the amylase and prevents further hydrolysis of the starch.

* After a time the food is passed into the small intestine, where it mixes with the secretion from the pancreas called pancreatic juice.

* The pancreatic juice contains **pancreatic amylase**. This continues the hydrolysis of any remaining starch to maltose. Alkaline salts are produced by both the pancreas and the intestinal wall to maintain the pH at around neutral so that the amylase can function.

* Muscles in the intestine wall push the food along the ileum. Its epithelial lining produces the disaccharidase **maltase**. Maltase is not released into the lumen of the ileum but is part of to the cell-surface membranes of the epithelial cells that line the ileum. It is therefore referred to as a **membrane-bound disaccharidase**. The maltase hydrolyses the maltose from starch breakdown into α-glucose.

In addition to the digestion of maltose described above, there are two other common disaccharides in the diet that are hydrolysed – sucrose and lactose.

Sucrose is found in many natural foods, especially fruits. Lactose is found in milk, and hence in milk products, such as yoghurt and cheese. Each disaccharide is hydrolysed by a membrane-bound disaccharidase as follows:

* **Sucrase** hydrolyses the single glycosidic bond in the sucrose molecule. This hydrolysis produces the two monosaccharides glucose and fructose.

* **Lactase** hydrolyses the single glycosidic bond in the lactose molecule. This hydrolysis produces the two monosaccharides glucose and galactose.

Lipid digestion

Lipids are hydrolysed by enzymes called **lipases**. Lipases are enzymes produced in the pancreas that hydrolyse the ester bond found in triglycerides to form fatty acids and monoglycerides. A monoglyceride is a glycerol molecule with a single fatty acid molecule attached. Lipids (fats and oils) are firstly split up into tiny droplets called **micelles** (Topic 6.10) by **bile salts**, which are produced by the liver. This process is called **emulsification** and increases the surface area of the lipids so that the action of lipases is speeded up.

Hint

Enzyme names usually end in '-ase' and start with the the first part of the name of their substrate (the substance on which they act). Hence maltase hydrolyses maltose, and sucrase hydrolyses sucrose.

Protein digestion

Proteins are large, complex molecules that are hydrolysed by a group of enzymes called **peptidases** (proteases). There are a number of different peptidases:

- **Endopeptidases** hydrolyse the peptide bonds between amino acids in the central region of a protein molecule forming a series of peptide molecules.
- **Exopeptidases** hydrolyse the peptide bonds on the terminal amino acids of the peptide molecules formed by endopeptidases. In this way they progressively release dipeptides and single amino acids.
- **Dipeptidases** hydrolyse the bond between the two amino acids of a dipeptide. Dipeptidases are membrane-bound, being part of the cell-surface membrane of the epithelial cells lining the ileum.

Summary questions

1 Define hydrolysis.
2 List **two** structures that produce amylase.
3 Suggest why the stomach does not have villi or microvilli.
4 Name the final product of starch digestion in the gut.
5 List **three** enzymes produced by the epithelium of the ileum.

Lactose intolerance

Milk is the only food of human babies and so they produce a relatively large amount of lactase, the enzyme that hydrolyses lactose, the sugar in milk. As milk forms a less significant part of the diet in adults, the production of lactase diminishes as children get older. This reduction can be so great in some adults that they produce little, or no, lactase at all.

This was not a problem to our ancestors but can be to humans of today. Humans that produce no lactase cannot hydrolyse the lactose they consume. When the undigested lactose reaches the large intestines, microorganisms hydrolyse it. This gives rise to small soluble molecules and a large volume of gas. This can result in diarrhoea because the soluble molecules lower the water potential of the material in the colon. The condition is known as lactose intolerance. Some people with the condition cannot consume milk or milk products at all while others can consume them only in small amounts.

▲ **Figure 3** *Milk and milk products*

1 a Suggest the process by which micoorganisms produce 'a large volume of gas' in lactose intolerant individuals.

   b Suggest a reason why this gas is unlikely to be carbon dioxide.

2 Suggest an explanation why lactose intolerance is a problem for modern day humans but wasn't for our ancestors.

3 Explain how the lowering of water potential in the colon can cause diarrhoea.

6.10 Absorption of the products of digestion

We have seen in Topic 6.9 how enzymes hydrolyse carbohydrates, fats and proteins. The products of this hydrolysis are monosaccharides, amino acids, monoglycerides and fatty acids. We will now see how these products are absorbed by the ileum.

Structure of the ileum

The ileum is adapted to the function of absorbing the products of digestion. The wall of the ileum is folded and possesses finger-like projections, about 1 mm long, called **villi** (Figure 2). They have thin walls, lined with epithelial cells on the other side of which is a rich network of blood capillaries. The villi considerably increase the surface area of the ileum and therefore accelerate the rate of absorption.

Villi are situated at the interface between the **lumen** (cavity) of the intestines (in effect outside the body) and the blood and other tissues inside the body. They are part of a specialised exchange surface adapted for the absorption of the products of digestion. Their properties increase the efficiency of absorption in the following ways:

- They increase the surface area for **diffusion**.
- They are very thin walled, thus reducing the distance over which diffusion takes place.
- They contain muscle and so are able to move. This helps to maintain diffusion gradients because their movement mixes the contents of the ileum. This ensures that, as the products of digestion are absorbed from the food adjacent to the villi, new material rich in the products of digestion replaces it.
- They are well supplied with blood vessels so that blood can carry away absorbed molecules and hence maintain a diffusion gradient.
- The epithelial cells lining the villi possess **microvilli** (Figure 1). These are finger-like projections of the cell-surface membrane that further increase the surface area for absorption.

Learning objectives

→ Describe the structure of the ileum.

→ Explain how the ileum is adapted for the function of absorption.

→ Explain how monosaccharides and amino acids are absorbed.

→ Explain how triglycerides are absorbed.

Specification reference: 3.3.3

Synoptic link

You will better understand the contents of this Topic if you first read through Topics 4.2, 4.5 and 7.6.

◄ **Figure 1** *Light micrograph of a section through a villus in the small intestine. Villi are projections that increase the surface area for the absorption of food. They are covered in microvilli (smaller, finger-like projections) that further increase this surface area*

▲ **Figure 2** *False-colour SEM of villi (brown) in the lining of the ileum*

Absorption of amino acids and monosaccharides

The digestion of proteins produces amino acids, while that of carbohydrates produces monosaccharides such as glucose, fructose and galactose. The methods of absorbing these products are the same, namely diffusion and co-transport. We saw how glucose and amino acids are absorbed in the ileum by these processes in Topic 4.2 and 4.5.

Absorption of triglycerides

Once formed during digestion, monoglycerides and fatty acids remain in association with the bile salts that initially emulsified the lipid droplets (see Topic 6.9). The structures formed are called **micelles**. They are tiny, being around 4–7 nm in diameter. Through the movement of material within the lumen of the ileum, the micelles come into contact with the epithelial cells lining the villi of the ileum. Here the micelles break down, releasing the monoglycerides and fatty acids. As these are non-polar molecules, they easily diffuse across the cell-surface membrane into the epithelial cells.

Once inside the epithelial cells, monoglycerides and fatty acids are transported to the endoplasmic reticulum where they are recombined to form triglycerides. Starting in the endoplasmic reticulum and continuing in the Golgi apparatus, the triglycerides associate with cholesterol and lipoproteins to form structures called **chylomicrons**. Chylomicrons are special particles adapted for the transport of lipids.

Chylomicrons move out of the epithelial cells by **exocytosis**. They enter lymphatic capillaries called **lacteals** that are found at the centre of each villus. The process is illustrated in Figure 2.

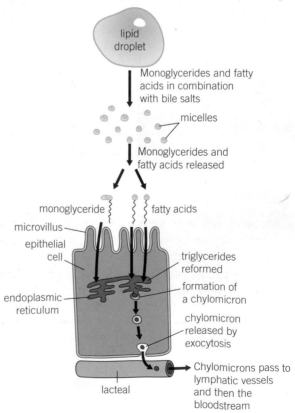

▶ **Figure 2** *The absorption of triglycerides*

From here, the chylomicrons pass, via lymphatic vessels, into the blood system. The triglycerides in the chylomicrons are hydrolysed by an enzyme in the endothelial cells of blood capillaries (see Topic 7.6) from where they diffuse into cells.

Absorption of fatty acids

Bile salts play a role in the digestion and absorption of fatty acids. One end of the bile salt molecule is soluble in fat (lipophilic) but not in water (hydrophobic). The other end is soluble in water (hydrophilic) but not in fat (lipophobic). Bile salt molecules therefore arrange themselves with their lipophilic ends in fat droplets, leaving their lipophobic ends sticking out. In this way they prevent fat droplets from sticking to each other to form large droplets, leaving only tiny ones (micelles). It is in this form that fatty acids reach the epithelial cells of the ileum where they break down, releasing the fatty acids for absorption.

An experiment was carried out to investigate the absorption of fatty acids. Six sections of intestine were filled with a fatty acid called oleic acid. To each section were added different mixtures of other contents as shown in Table 1.

Iodoacetate inhibits an enzyme involved in glycolysis – a stage of the respiratory process in cells that involves phosphorylation.

▼ Table 1

Contents of section of intestine					Relative amounts of oleic acid absorbed in 10 hours
Bile salts	Glycerol	Phosphate	Glycerol phosphate	Iodoacetate	
✓	✗	✗	✗	✗	2.9
✓	✗	✓	✗	✗	1.1
✓	✓	✗	✗	✗	2.6
✓	✓	✓	✗	✗	5.8
✓	✗	✗	✓	✗	8.5
✓	✗	✗	✓	✓	0.0

✓ = substance present ✗ = substance absent

From the information in Table 1:

1 √x̄ List **three** pieces of evidence that support the idea that the absorption of fatty acids in the intestine is increased if they are combined with a compound of glycerol and phosphate.

2 √x̄ Recognise the evidence supporting the view that the absorption of fatty acids involves phosphorylation.

Summary questions

1 List **three** organelles that you would expect to be numerous and/or well developed in an epithelial cell of the ileum, giving a reason for your choice in each case.

2 Name the other chemical that moves across epithelial cells with glucose molecules during co-transport.

3 In addition to having microvilli, state **one** other feature of the epithelial cells of the ileum that would increase the rate of absorption of amino acids.

Maths link √x̄

MS 1.3, see Chapter 22.

Practice questions: Chapter 6

1 **(a)** Flatworms are small animals that live in water. They have no specialised gas exchange or circulatory systems. The drawing shows one type of flatworm.

1 mm

   (i)   Name the process by which oxygen reaches the cells inside the body of this flatworm. *(1 mark)*

   (ii) The body of a flatworm is adapted for efficient gas exchange between the water and the cells inside the body. Using the diagram, explain how two features of the flatworm's body allow efficient gas exchange. *(2 marks)*

   (b) **(i)**   A leaf is an organ. What is an organ? *(1 mark)*

   (ii) Describe how carbon dioxide in the air outside a leaf reaches mesophyll cells inside the leaf. *(3 marks)*

AQA June 2012

2 **(a)** The diagram shows the structure of the human gas exchange system.

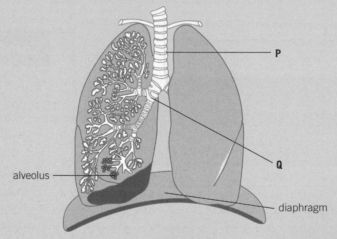

alveolus

diaphragm

P

Q

Name organs
P
Q   *(1 mark)*

   (b) Explain how downward movement of the diaphragm leads to air entering the lungs. *(2 marks)*

AQA Jan 2013

3 The diagram shows the position of the diaphragm at times **P** and **Q**.

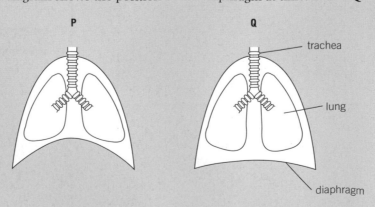

P   Q

trachea

lung

diaphragm

(a) Describe what happens to the diaphragm between times **P** and **Q** to bring about the change in its shape. *(2 marks)*

(b) Air moves into the lungs between times P and Q. Explain how the diaphragm causes this. *(3 marks)*

(c) Describe how oxygen in air in the alveoli enters the blood in capillaries. *(2 marks)*

AQA June 2012

4   Insects such as beetles obtain oxygen by drawing air into their tracheae through spiracles. Diving beetles live in ponds. They carry a bubble of air under their wing cases when they swim underwater. The bubble supplies air to the spiracles. When the bubble has been used up, the beetle comes to the surface to collect a new bubble.

An investigation was carried out into the effect of temperature on diving beetles. Three beetles, **A**, **B** and **C**, of the same species, were observed in thermostatically-controlled water baths. The number of times each beetle surfaced to renew its air bubble was counted at three different temperatures.

The results are shown in Table 1 below.

temperature / °C	number of times air bubble was renewed per hour		
	beetle A	beetle B	beetle C
10	10	12	8
20	18	22	18
30	44	48	38

(a) Calculate the mean number of times the air bubble was renewed per hour at each temperature. *(1 mark)*

(b) Sketch a graph to show the relationship between temperature and the mean number of times the air bubble was renewed per hour and name the shape of the line obtained. *(2 marks)*

(c) The number of times the air bubble is renewed per hour is related to a beetle's need for oxygen to carry out aerobic respiration, which is catalysed by enzymes. Explain what the data reveal about the size of the effect of each 10°C rise in temperature on the rate of respiration. *(2 marks)*

5   Forced expiratory volume (FEV) is the greatest volume of air a person can breathe out in 1 second.

Forced vital capacity (FVC) is the greatest volume of air a person can breathe out in a single breath. Figure 2 shows results for the volume of air breathed out by three groups of people, **A**, **B** and **C**. Group **A** had healthy lungs. Groups **B** and **C** had different lung conditions that affect breathing.

(a) Calculate the percentage drop in FEV for group **C** compared with the healthy people. *(1 mark)*

(b) Asthma affects bronchioles and reduces flow of air in and out of the lungs.
Fibrosis does not affect bronchioles; it reduces the volume of the lungs.
Which group, **B** or **C**, was the one containing people with fibrosis of their lungs?
Use the information provided and evidence from **Figure 2** to explain your answer.

(3 marks)

AQA SAMS A LEVEL PAPER 1

6 An animal cell takes in oxygen over its surface area but uses oxygen in proportion to its volume. Size and shape affect the ratio of surface area to volume of a cell, and therefore affect the efficiency of oxygen uptake.

(a) Complete the table to compare the surface area to volume ratios of the four model cells described. *(4 marks)*

model cell description	surface area / μm^2	volume / μm^3	ratio of surface area to volume
cube, side length 4 μm	96		
sphere, diameter 4 μm	50.3	33.5	
cube, side length 6 μm		216	
sphere, diameter 6 μm			1:1

(b) Summarise what the results show about the effect of size and shape on the ability of a cell to obtain enough oxygen for its needs. *(2 marks)*

7 (a) Describe how you would use a simple respirometer to measure the oxygen uptake of 5g of maggots. *(5 marks)*

(b) A student takes respirometer readings by measuring the distance moved by the marker fluid along a capillary tube in ten minutes. Explain what calculations need to be performed to obtain an hourly oxygen uptake rate per gram of maggots. *(3 marks)*

8 Breathing out as hard as you can is called forced expiration.

(a) Describe and explain the mechanism that causes forced expiration. *(4 marks)*

Two groups of people volunteered to take part in an experiment.
• People in group **A** were healthy.
• People in group **B** were recovering from an asthma attack.

Each person breathed in as deeply as they could. They then breathed out by forced expiration.

A scientist measured the volume of air breathed out during forced expiration by each person.

Forced expiration volume (FEV) is the volume of air a person can breathe out in 1 second.

(b) Using data from the first second of forced expiration, calculate the percentage decrease in the FEV for group **B** compared with group **A**. *(1 mark)*

(c) The people in group **B** were recovering from an asthma attack. Explain how an asthma attack caused the drop in the mean FEV shown in **Figure 4**. *(4 marks)*

AQA SAMS AS PAPER

In the last chapter we looked at how substances are exchanged between the internal and external environments. This chapter looks at how these substances are distributed throughout an organism. Before considering mass transport systems, it begins by looking at an important group of molecules that are highly adapted for transporting oxygen – the haemoglobins.

Haemoglobin molecules

The haemoglobins are a group of chemically similar molecules found in a wide variety of organisms. In Topic 1.6 we investigated the structure of proteins and how their shape is important to their functions. Haemoglobins are protein molecules with a quaternary structure that has evolved to make it efficient at loading oxygen under one set of conditions but unloading it under a different set of conditions. The structure of a haemoglobin molecule is shown in Figure 1. It is made up as follows:

- **primary structure**, sequence of amino acids in the four polypeptide chains
- **secondary structure**, in which each of these polypeptide chains is coiled into a helix
- **tertiary structure**, in which each polypeptide chain is folded into a precise shape – an important factor in its ability to carry oxygen
- **quaternary structure**, in which all four polypeptides are linked together to form an almost spherical molecule. Each polypeptide is associated with a haem group – which contains a ferrous (Fe^{2+}) ion. Each Fe^{2+} ion can combine with a single oxygen molecule (O_2), making a total of four O_2 molecules that can be carried by a single haemoglobin molecule in humans.

Loading and unloading oxygen

The process by which haemoglobin binds with oxygen is called **loading**, or **associating**. In humans this takes place in the lungs.

The process by which haemoglobin releases its oxygen is called **unloading**, or **dissociating**. In humans this takes place in the tissues.

Haemoglobins with a high affinity for oxygen take up oxygen more easily, but release it less easily. Haemoglobins with a low affinity for oxygen take up oxygen less easily, but release it more easily.

The role of haemoglobin

The role of haemoglobin is to transport oxygen. To be efficient at transporting oxygen, haemoglobin must:

- readily associate with oxygen at the surface where gas exchange takes place
- readily dissociate from oxygen at those tissues requiring it.

β-polypeptide

α-polypeptide

each chain is attached to a haem group that can combine with oxygen

▲ **Figure 1** *Quaternary structure of a haemoglobin molecule*

▲ **Figure 2** *Computer graphic representation of a haemoglobin molecule showing two pairs of polypeptide chains (orange and blue) associated with a haem group (red)*

These two requirements may appear to contradict each other, but they are achieved by a remarkable property of haemoglobin. It changes its affinity (chemical attraction) for oxygen under different conditions (Table 1). It achieves this because its shape changes in the presence of certain substances, such as carbon dioxide. In the presence of carbon dioxide, the new shape of the haemoglobin molecule binds more loosely to oxygen. As a result haemoglobin releases its oxygen.

▼ **Table 1** *Affinity of haemoglobin for oxygen under different conditions*

Region of body	Oxygen concentration	Carbon dioxide concentration	Affinity of haemoglobin for oxygen	Result
gas exchange surface	high	low	high	oxygen is associated
respiring tissues	low	high	low	oxygen is dissociated

Why are there different haemoglobins?

Scientists long ago observed that many organisms possessed haemoglobin. They proposed that it carried oxygen from the gas-exchange surface to the tissues that required it for respiration. If so, this meant that it must readily combine with oxygen. Consequently they investigated the ability of haemoglobin from different organisms to combine with oxygen. Results showed that there were different types of haemoglobins. These exhibited different properties relating to the way they took up and released oxygen.

Why do different haemoglobins have different affinities for oxygen?

The answer, scientists discovered, lies in the shape of the molecule. Each species produces a haemoglobin with a slightly different amino acid sequence. The haemoglobin of each species therefore has a slightly different tertiary and quaternary structure and hence different oxygen binding properties. Depending on its structure haemoglobin molecules range from those that have a high affinity for oxygen to those that have a low affinity for oxygen.

Summary questions

1 Describe the quaternary structure of haemoglobin.

2 Explain how DNA leads to different haemoglobin molecules having different affinities for oxygen.

3 When the body is at rest, only one of the four oxygen molecules carried by haemoglobin is normally released into the tissues. Suggest why this could be an advantage when the organism becomes more active.

4 Carbon monoxide occurs in car exhaust fumes. It binds permanently to haemoglobin in preference to oxygen. Suggest a reason why a person breathing in car-exhaust fumes might lose consciousness.

Having looked at haemoglobin in topic 7.1, let us now consider its properties. How does it load and unload oxygen and what effect does carbon dioxide have on this process?

Oxygen dissociation curves

When haemoglobin is exposed to different partial pressures of oxygen, it does not bind the oxygen evenly. The graph of the relationship between the saturation of haemoglobin with oxygen and the partial pressure of oxygen is known as the **oxygen dissociation curve** (see Figure 1). The explanation for the shape of the oxygen dissociation curve is as follows:

- The shape of the haemoglobin molecule makes it difficult for the first oxygen molecule to bind to one of the sites on its four polypeptide subunits because they are closely united. Therefore at low oxygen concentrations, little oxygen binds to haemoglobin. The gradient of the curve is shallow initially.

- However, the binding of this first oxygen molecule changes the quaternary structure of the haemoglobin molecule, causing it to change shape. This change makes it easier for the other subunits to bind to an oxygen molecule. In other words, the binding of the first oxygen molecule induces the other subunits to bind to an oxygen molecule.

- It therefore takes a smaller increase in the partial pressure of oxygen to bind the second oxygen molecule than it did to bind the first one. This is known as **positive cooperativity** because binding of the first molecule makes binding of the second easier and so on. The gradient of the curve steepens.

- The situation changes, however, after the binding of the third molecule. While in theory it is easier for haemoglobin to bind the fourth oxygen molecule, in practice it is harder. This is simply due to probability. With the majority of the binding sites occupied, it is less likely that a single oxygen molecule will find an empty site to bind to. The gradient of the curve reduces and the graph flattens off.

We saw in Topic 7.1 that there are different types of haemoglobin molecules in different species, each with a different shape and hence a different affinity for oxygen. In addition, the shape of any one type of haemoglobin molecule can change under different conditions. These facts both mean that there are a large number of different oxygen dissociation curves. They all have a roughly similar shape but differ in their position on the axes.

The many different oxygen dissociation curves are better understood if two facts are always kept in mind:

- The further to the left the curve, the greater is the affinity of haemoglobin for oxygen (so it loads oxygen readily but unloads it less easily).

- The further to the right the curve, the lower is the affinity of haemoglobin for oxygen (so it loads oxygen less readily but unloads it more easily).

Learning objectives

→ Describe the nature of an oxygen dissociation curve.

→ Explain the effect of carbon dioxide concentration on the curve and the reasons why.

→ Explain how the properties of the haemoglobins in different organisms relate to the environment and way of life of the organism concerned.

Specification reference: 3.3.4.1

Hint

Haemoglobin that is 100% saturated with oxygen has the maximum number of oxygen molecules it can bind to. If 50% saturated, it has half the maximum number it can bind to.

Maths link √x̄

MS 3.1, see Chapter 22.

▲ **Figure 1** *Oxygen dissociation curve for adult human haemoglobin*

Hint

Measuring oxygen concentration
The amount of a gas that is present in a mixture of gases is measured by the pressure it contributes to the total pressure of the gas mixture. This is known as the **partial pressure** of the gas and, in the case of oxygen, is written as pO_2. It is measured in kiloPascals *kPa*. Normal atmospheric pressure is 100 kPa. As oxygen makes up 21% of the atmosphere, its partial pressure is normally 21 kPa.

▲ **Figure 2** *The effect of carbon dioxide concentration on the oxygen dissociation curve*

Effects of carbon dioxide concentration

Haemoglobin has a reduced affinity for oxygen in the presence of carbon dioxide. The greater the concentration of carbon dioxide, the more readily the haemoglobin releases its oxygen (the Bohr effect). This explains why the behaviour of haemoglobin changes in different regions of the body.

*   At the gas-exchange surface (e.g., lungs), the concentration of carbon dioxide is low because it diffuses across the exchange surface and is excreted from the organism. The affinity of haemoglobin for oxygen is increased, which, coupled with the high concentration of oxygen in the lungs, means that oxygen is readily loaded by haemoglobin. The reduced carbon dioxide concentration has shifted the oxygen dissociation curve to the left (Figure 2).
*   In rapidly respiring tissues (e.g., muscles), the concentration of carbon dioxide is high. The affinity of haemoglobin for oxygen is reduced, which, coupled with the low concentration of oxygen in the muscles, means that oxygen is readily unloaded from the haemoglobin into the muscle cells. The increased carbon dioxide concentration has shifted the oxygen dissociation curve to the right (Figure 2).

We have seen that the greater the concentration of carbon dioxide, the more readily haemoglobin releases its oxygen. This is because dissolved carbon dioxide is acidic and the low pH causes haemoglobin to change shape. Let us see how this works in the transport of oxygen by haemoglobin.

Loading, transport and unloading of oxygen

*   At the gas-exchange surface carbon dioxide is constantly being removed.
*   The pH is slightly raised due to the low concentration of carbon dioxide.
*   The higher pH changes the shape of haemoglobin into one that enables it to load oxygen readily.
*   This shape also increases the affinity of haemoglobin for oxygen, so it is not released while being transported in the blood to the tissues.
*   In the tissues, carbon dioxide is produced by respiring cells.
*   Carbon dioxide is acidic in solution, so the pH of the blood within the the tissues is lowered.
*   The lower pH changes the shape of haemoglobin into one with a lower affinity for oxygen.
*   Haemoglobin releases its oxygen into the respiring tissues.

The above process is a flexible way of ensuring that there is always sufficient oxygen for respiring tissues. The more active a tissue, the more oxygen is unloaded. This works as follows:

The higher the rate of respiration → the more carbon dioxide the tissues produce → the lower the pH → the greater the haemoglobin shape change → the more readily oxygen is unloaded → the more oxygen is available for respiration.

In humans, haemoglobin normally becomes saturated with oxygen as it passes through the lungs. In practice not all haemoglobin molecules are loaded with their maximum four oxygen molecules. As a consequence, the overall saturation of haemoglobin at atmospheric pressure is normally around 97%. When this haemoglobin reaches a tissue with a low respiratory rate, only one of these molecules will normally be released. The blood returning to the lungs will therefore contain haemoglobin that is still 75 per cent saturated with oxygen. If a tissue is very active, for example, an exercising muscle, then three oxygen molecules will usually be unloaded from each haemoglobin molecule. These events are shown in Figure 3.

Different species have different types of haemoglobin, each with its own different oxygen dissociation curve. These different types have evolved within species as adaptations to different environments and conditions. For example, species of animals that live in an environment with a lower partial pressure of oxygen have evolved haemoglobin that has a higher affinity for oxygen than the haemoglobin of animals that live where the partial pressure of oxygen is higher.

Take for example the lugworm, an animal that lives on the seashore.

The lugworm is not very active, spending almost all its life in a U-shaped burrow. Most of the time the lugworm is covered by sea water, which it circulates through its burrow. Oxygen diffuses into the lugworm's blood from the water and it uses haemoglobin to transport oxygen to its tissues.

When the tide goes out, the lugworm can no longer circulate a fresh supply of oxygenated water through its burrow. As a result, the water in the burrow contains progressively less oxygen as the lugworm uses it up. The lugworm has to extract as much oxygen as possible from the water in the burrow if it is to survive until the tide covers it again. Figure 4 shows the oxygen dissociation curve of lugworm haemoglobin compared to that of adult human haemoglobin.

The dissociation curve is shifted far to the left of that of a human. This means that the haemoglobin of the lugworm is fully loaded with oxygen even when there is little available in its environment.

Another example is the llama. It is an animal that lives at high altitudes. At these altitudes the atmospheric pressure is lower and so the partial pressure of oxygen is also lower. It is therefore difficult to load haemoglobin with oxygen. Llamas also have a type of haemoglobin that has a higher affinity for oxygen than human haemoglobin. In other words it is shifted to the left of that of human haemoglobin.

Key
— Haemoglobin molecule that is fully loaded with oxygen in the lungs
— Haemoglobin molecule in a resting tissue unloads only 25% of its oxygen
— Haemoglobin molecule in an active tissue unloads 75% of its oxygen

▲ **Figure 3** *The loading and unloading of a haemoglobin with oxygen*

Maths link √x̄

MS 3.1 and 3.4, see Chapter 22.

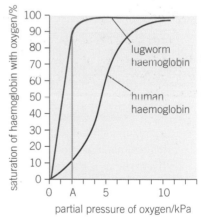

▲ **Figure 4** *Comparison of the oxygen dissociation curves of lugworm and human haemoglobin*

▲ **Figure 5** *Three lugworms lying on sand (left); lugworm casts at the entrances to their burrows (right)*

Summary questions √x̄

1 Study Figure 3 on the previous page and answer the following questions:

   a State at what partial pressure of oxygen the haemoglobin is 50% saturated with oxygen.

   b Determine the percentage saturation of haemoglobin with oxygen when the partial pressure of oxygen is 9 kPa.

   c In an exercising muscle the partial pressure of oxygen is 4 kPa while in the lungs it is 12 kPa. Calculate the percentage of the oxyhaemoglobin from the lungs that will have released its oxygen to an exercising muscle.

2 a Describe the effect of increased carbon dioxide concentration on oxygen dissociation.

   b State how this changes the saturation of haemoglobin with oxygen.

3 A rise in temperature shifts the oxygen dissociation curve to the right. Suggest how this enables an exercising muscle to work more efficiently.

4 In Figure 4, line A is drawn at a partial pressure of oxygen of 2 kPa. This is the partial pressure of oxygen found in lugworm burrows after the sea no longer covers them. Use figures from the graph to explain why a lugworm can survive at these concentrations of oxygen while a human could not.

5 Haemoglobin usually loads oxygen less readily when the concentration of carbon dioxide is high (the Bohr effect). The haemoglobin of lugworms does not exhibit this effect. Explain why to do so could be harmful.

6 In terms of obtaining oxygen, suggest a reason why lugworms are not found higher up the seashore.

Activity counts

Flight in birds and swimming in fish are both energy-demanding processes. The muscles that move a bird's wings are powerful and require a lot of oxygen to enable them to respire at a sufficient rate to keep the body airborne. Flight muscles have a very high metabolic rate and, during flight, much of the blood pumped by the heart goes to these muscles. While birds use a great deal of energy opposing gravity in a medium that gives little support, fish have a different problem. They expend considerable energy swimming in a medium that is very dense and therefore difficult to move through.

1 Suggest whether the oxygen dissociation curve of a pigeon is shifted to the right or left of the curve for a human. Explain your answer.

2 The mackerel is a type of fish that swims freely in the surface waters of the sea. These fish rely on their ability to swim very fast in order to escape from predators. The plaice is a marine fish that uses a different strategy. These fish spend much of their lives stationary or moving very slowly on the sea bed, where they are camouflaged by their skin colour. The two fish are of relatively similar mass. Sketch a graph to show what you would expect to be the relative positions of the oxygen dissociation curves of these two fish.

▲ **Figure 6** *Mackerel (top) live in surface waters and swim rapidly. Plaice (bottom) live on the sea bed and move very slowly*

Size matters $\sqrt{x}$

Mice are small mammals and therefore have a large surface area to volume ratio. As a result they tend to lose heat rapidly when the environmental temperature is lower than their body temperature. Figure 7 shows the oxygen dissociation curve for the haemoglobin of a mouse compared to that of adult human haemoglobin.

▲ **Figure 7** *Oxygen dissociation curves of mouse and human haemoglobin*

1 The partial pressure of oxygen at which haemoglobin is 50 per cent saturated is known as the unloading pressure. Calculate the difference between the unloading pressure of human haemoglobin and that of mouse haemoglobin.

2 The oxygen dissociation curve of the mouse is shifted to the right of that for a human.
   a Explain what difference this makes to the way oxygen is unloaded from mouse haemoglobin compared to human haemoglobin.

   b Suggest an advantage this has for the maintenance of body temperature in mice.

   c The position of the oxygen dissociation curve for a mouse means that its haemoglobin loads oxygen less readily than human haemoglobin. Given that the partial pressure of oxygen in air is normally 21 kPa, use the graph to explain why this is of no disadvantage to the mouse.

3 Sketch a graph to show the shapes and relative positions of the oxygen dissociation curves of the following mammals:
   a a human

   b an elephant

   c a shrew.

4 Ice fish live in the Antarctic and are the only vertebrates to completely lack haemoglobin. Suggest **one** reason why they can survive in the seas around Antarctica without haemoglobin in their blood.

Learning objectives

→ Explain why large organisms move substances around their bodies.

→ Describe the features of the transport systems of large organisms.

→ Describe the circulatory system of insects.

→ Describe the circulatory system of fish.

→ Describe the circulatory system of mammals.

→ Explain the relative efficiency of different circulatory systems.

Specification reference: 3.3.4.1

Synoptic link

A useful starting point to this topic would be to revise information on diffusion (Topic 4.2) and surface area to volume ratios (Topic 6.1).

▲ **Figure 1** *Large organisms require a transport system to take materials from exchange surfaces to the cells that need them*

Diffusion is fast enough for transport over short distances (see Topic 4.2). The efficient supply of materials over larger distances requires a mass transport system.

Why large organisms have a transport system

All organisms exchange materials between themselves and their environment. We have seen that in small organisms this exchange takes place over the surface of the body (see Topic 6.2). However, with increasing size, the surface area to volume ratio decreases to a point where the needs of the organism cannot be met by the body surface alone (see Topic 6.1). Specialist exchange surfaces are therefore required to absorb nutrients and respiratory gases, and remove excretory products. These exchange surfaces are located in specific regions of the organism. A transport system is required to take materials from cells to exchange surfaces and from exchange surfaces to cells. Materials have to be transported between exchange surfaces and the environment. They also need to be transported between different parts of the organism. As organisms have evolved into larger and more complex structures, the tissues and organs of which they are made have become more specialised and dependent upon one another. This makes a transport system all the more essential.

Whether or not there is a specialised transport medium, and whether or not it is circulated by a pump, depends on two factors:

*   the surface area to volume ratio,
*   how active the organism is.

The lower the surface area to volume ratio, and the more active the organism, the greater is the need for a specialised transport system with a pump.

Features of transport systems

Any large organism encounters the same problems in transporting materials within itself. Not surprisingly, the transport systems of many organisms have many common features:

*   A suitable medium in which to carry materials, for example blood. This is normally a liquid based on water because water readily dissolves substances and can be moved around easily, but can be a gas such as air breathed in and out of the lungs.

*   A form of mass transport in which the transport medium is moved around in bulk over large distances – more rapid than diffusion.

*   A closed system of tubular vessels that contains the transport medium and forms a branching network to distribute it to all parts of the organism.

*   A mechanism for moving the transport medium within vessels. This requires a pressure difference between one part of the system and another.

It is achieved in two main ways:

a   Animals use muscular contraction either of the body muscles or of a specialised pumping organ, such as the heart (see Topic 7.4).

b Plants rely on natural, passive processes such as the evaporation of water (see Topic 7.8).

- A mechanism to maintain the mass flow movement in one direction, for example, valves.
- A means of controlling the flow of the transport medium to suit the changing needs of different parts of the organism.
- A mechanism for the mass flow of water or gases, for example, intercostal muscles and diaphragm during breathing in mammals.

Circulatory systems in mammals

Mammals have **a closed, double circulatory system** in which blood is confined to vessels and passes twice through the heart for each complete circuit of the body (Figure 2). This is because, when blood is passed through the lungs, its pressure is reduced. If it were to pass immediately to the rest of the body its low pressure would make circulation very slow. Blood is therefore returned to the heart to boost its pressure before being circulated to the rest of the tissues. As a result, substances are delivered to the rest of the body quickly, which is necessary as mammals have a high body temperature and hence a high rate of **metabolism**. The vessels that make up the circulatory system of a mammal are divided into three types: arteries, veins and capillaries. We will look in more detail at these in Topic 7.6.

The arrangement of the main arteries and veins that make up the circulatory system of a mammal is shown in Figure 3.

Although a transport system is used to move substances longer distances, the final part of the journey to cells is by diffusion. The final exchange from blood vessels into cells is rapid because it takes place over a large surface area, across short distances and there is a steep diffusion gradient.

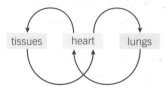

▲ **Figure 2** *Double circulation of a mammal*

▲ **Figure 3** *Plan of the mammalian circulatory system*

Summary questions

1 Name the blood vessel in each of the following descriptions:

   a joins the right ventricle of the heart to the capillaries of the lungs

   b carries oxygenated blood away from the heart

   c carries deoxygenated blood away from the kidney

   d the first main blood vessel that an oxygen molecule reaches after being absorbed from an alveolus

   e has the highest blood pressure.

2 State **two** factors that make it more likely that an organism will have a circulatory pump such as the heart.

3 State the main advantage of the double circulation found in mammals.

The heart is a muscular organ that lies in the thoracic cavity behind the sternum (breastbone). It operates continuously and tirelessly throughout the life of an organism.

Structure of the human heart

The human heart is really two separate pumps lying side by side. The pump on the left deals with oxygenated blood from the lungs, while the one on the right deals with deoxygenated blood from the body. Each pump has two chambers:

- The **atrium** is thin-walled and elastic and stretches as it collects blood.
- The **ventricle** has a much thicker muscular wall as it has to contract strongly to pump blood some distance, either to the lungs or to the rest of the body.

Why have two separate pumps? Why not just pump the blood through the lungs to collect oxygen and then straight to the rest of the body before returning it to the heart? The problem with such a system is that the blood has to pass through tiny capillaries in the lungs in order to present a large surface area for the exchange of gases (see Topic 6.8). In doing so, there is a very large drop in pressure and so the blood flow to the rest of the body would be very slow. This drop in pressure is illustrated in Figure 1. Mammals therefore have a system in which the blood is returned to the heart to increase its pressure before it is distributed to the rest of the body. It is essential to keep the oxygenated blood in the pump on the left side separate from the deoxygenated blood in the pump on the right.

The right ventricle pumps blood only to the lungs, and it has a thinner muscular wall than the left ventricle. The left ventricle, in contrast, has a thick muscular wall, enabling it to contract to create enough pressure to pump blood to the rest of the body. Although the two sides of the heart are separate pumps and, after birth, there is no mixing of the blood in each of them, they nevertheless pump in time with each other. Both atria contract together and then both ventricles contract together, pumping the same volume of blood.

Between each atrium and ventricle are valves that prevent the backflow of blood into the atria when the ventricles contract. There are two valves:

- the **left atrioventricular (bicuspid) valve**
- the **right atrioventricular (tricuspid) valve**

Each of the four chambers of the heart is connected to large blood vessels that carry blood towards or away from the heart. The ventricles pump blood away from the heart and into the arteries. The atria receive blood from the veins.

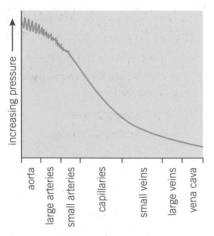

▲ **Figure 1** *Pressure changes in blood vessels*

Vessels connecting the heart to the lungs are called **pulmonary** vessels. The vessels connected to the four chambers are therefore as follows:

- The **aorta** is connected to the left ventricle and carries oxygenated blood to all parts of the body except the lungs.
- The **vena cava** is connected to the right atrium and brings deoxygenated blood back from the tissues of the body (except the lungs).
- The **pulmonary artery** is connected to the right ventricle and carries deoxygenated blood to the lungs, where its oxygen is replenished and its carbon dioxide is removed. Unusually for an artery, it carries deoxygenated blood.
- The **pulmonary vein** is connected to the left atrium and brings oxygenated blood back from the lungs. Unusually for a vein, it carries oxygenated blood.

The structure of the heart and its associated blood vessels is shown in Figure 2.

▲ Figure 2 Section through the human heart

Supplying the heart muscle with oxygen

Although oxygenated blood passes through the left side of the heart, the heart does not use this oxygen to meet its own great respiratory needs. Instead, the heart muscle is supplied by its own blood vessels, called the **coronary arteries**, which branch off the aorta shortly after it leaves the heart. Blockage of these arteries, for example by a blood clot, leads to **myocardial infarction**, or heart attack, because an area of the heart muscle is deprived of blood and therefore oxygen also. The muscle cells in this region are unable to respire (aerobically) and so die.

Hint

An easy way to recall which heart chambers are attached to which type of blood vessel is to remember that A and V always go together. Hence: Atria link to Veins and Arteries link to Ventricles.

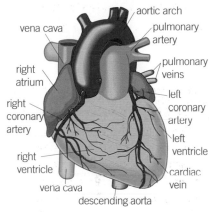

▲ Figure 3 External appearance of the human heart showing the blood supply to the heart muscle

Summary questions

1 Name the blood vessel that supplies the heart muscle with oxygenated blood.

2 State whether the blood in each of the following structures is oxygenated or deoxygenated:
   a vena cava
   b pulmonary artery
   c left atrium.

3 List the correct sequence of four main blood vessels and four heart chambers that a red blood cell passes through on its journey from the lungs, though the heart and body, and back again to the lungs.

4 Suggest why it is important to prevent mixing of the blood in the two sides of the heart.

Risk factors associated with cardiovascular disease

There are a number of factors that separately increase the risk of an individual suffering from cardiovascular disease. When combined together, four or five of these factors produce a disproportionately greater risk (Figure 4). These risk factors include the following.

▲ **Figure 4** *The combined impact of six risk factors on the likelihood of a 70-year-old man experiencing a stroke in the next ten years*

Smoking

Smokers are between two and six times more likely to suffer from heart disease than non-smokers. Giving up smoking is the single most effective way of increasing life expectancy. There are two main constituents of tobacco smoke that increase the likelihood of heart disease:

- **Carbon monoxide** combines easily, but irreversibly, with the haemoglobin in red blood cells to form carboxyhaemoglobin. It thereby reduces the oxygen-carrying capacity of the blood. To supply the equivalent quantity of oxygen to the tissues, the heart works harder. This can lead to raised blood pressure that increases the risk of coronary heart disease and strokes. In addition, the reduction in the oxygen-carrying capacity of the blood means that it may be insufficient to supply the heart muscle during exercise. This leads to chest pain (angina) or, in severe cases, a myocardial infarction (heart attack).

- **Nicotine** stimulates the production of the hormone adrenaline, which increases heart rate and raises blood pressure. As a consequence there is a greater risk of smokers suffering coronary heart disease or a stroke. Nicotine also makes the platelets in the blood more

'sticky', and this leads to a higher risk of thrombosis and hence of strokes or myocardial infarction.

High blood pressure

If your genes cause you to have a high blood pressure, altering your lifestyle will not change this fact. Lifestyle factors such as excessive prolonged stress, certain diets and lack of exercise, increase the risk of high blood pressure. These are factors over which the individual can exert control. High blood pressure increases the risk of heart disease for the following reasons:

- As there is already a higher pressure in the arteries, the heart must work harder to pump blood into them and is therefore more prone to failure.

- Higher blood pressure within the arteries means that they are more likely to develop an aneurysm (weakening of the wall) and burst, causing haemorrhage.

- To resist the higher pressure within them, the walls of the arteries tend to become thickened and may harden, restricting the flow of blood.

> ### Hint
>
> Always remember that risk factors increase the *probability* of getting heart disease, but they do not mean that someone will certainly get it. Heavy smokers, with high blood pressure and high blood cholesterol, may never develop heart disease, they are just more likely to (see Figure 5).

Blood cholesterol

Cholesterol is an essential component of membranes. As such, it is an essential biological molecule which must be transported in the blood. It is carried in the plasma as tiny spheres of lipoproteins (lipid and protein). There are two main types:

- **high-density lipoproteins (HDLs)**, which remove cholesterol from tissues and transport it to the liver for excretion. They help protect arteries against heart disease

- **low-density lipoproteins (LDLs)**, which transport cholesterol from the liver to the tissues, including the artery walls, which they infiltrate, leading to the development of atheroma, which may lead to heart disease.

Diet

There are a number of aspects of diet that increase the risk of heart disease, both directly and indirectly:

- **High levels of salt** raise blood pressure.
- **High levels of saturated fat** increase low-density lipoprotein levels and hence blood cholesterol concentration.

By contrast, foods that act as **antioxidants**, for example, vitamin C, reduce the risk of heart disease, and so does non-starch polysaccharide (dietary fibre).

A calculated risk

Figure 5 shows the effect of three of the above risk factors on the chance of heart attack in American men. Study the data and answer the questions:

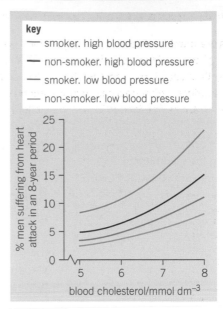

▲ **Figure 5** *Effects of blood pressure, smoking and blood cholesterol on the risk of heart attack in American men*

1 A smoker with high blood pressure wishes to reduce his risk of heart attack. If he could only alter one factor, would he be better giving up smoking or reducing his blood pressure? Explain your answer.

2 A non-smoker with high blood pressure has a blood cholesterol level of 5 mmol dm^{-3}. Over a period of 3 years this concentration increases to 8 mmol dm^{-3}. Calculate how many times greater his risk of heart disease is. Show your working.

3 Two non-smoking men with low blood pressure both have a blood cholesterol level of 5 mmol dm^{-3}. One of them starts to smoke and the blood cholesterol level of the other increases to 7 mmol dm^{-3}. State which man is now at the greater risk from heart disease. Explain your answer.

The heart undergoes a sequence of events that is repeated in humans around 70 times each minute when at rest. This is known as the **cardiac cycle**. There are two phases to the beating of the heart: contraction (systole) and relaxation (diastole). Contraction occurs separately in the ventricles and the atria and is therefore described in two stages. For some of the time, relaxation takes place simultaneously in all chambers of the heart and is therefore treated as a single phase in the account below, which is illustrated in Figure 1.

Relaxation of the heart (diastole)

Blood returns to the atria of the heart through the pulmonary vein (from the lungs) and the vena cava (from the body). As the atria fill, the pressure in them rises. When this pressure exceeds that in the ventricles, the atrioventricular valves open allowing the blood to pass into the ventricles. The passage of blood is aided by gravity. The muscular walls of both the atria and ventricles are relaxed at this stage. The relaxation of the ventricle walls causes them to recoil and reduces the pressure within the ventricle. This causes the pressure to be lower than that in the aorta and the pulmonary artery, and so the semi-lunar valves in the aorta and the pulmonary artery close, accompanied by the characteristic 'dub' sound of the heart beat.

1. Blood enters atria and ventricles from pulmonary veins and vena cava.

semi-lunar valves closed

left and right atrioventricular valves open

relaxation of ventricles allows blood to enter from atria

Relaxation of heart (diastole)
Atria are relaxed and fill with blood. Ventricles are also relaxed.

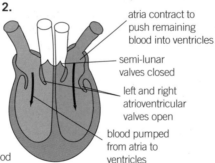

2.

atria contract to push remaining blood into ventricles

semi-lunar valves closed

left and right atrioventricular valves open

blood pumped from atria to ventricles

Contraction of atria (atrial systole)
Atria contract, pushing blood into the ventricles. Ventricles remain relaxed.

3. Blood pumped into pulmonary arteries and the aorta.

semi-lunar valves open

left and right atrioventricular valves closed

ventricles contract and walls thicken

Contraction of ventricles (ventricular systole)
Atria relax. Ventricles contract, pushing blood away from heart through pulmonary arteries and the aorta.

▲ **Figure 1** *The cardiac cycle*

Link

A level students will also learn about how the cardiac cycle is controlled in Topic 14.5 Control of heart rate.

Contraction of the atria (atrial systole)

The contraction of the atrial walls, along with the recoil of the relaxed ventricle walls, forces the remaining blood into the ventricles from the atria. Throughout this stage the muscle of the ventricle walls remains relaxed.

Contraction of the ventricles (ventricular systole)

After a short delay to allow the ventricles to fill with blood, their walls contract simultaneously. This increases the blood pressure within them, forcing shut the atrioventricular valves and preventing backflow of blood into the atria. The 'lub' sound of these valves closing is a characteristic of the heart beat. With the atrioventricular valves closed, the pressure in the ventricles rises further. Once it exceeds that in the aorta and pulmonary artery, blood is forced from the ventricles into these vessels. The ventricles have thick muscular walls which mean they contract forcefully. This creates the high pressure necessary to pump blood around the body. The thick wall of the left ventricle has to pump blood to the extremities of the body while the relatively thinner wall of the right ventricle, has to pump blood to the lungs.

Valves in the control of blood flow

Blood is kept flowing one direction through the heart and around the body by the pressure created by the heart muscle. Blood will always move from a region of higher pressure to one of lower pressure. There are, however, situations within the circulatory system when pressure differences would result in blood flowing in the opposite direction from that which is desirable. In these circumstances, valves are used to prevent any unwanted backflow of blood.

Valves in the cardiovascular system are designed so that they open whenever the difference in blood pressure either side of them favours the movement of blood in the required direction. When pressure differences are reversed, that is, when blood would tend to flow in the opposite direction to that which is desirable, the valves are designed to close. Examples of such valves include:

- **Atrioventricular valves** between the left atrium and ventricle and the right atrium and ventricle. These prevent backflow of blood when contraction of the ventricles means that ventricular pressure exceeds atrial pressure. Closure of these valves ensures that, when the ventricles contract, blood within them moves to the aorta and pulmonary artery rather than back to the atria.
- **Semi-lunar valves** in the aorta and pulmonary artery. These prevent backflow of blood into the ventricles when the pressure in these vessels exceeds that in the ventricles. This arises when the elastic walls of the vessels recoil increasing the pressure within them and when the ventricle walls relax reducing the pressure within the ventricles.
- **Pocket valves** in veins (see Topic 7.6) that occur throughout the venous system. These ensure that when the veins are squeezed, e.g. when skeletal muscles contract, blood flows back towards the heart rather than away from it.

The design of these valves is basically the same. They are made up of a number of flaps of tough, but flexible, fibrous tissue, which are cusp-shaped, in other words like deep bowls. When pressure is greater on the convex side of these cusps, rather than on the concave side, they move apart to let blood pass between the cusps. When pressure is greater on the concave side than on the convex side, blood collects within the 'bowl' of the cusps. This pushes them together to form a tight fit that prevents the passage of blood (Figure 2).

Pressure and volume changes of the heart

Mammals have a closed circulatory system, in other words the blood is confined to vessels, and this allows the pressure within them to be maintained and regulated. Figure 4 illustrates the pressure and volume changes, and associated valve movements, that take place in the heart during a typical cardiac cycle.

a *Valve open*

cusp of valve

higher blood pressure above valve forces it open

string-like tendons

pillar muscles

Lower blood pressure beneath valve

b *Valve closed*

lower blood pressure cannot open valve

higher blood pressure beneath valve forces it closed

cusps of valves fit closely together

▲ **Figure 2** *Action of the valves*

semi-lunar valves

▲ **Figure 3** *False-colour SEM of the semi-lunar valve of the aorta*

Maths link $\sqrt{x}$

MS 2.4, see Chapter 22.

Study tip

Do not mix up *cardiac output and pulmonary ventilation* (see Topic 6.7, The mechanism of breathing). While both measure volumes, the first involves blood and the heart (cardiac) and the second involves air and the lungs (pulmonary).

Cardiac output

Cardiac output is the volume of blood pumped by one ventricle of the heart in one minute. It is usually measured in $dm^3 min^{-1}$ and depends upon two factors:

- the heart rate (the rate at which the heart beats)
- the stroke volume (volume of blood pumped out at each beat).

Cardiac output = heart rate × stroke volume

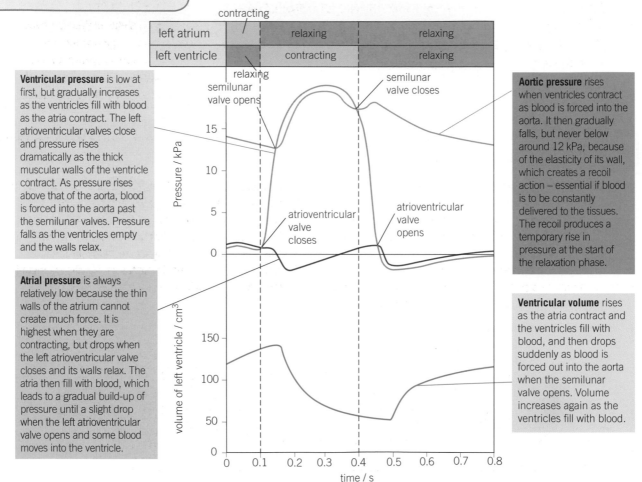

Ventricular pressure is low at first, but gradually increases as the ventricles fill with blood as the atria contract. The left atrioventricular valves close and pressure rises dramatically as the thick muscular walls of the ventricle contract. As pressure rises above that of the aorta, blood is forced into the aorta past the semilunar valves. Pressure falls as the ventricles empty and the walls relax.

Atrial pressure is always relatively low because the thin walls of the atrium cannot create much force. It is highest when they are contracting, but drops when the left atrioventricular valve closes and its walls relax. The atria then fill with blood, which leads to a gradual build-up of pressure until a slight drop when the left atrioventricular valve opens and some blood moves into the ventricle.

Aortic pressure rises when ventricles contract as blood is forced into the aorta. It then gradually falls, but never below around 12 kPa, because of the elasticity of its wall, which creates a recoil action – essential if blood is to be constantly delivered to the tissues. The recoil produces a temporary rise in pressure at the start of the relaxation phase.

Ventricular volume rises as the atria contract and the ventricles fill with blood, and then drops suddenly as blood is forced out into the aorta when the semilunar valve opens. Volume increases again as the ventricles fill with blood.

▲ **Figure 4** *Pressure and volume changes, and associated valve movements, in the left side of the heart during the cardiac cycle*

Summary questions

1 Name the chamber of the heart that produces the greatest pressure.

2 State whether each of the following statements is true or false.

  a The left and right ventricles contract together.

  b Veins have pocket valves.

  c Semi-lunar valves occur between the atria and ventricles.

  d ⓥ𝓍 If a person's cardiac output is 4.9 $dm^3 min^{-1}$ and their heart rate is 70 beats a minute, then their stroke volume is 0.7 dm^3.

3   In each case, state what is being described.

   a   On contraction it forces blood into the ventricles.

   b   The relaxation phase of the heart.

   c   Structures that prevent flow of blood from the aorta into the left ventricle.

4   After a period of training, the heart rate is often decreased when at rest although the cardiac output is unchanged. Suggest an explanation for this.

5   $\sqrt{x}$ Use Figure 4 to calculate the heart rate in beats per minute. Show your working.

6   $\sqrt{x}$ If a person has a stroke volume of $0.065\,dm^3$ and a cardiac output of $5.2\,dm^3\,min^{-1}$, calculate their heart rate.

Hint

Two facts will help you to understand the rather complex graph shown in Figure 4.

- Pressure and volume within a closed container are inversely related. When pressure increases, volume decreases, and vice versa.

- Blood, like all fluids, moves from a region where its pressure is greater to one where it is lower, i.e. it moves down a pressure gradient.

Electrocardiogram

During the cardiac cycle, the heart undergoes a series of electrical current changes. These are related to the waves of electrical activity created by the sinoatrial node and the heart's response to these. If displayed on a cathode ray oscilloscope, these changes can produce a trace known as an **electrocardiogram**. Doctors can use this trace to provide a picture of the heart's electrical activity and hence its health. In a normal electrocardiogram (ECG) there is a pattern of large peaks and small troughs that repeat identically at regular intervals. An ECG produced during a heart attack shows less pronounced peaks and larger troughs that are repeated in a similar, but not identical, way. During a condition called fibrillation, the heart muscle contracts in a disorganised way that is reflected in an irregular ECG.

a

b

c

▲ **Figure 6**   *Three different electrocardiogram (ECG) traces*

1   The three ECG traces shown in Figure 6 represent an ECG trace for:

- a normal heart

- a heart in fibrillation

- during a heart attack

Using the letters a, b and c, suggest which trace corresponds to which heart condition. Give reasons for your answers.

Maths link

Maths skill 1.3, see Chapter 22.

In Topic 7.3 we saw that, in larger organisms, materials are transported around the body by the blood that is confined to blood vessels. This topic looks in more detail at these vessels.

Structure of blood vessels

There are different types of blood vessels:

- **Arteries** carry blood away from the heart and into arterioles.
- **Arterioles** are smaller arteries that control blood flow from arteries to capillaries.
- **Capillaries** are tiny vessels that link arterioles to veins.
- **Veins** carry blood from capillaries back to the heart.

Arteries, arterioles and veins all have the same basic layered structure. From the outside inwards, these layers are:

- **tough fibrous outer layer** that resists pressure changes from both within and outside
- **muscle layer** that can contract and so control the flow of blood
- **elastic layer** that helps to maintain blood pressure by stretching and springing back (recoiling)
- **thin inner lining (endothelium)** that is smooth to reduce friction and thin to allow diffusion
- **lumen** that is not actually a layer but the central cavity of the blood vessel through which the blood flows.

What differs between each type of blood vessel is the relative proportions of each layer. These differences are shown in Figure 1. Arterioles are not included because they are similar to arteries. They differ from arteries in being smaller in diameter and having a relatively larger muscle layer and lumen. The differences in structure are related to the differences in the function that each type of vessel performs.

Study tip

Arteries, arterioles and veins carry out transport not exchange; only capillaries carry out exchange.

Study tip

The elastic tissue of arteries will stretch and recoil. It is not muscle and will not contract and relax.

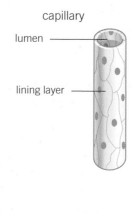

▲ **Figure 1** *Comparison of arteries, veins and capillaries*

Artery structure related to function

The function of arteries is to transport blood rapidly under high pressure from the heart to the tissues. Their structure is adapted to this function as follows:

- **The muscle layer is thick compared to veins.** This means smaller arteries can be constricted and dilated in order to control the volume of blood passing through them.

- **The elastic layer is relatively thick compared to veins** because it is important that blood pressure in arteries is kept high if blood is to reach the extremities of the body. The elastic wall is stretched at each beat of the heart (systole). It then springs back when the heart relaxes (diastole) in the same way as a stretched elastic band. This stretching and recoil action helps to maintain high pressure and smooth pressure surges created by the beating of the heart.

- **The overall thickness of the wall is great.** This also resists the vessel bursting under pressure.

- **There are no valves** (except in the arteries leaving the heart) because blood is under constant high pressure due to the heart pumping blood into the arteries. It therefore tends not to flow backwards.

▲ **Figure 2** *Artery (left) and vein (right)*

Arteriole structure related to function

Arterioles carry blood, under lower pressure than arteries, from arteries to capillaries. They also control the flow of blood between the two. Their structure is related to these functions as follows:

- **The muscle layer is relatively thicker than in arteries.** The contraction of this muscle layer allows constriction of the lumen of the arteriole. This restricts the flow of blood and so controls its movement into the capillaries that supply the tissues with blood.

- **The elastic layer is relatively thinner than in arteries** because blood pressure is lower.

Vein structure related to function

Veins transport blood slowly, under low pressure, from the capillaries in tissues to the heart. Their structure is related to this function as follows:

- **The muscle layer is relatively thin** compared to arteries because veins carry blood away from tissues and therefore their constriction and dilation cannot control the flow of blood to the tissues.

- **The elastic layer is relatively thin** compared to arteries because the low pressure of blood within the veins will not cause them to burst and pressure is too low to create a recoil action.

- **The overall thickness of the wall is small** because there is no need for a thick wall as the pressure within the veins is too low to create any risk of bursting. It also allows them to be flattened easily, aiding the flow of blood within them (see below).

- **There are valves at intervals throughout** to ensure that blood does not flow backwards, which it might otherwise do because the pressure is so low. When body muscles contract, veins are compressed, pressurising the blood within them. The valves ensure that this pressure directs the blood in one direction only: towards the heart (Figure 3).

wall of vein

valve flap

valve open

Blood flowing towards the heart passes easily through the valves.

blood flowing away from the heart pushes flaps of valve closed

valve closed

Blood flowing away from the heart pushes valves closed and so blood is prevented from flowing any further in this direction.

▲ **Figure 3** *Action of valves in veins in ensuring one-way flow of blood*

▲ **Figure 4** *False-colour SEM of a section through a capillary with red blood cells passing through it*

▲ **Figure 5** *Resin cast of a capillary network from the large intestine*

Capillary structure related to function

The function of capillaries (Figures 4 and 5) is to exchange metabolic materials such as oxygen, carbon dioxide and glucose between the blood and the cells of the body. The flow of blood in capillaries is much slower. This allows more time for the exchange of materials.

The structure of capillaries is related to their function as follows:

- **Their walls consist mostly of the lining layer**, making them extremely thin, so the distance over which diffusion takes place is short. This allows for rapid diffusion of materials between the blood and the cells.

- **They are numerous and highly branched**, thus providing a large surface area for exchange.

- **They have a narrow diameter** and so permeate tissues, which means that no cell is far from a capillary and there is a short diffusion pathway.

- **Their lumen is so narrow** that red blood cells are squeezed flat against the side of a capillary. This brings them even closer to the cells to which they supply oxygen. This again reduces the diffusion distance.

- **There are spaces between the lining (endothelial) cells** that allow white blood cells to escape in order to deal with infections within tissues.

Although capillaries are small, they cannot serve every single cell directly. Therefore the final journey of metabolic materials is made in a liquid solution that bathes the tissues. This liquid is called **tissue fluid**.

Tissue fluid and its formation

Tissue fluid is a watery liquid that contains glucose, amino acids, fatty acids, ions in solution and oxygen. Tissue fluid supplies all of these substances to the tissues. In return, it receives carbon dioxide and other waste materials from the tissues. Tissue fluid is therefore the means by which materials are exchanged between blood and cells and, as such, it bathes all the cells of the body. It is the immediate environment of cells and is, in effect, where they live. Tissue fluid is formed from blood plasma, and the composition of blood plasma is controlled by various homeostatic systems. As a result tissue fluid provides a mostly constant environment for the cells it surrounds.

Formation of tissue fluid

Blood pumped by the heart passes along arteries, then the narrower arterioles and, finally, the even narrower capillaries. Pumping by the heart creates a pressure, called **hydrostatic pressure**, at the arterial end of the capillaries. This hydrostatic pressure causes tissue fluid to move out of the blood plasma. The outward pressure is, however, opposed by two other forces:

- hydrostatic pressure of the tissue fluid outside the capillaries, which resists outward movement of liquid

- the lower **water potential** of the blood, due to the plasma proteins, that causes water to move back into the blood within the capillaries.

However, the combined effect of all these forces is to create an overall pressure that pushes tissue fluid out of the capillaries at the arterial end. This pressure is only enough to force small molecules out of the capillaries, leaving all cells and proteins in the blood because these are too large to cross the membranes. This type of filtration under pressure is called **ultrafiltration**.

Return of tissue fluid to the circulatory system

Once tissue fluid has exchanged metabolic materials with the cells it bathes, it is returned to the circulatory system. Most tissue fluid returns to the blood plasma directly via the capillaries. This return occurs as follows:

- The loss of the tissue fluid from the capillaries reduces the hydrostatic pressure inside them.

- As a result, by the time the blood has reached the venous end of the capillary network its hydrostatic pressure is usually lower than that of the tissue fluid outside it.

- Therefore tissue fluid is forced back into the capillaries by the higher hydrostatic pressure outside them.

- In addition, the plasma has lost water and still contains proteins. It therefore has a lower water potential than the tissue fluid.

- As a result, water leaves the tissue by osmosis down a water potential gradient.

The tissue fluid has lost much of its oxygen and nutrients by diffusion into the cells that it bathed, but it has gained carbon dioxide and waste materials in return. These events are summarised in Figure 6.

▲ **Figure 6** *Formation and return of tissue fluid*

Not all the tissue fluid can return to the capillaries; the remainder is carried back via the lymphatic system. This is a system of vessels that begin in the tissues. Initially they resemble capillaries (except that they have dead ends), but they gradually merge into larger vessels that form a network throughout the body. These larger vessels drain their contents back into the bloodstream via two ducts that join veins close to the heart.

▲ Figure 7 *Formation and return of tissue fluid to the bloodstream*

Maths link $\sqrt{x}$

MS 1.3, see Chapter 22.

The contents of the lymphatic system (lymph) are not moved by the pumping of the heart. Instead they are moved by:

- **hydrostatic pressure** of the tissue fluid that has left the capillaries
- **contraction of body muscles** that squeeze the lymph vessels – valves in the lymph vessels ensure that the fluid inside them moves away from the tissues in the direction of the heart.

A summary of the methods of tissue fluid formation and its return to the bloodstream is shown in Figure 7.

Summary questions

1 State **one** advantage of having:

   a thick elastic tissue in the walls of arteries

   b relatively thick muscle walls in arterioles

   c valves in veins

   d only a lining layer in capillaries.

2 Table 1 shows the mean wall thickness of different blood vessels in a mammal. Suggest the letter that is most likely to refer to **a** the aorta, **b** a capillary, **c** a vein, **d** an arteriole and **e** the renal artery.

3 State what forces tissue fluid out of the blood plasma in capillaries and into the surrounding tissues.

4 List the **two** routes by which tissue fluid returns to the bloodstream.

▼ Table 1

Blood vessel	Mean wall thickness/mm
A	1.000
B	0.001
C	2.000
D	0.500
E	0.030

Blood flow in various blood vessels $\sqrt{x}$

The graph in Figure 8 shows certain features of the flow of blood from and to the heart through a variety of blood vessels.

1 Describe the changes in the rate of blood flow as blood passes from the aorta to the vena cava.

2 Explain why blood pressure in region A fluctuates up and down.

3 Explain why the rate of blood flow decreases between the aorta and capillaries.

4 Explain how the rate of blood flow in the capillaries increases the rate of exchange of metabolic materials.

5 Explain why the structure of capillaries increases the efficiency of the exchange of metabolic substances.

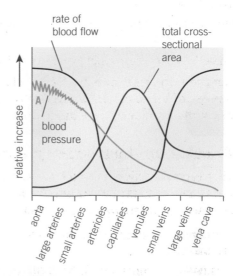

▲ Figure 8 *Flow of blood to and from the heart*

In plants water is absorbed by the roots through extensions called root hairs. In flowering plants the vast majority of water is transported through hollow, thick-walled tubes called **xylem vessels**. The main force that pulls water through the xylem vessels in the stem of a plant is the evaporation of water from leaves – a process called **transpiration**. The energy for this is supplied by the sun and the process is therefore passive. It is therefore logical to begin an explanation of how water moves through the xylem from the point where water evaporates from the surfaces of cells surrounding the stomatal air space and water vapour diffuses out of the stomatal pore.

Movement of water out through stomata

The humidity of the atmosphere is usually less than that of the air spaces next to the stomata (see Figure 2). As a result there is a water potential gradient from the air spaces through the stomata to the air. Provided the stomata are open, water vapour molecules diffuse out of the air spaces into the surrounding air. Water lost by diffusion from the air spaces is replaced by water evaporating from the cell walls of the surrounding mesophyll cells. By changing the size of the stomatal pores, plants can control their rate of transpiration.

Movement of water across the cells of a leaf

Water is lost from mesophyll cells by evaporation from their cell walls to the air spaces of the leaf. This is replaced by water reaching the mesophyll cells from the xylem either via cell walls or via the cytoplasm. In the case of the cytoplasmic route the water movement occurs because:

*   mesophyll cells lose water to the air spaces by evaporation due to heat supplied by the sun
*   these cells now have a lower **water potential** and so water enters by **osmosis** from neighbouring cells
*   the loss of water from these neighbouring cells lowers their water potential
*   they, in turn, take in water from their neighbours by osmosis.

In this way, a water potential gradient is established that pulls water from the xylem, across the leaf mesophyll, and finally out into the atmosphere. These events are summarised in Figure 2 on the next page.

Movement of water up the stem in the xylem

The main factor that is responsible for the movement of water up the xylem, from the roots to the leaves, is cohesion–tension. The movement of water up the stem occurs as follows:

*   Water evaporates from mesophyll cells due to heat from the sun leading to transpiration.
*   Water molecules form hydrogen bonds between one another and hence tend to stick together. This is known as **cohesion**.
*   Water forms a continuous, unbroken column across the mesophyll cells and down the xylem.

Learning objectives

→ Define what transpiration is.
→ Explain how water moves through the leaf.
→ Explain how water moves up the xylem.
→ Explain the cohesion–tension theory of water transport.

Specification reference: 3.3.4.2

▲ **Figure 1** *False-colour SEM showing hollow, tubular xylem vessels adapted to transport water*

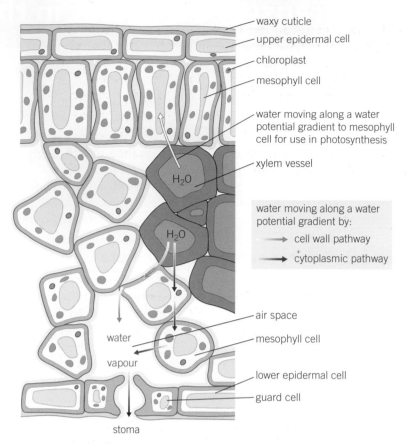

Link

The leaf is the site of photosynthesis and A level students will learn more about its role in this process in Topic 11.1 An overview of photosynthesis

waxy cuticle

upper epidermal cell

chloroplast

mesophyll cell

water moving along a water potential gradient to mesophyll cell for use in photosynthesis

xylem vessel

water moving along a water potential gradient by:

→ cell wall pathway

→ cytoplasmic pathway

air space

mesophyll cell

lower epidermal cell

guard cell

water vapour

stoma

▲ **Figure 2** *Movement of water across leaf*

▲ **Figure 3** *Section through a leaf showing the tissues involved in the movement of water*

- As water evaporates from the mesophyll cells in the leaf into the air spaces beneath the stomata, more molecules of water are drawn up behind it as a result of this cohesion.
- A column of water is therefore pulled up the xylem as a result of transpiration. This is called the **transpiration pull**.
- Transpiration pull puts the xylem under tension, that is, there is a negative pressure within the xylem, hence the name **cohesion–tension theory**.

Such is the force of the transpiration pull that it can easily raise water up the 100 m or more of the tallest trees. There are several pieces of evidence to support the cohesion–tension theory. These include:

- Change in the diameter of tree trunks according to the rate of transpiration. During the day, when transpiration is at its greatest, there is more tension (more negative pressure) in the xylem. This pulls the walls of the xylem vessels inwards and causes the trunk to shrink in diameter. At night, when transpiration is at its lowest, there is less tension in the xylem and so the diameter of the trunk increases.
- If a xylem vessel is broken and air enters it, the tree can no longer draw up water. This is because the continuous column of water is broken and so the water molecules can no longer stick together.
- When a xylem vessel is broken, water does not leak out, as would be the case if it were under pressure. Instead air is drawn in, which is consistent with it being under tension.

Transpiration pull is a passive process and therefore does not require metabolic energy to take place. Indeed, the xylem vessels through which the water passes are dead and so cannot actively move the water. Xylem vessels have no end walls which means that xylem forms a series of continuous, unbroken tubes from root to leaves, which is essential to the cohesion–tension theory of water flow up the stem. Energy is nevertheless needed to drive the process of transpiration. This energy is in the form of heat that evaporates water from the leaves and it ultimately comes from the sun.

Figure 4 summarises the movement of water from the soil, through the plant, and into the atmosphere.

▲ **Figure 4** *Summary of water transport through a plant*

 Hug a tree √x̄

If you put your arms around a suitably sized tree trunk in the middle of the day your fingers will just touch on the far side of the tree. Try to hug the same tree at night and your fingers will probably no longer meet. The graph in Figure 5 shows why. It shows the rate of water flow up a tree and the diameter of the tree trunk over a 24-hour period.

1 At what time of day is transpiration rate greatest? Explain your answer using information in Figure 5.
2 Describe the changes in the rate of flow of water during the 24-hour period.
3 Explain in terms of the cohesion–tension theory the changes in the rate of flow of water during the 24-hour period.
4 Explain the changes in the diameter of the tree trunk over the 24-hour period.
5 If the tree was sprayed with ammonium sulfamate, a herbicide that kills living cells, the rate of water flow would be unchanged. Explain why.

Summary question

1 State the most suitable word, or words, represented by **a–f** in the passage below.

Water leaves from the air spaces in a plant by a process called **a**. This takes place mainly through pores called **b** in the epidermis of the leaf. Water evaporates into the air spaces from mesophyll cells. As a result these cells have a **c** water potential and so draw water by **d** from neighbouring cells. In this way, a water potential gradient is set up that draws water from the xylem. Water is pulled up the xylem because water molecules stick together – a phenomenon called **e**. During the night the diameter of a tree trunk **f**.

Maths link √x̄

MS 1.3, see Chapter 22.

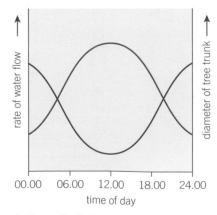

▲ **Figure 5** *Variation of rate of water flow and diameter of a tree trunk*

Measurement of water uptake using a potometer

It is almost impossible to measure transpiration because it is extremely difficult to condense and collect all the water vapour that leaves all the parts of a plant. What we can easily measure, however, is the amount of water that is taken up in a given time by a part of the plant such as a leafy shoot. About 99% of the water taken up by a plant is lost during transpiration, which means that the rate of uptake is almost the same as the rate at which transpiration is occurring. We can then measure water uptake by the same shoot under different conditions, e.g. various humidities, wind speeds or temperatures. In this way we get a reasonably accurate measure of the effects of these conditions on the rate of transpiration.

The rate of water loss in a plant can be measured using a potometer (Figure 6). The experiment is carried out in the following stages:

- A leafy shoot is cut under water. Care is taken not to get water on the leaves.

- The potometer is filled completely with water, making sure there are no air bubbles.

- Using a rubber tube, the leafy shoot is fitted to the potometer under water.

- The potometer is removed from under the water and all joints are sealed with waterproof jelly.

- An air bubble is introduced into the capillary tube.

- The distance moved by the air bubble in a given time is measured a number of times and the mean is calculated.

- Using this mean value, the volume of water lost is calculated.

- The volume of water lost against the time in minutes can be plotted on a graph.

- Once the air bubble nears the junction of the reservoir tube and the capillary tube, the tap on the reservoir is opened and the syringe is pushed down until the bubble is pushed back to the start of the scale on the capillary tube. Measurements then continue as before.

- The experiment can be repeated to compare the rates of water uptake under different conditions, for example at different temperatures, humidity, light intensity, or the differences in water uptake between different species under the same conditions.

1 From your knowledge of how water moves up the stem, suggest a reason why each of the following procedures is carried out:
   a The leafy shoot is cut under water rather than in the air.
   b All joints are sealed with waterproof jelly.
2 State what assumption must be made if a potometer is used to measure the rate of transpiration.
3 The volume of water taken up in a given time can be calculated using the formula $\pi r^2 l$ (where $\pi = 3.142$, r = radius of the capillary tube, and l = the distance moved by the air bubble). In an experiment the mean distance moved by the air bubble in a capillary tube of radius 0.5 mm during 1 min was 15.28 mm. Calculate the rate of water uptake in $cm^3\,h^{-1}$. Show your working.
4 If a potometer is used to compare the transpiration rates of two different species of plant, suggest one feature of both plant shoots that should, as far as possible, be kept the same.
5 Suggest reasons why the results obtained from a laboratory potometer experiment may not be representative of the transpiration rate of the same plant in the wild.

reservoir of water

syringe

cut end of shoot

tap

rubber tube

capillary tube   air bubble

millimetre scale

▲ **Figure 6** *A potometer*

Specialised plant cells

The root hair cell

Figure 7 shows the structure of a root hair cell. Each root hair is an extension of a root epidermal cell. Root hairs are the exchange surfaces in plants that are responsible for the absorption of water by osmosis and mineral ions by active transport.

1 State **two** features shown in Figure 7 that suit a root hair cell to its function of absorbing water and mineral ions.

2 Define osmosis.

3 Explain in terms of water potential how water might be absorbed into a root hair cell.

4 Suggest the name of an organelle that you might expect to occur in large numbers in a root hair cell, giving a reason for your answer.

Xylem vessels

Xylem vessels vary in appearance, depending on the type and amount of thickening of their cell walls. As they mature, their walls incorporate a substance called lignin and the cells die. The lignin often forms rings or spirals around the vessel. The structure of xylem vessels is shown in Figure 8.

5 The process of transporting water in plants in the transpiration stream involves water being pulled up the plant, which causes a negative pressure in the xylem vessels. Explain how xylem vessels are adapted to cope with this.

6 Name **two** other features shown in Figure 8 that suit xylem vessels to their function of transporting water up a plant.

7 Suggest one advantage of xylem vessels being dead cells in order to carry out their function effectively.

8 Suggest another possible feature of lignin, other than its mechanical strength, that would be useful in ensuring that water was carried up the plant.

9 The thickening of the cell wall in xylem vessels is often spiral. Suggest **three** advantages to the plant of having this arrangement rather than continuous thickening.

cellulose cell wall

cytoplasm

vacuole

nucleus

▲ **Figure 7** *A root hair cell (top); root hairs on radish seedlings are for the absorption of water and mineral ions (bottom)*

longitudinal section

transverse section

lignified wall of xylem vessel

spiral thickening in xylem vessel

hollow centre of xylem vessel

▲ **Figure 8** *Xylem vessels seen in longitudinal and transverse section*

The process by which organic molecules and some mineral ions are transported from one part of a plant to another is called **translocation**. In flowering plants, the tissue that transports biological molecules is called phloem. Phloem is made up of sieve tube elements, long thin structures arranged end to end. Their end walls are perforated to form sieve plates. Associated with the sieve tube elements are cells called companion cells. The structure of phloem is shown in Figure 1.

Having produced sugars during photosynthesis, the plant transports them from the sites of production, known as **sources**, to the places where they will be used directly or stored for future use – known as **sinks**. As sinks can be anywhere in a plant – sometimes above and sometimes below the source – it follows that the translocation of molecules in phloem can be in either direction. Organic molecules to be transported include sucrose and amino acids. The phloem also transports inorganic ions such as potassium, chloride, phosphate and magnesium ions.

Mechanism of translocation

It is accepted that materials are transported in the phloem and that the rate of movement is too fast to be explained by diffusion. What is in doubt is the precise mechanism by which translocation is achieved. Current thinking favours the **mass flow theory**, a theory that can be divided into three phases:

1. Transfer of sucrose into sieve elements from photosynthesising tissue

- Sucrose is manufactured from the products of photosynthesis in cells with chloroplasts.

- The sucrose diffuses down a concentration gradient by facilitated diffusion from the photosynthesising cells into companion cells.

- Hydrogen ions are actively transported from companion cells into the spaces within cell walls using ATP.

- These hydrogen ions then diffuse down a concentration gradient through carrier proteins into the sieve tube elements.

- Sucrose molecules are transported along with the hydrogen ions in a process known as co-transport (Topic 4.5). The protein carriers are therefore also known as **co-transport proteins**.

2. Mass flow of sucrose through sieve tube elements

Mass flow is the bulk movement of a substance through a given channel or area in a specified time. Mass flow of sucrose through sieve tube elements takes place as follows:

- The sucrose produced by photosynthesising cells (source) is actively transported into the sieve tubes as described above.

- This causes the sieve tubes to have a lower (more negative) water potential.

- As the xylem has a much higher (less negative) water potential (see Topic 7.7), water moves from the xylem into the sieve tubes by osmosis, creating a high hydrostatic pressure within them.

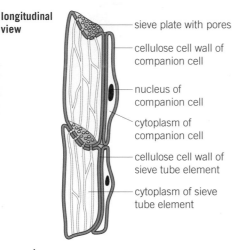

longitudinal view

- sieve plate with pores
- cellulose cell wall of companion cell
- nucleus of companion cell
- cytoplasm of companion cell
- cellulose cell wall of sieve tube element
- cytoplasm of sieve tube element

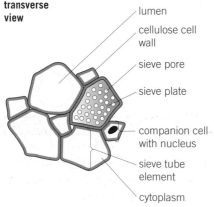

transverse view

- lumen
- cellulose cell wall
- sieve pore
- sieve plate
- companion cell with nucleus
- sieve tube element
- cytoplasm

▲ **Figure 1** *Phloem as seen under a light microscope*

- At the respiring cells (sink), sucrose is either used up during respiration or converted to starch for storage.
- These cells therefore have a low sucrose content and so sucrose is actively transported into them from the sieve tubes lowering their water potential.
- Due to this lowered water potential, water also moves into these respiring cells, from the sieve tubes, by osmosis.
- The hydrostatic pressure of the sieve tubes in this region is therefore lowered.
- As a result of water entering the sieve tube elements at the source and leaving at the sink, there is a high hydrostatic pressure at the source and a low one at the sink.
- There is therefore a mass flow of sucrose solution down this hydrostatic gradient in the sieve tubes.

▲ **Figure 2** *Colourised SEM of sieve plates*

While mass flow is a passive process, it occurs as a result of the active transport of sugars. Therefore the process as a whole is active which is why it is affected by, for example, temperature and metabolic poisons. A model of this theory is shown in Figure 3 and the evidence for and against the mass flow theory is listed in Table 1.

Cell A – has membrane permeable only to water. It contains a solution with a high concentration of sucrose. it represents the leaf of a plant.

glass tube linking cell A to cell B represents the phloem

direction of flow of sucrose solution

Cell B – has a membrane permeable only to water. It contains a solution with little sucrose. It represents respiring or storage regions of the plant, e.g., roots.

Water enters cell A by osmosis because it has a lower water potential than the water surrounding it.

direction of water flow

Water leaves because it is forced out due to the high pressure potential created in A.

glass tube linking the two vessels – represents the xylem

Provided sucrose is continually produced in A (leaf) and continually removed at B (e.g., root), the mass flow of sucrose from A to B continues.

▲ **Figure 3** *Model illustrating the movement of sucrose by mass flow in phloem*

▼ **Table 1** *Evidence for and against the mass flow theory*

Evidence supporting the mass flow hypothesis	Evidence questioning the mass flow hypothesis
<ul><li>there is a pressure within sieve tubes, as shown by sap being released when they are cut.</li><li>the concentration of sucrose is higher in leaves (source) than in roots (sink).</li><li>downward flow in the phloem occurs in daylight, but ceases when leaves are shaded, or at night.</li><li>increases in sucrose levels in the leaf are followed by similar increases in sucrose levels in the phloem a little later.</li><li>metabolic poisons and/or lack of oxygen inhibit translocation of sucrose in the phloem.</li><li>companion cells possess many mitochondria and readily produce ATP.</li></ul>	<ul><li>the function of the sieve plates is unclear, as they would seem to hinder mass flow (it has been suggested that they may have a structural function, helping to prevent the tubes from bursting under pressure).</li><li>not all solutes move at the same speed – they should do so if movement is by mass flow.</li><li>sucrose is delivered at more or less the same rate to all regions, rather than going more quickly to the ones with the lowest sucrose concentration, which the mass flow theory would suggest.</li></ul>

3. Transfer of sucrose from the sieve tube elements into storage or other sink cells

The sucrose is actively transported by companion cells, out of the sieve tubes and into the sink cells.

The process of translocation of sucrose in phloem is illustrated in Figure 4.

▲ **Figure 4** *Movement of sucrose from source to sink through the phloem of a plant*

Summary questions

State the most suitable word or words represented by the letters **a**–**k** in the passage below.

Transport of sucrose in plants occurs in the tissue called **a**, from places where it is produced, known as **b**, to places where it is used up or stored, called **c**. One theory of how it is translocated is called the **d** theory. Initially the sucrose is transferred into **e** elements by the process of **f**. The sucrose is produced by **g** cells that therefore have a **h** water potential due to this sucrose. Water therefore moves into them from the nearby **i** tissue that has a **j** water potential. The opposite occurs in those cells (sinks) using up sucrose, and water therefore leaves them by the process of **k**.

We have seen that water is carried in xylem while sugars and amino acids are carried in phloem. How can we be sure that this is the case? This topic looks at some of the evidence and how it is obtained.

Ringing experiments

Woody stems have an outer protective layer of bark on the inside of which is a layer of phloem that extends all round the stem. Inside the phloem layer is xylem (Figure 1).

At the start of a ringing experiment, a section of the outer layers (protective layer and phloem) is removed around the complete circumference of a woody stem while it is still attached to the rest of the plant. After a period of time, the region of the stem immediately above the missing ring of tissue is seen to swell (Figure 1). Samples of the liquid that has accumulated in this swollen region are found to be rich in sugars and other dissolved organic substances. Some non-photosynthetic tissues in the region below the ring (towards the roots) are found to wither and die, while those above the ring continue to grow.

These observations suggest that removing the phloem around the stem has led to:

- the sugars of the phloem accumulating above the ring, leading to swelling in this region
- the interruption of flow of sugars to the region below the ring and the death of tissues in this region.

The conclusion drawn from this type of ringing experiment is that phloem, rather than xylem, is the tissue responsible for translocating sugars in plants. As the ring of tissue removed had not extended into the xylem, its continuity had not been broken. If it were the tissue responsible for translocating sugars you would not have expected sugars to accumulate above the ring nor tissues below it to die.

Tracer experiments

Radioactive **isotopes** are useful for tracing the movement of substances in plants. For example the isotope ^{14}C can be used to make radioactively labelled carbon dioxide ($^{14}CO_2$). If a plant is then grown in an atmosphere containing $^{14}CO_2$, the ^{14}C isotope will be incorporated into the sugars produced during photosynthesis. These radioactive sugars can then be traced as they move within the plant using autoradiography. In our example, this involves taking thin cross-sections of the plant stem and placing them on a piece of X-ray film. The film becomes blackened where it has been exposed to the radiation produced by the ^{14}C in the sugars. The blackened regions are found to correspond to where phloem tissue is in the stem. As the other tissues do not blacken the film, it follows that they do not carry sugars and that phloem alone is responsible for their translocation.

Learning outcomes

→ Describe the use of ringing experiments to investigate transport in plants.

→ Describe the use of tracer experiments to investigate transport in plants.

→ Explain the evidence that translocation of organic molecules occurs in the phloem.

Specification reference: 3.3.4.2

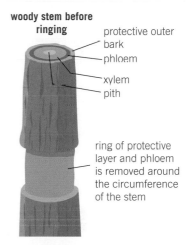

woody stem before ringing

protective outer bark
phloem
xylem
pith

ring of protective layer and phloem is removed around the circumference of the stem

woody stem after ringing

to the leaves

stem immediately above the ring swells due to accumulation of sugars

to the roots

▲ **Figure 1** *Ringing of a woody stem and its results*

Evidence that translocation of organic molecules occurs in phloem

The techniques described are only two of the pieces of evidence that support the view that translocation of organic molecules such as sugars takes place in phloem. A more complete list of evidence is given below.

- When phloem is cut, a solution of organic molecules flow out.

- Plants provided with radioactive carbon dioxide can be shown to have radioactively labelled carbon in phloem after a short time.

- Aphids are a type of insect that feed on plants. They have needle-like mouthparts which penetrate the phloem. They can therefore be used to extract the contents of the sieve tubes. These contents show daily variations in the sucrose content of leaves that are mirrored a little later by identical changes in the sucrose content of the phloem Figure 2.

- The removal of a ring of phloem from around the whole circumference of a stem leads to the accumulation of sugars above the ring and their disappearance from below it.

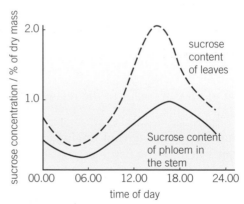

▲ **Figure 2** *Diurnal variation in sucrose content of leaves and phloem*

Using radioactive tracers to find which tissue transports minerals ⚠ √x̄

In an experiment to determine whether minerals are transported in xylem or phloem, a plant was grown in a pot. One branch (Y) on each plant had a 225 mm section of its phloem and xylem separated by inserting strips of impervious wax paper between them as shown in Figure 3. A 225 mm section of another branch (X) of the same plant that had *not* had its xylem and phloem separated by wax paper was used as a control.

Summary questions

1 **a** Suggest what difference there would be between the results of a ringing experiment carried out in the summer and one carried out in the winter.

   b Explain the reason for the difference you have suggested.

2 Squirrels sometimes strip sections of bark from around branches. Explain why these branches might die.

3 Suggest how a branch with a complete ring of phloem stripped from it by squirrels might still survive.

4 Explain why squirrels are unlikely to cause the death of a large mature tree by stripping some bark from its trunk.

5 Study Figure 2 and suggest an explanation for:

   a Why there is a time lag between the maximum sucrose content in the leaves and the phloem in the stem and roots.

   b Why the sucrose concentration in the phloem in the stem is lower than that in the leaves.

▲ **Figure 3** *Portion of branch of plant showing how xylem and phloem are separated by wax paper and where samples were taken*

The plant was watered with a solution that contained radioactive potassium (^{42}K). After 5 hours absorbing radioactive ^{42}K, sections of the experimental branch were tested for the quantity of ^{42}K in the xylem and phloem. The sections tested are labelled on Figure 3.

The equivalent positions on the control branch were also tested for ^{42}K.

The results are shown in Table 1.

▼ **Table 1**

| Section of stem | Percentage of total ^{42}K | | | |
| | Branch X (phloem and xylem together) | | Branch Y (phloem and xylem separated) | |
	Phloem	Xylem	Phloem	Xylem
1	53	47	53	47
2			09	91
3	56	44	01	99
4			15	85
5	52	48	59	41

1 Draw a conclusion from the data in the table.
2 Justify your conclusion with supporting evidence.
3 Explain the fact that the levels of ^{42}K are similar in the xylem and phloem of branch Y in sections 1 and 5.
4 The control (branch X) was an identical length of a different branch that had not had wax paper placed between the xylem and phloem. Suggest a way in which this control could have been improved. Explain why the change you suggest is an improvement.

1   Lugworms live in mud where the partial pressure of oxygen is low. The graph shows oxygen dissociation curves for a lugworm and for a human.

(a)   Explain the advantage to the lugworm of having haemoglobin with a dissociation curve in the position shown.   (*2 marks*)
(b)   In humans, substances move out of the capillaries to form tissue fluid. Describe how this tissue fluid is returned to the circulatory system.   (*3 marks*)
                                                                                                    AQA June 2011

2   The table shows pressure changes in the left side of the heart during one cardiac cycle.

Time / s	Blood pressure / kPa	
	Left atrium	Left ventricle
0.1	0.7	0.3
0.2	1.0	2.0
0.3	0.1	12.5
0.4	0.2	15.3
0.5	1.0	4.5
0.6	0.5	1.0
0.7	0.6	0.3
0.8	0.7	0.3

(a)   Between which times is the valve between the atrium and the ventricle closed? Explain your answer.   (*2 marks*)
(b)   The maximum pressure in the ventricle is much higher than that in the atrium. Explain what causes this.   (*2 marks*)
(c)   Use the information in the table to calculate the heart rate in beats per minute.   (*1 mark*)
                                                                                                    AQA June 2011

3   (a)   (i)   An arteriole is described as an organ. Explain why.   (*1 mark*)
          (ii)   An arteriole contains muscle fibres. Explain how these muscle fibres reduce blood flow to capillaries.   (*2 marks*)
    (b)   (i)   A capillary has a thin wall. This leads to rapid exchange of substances between the blood and tissue fluid. Explain why.   (*1 mark*)
          (ii)   Blood flow in capillaries is slow. Give the advantage of this.   (*1 mark*)
    (c)   Kwashiorkor is a disease caused by a lack of protein in the blood. This leads to a swollen abdomen due to a build up of tissue fluid.
          Explain why a lack of protein in the blood causes a build up of tissue fluid.   (*3 marks*)
                                                                                                    AQA Jan 2013

4   (a)   Scientists measured the rate of water flow and the pressure in the xylem in a small branch. Their results are shown in the graph.

(i) Use your knowledge of transpiration to explain the changes in the rate of flow in the xylem shown in the graph. *(3 marks)*

(ii) Explain why the values for the pressure in the xylem are negative. *(1 mark)*

(b) Doctors measured the thickness of the walls of three blood vessels in a large group of people. Their results are given in the table.

Name of vessel	Mean wall thickness / mm (± standard deviation)
Aorta	5.7 ± 1.2
Pulmonary artery	1.0 ± 0.2
Pulmonary vein	0.5 ± 0.2

(i) Explain the difference in thickness between the pulmonary artery and the pulmonary vein. *(1 mark)*

(ii) The thickness of the aorta wall changes all the time during each cardiac cycle. Explain why. *(3 marks)*

(iii) Which of the three blood vessels shows the greatest variation in wall thickness? Explain your answer. *(1 mark)*

(c) Describe how tissue fluid is formed **and** how it is returned to the circulatory system. *(6 marks)*

AQA June 2012

5 Explain how water enters xylem from the endodermis in the root and is then transported to the leaves. *(6 marks)*

AQA June 2013

6 The table shows measurements of pulse rate, systolic blood pressure and diastolic blood pressure of an individual after sitting in a chair or walking fast or running.

level of activity	pulse rate / beats min^{-1}	systolic pressure / kPa	diastolic pressure / kPa	stroke volume / ml
sitting	62	15.5	10.4	55
after walking	58	19.2	10.7	74
after running	106	23.8	11.1	88

(a) Calculate the changes in systolic pressure as the level of activity increases. *(2 marks)*

(b) Explain the difference in the effect of exercise on systolic and diastolic pressure. *(3 marks)*

(c) Calculate the cardiac output of this individual after running. Give you answer in dm^3 min^{-1}. *(1 mark)*

(d) Predict and explain the effect on potential cardiac output of daily exercise sessions over a six month period. *(2 marks)*

Section 3 Summary

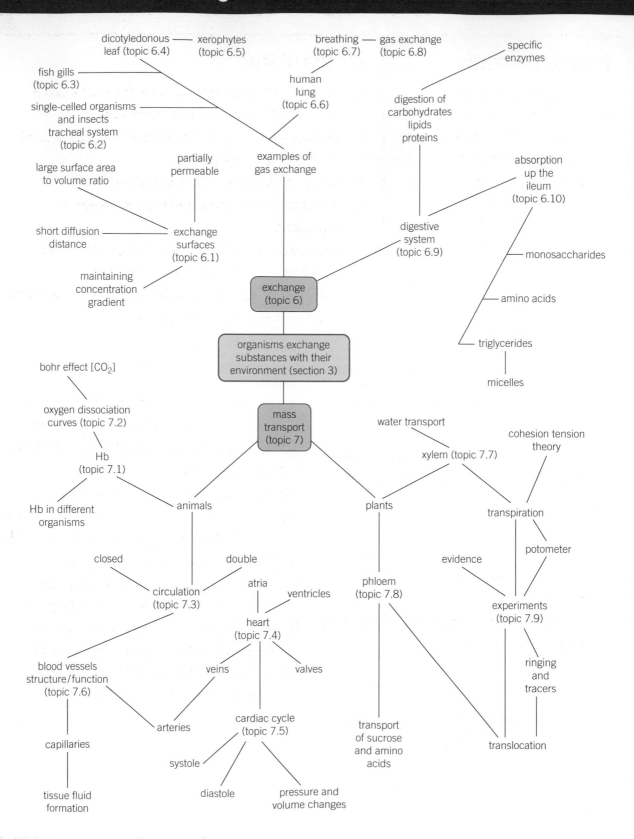

Practical skills

In this section you have met the following practical skills:

- Evaluating the results of scientific experiments
- Using appropriate apparatus, such as a potometer, to obtain quantitative measurements
- Commenting on experimental design and suggesting improvements.

Maths skills

In this section you have met the following maths skills:

- Calculating surface area to volume ratios
- Changing the subject in pulmonary ventilation and cardiac output equations
- Using appropriate units in calculations
- Substituting values in, and solving, algebraic equations
- Interpreting graphs and translating information between graphical and numerical forms
- Recognising expressions in decimal and standard forms
- Understanding simple probability
- Interpreting bar charts.

Extension tasks

Using the knowledge that you have gained from this section make a comprehensive list of each of the following:

a The general features of all transport systems.

b The differences between transport systems in plants and transport systems in animals.

c An explanation for each of the differences you have listed under b).

Figure 1 shows you how to take a person's pulse at the wrist. Each pulse is equivalent to a single heartbeat.

You should take each person's pulse for 30 seconds and double the reading to give the number of heart beats per minute (heart rate).

Find out what is meant by the 'recovery heart rate' and then design and carry out an experiment to determine any difference between the recovery heart rate of people who exercise frequently and those who do not. Draw conclusions from your results and suggest an explanation for them.

Section 3 Practice questions

1   Large insects contract muscles associated with the abdomen to force air in and out of the spiracles. This is known as 'abdominal pumping'. The table shows the mean rate of abdominal pumping of an insect before and during flight.

Stage of flight	Mean rate of abdominal pumping / dm^3 of air kg^{-1} $hour^{-1}$
Before	42
During	186

(a) Calculate the percentage increase in the rate of abdominal pumping before and during flight.
     Show your working.   (2 marks)

(b) Abdominal pumping increases the efficiency of gas exchange between the tracheoles and muscle tissue of the insect. Explain why.   (2 marks)

(c) Abdominal pumping is an adaptation not found in many small insects.
     These small insects obtain sufficient oxygen by diffusion.
     Explain how their small size enables gas exchange to be efficient without the need for abdominal pumping.   (1 mark)
     The graph shows the concentration of oxygen inside the tracheoles of an insect when at rest. It also shows when the spiracles are fully open.

(d) Use the graph to calculate the frequency of spiracle opening. Show your working.   (2 marks)

(e) The insect opens its spiracles at a lower frequency in very dry conditions.
     Suggest **one** advantage of this.   (1 mark)

(f) The ends of tracheoles connect directly with the insect's muscle tissue and are filled with water. When flying, water is absorbed into the muscle tissue.
     Removal of water from the tracheoles increases the rate of diffusion of oxygen between the tracheoles and muscle tissue. Suggest **one** reason why.   (1 mark)
     AQA June 2013

2   (a) Describe how proteins are digested in the human gut.   (4 marks)
     AQA SAMS A LEVEL PAPER 1

   (b) The enzyme lipase catalyses the digestion of lipids. The optimum pH of one version of this enzyme is 4.7. Calculate the concentration of hydrogen ions in a solution of pH 4.7.   (2 marks)

3   Newborn babies can be fed with breast milk or with formula milk. Both types of milk contain carbohydrates, lipids and proteins.

• Human breast milk also contains a bile-activated lipase. This enzyme is thought to be inactive in milk but activated by bile in the small intestine of the newborn baby.

• Formula milk does not contain a bile-activated lipase.

Scientists investigated the benefits of breast milk compared with formula milk.
The scientists used kittens (newborn cats) as model organisms in their laboratory investigation.

(a)   Other than ethical reasons, suggest **two** reasons why they chose to use cats as model organisms.   (2 marks)

(b)   Before starting their experiments, the scientists confirmed that, like human breast milk, cat's milk also contained bile-activated lipase. To do this, they added bile to cat's milk and monitored the pH of the mixture.
Explain why monitoring the pH of the mixture could show whether the cat's milk contained lipase.   (2 marks)

The scientists then took 18 kittens. Each kitten had been breastfed by its mother for the previous 48 hours.
The scientists divided the kittens randomly into three groups of six.

• The kittens in group **1** were fed formula milk.

• The kittens in group **2** were fed formula milk plus a supplement containing bile-activated lipase.

• The kittens in group **3** were fed breast milk taken from their mothers.

Each kitten was fed 2 cm³ of milk each hour for 5 days.
The scientists weighed the kittens at the start of the investigation and on each day for 5 days.

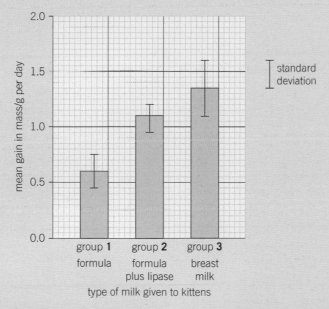

▲ **Figure 5** shows the scientists' results.

(c)   What can you conclude from **Figure 5** about the importance of bile-activated lipase in breast milk?   (3 marks)

AQA SAMS AS PAPER 2

Section 4
Genetic information, variation and relationships between organisms

Introduction

A look at the living world around us is enough to demonstrate the striking variety of life. This biodiversity is reflected not just in the multiplicity of different species, but also in the range of different individuals within a single species. Then there is the variety of different tissues, organs and cells that make up one individual. All this diversity is primarily the result of the genes that each organism possesses. It is not entirely genetic however, the environment plays its part by modifying the characteristics determined by an individual's genes.

A gene is found on a specific region of a DNA molecule known as the locus. DNA is a remarkable molecule, carrying vast amounts of information in the form of its sequence of nucleotide bases. The base sequence of each gene carries coded genetic information that determines, in turn, the sequence of amino acids in an organism's proteins. The genetic code is universal, being the same for all living organisms and therefore provides indirect evidence that organisms have evolved from one another.

DNA is a very stable molecule. It has to be if it is to reliably transfer information from one generation to the next in a way that faithfully reproduces the characteristics of the parents in the offspring. How then has the genetic diversity that is so evident arisen if DNA is not easily altered? There are a number of mechanisms. The process of sexual reproduction introduces variety in a number of ways, not least the combining of two different sets of genes – one from each parent. In addition, meiosis at some point in the life cycle leads to a random shuffling of the chromosomes. Then there are spontaneous random changes to DNA called mutations. Despite being rare, they are the basis of genetic change and evolution.

From the diverse range of individuals in a population of a species there will be some that are better adapted to the particular conditions that exist at the time they are alive. These individuals will be more likely to survive and breed and therefore pass on their alleles to the next generation. In this way populations may evolve by natural selection into new species.

Measuring diversity within a species can be achieved by comparing differences in the sequence of nucleotide bases in an individual's DNA or in the sequence of amino acids in the proteins that this DNA codes for. Measuring biodiversity within a community can be achieved by using an index of diversity or species richness.

Working scientifically

The study of genetic information, variation and relationships between organisms provides scope to perform practical work and to develop practical skills. A required practical activity is the use of aseptic techniques to investigate the effect of antimicrobial substances on bacterial growth.

A range of mathematical skills will be needed in your work, in particular the ability to use power and logarithmic functions of a calculator, use a logarithmic scale, find arithmetic means, understand measures of dispersion including standard deviation and substitute values in algebraic equations.

What you already know

The material in this unit is intended to be self-explanatory, but there are certain facts from GCSE that will help your understanding of this section. These facts include:

- ☐ Petri dishes and culture media must be sterilised before use to kill unwanted microorganisms.
- ☐ Inoculating loops are used to transfer microorganisms to culture media and they must be sterilised by passing them through a flame.
- ☐ The lid of the Petri dish should be secured with adhesive tape to prevent microorganisms from the air contaminating the culture.
- ☐ The type of cell division in which a cell divides to form gametes is called meiosis.
- ☐ Sexual reproduction gives rise to variation because, when gametes fuse, one of each pair of alleles comes from each parent.
- ☐ Chromosomes are made up of large molecules of DNA (deoxyribonucleic acid), which has a double helix structure.
- ☐ A gene is a small section of DNA.
- ☐ Each gene codes for a particular combination of amino acids that make a specific protein.
- ☐ Genetic variation is due to a population having a wide range of alleles that control their characteristics.
- ☐ In each population, the alleles that control the characteristics which help the organism to survive are selected.
- ☐ Quantitative data on the distribution of organisms can be obtained by random sampling with quadrats and sampling along a transect.

Once it had been established that DNA was the means by which genetic information was passed from generation to generation, scientists puzzled as to exactly how DNA determined the features of organisms. Before we look at this problem, we need first to be clear about what is meant by a gene.

What is a gene?

A gene is a section of DNA that contains the coded information for making polypeptides and functional RNA. The coded information is in the form of a specific sequence of bases along the DNA molecule. Polypeptides make up proteins and so genes determine the proteins of an organism. Enzymes are proteins. As enzymes control chemical reactions they are responsible for an organism's development and activities. In other words genes, along with environmental factors, determine the nature and development of all organisms. A gene is a section of DNA located at a particular position, called a **locus**, on a DNA molecule. The gene is a base sequence of DNA that codes for:

- the amino acid sequence of a polypeptide
- or a functional RNA, including ribosomal RNA and transfer RNAs (Topic 8.3).

One DNA molecule carries many genes.

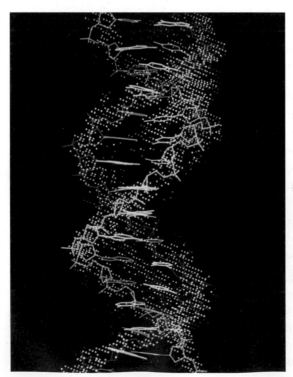

▲ **Figure 1** *Computer graphics representation of a short section of DNA*

The genetic code

In trying to discover how DNA bases coded for amino acids, scientists suggested that there must be a minimum of three bases that coded for each amino acid. Their reasoning was as follows:

- Only 20 different amino acids regularly occur in proteins.
- Each amino acid must have its own code of bases on the DNA.
- Only four different bases (adenine, guanine, cytosine and thymine) are present in DNA.
- If each base coded for a different amino acid, only four different amino acids could be coded for.
- Using a pair of bases, 16 (4^2) different codes are possible, which is still inadequate.
- Three bases produce 64 (4^3) different codes, more than enough to satisfy the requirements of 20 amino acids.

As the code has three bases for each amino acid, each one is called a triplet. As there are 64 possible triplets and only 20 amino acids, it follows that some amino acids are coded for by than one triplet.

Features of the genetic code

Further experiments have revealed the following features of the genetic code.

- A few amino acids are coded for by only a single triplet.
- The remaining amino acids are coded for by between two and six triplets each.
- The code is known as a '**degenerate code**' because most amino acids are coded for by more than one triplet.
- A triplet is always read in one particular direction along the DNA strand.
- The start of a DNA sequence that codes for a polypeptide is always the same triplet. This codes for the amino acid methionine. If this first methionine molecule does not form part of the final polypeptide, it is later removed.
- Three triplets do not code for any amino acid. These are called 'stop codes' and mark the end of a polypeptide chain. They act in much the same way as a full stop at the end of a sentence.
- The code is **non-overlapping**, in other words each base in the sequence is read only once. Thus six bases numbered 123456 are read as triplets 123 and 456, rather than as triplets 123, 234, 345, 456.
- The code is **universal**, with a few minor exceptions each triplet codes for the same amino acid in all organisms. This is indirect evidence for evolution.

Much of the DNA in eukaryotes does not code for polypeptides. For example, between genes there are non-coding sequences made up of multiple repeats of base sequences. Even within genes, only certain sequences code for amino acids. These coding sequences are called **exons**. Within the gene these exons are separated by further non-coding sequences called **introns**. Some genes code for ribosomal RNA and transfer RNAs.

Summary questions

1 Describe what a gene is.

2 Calculate how many bases are required to code for a chain of six consecutive amino acids.

3 Explain how a change in one base along a DNA molecule may result in an enzyme becoming non-functional.

4 A section of DNA has the following sequence of bases along it: TAC GCT CCG CTG TAC. All of the bases are part of the code for amino acids. The first base in the sequence is the start of the code.

a Calculate the number of amino acids that the section of DNA codes for.

b Determine which two sequences code for the same amino acid.

c It is possible that this sequence codes for many different amino acids or many copies of the same amino acid. From your knowledge of the genetic code explain how this can happen.

Interpreting the genetic code

Table 1 is a genetic code table showing the amino acids that each codon (set of three nucleotides in mRNA) is translated into during protein synthesis. An amino acid is indicated by three letters of its name, for example arg = **arg**inine, ile = **isole**ucine. To find the code for any amino acid you find the relevant three letters (usually the first three) of its name in Table 1 and then read:

▼ **Table 1** *The genetic code. The base sequences shown are those on mRNA*

First position	Second position				Third position
	U	C	A	G	
U	Phe	Ser	Tyr	Cys	U
	Phe	Ser	Tyr	Cys	C
	Leu	Ser	Stop	Stop	A
	Leu	Ser	Stop	Trp	G
C	Leu	Pro	His	Arg	U
	Leu	Pro	His	Arg	C
	Leu	Pro	Gln	Arg	A
	Leu	Pro	Gln	Arg	G
A	Ile	Thr	Asn	Ser	U
	Ile	Thr	Asn	Ser	C
	Ile	Thr	Lys	Arg	A
	Met	Thr	Lys	Arg	G
G	Val	Ala	Asp	Gly	U
	Val	Ala	Asp	Gly	C
	Val	Ala	Glu	Gly	A
	Val	Ala	Glu	Gly	G

- the first base in the sequence from the column on the left
- the second base in the sequence from the row at the top
- the third base in the sequence from the column to the right.

You can also use the table to find an amino acid that is coded for by a particular codon. For example, UGC codes for the amino acid Cys (cysteine):

- the first letter (U) is in the column on the left
- the second letter (G) is in the row at the top
- the third letter (C) is in the column to the right.

You will notice that most amino acids have more than one codon(s), for example alanine (ala) has four codon(s) GCU, GCC, GCA and GCG.

Using Table 1, answer the following questions. In each case identify amino acids by their three-letter codon(s).

1 List the two amino acids that have only one codon and state what it is in each case.
2 Name the amino acids that have each of the following codon.
   a CUC
   b AAA
   c GAU
3 For each of the following base sequences on a DNA molecule, deduce the sequence of amino acids in the order in which they would occur in the resultant polypeptide.
   a ATGCGTTAAGGCAGT
   b GCTAAGTTTCCAGAT

In Topic 3.6 we saw that, according to their organisation, there are two types of cell: **prokaryotic cells** and **eukaryotic cells**. We looked at some of the differences between the two. These differences extend to their DNA:

- In prokaryotic cells, such as bacteria, the DNA molecules are shorter, form a circle and are not associated with protein molecules. Prokaryotic cells therefore do not have chromosomes.

- In eukaryotic cells, the DNA molecules are longer, form a line (are linear) rather than a circle and occur in association with proteins called **histones** to form structures called **chromosomes**. The mitochondria and chloroplasts of eukaryotic cells also contain DNA which, like the DNA of prokaryotic cells, is short, circular and not associated with proteins.

Chromosome structure

Chromosomes are only visible as distinct structures when a cell is dividing. For the rest of the time they are widely dispersed throughout the nucleus. When they first become visible at the start of cell division chromosomes appear as two threads, joined at a single point (Figure 1). Each thread is called a **chromatid** because DNA has already replicated to give two identical DNA molecules. The DNA in chromosomes is held by histones. The considerable length of DNA found in each cell (around 2 m in every human cell) is highly coiled and folded as illustrated in Figure 2. Let us look carefully at this diagram.

Learning objectives

→ Distinguish between the DNA in prokaryotic cells and the DNA in eukaryotic organisms.

→ Describe the structure of a chromosome.

→ Explain how genes are arranged on a DNA molecule.

→ Describe the nature of homologous chromosomes.

→ Explain what is meant by an allele.

Specification reference: 3.4.1

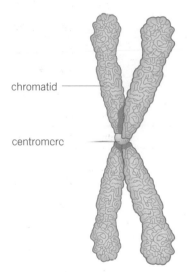

▲ **Figure 1** *Structure of a chromosome*

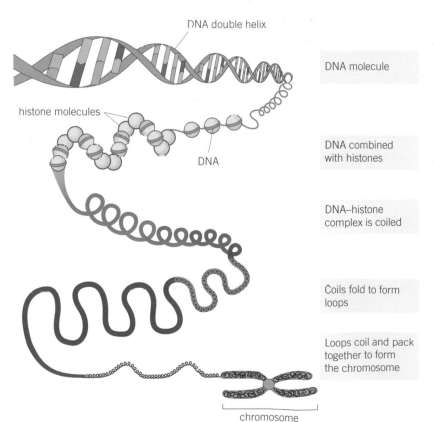

DNA double helix

DNA molecule

histone molecules

DNA

DNA combined with histones

DNA–histone complex is coiled

Coils fold to form loops

Loops coil and pack together to form the chromosome

chromosome

▲ **Figure 2** *How DNA is packed into a chromosome*

Study tip

Do not confuse the two threads (chromatids) of a chromosome with the two strands of the DNA double helix.

Hint

It is often mistakenly thought that humans have just 46 chromosomes throughout the body, rather than 46 in every single cell (except sperm and eggs).

We already know that DNA is a double helix. From the diagram we will see that this helix is wound around histones to fix it in position. This DNA–histone complex is then coiled. The coil, in turn, is looped and further coiled before being packed into the chromosome. In this way a lot of DNA is condensed into a single chromosome. If we follow the diagram carefully we will see that a chromosome contains just a single molecule of DNA, although this is very long. This single DNA molecule has many genes along its length (see Topic 8.1). Each gene occupies a specific position (locus) along the DNA molecule.

Although the number of chromosomes is always the same for normal individuals of a species, it varies from one species to another. For example, while humans have 46 chromosomes, potato plants have 48 and dogs have 78. In most species, there is an even number of chromosomes in the cells of adults.

Homologous chromosomes

Sexually produced organisms, such as humans, are the result of the fusion of a sperm and an egg, each of which contributes one complete set of chromosomes to the offspring. Therefore, one of each pair is derived from the chromosomes provided by the mother in the egg (maternal chromosomes) and the other is derived from the chromosomes provided by the father in the sperm (paternal chromosomes). These are known as **homologous pairs** and the total number is referred to as the **diploid** number. In humans this is 46.

A homologous pair is always two chromosomes that carry the same genes but not necessarily the same alleles of the genes.

▲ **Figure 3** *False-colour SEM of a group of human chromosomes*

For instance, a homologous pair of chromosomes may each possess genes for tongue rolling and blood group, but one chromosome may carry the allele for non-roller and blood group A, while the other carries the allele for roller and blood group B. During **meiosis**, the halving of the number of chromosomes is done in a manner which ensures that each daughter cell receives one chromosome from each homologous pair. In this way each cell receives one gene for each characteristic of the organism. When these haploid cells combine, the diploid state, with paired homologous chromosomes, is restored.

What is an allele?

An **allele** is one of a number of alternative forms of a gene. We have seen that genes are sections of DNA that contain coded information in the form of specific sequences of bases. Each gene exists in two, occasionally more, different forms. Each of these forms is called an allele. Each individual inherits one allele from each of its parents. These two alleles may be the same or they may be different. When they are different, each allele has a different base sequence, therefore a different amino acid sequence, so produces a different polypeptide.

Any changes in the base sequence of a gene produces a new allele of that gene (=mutation) and results in a different sequence of amino acids being coded for. This different amino acid sequence will lead

Hint

Do not confuse genes and alleles. A *gene* refers to a particular characteristic such as blood groups. Genes can exist in two or more different forms called **alleles**.

to the production of a different polypeptide, and hence a different protein. Sometimes this different protein may not function properly or may not function at all. When the protein produced is an enzyme, it may have a different shape. The new shape may not fit the enzyme's substrate. As a result the enzyme may not function and this can have serious consequences for the organism.

Synoptic link

To remind you how the shape of an enzyme is important to the way it works look back at Topic 1.7. We shall also learn more about the importance of meiosis in producing genetic variation in Topic 9.2

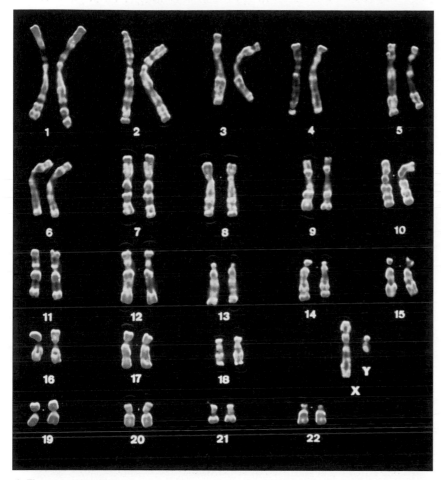

▲ **Figure 4** *The 46 chromosomes of a human showing them in their 22 pairs, as well as the X and Y sex chromosomes*

Link

A level students will learn about the roles of chromosomes and alleles in inheritance in Chapter 17 Inherited change.

Summary questions

1 Contrast the DNA of a prokaryotic cell with that of a eukaryotic cell.

2 State the function of the protein found in chromosomes.

3 Explain how the considerable length of a DNA molecule is compacted into a chromosome.

4 $\sqrt{x}$ Suppose the total length of all the DNA in a single human muscle cell is 2.3 m.

   a If all the DNA were distributed equally between the chromosomes, calculate the mean length of DNA in each one.

   b Calculate in mm the length of DNA in a human brain cell.

Learning objectives

→ Describe what the genetic code is and its main features

→ Describe the structure of ribonucleic acid (RNA).

→ Describe the structure and the role of messenger RNA (mRNA).

→ Describe the structure and the role of transfer RNA (tRNA).

Specification reference: 3.4.1

In Topic 8.1 we learned about the importance of DNA and how it carries coded information for the sequence of amino acids that make up a protein. In this topic, we will turn our attention to exactly how DNA triplets are used to make the proteins that they code for.

Transferring the coded information

We know that the sequence of nucleotide bases in DNA determines the sequence of amino acids in the proteins of an organism. In eukaryotic cells DNA is largely confined to the nucleus. However, the synthesis of proteins takes place in the cytoplasm. So how is the coded information on the DNA in the nucleus transferred to the cytoplasm where it is translated into proteins? The answer is that sections of the DNA code are transcribed onto a single-stranded molecule called ribonucleic acid (RNA).

There are a number of types of RNA. The one that transfers the DNA code from the nucleus to the cytoplasm acts as a type of messenger and is hence given the name **messenger RNA**, or **mRNA** for short. This mRNA is small enough to leave the nucleus through the nuclear pores and to enter the cytoplasm, where the coded information that it contains is used to determine the sequence of amino acids in the proteins which are synthesised there.

The term **codon** refers to the sequence of three bases on mRNA that codes for a single amino acid.

There are two other terms that are relevant:

- **Genome** – the complete set of genes in a cell, including those in mitochondria and/or chloroplasts.
- **Proteome** – the full range of proteins produced by the genome. This is sometimes called the **complete proteome**, in which case the term proteome refers to the proteins produced by a given type of cell under a certain set of conditions.

We saw in Topic 2.1 that DNA is composed of two nucleotide chains wound around each other (double helix). We will now look at the structure of a related molecule that is usually made up of a single nucleotide chain: **ribonucleic acid (RNA)**.

Ribonucleic acid (RNA) structure

Ribonucleic acid (RNA) is a **polymer** made up of repeating mononucleotide sub-units (Figure 1). It forms a single strand in which each nucleotide is made up of:

- the **pentose sugar** ribose
- one of the organic bases adenine (A), guanine (G), cytosine (C) and uracil (U)
- a phosphate group.

▲ **Figure 1** Section of ribonucleic acid (RNA) molecule

The two types of RNA that are important in protein synthesis are:

- messenger RNA (mRNA)
- transfer RNA (tRNA).

Messenger RNA (mRNA)

Consisting of thousands of mononucleotides, mRNA is a long strand that is arranged in a single helix. The base sequence of mRNA is determined by the sequence of bases on a length of DNA in a process called transcription. There is a great variety of different types of mRNA. Once formed, mRNA leaves the nucleus via pores in the nuclear envelope and enters the cytoplasm, where it associates with the ribosomes. There it acts as a template for protein synthesis. Its structure is suited to this function because it possesses information in the form of codons (three bases that are complementary to a triplet in DNA). The sequence of codons determines the amino acid sequence of a specific polypeptide that will be made.

Transfer RNA (tRNA)

Transfer RNA (tRNA) is a relatively small molecule that is made up of around 80 nucleotides. It is a single-stranded chain folded into a clover-leaf shape, with one end of the chain extending beyond the other. This is the part of the tRNA molecule to which an amino acid can easily attach. There are many types of tRNA, each of which binds to a specific amino acid. At the opposite end of the tRNA molecule is a sequence of three other organic bases, known as the **anticodon**. Given that the genetic code is degenerate there must be as many tRNA molecules as there are coding triplets. However, each tRNA is specific to one amino acid and has an anticodon that is specific to that amino acid.

Synoptic link

A third type of RNA, ribosomal RNA, was covered in Topic 3.4.

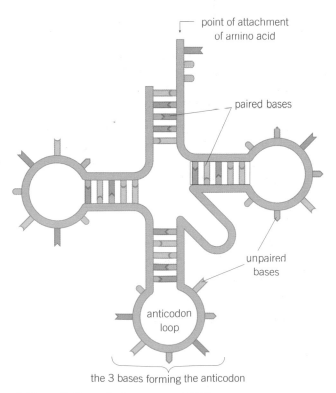

point of attachment of amino acid

paired bases

unpaired bases

anticodon loop

the 3 bases forming the anticodon

▲ **Figure 3** *Clover-leaf structure of tRNA*

▲ **Figure 2** *Computer artwork of a tRNA molecule*

You will recall from Topic 8.1 that the organic bases in DNA pair up in a precise way, for example, guanine pairs with cytosine, and adenine pairs with thymine. These are known as complementary base pairs. In RNA, however, the base thymine is always replaced by a similar base called uracil. RNA can join with both DNA and other RNA molecules by complementary base pairing. The complementary base pairings that RNA forms are therefore:

• guanine with cytosine

• adenine with uracil (in RNA) or thymine (in DNA).

During protein synthesis, an anticodon pairs with the three complementary organic bases that make up the codon on mRNA. The tRNA structure (Figure 3), with its end chain for attaching amino acids and its anticodon for complementary base pairing with the codon of the mRNA, is structurally suited to its role of lining up amino acids on the mRNA template during protein synthesis.

Comparison of DNA, messenger RNA and transfer RNA

Table 1 compares the structure, function and composition of DNA, mRNA and tRNA.

▼ **Table 1** *Comparison of DNA, mRNA and tRNA*

DNA	Messenger RNA	Transfer RNA
double polynucleotide chain	single polynucleotide chain	single polynucleotide chain
largest molecule of the three	molecule is smaller than DNA but larger than tRNA	smallest molecule of the three
double-helix molecule	single-helix molecule (except in a few viruses)	clover-shaped molecule
pentose sugar is deoxyribose	pentose sugar is ribose	pentose sugar is ribose
organic bases are adenine, guanine, cytosine and thymine	organic bases are adenine, guanine, cytosine and uracil	organic bases are adenine, guanine, cytosine and uracil
found mostly in the nucleus	manufactured in the nucleus but found throughout the cell	manufactured in the nucleus but found throughout the cell
quantity is constant for all cells of a species (except gametes)	quantity varies from cell to cell and with level of metabolic activity	quantity varies from cell to cell and with level of metabolic activity
chemically very stable	Less stable than DNA or tRNA, individual molecules are usually broken down in cells within a few days.	chemically more stable than mRNA but less stable than DNA

1 Table 1 states that, for DNA, the 'quantity is constant for all cells of a species (except gametes)'.
    a State how the quantity in a gamete differs from that in a body cell.
    b Explain the significance of the difference you have described.
2 Explain an advantage of:
    a DNA being a chemically stable molecule
    b mRNA being broken down relatively quickly.

8.4 Polypeptide synthesis – transcription and splicing

We saw in Topic 1.6 that proteins are made up of one or more polypeptides. Proteins, especially enzymes, are essential to all aspects of life. Every organism needs to make its own, unique, proteins. The biochemical machinery in the cytoplasm of each cell has the capacity to make every protein from just 20 amino acids. Exactly which proteins it manufactures depends upon the instructions that are provided, at any given time, by the DNA in the cell's nucleus. The basic process is as follows.

- DNA provides the instructions in the form of a long sequence of bases.
- A complementary section of part of this sequence is made in the form of a molecule called pre-mRNA – a process called **transcription**.
- The pre-mRNA is spliced to form mRNA.
- The mRNA is used as a template to which complementary tRNA molecules attach and the amino acids they carry are linked to form a polypeptide – a process called **translation**.

The process can be likened to a bakery, where the basic equipment and ovens (cell organelles) can manufacture any variety of cake (protein) from relatively few basic ingredients (amino acids). Which particular variety of cake is made depends on the recipe (genetic code) that the baker uses on any particular day. By choosing different recipes at different times, rather than making everything all the time, the baker can meet seasonal demands, adapt to changing customer needs and avoid waste.

DNA replication can be likened to the publication of many copies of a recipe book (genome); making a photocopy of a recipe to use in the bakery is therefore transcription. Making the cakes, using the photocopied recipe, is translation. If the book is not removed from the library, many copies of the recipe can be made, and the same cakes can be produced in many places at the same time or over many years.

Transcription

Transcription is the process of making pre-mRNA using part of the DNA as a template. The process, which is illustrated in Figure 1, is as follows.

- An enzyme acts on a specific region of the DNA causing the two strands to separate and expose the **nucleotide** bases in that region.
- The nucleotide bases on one of the two DNA strands, known as the **template strand**, pair with their complementary nucleotides from the pool which is present in the nucleus. The enzyme RNA polymerase then moves along the strand and joins the nucleotides together to form a pre mRNA molecule.

Learning objectives

→ Explain how pre-messenger RNA is produced from DNA in the process called transcription

→ Describe how pre-messenger RNA is modified to form messenger RNA.

Specification reference: 3.4.2

Hint

The bakery analogy is to help understanding but this should not be used when you are writing a scientific explanation.

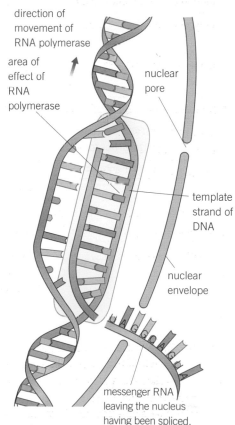

direction of movement of RNA polymerase

area of effect of RNA polymerase

nuclear pore

template strand of DNA

nuclear envelope

messenger RNA leaving the nucleus having been spliced.

▲ **Figure 1** *Summary of transcription*

Link

A level students will also look at how transcription and translation are controlled in Topic 20.3 Regulation of transcription and translation.

Summary questions

1 Describe the role of RNA polymerase in transcription.

2 State which other enzyme is involved in transcription and describe its role.

3 Explain why splicing of pre-mRNA is necessary.

4 A sequence of bases along the template strand of DNA is ATGCAAGTCCAG.

   a Deduce the sequence of bases on a pre-messenger RNA molecule that has been transcribed from this part of the DNA molecule.

   b Calculate how many amino acids the sequence codes for.

5 A gene is made up of 756 base pairs. The mRNA that is transcribed from this gene is only 524 nucleotides long. Explain why there is this difference.

- In this way an exposed guanine base on the DNA binds to the cytosine base of a free nucleotide. Similarly, cytosine links to guanine, and thymine joins to adenine. The exception is adenine, which links to uracil rather than thymine.
- As the RNA polymerase adds the nucleotides one at a time to build a strand of pre-mRNA, the DNA strands rejoin behind it. As a result, only about 12 base pairs on the DNA are exposed at any one time.
- When the RNA polymerase reaches a particular sequence of bases on the DNA that it recognises as a 'stop' triplet code, it detaches, and the production of pre-mRNA is then complete.

Splicing of pre-mRNA

In prokaryotic cells, transcription results directly in the production of mRNA from DNA. In **eukaryotic cells** transcription results in the production of pre-mRNA, which is then spliced to form mRNA. The DNA of a gene eukaryotic cells is made up of sections called exons that code for proteins and sections called introns that do not. These intervening introns would prevent the synthesis of a polypeptide. In the pre-mRNA of eukaryotic cells. The base sequences corresponding to the introns are removed and the functional exons are joined together during a process called **splicing**. As most prokaryotic cells do not have introns, splicing of their DNA is unnecessary. The process of splicing is shown in Figure 2.

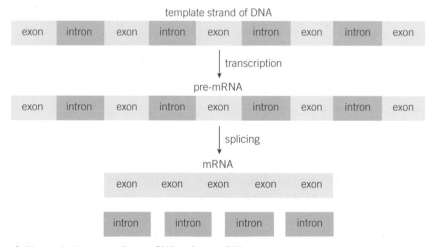

▲ **Figure 2** *Splicing of pre-mRNA to form mRNA*

The mRNA molecules are too large to diffuse out of the nucleus and so, once they have been spliced, they leave via a nuclear pore. Outside the nucleus, the mRNA is attracted to the ribosomes to which it becomes attached, ready for the next stage of the process: translation.

8.5 Polypeptide synthesis – translation

In Topic 8.4 we looked at how the triplet code of DNA is transcribed into a sequence of **codons** (genetic code) on messenger RNA (mRNA). The next stage is to translate the codons on the mRNA into a sequence of amino acids that make up a polypeptide.

There are about 60 different tRNAs. A particular tRNA has a specific **anticodon** and attaches to a specific amino acid. Each amino acid therefore has one or more tRNA molecule, with its own anticodon of bases.

Synthesising a polypeptide

Once mRNA has passed out of the nuclear pore it determines the synthesis of a polypeptide. The following explanation of how a polypeptide is made is illustrated in Figures 3 and 4. (The information given in brackets below is only to help you follow the process and does not need to be learned.)

- A ribosome (Figure 4, part 1) becomes attached to the starting codon (AUG) at one end of the mRNA molecule.
- The tRNA molecule with the complementary anticodon sequence (UAC) moves to the ribosome and pairs up with the codon on the mRNA. This tRNA carries a specific amino acid (methionine).
- A tRNA molecule with a complementary anticodon (UGC) pairs with the next codon on the mRNA (ACG). This tRNA molecule carries another amino acid (threonine).
- The ribosome moves along the mRNA, bringing together two tRNA molecules at any one time, each pairing up with the corresponding two codons on the mRNA.
- The two amino acids (methonine and threonine) on the tRNA are joined by a **peptide bond** using an enzyme and ATP which is hydrolysed to provide the required energy.
- The ribosome moves on to the third codon (GAU) in the sequence on the mRNA, thereby linking the amino acids (threonine and aspartic acid) on the second and third tRNA molecules (Figure 4, part 2).
- As this happens, the first tRNA is released from its amino acid (methionine) and is free to collect another amino acid (methionine) from the amino acid pool in the cell.
- The process continues in this way, with up to 15 amino acids being added each second, until a polypeptide chain is built up (Figure 4, part 3).
- Up to 50 ribosomes can pass immediately behind the first, so that many identical polypeptides can be assembled simultaneously (Figure 3).
- The synthesis of a polypeptide continues until a ribosome reaches a stop codon. At this point, the ribosome, mRNA and the last tRNA molecule all separate and the polypeptide chain is complete.

In summary, the DNA sequence of triplets that make up a gene determine the sequence of codons on mRNA. The sequence of codons on mRNA determine the order in which the tRNA molecules line up.

Learning objectives

→ Explain how a polypeptide is synthesised during the process of translation.

→ Describe the roles of messenger RNA and transfer RNA in translation.

Specification reference: 3.4.2

point of attachment of amino acid

Anticodon – this sequence of ACG means that the amino acid cysteine will attach to the other end of this tRNA molecule. This anticodon will combine with the codon UGC on a mRNA molecule during the formation of a polypeptide. The mRNA codon UGC therefore translates into the amino acid cysteine.

▲ **Figure 1** *Simplified structure of one type of tRNA*

large sub-unit

small sub-unit

▲ **Figure 2** *Structure of a ribosome. The smaller sub-unit fits into a depression on the surface of the larger one*

Hint

Remember that there is no thymine in any RNA molecule. It is uracil in RNA that pairs with adenine.

▲ **Figure 3** *Polypeptide formation*

▲ **Figure 4** *Translation*

They, in turn, determine the sequence of amino acids in the polypeptide. In this way genes precisely determine which proteins a cell manufactures. As many of these proteins are enzymes, genes effectively control the activities of cells.

Assembling a protein

Sometimes a single polypeptide chain is a functional protein. Often, a number of polypeptides are linked together to give a functional protein (quaternary structure). What happens to the polypeptide next depends upon the protein being made, but usually involves the following:

- The polypeptide is coiled or folded, producing its secondary structure.
- The secondary structure is folded, producing the tertiary structure.
- Different polypeptide chains, along with any non-protein groups, are linked to form the quaternary structure.

Summary questions

1 Name the cell organelle involved in translation.

2 A codon found on a section of mRNA has the sequence of bases AUC. List the sequence of bases found on:

   a   the tRNA anticodon that attaches to this codon;

   b   the template strand of DNA that formed the mRNA codon.

3 Describe the role of tRNA in the process of translation.

4 A strand of mRNA has 64 codons but the protein produced from it has only 63 amino acids. Suggest a reason for this difference.

Protein synthesis

▲ Figure 5

Figure 5 shows the formation of part of a polypeptide along a section of eight codons. Codons 4 and 5 have been left blank. Using Figure 5 and Table 1 answer the following questions.

1 Name the structures X and Y.
2 Recall the chemical group shown on the end of the polypeptide chain.
3 Determine the anticodon sequence on tRNA molecule 4.
4 Deduce the sequence of the first five amino acids in the polypeptide.
5 Determine the sequence of bases on that portion of DNA from which codons 1 – 3 are transcribed.
6 A DNA mutation results in the base cytosine being replaced by uracil in codon 8. Explain the significance of this change.
7 Another mutant form of a gene causes the inversion (reversal) of the code for the amino acid glutamine (Glu).
   a Consider all possible outcomes from this change and explain the effect on the polypeptide in each case.
   b If the polypeptide formed from this mutant gene forms part of an enzyme, suggest **two** reasons why it might fail to function. Explain your answer.

▼ Table 1 *The base sequences shown are those on mRNA*

First position	Second position				Third position
	U	C	A	G	
U	Phe	Ser	Tyr	Cys	U
	Phe	Ser	Tyr	Cys	C
	Leu	Ser	Stop	Stop	A
	Leu	Ser	Stop	Trp	G
C	Leu	Pro	His	Arg	U
	Leu	Pro	His	Arg	C
	Leu	Pro	Gln	Arg	A
	Leu	Pro	Gln	Arg	G
A	Ile	Thr	Asn	Ser	U
	Ile	Thr	Asn	Ser	C
	Ile	Thr	Lys	Arg	A
	Met	Thr	Lys	Arg	G
G	Val	Ala	Asp	Gly	U
	Val	Ala	Asp	Gly	C
	Val	Ala	Glu	Gly	A
	Val	Ala	Glu	Gly	G

Cracking the code

How exactly did scientists decipher which amino acid was coded for by which codon? Nirenberg and others did so by making synthetic mRNA and using this to make polypeptides.

The basic stages of the experiments were as follows:

- Cell extracts with the necessary components to make polypeptides were obtained and treated with DNase.
- Synthetic mRNA was added to the extract and all 20 amino acids attached to their appropriate tRNA.
- One amino acid was radioactively labelled with carbon 14 (^{14}C) while the remaining 19 had normal, non-radioactive carbon 12 (^{12}C).
- The extracts were incubated and the polypeptide produced was later extracted.
- The radioactivity level of the polypeptide produced in each case was measured.

> **1** Suggest a reason why DNase was added to the cell extract.

- In one experiment the radioactive amino acid was phenylalanine and four mixtures, differing only in their mRNA, were set up as follows:
  - mRNA made up of a chain of nucleotides containing only the base adenine = poly A
  - mRNA made up of a chain of nucleotides containing only the base uracil = poly U
  - mRNA made up of a chain of nucleotides containing only the base cytosine = poly C
  - no mRNA was present.

The results are shown in Table 2.

▼ **Table 2** *Results of experiment using radioactively labelled phenylalanine*

Type of synthetic mRNA	Radioactivity / counts min^{-1}
poly A	50
poly U	39 800
poly C	38
none	44

> **2** State **one** codon for the amino acid phenylalanine that is suggested by the results of this experiment. Explain your answer.
> **3** Explain why a mixture without any synthetic RNA was used.

Using this method Nirenberg deciphered 47 of the 64 possible codons in the genetic code. The remaining 17 codons, however, gave ambiguous results. This led another scientist, called Khorana, to devise a different technique. He formed very long mRNA molecules that had a repeating sequence of nucleotide bases, such as GUGUGUGUGUGUG. The polypeptide produced by this mRNA was made up of alternating cysteine and valine amino acids. The question was, what was the codon for each amino acid?

> **4** Suggest why it is not possible to say what the codon is for each amino acid.

From Nirenberg's earlier experiments, Khorana knew that UGU was a codon for cysteine. This meant that GUG was a codon for valine. By analysing the results of similar experiments using specific sequences of mRNA he was able to decipher the complete genetic code and to show that the code was degenerate. Further experiments by Nirenberg verified these findings.

> **5** The genetic code can be described as degenerate but not ambiguous. Discuss this statement.
> **6** Despite all these experiments, it was still not possible to find the amino acids coded for by certain codons. Explain why not.

1 The diagram shows a short sequence of DNA bases.

T T T G T A T A C T A G T C T A C T T C G T T A A T A

(a) (i) What is the maximum number of amino acids for which this sequence of DNA bases could code? *(1 mark)*

(ii) The number of amino acids coded for could be fewer than your answer to part **(a)(i)**.
Give **one** reason why. *(1 mark)*

(b) Explain how a change in the DNA base sequence for a protein may result in a change in the structure of the protein. *(3 marks)*

(c) A piece of DNA consisted of 74 base pairs. The two strands of the DNA, strands **A** and **B**, were analysed to find the **number** of bases of each type that were present. Some of the results are shown in the table.

	Number of bases			
	C	G	A	T
Strand **A**	26			
Strand **B**	19		9	

Complete the table by writing in the missing values. *(2 marks)*

AQA June 2011

2 **Figure 1** shows a short section of a DNA molecule.

(a) Name parts **R** and **Q**.
(i) **R**
(ii) **Q** *(2 marks)*
(b) Name the bonds that join **A** and **B**. *(1 mark)*
(c) Ribonuclease is an enzyme. It is 127 amino acids long.
What is the minimum number of DNA bases needed to code for ribonuclease? *(1 mark)*
(d) **Figure 2** shows the sequence of DNA bases coding for seven amino acids in the enzyme ribonuclease.

GTTTACTACTCTTCTTCTTTA

▲ **Figure 2**

The number of each type of amino acid coded for by this sequence of DNA bases is shown in the table.

Amino acid	Number present
Arg	3
Met	2
Gln	1
Asn	1

Use the table and **Figure 2** to work out the sequence of amino acids in this part of the enzyme. Write your answer in the boxes below.

Gln						

(1 mark)

(e) Explain how a change in a sequence of DNA bases could result in a non-functional enzyme. *(3 marks)*

AQA Jan 2010

3 (a) What name is used for the non-coding sections of a gene? *(1 mark)*

Figure 1 shows a DNA base sequence. It also shows the effect of two mutations on this base sequence. **Figure 2** shows DNA triplets that code for different amino acids.

Original DNA base sequence	A	T	T	G	G	C	G	T	G	T	C	T
Amino acid sequence												
Mutation **1** DNA base sequence	A	T	T	G	G	A	G	T	G	T	C	T
Mutation **2** DNA base sequence	A	T	T	G	G	C	C	T	G	T	C	T

▲ **Figure 1**

DNA triplets	Amino acid
GGT, GGC, GGA, GGG	Gly
GTT, GTA, GTG, GTC	Val
ATC, ATT, ATA	Ile
TCC, TCT, TCA, TCG	Ser
CTC, CTT, CTA, CTG	Leu

▲ **Figure 2**

(b) Complete **Figure 1** to show the sequence of amino acids coded for by the original DNA base sequence. *(1 mark)*

(c) Some gene mutations affect the amino acid sequence. Some mutations do not. Use the information from **Figure 1** and **Figure 2** to explain

(i) whether mutation 1 affects the amino acid sequence *(2 marks)*

(ii) how mutation 2 could lead to the formation of a non-functional enzyme. *(3 marks)*

(d) Gene mutations occur spontaneously.

(i) During which part of the cell cycle are gene mutations most likely to occur? *(1 mark)*

(ii) Suggest an explanation for your answer. *(1 mark)*

AQA June 2010

4 The diagram shows part of a pre-mRNA molecule.

A   U   C   C   G   U

part X

(a) (i) Name the two substances that make up part **X**. *(1 mark)*

(ii) Give the sequence of bases on the DNA strand from which this pre-mRNA has been transcribed. *(1 mark)*

(b) (i) Give **one** way in which the structure of an mRNA molecule is different from the structure of a tRNA molecule. *(1 mark)*

(ii) Explain the difference between pre-mRNA and mRNA. *(1 mark)*

(c) The table shows the percentage of different bases in two pre-mRNA molecules.

The molecules were transcribed from the DNA in different parts of a chromosome.

Part of chromosome	Percentage of base			
	A	G	C	U
Middle	38	20	24	
End	31	22	26	

(i) Complete the table by writing the percentage of uracil (U) in the
appropriate boxes.   (*1 mark*)

(ii) Explain why the percentages of bases from the middle part of the
chromosome and the end part are different.   (*2 marks*)

AQA June 2011

5 **Figure 3** represents one process that occurs during protein synthesis.

A U G C C G U A C C G A C U

(a) Name the process shown.   (*1 mark*)
(b) Identify the molecule labelled **Q**.   (*1 mark*)
(c) In **Figure 3**, the first codon is AUG. Give the base sequence of the
complementary DNA base sequence and the missing anticodon.   (*2 marks*)

Table 1 shows the base triplets that code for two amino acids.

▼ **Table 1**

Amino acid	Encoding base triplet
Aspartic acid	GAC, GAU
Proline	CCA, CCG, CCC, CCU

Aspartic acid and proline are both amino acids.
(d) Describe how two amino acids differ from one another. You may use
a diagram to help your description.   (*1 mark*)
(e) Deletion of the sixth base (G) in the sequence shown in **Figure 3** would change
the nature of the protein produced but substitution of the same base would not.
Use the information in **Table 1** and your own knowledge to explain why. (*3 marks*)

AQA SAMS PAPER 2

6 **Table 6** lists the chromosome numbers and genome sizes of four plant species.
One Mbp (mega base pair) is equal to 1 000 000 base pairs of DNA.

▼ **Table 6**

name	chromosome number(s)	genome size (Mbp)
Amborella	2n = 26	870
sweet rush	2n = 18	392
monkey flower	2n = 28	430

(a) Calculate the mean number of base pairs per chromosome in sweet rush,
expressing your answer in standard form.   (*2 marks*)
(b) Modern techniques allow the genomes to be sequenced, but only
750 base pairs can be read at a time.

Calculate the smallest number of DNA fragments that would need to be
made to sequence the DNA of monkey flower.   (*2 marks*)

9 Genetic diversity and adaptation
9.1 Gene mutation

Learning objectives

→ Describe gene mutations.

→ Explain how deletion and substitution of bases result in different amino acid sequences in polypeptides.

→ Explain why some mutations do not result in a changed amino acid sequence.

→ Describe what chromosome mutations are.

Specification reference: 3.4.3

Synoptic link

You will find it easier to follow this topic if you first revise DNA structure and replication (Topics 2.1 and 2.2), genes and the triplet code (Topic 8.1) as well as chromosome structure (Topic 8.2).

Hint

The various gene mutations are illustrated by specific examples that name bases and amino acids. These are only to illustrate the points being made and do not need to be remembered.

Any change to the quantity or the base sequence of the DNA of an organism is known as a **mutation**. Mutations occurring during the formation of gametes may be inherited, often producing sudden and distinct differences between individuals. Any change to one or more nucleotide bases, or a change in the sequence of the bases, in DNA is known as a **gene mutation**.

We have seen that a sequence of triplets on DNA is transcribed into mRNA and is then translated into a sequence of amino acids that make up a polypeptide. It follows that any changes to one or more bases in the DNA triplets could result in a change in the amino acid sequence of the polypeptide. Gene mutations can arise spontaneously during DNA replication and include base substitution and base deletion.

Substitution of bases

The type of gene mutation in which a nucleotide in a DNA molecule is replaced by another nucleotide that has a different base is known as a substitution. As an example, consider the DNA triplet of bases, guanine–thymine–cytosine (GTC) that codes for the amino acid glutamine. If the final base, cytosine, is replaced by guanine, then GTC becomes GTG. GTG is one of the DNA triplet that codes for the amino acid histidine and this then replaces the original amino acid: glutamine. The polypeptide produced will differ in a single amino acid. The significance of this difference will depend upon the precise role of the original amino acid. If it is important in forming bonds that determine the tertiary structure of the final protein, then the replacement amino acid may not form the same bonds. The protein may then be a different shape and therefore not function properly. For example, if the protein is an enzyme, its active site may no longer fit the substrate and it will no longer catalyse the reaction.

The effect of the mutation is different if the new triplet of bases still codes for the same amino acid as before. This is due to the degenerate nature of the genetic code, in which most amino acids have more than one codon. For instance, if the third base in our example is replaced by thymine, then GTC becomes GTT. However, as both DNA triplets code for glutamine, there is no change in the polypeptide produced and so the mutation will have no effect.

Deletion of bases

A gene mutation by deletion arises when a nucleotide is lost from the normal DNA sequence. The loss of a single nucleotide from the thousands in a typical gene may seem a minor change but the consequences can be considerable. Usually the amino acid sequence of the polypeptide is entirely different and so the polypeptide is unlikely to function correctly. This is because the sequence of bases in DNA is read in units of three bases (triplet). One deleted nucleotide causes all triplets in a sequence to be read differently because each has been shifted to the left by one base as shown in Figure 1.

Link

More detail on mutations will be provided to those students studying A level in Topic 20.1 Gene mutations.

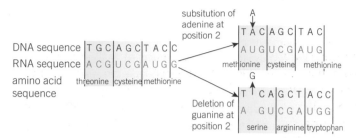

▲ **Figure 1** *Effect of substitution and deletion mutations on amino acid sequence.*

Chromosome mutations

Changes in the structure or number of whole chromosomes are called **chromosome mutations**.

Chromosome mutations can arise spontaneously and take two forms:

- **Changes in whole sets of chromosomes** occur when organisms have three or more sets of chromosomes rather than the usual two. This condition is called **polyploidy** and occurs mostly in plants.

- **Changes in the number of individual chromosomes**. Sometimes individual homologous pairs of chromosomes fail to separate during **meiosis** (see Topic 9.2). This is known as **non-disjunction** and usually results in a gamete having either one more or one fewer chromosome. On fertilisation with a gamete that has the normal complement of chromosomes, the resultant offspring have more or fewer chromosomes than normal in all their body cells. An example of a non-disjunction in humans is Down's syndrome, where individuals have an additional chromosome 21.

✚ Hybridisation and polyploidy

Around 10 000 years ago, in regions of the Middle East known as the 'fertile crescent', groups of humans are thought to have gathered the grain of the wild **einkorn wheat** and **emmer wheat** using sickles made of antlers. In time, these farmers began to sow their crops. Although they knew nothing of genetics, they will naturally have selected the seed from the varieties that suited their needs rather than types that did not. Continued selection over the intervening years has produced the modern high-yielding varieties of wheat that we cultivate today. These modern varieties are the result of a process known as **hybridisation**. Hybridisation is combining the genes of different varieties or species of organisms to produce a **hybrid**. Sometimes this is followed by organisms that have additional complete sets of chromosomes (polyploidy). How then can polyploidy arise?

Polyploidy can come about in several different ways. One is for the chromosomes not to separate into two distinct sets during meiosis. Gametes could then be produced that have both sets, in other words they

▲ **Figure 2** *Einkorn wheat (Triticum urartu)*

▲ **Figure 3** *Goat grass* (Aegilops speltoides)

▲ **Figure 4** *Emmer wheat* (Triticum turgidum)

are diploid rather than haploid. If these fused with one another the offspring could have four sets of chromosomes – they would be tetraploid. Alternatively, if a diploid gamete fused with a haploid gamete, the offspring would have three sets of chromosomes – they would be triploid.

Sometimes hybrids can be formed by combining sets of chromosomes from two different species. For example, by cross-pollination between two closely related plants leading to successful fertilisation. These hybrids are usually sterile. If, however, the hybrid has a chromosome number that is a multiple of the original chromosome number, a new fertile species can arise because chromosomes have homologous partners and so meiosis is possible. Modern wheat plants have arisen due to a combination of both these forms of polyploidy.

Hybridisation in wheat

Einkorn wheat (*Triticum urartu*) is a wild form of wheat that has 14 chromosomes. It is thought that around 500 000 years ago it hybridised by cross-pollination with *Aegilops speltoides*, a type of goat grass that also had 14 chromosomes. This gave rise to a new hybrid species called emmer wheat (*Triticum turgidum*). This species therefore had four sets of chromosomes (28 chromosomes) and was therefore tetraploid. Emmer and einkorn wheat were the first ones harvested by humans around 10 000 years ago. Sometime shortly after, further hybridisation occurred when the tetraploid emmer wheat with its 28 chromosomes cross-pollinated with another type of goat grass (*Aegilops tauschii*) with 14 chromosomes. The new hybrid species now had 42, or six sets of chromosomes, in other words it was hexaploid. This species is *Triticum aestivum* or bread wheat, which is the type of wheat grown today.

▲ **Figure 5** *Bread wheat* (Triticum aestivum)

The development of modern wheat is summarised in Figure 6.

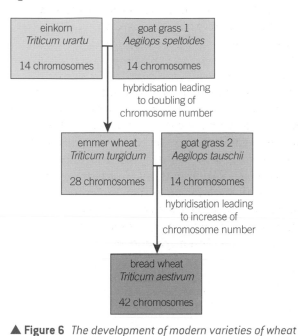

▲ **Figure 6** *The development of modern varieties of wheat by hybridisation*

1 Explain why hybrids formed by combining sets of chromosomes from two different species are often sterile.

2 Selective breeding has led to strains of wheat with shorter stems. Suggest an advantage to farmers of these strains of wheat.

3 Explain why *Triticum urartu* and *Triticum turgidum* are classified as different species.

Summary questions

1 The following is a sequence of 12 nucleotides within a much longer mRNA molecule: AUGCAUGUUACU. Following a gene mutation the same 12-nucleotide portion of the mRNA molecule is AUGCUGUUACUG. Name the type of gene mutation that has occurred. Show your reasoning.

2 Explain why a deletion gene mutation is more likely to result in a change to an organism than a substitution gene mutation.

3 Explain why a mutation that is transcribed onto mRNA may not result in any change to the polypeptide that it codes for.

4 Errors in transcription occur about 100 000 times more often than errors in DNA replication. Explain why errors in DNA replication can be far more damaging than errors in transcription.

life cycle of most animals

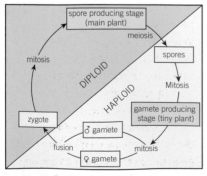

life cycle of fern

▲ **Figure 1** *A comparison of the life cycles of most animals with that of a fern*

Cell division occurs in one of two ways:

- **Mitosis** produces two daughter cells with the same number of **chromosomes** as the parent cell and as each other.
- **Meiosis** usually produces four daughter cells, each with half the number of chromosomes as the parent cell.

Importance of meiosis

In sexual reproduction two **gametes** fuse to give rise to new offspring. If each gamete had a full set of chromosomes (**diploid** number) then the cell that they produce has double this number. In humans, the diploid number of chromosomes is 46, which means that this cell would have 92 chromosomes. This doubling of the number of chromosomes would continue at each generation. It follows that, in order to maintain a constant number of chromosomes in the adults of a species, the number of chromosomes must be halved at some stage in the life cycle. This halving occurs as a result of meiosis. In most animals meiosis occurs in the formation of gametes. In some plants such as ferns, however, gametes are produced by mitosis. In the fern life cycle meiosis occurs in the formation of spores (Figure 1).

Every diploid cell of an organism has two complete sets of chromosomes: one set provided by each parent. During meiosis, homologous pairs of chromosomes separate, so that only one chromosome from each pair enters a daughter cell. This is known as the **haploid** number of chromosomes which, in humans, is 23. When two haploid gametes fuse at fertilisation, the diploid number of chromosomes is restored.

The process of meiosis

Meiosis involves two nuclear divisions that normally occur immediately one after the other:

1   In the **first division (meiosis 1) homologous chromosomes** pair up and their **chromatids** wrap around each other. Equivalent portions of these chromatids may be exchanged in a process called **crossing over**. We shall see the significance of this later. By the end of this division the homologous pairs have separated, with one chromosome from each pair going into one of the two daughter cells.

2   In the **second meiotic division (meiosis 2)** the chromatids move apart. At the end of meiosis 2, four cells have usually been formed. In humans, each of these cells contains 23 chromosomes.

This is summarised in Figure 1.

In addition to halving the number of chromosomes, meiosis also produces genetic variation among the offspring, which may lead to adaptations that improve survival chances. Meiosis brings about this genetic variation in the following two ways:

- independent segregation of homologous chromosomes
- new combinations of maternal and paternal alleles by crossing over.

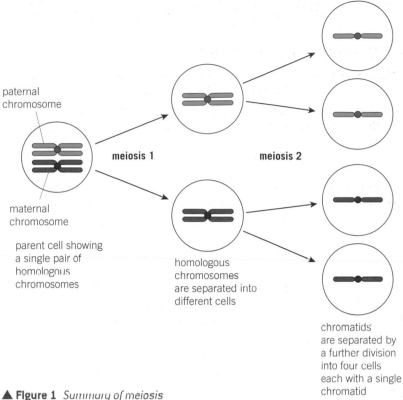

▲ **Figure 1** *Summary of meiosis*

Before we look at these two processes in more detail, let us recall the meaning of three important terms:

- **gene** – a length of DNA that codes for a polypeptide
- **locus** – the position of a gene on a chromosome or DNA molecule
- **allele** – one of the different forms of a particular gene.
- **homologous chromosomes** – a pair of chromosomes, one maternal and one paternal, that have the same gene loci.

Independent segregation of homologous chromosomes

During meiosis 1, each chromosome lines up alongside its homologous partner (see Figure 2). In humans, for example, this means that there will be 23 homologous pairs of chromosomes lying side by side. When these homologous pairs arrange themselves in this line they do so at random. One of each pair will pass to each daughter cell. Which one of the pair goes into the daughter cell, and with which one of any of the other pairs, depends on how the pairs are lined up in the parent cell. Since the pairs are lined up at random, the combination of chromosomes of maternal and paternal origin that go into the daughter cell at meiosis 1 is also a matter of chance. This is called **independent segregation**.

Variety from new genetic combinations

Each member of a homologous pair of chromosomes has exactly the same genes and therefore determines the same characteristics (e.g., tongue rolling and blood group). However, the alleles of these genes may differ (e.g., they may code for rollers or non-rollers, or blood group A or B). The independent assortment, of these chromosomes therefore produces new genetic combinations. An example is shown in Figure 2.

> **Hint**
>
> Imagine your chromosomes as two packs of 23 cards, red and blue, in which the cards are labelled from A to W. You were given the red pack by your mother and the blue pack by your father. Independent segregation is like dealing a card, of each letter in turn, at random from either of these two packs. Your final hand of 23 cards could contain any proportion of red and blue cards. In fact there are 2^{23} (over 8 million) different possible combinations.

Link

A level students will be provided with more detail on the effects of independent segregation of chromosomes in Topic 17.4 Dihybrid inheritance.

In this diagram we look at just two homologous pairs. The stages shown on the figure are:

- **Stage 1.** One of the pair of chromosomes includes the gene for tongue rolling and carries one allele for roller and one for non-roller. The other chromosome includes the gene for blood group and carries the allele for blood group A and the allele for blood group B. There are two possible arrangements, P and Q, of the two chromosomes at the start of meiosis. Both are equally probable, but each produces a different outcome in terms of the characteristics that may be passed on via the gametes.
- **Stage 2.** At the end of meiosis 1, the homologous chromosomes have segregated into two separate cells.
- **Stage 3.** At the end of meiosis 2, the chromosomes have segregated into chromatids producing four gametes for each arrangement. The actual gametes are different, depending on the original arrangement (P or Q) of the chromosomes at stage 1.

Arrangement P produces the following types of gamete with alleles for:

- roller and blood group B
- non-roller and blood group A.

Arrangement Q produces the following types of gamete with alleles for:

- non-roller and blood group B
- roller and blood group A.

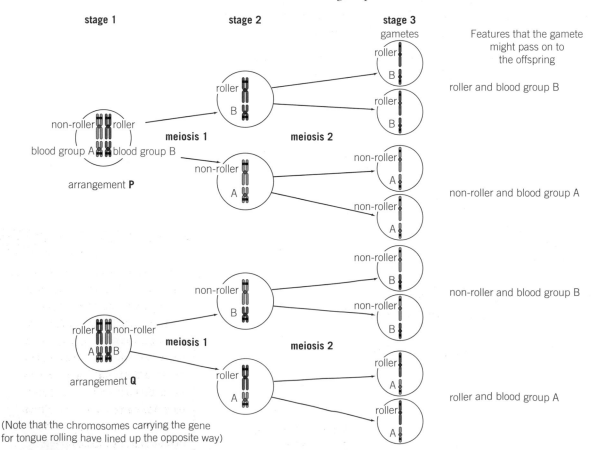

(Note that the chromosomes carrying the gene for tongue rolling have lined up the opposite way)

▲ **Figure 2** *Genetic variation produced as a result of independent segregation of chromosomes during meiosis. This diagram illustrates the independent segregation of two features, tongue rolling and blood group, that are carried on separate chromosomes*

Where the cells produced in meiosis are gametes these will be genetically different as a result of the different combinations of the maternal and paternal chromosomes/alleles they contain. These haploid gametes fuse randomly at fertilisation. The haploid gametes produced by meiosis fuse to restore the diploid state. Each gamete has a different make-up and their random fusion therefore produces variety in the offspring. Where the gametes come from different parents (as is usually the case) two different genetic make-ups are combined and even more variety results.

Genetic recombination by crossing over

We saw above that, during meiosis 1, each chromosome lines up alongside its homologous partner. The following events then take place:

- The chromatids of each pair become twisted around one another.
- During this twisting process tensions are created and portions of the chromatids break off.
- These broken portions might then rejoin with the chromatids of its homologous partner.
- Usually it is the equivalent portions of homologous chromosomes that are exchanged.
- In this way new genetic combinations of maternal and paternal alleles are produced (Figure 3).

The chromatids cross over one another many times and so the process is known as **crossing over**. The broken-off portions of chromatid recombine with another chromatid, so this process is called **recombination**.

The effect of this recombination by crossing over on the cells produced at the end of meiosis is illustrated in Figure 4. Compare the four cells

Chromatids of homologous chromosomes twist around one another, crossing over many times

Simplified representation of a single cross over

point of breakage

Result of a single cross over showing equivalent portions of the chromatid having been exchanged

▲ **Figure 3** *Crossing over*

meiosis 1   meiosis 2

Parent cell with homologous chromosomes after crossing over and recombination

All four cells produced have a different genetic composition

▲ **Figure 4** *Genetic variation as a result of recombination by crossing over*

Hint

Imagine your packs of blue and red cards again. Recombination is like taking a red card and a blue card, each with the same number, and tearing an identical portion from each card and attaching it to the other card, so that you have new cards that are part red and part blue. You can do this in an almost infinite number of ways and all before you start to deal them as before!

that result with those shown in Figure 1. If there is no recombination by crossing over only two different types of cell are produced. However, if recombination does occur, four different cell types are produced. Crossing over therefore increases genetic variety even further.

Possible chromosome combinations following meiosis

Homologous pairs of chromosomes line up at the equator of a cell during meiosis I. Either one of a pair can pass into each daughter cell (independent segregation) and so there are a large number of possible combinations of chromosomes in any daughter cell. It is possible to make a mathematical calculation based on the number of chromosomes in an organism to determine the number of possible combinations of chromosomes for each daughter cell. The formula is:

2^n where n = the number of pairs of homologous chromosomes.

So an organism with 4 homologous pairs of chromosomes can produce 2^4 or 16 possible different combinations of chromosomes of maternal and paternal origin in its daughter cells as a result of meiosis.

We have also seen that variety is further increased through the random pairing of male and female gametes. Where the gametes come from different parents two different genetic complements with different alleles are combined, providing yet more variety. Again, we can calculate this mathematically using the formula:

$(2^n)^2$ where n = the number of pairs of homologous chromosomes.

Using our example of an organism with 4 homologous pairs of chromosomes, there are $(2^n)^2$ or 256 different combinations of chromosomes in the offspring produced as the result of sexual reproduction.

These calculations are based on chromosomes staying intact throughout meiosis. In practice we know that crossing over between chromatids during meiosis I exchanges sections of chromosomes between homologous pairs in the process called recombination. As recombination occurs each time gametes are made, it will greatly increase the number of possible chromosome combinations in the gametes.

Maths link √x̄

MS 0.5 and 1.4, see Chapter 22.

Summary questions

1 A cell is examined and found to have 27 chromosomes. Is it likely to be haploid or diploid? Explain your answer.

2 State **two** ways in which meiosis leads to an increase in genetic variety.

3 Study Figure 2. Imagine that both alleles of the gene on the smaller chromosome are for blood group A (rather than blood groups A and B). List all the different combinations of alleles in the gametes.

4 A mule is a cross between a horse (64 chromosomes) and a donkey (62 chromosomes). Mules therefore have 63 chromosomes. From your knowledge of meiosis, suggest why mules are sterile.

5 √x̄ Calculate the number of possible chromosome combinations produced from the fertilisation of two gametes, each of which contains five chromosomes (assume there is no crossing over).

Worked example

Calculate the number of possible chromosome combinations produced from the fertilisation of two gametes from separate individuals whose diploid number is 12 (assume no crossing over).

As you are looking at the possible chromosomes in the offspring (after fertilisation), you must use the formula:

$(2^n)^2$, where n = the number of pairs of homologous chromosomes.

First you need to find the value of n.

You are told the diploid number is 12.

Therefore the number of *pairs* of homologous chromosomes is $12 \div 2 = 6$.

Substituting in the formula you get: $(2^6)^2 = 4096$.

Organisms are varied. Around 1.8 million **species** of organisms on Earth have been identified and named. Many more are unnamed or undiscovered. Estimates of the total number of species on this planet today range from 5 million to 100 million. All of these species are different.

Even between members of the same species there are a multitude of differences. Almost every one of the 7.3 billion people alive in 2014 are similar enough to be recognised as humans and yet different enough to be distinguished from one another. What makes us and other species similar and yet different?

Genetic diversity

We saw Topic 8.1, that it is DNA which determines the considerable variety of proteins that make up each organism. Therefore genetic similarities and differences between organisms may be defined in terms of variation in DNA. Hence it is differences in DNA that lead to the vast genetic diversity we find on Earth.

We also saw in Topic 8.1 that a section of DNA that codes for one polypeptide is called a gene. All members of the same species have the same genes. For example, all humans have a gene for blood group, just as all snapdragons (*Antirrhinum majus*) have a gene for petal colour. Which blood group humans have depends on which two **alleles** of the gene they possess. Likewise, the colour of a snapdragon's petals depends on which two alleles for petal colour it possesses. Organisms of the same species differ in their combination of alleles, not their genes.

Genetic diversity is described as the total number of different alleles in a population. A population is a group of individuals of the same species that live in the same place and can interbreed. A species consists of one, or more, populations. The greater the number of different alleles that all members of a species possess, the greater the genetic diversity of that species. Genetic diversity is reduced when a species has fewer different alleles. The greater the genetic diversity, the more likely that some individuals in a population will survive an environmental change. This is because of a wider range of alleles and therefore a wider range of characteristics. This gives a greater probability that some individual will possess a characteristic that suits it to the new environmental conditions. Genetic diversity is a factor that enables natural selection to occur.

Natural selection in the evolution of populations

Not all alleles of a population are equally likely to be passed to the next generation. This is because only certain individuals are reproductively successful and so pass on their alleles.

Reproductive success and allele frequency

Differences between the reproductive success of individuals affects **allele frequency** in populations. The process works like this:

- Within any population of a species there will be a **gene pool** containing a wide variety of alleles.

Learning objectives

→ Explain why organisms are different from one another.

→ Describe what factors influence genetic diversity.

→ Explain how reproductive success affects allele frequency within a gene pool.

→ Explain how genetic diversity enables natural selection.

Specification reference: 3.4.4

▲ **Figure 1** *Examples of genetic diversity (from top to bottom): anemone; lichens; mountain goat; fritillary butterfly*

Hint

Remember that an allele is one alternative form of a gene and, as such, is a length of DNA on one chromosome of a homologous pair.

Link

A more detailed account of natural selection is given in Topic 18.3 Natural selection, for those students studying A level.

- Random mutation of alleles within this gene pool may result in a new allele of a gene which in most cases will be harmful.

- However in certain environments, the new allele of a gene might give its possessor an advantage over other individuals in the population.

- These individuals will be better adapted and therefore more likely to survive in their competition with others.

- These individuals are more likely to obtain the available resources and so grow more rapidly and live longer. As a result, they will have a better chance of breeding successfully and producing more offspring.

- Only those individuals that reproduce successfully will pass on their alleles to the next generation.

- Therefore it is the new allele that gave the parents an advantage in the competition for survival that is most likely to be passed on to the next generation.

- As these new individuals also have the new, 'advantageous' allele, they in turn are more likely to survive, and so reproduce successfully.

- Over many generations, the number of individuals with the new, 'advantageous' allele will increase at the expense of the individuals with the 'less advantageous' alleles.

- Over time, the frequency of the new, 'advantageous' allele in the population increases while that of the 'non-advantageous' ones decreases.

It must be stressed that what is 'advantageous' depends upon the environmental conditions at any one time. For example, alleles for black body colour may be 'advantageous' as camouflage against a smoke-blackened wall, but 'non-advantageous' against a snowy landscape.

Summary questions

1 State whether each of the following is likely to *increase* or *decrease* genetic diversity:

   a increasing the variety of alleles within a population

   b breeding together closely related cats to develop varieties with longer fur

   c mutation (permanent change to the DNA) of an allele.

2 Explain how a difference in its DNA might lead to an organism having a different appearance and hence the species showing greater genetic diversity.

 Natural selection in action

The peppered moth, *Biston betularia* normally has a light colour that camouflages it against the light background of the lichen-covered trees on which it rests. In 1848 a dark (melanic) form of the peppered moth appeared in Manchester. At this time, most buildings, walls and trees in the city were blackened by the soot from 50 years of industrial development. Both types of the moth are shown in Figure 4 against this blackened background. By 1895, 98% of Manchester's population of the moth was of the black type.

▲ **Figure 2** *Light and dark varieties of the peppered moth*

1 Suggest an explanation that accounts for the change from the light to dark variety of the moth.

Selection is the process by which organisms that are better adapted to their environment tend to survive and breed, while those that are less well adapted tend not to. Every organism is subjected to a process of selection, based on its suitability for surviving the conditions that exist at the time. Different environmental conditions favour different characteristics in the population. Depending on which characteristics are favoured, selection will produce a number of different results.

- Selection may favour individuals that vary in one direction from the mean of the population. This is called **directional selection** and changes the characteristics of the population.
- Selection may favour average individuals. This is called **stabilising selection** and preserves the characteristics of a population.

Most charactcristics are influenced by more than one gene(**polygenes**). These types of characteristics are more influcnccd by the environment than ones determined by a single gene. The effect of the environment on polygenes produces individuals in a population that vary about the mean. When you plot this variation on a graph you get a **normal distribution curve**. The next part of this topic looks at how these two types of selection affect this curve.

Directional selection

If the environmental conditions change, the phenotypes (the observable physical and biochemical characteristics of an organism) that are best suited to thc ncw conditions are most likely to survive. Some individuals, which fall to either the left or right of the mean, will possess a phenotype more suited to the new conditions. These individuals will be more likely to survive and breed. They will therefore contribute more offspring (and thc allelcs these offspring possess) to the next generation than other individuals. Over time, the mean will then move in the direction of these individuals. To explain, let us take the example of the development of antibiotic resistance in bacteria.

Shortly after the discovery of antibiotics it became apparent that the effectiveness of some antibiotics at killing bacteria was reduced. It was found that these populations of bacteria had developed resistance to antibiotics such as penicillin. The resistance was not due to the development of tolerance to the antibiotic, but rather a chance mutation within the bacteria. We saw in Topic 9.1 that a mutation is a change in DNA that results in different characteristics, usually due to a change to some protein. As an example of directional selection, let us look at the case of resistance to penicillin:

- A spontaneous mutation occurred in the allele of a gene in a bacterium that enabled it to make a new protein. The new protein was an enzyme that broke down the antibiotic penicillin before it was able to kill the bacterium. The enzyme was given the name penicillinase.
- The bacterium happened, by chance, to be in a situation where penicillin was being used to treat an individual. In these circumstances, the mutation gave the bacterium an advantage in

Learning objectives

→ Describe what selection is

→ Describe the environmental factors which exert selection pressure

→ Explain what stabilising and directional selection are.

Specification reference: 3.4.4

Maths link $\sqrt{x}$

MS 3.1, see Chapter 22.

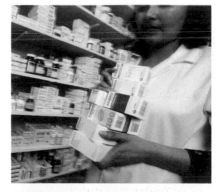

▲ **Figure 1** *One in every six prescriptions issued by doctors in the UK is for an antibiotic*

Link

A level students will expand their knowledge of types of selection in Topic 18.4 Effects of different forms of selection on evolution.

being able to use penicillinase to break down the antibiotic and so survive while the rest of the population of bacteria were killed by it.

- The bacterium that survived was able to divide by binary fission to build up a small population of penicillin-resistant bacteria.
- Members of this small penicillin-resistant population were more able to survive, and therefore multiply, in the presence of penicillin than members of the non-resistant population.
- The population of penicillin-resistant bacteria increased at the expense of the non-resistant population. Consequently the frequency of the allele that enabled the production of penicillinase increased in the population.
- The population's normal distribution curve shifted in the direction of a population having greater resistance to penicillin (see Figure 2).

While our example illustrates penicillin resistance, the process applies equally to any antibiotic.

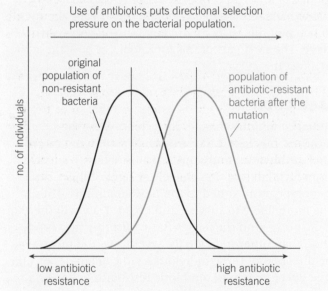

▲ **Figure 2** *Directional selection as exemplified by antibiotic resistance in bacteria*

It must be stressed that bacteria do not mutate because of the presence of antibiotics. Mutations occur randomly and are very rare. However as there are so many bacteria around, the total number of mutations is large. Many of these mutations will be of no advantage to a bacterium. Indeed most will be harmful, in which case the bacterium will probably die. Very occasionally a mutation will be advantageous. Even then it depends upon the circumstances. For example, a mutation that leads to the production of penicillinase is only an advantage when the bacterium is in the presence of penicillin. With continued use of antibiotics, there is a greater chance that the mutant population will out-compete, and replace, the original population.

Directional selection therefore results in phenotypes at one extreme of the population being selected for and those at the other extreme being selected against.

▲ **Figure 3** *The use of many antibiotics in hospitals increases the chance of multiple antibiotic resistance developing in bacteria*

Stabilising selection

If environmental conditions remain stable, it is the individuals with phenotypes closest to the mean that are favoured. These individuals are more likely to pass their alleles on to the next generation. Those individuals with phenotypes at the extremes are less likely to pass on their alleles. Stabilising selection therefore tends to eliminate the phenotypes at the extremes.

Let us look at the example of human birth weights.

Stabilising selection results in phenotypes around the mean of the population being selected for and those at both extremes being selected against. These events are summarised in Figure 4.

The body mass at birth of babies born at a hospital was measured over a 12-year period. In the graph in Figure 5 the percentage of births in the population (*y*-axis on the left) is plotted against birth mass of the infants as a histogram.

Over the same period, the infant mortality (death) rate was also recorded. The infant mortality rate is measured on a logarithmic scale (*y*-axis on the right) and plotted against infant body mass at birth as a line graph.

Looking at the histogram in Figure 4, you will see that the body mass of the babies at birth is within a relatively narrow range – mostly between 2.5 and 4.0 kg. The likely explanation for this can be found from looking at the line for infant mortality. This line climbs steeply where the birth weight is below 2.5 kg and again where it is above 4.0 kg. In other words, there is a much greater risk of infant death when the birth weight is outside the range 2.5–4.0 kg. Bigger isn't always better.

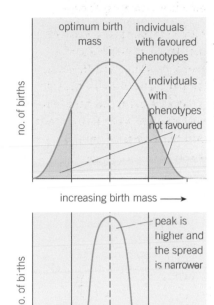

▲ **Figure 4** *Stabilising selection*

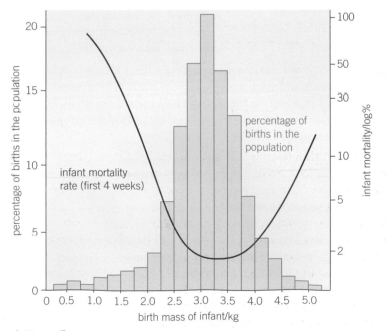

▲ **Figure 5**

1 What is selection?

2 Distinguish between directional selection and stabilising selection.

3 A severe cold spell in 1996 killed over 50% of swallows living on cliffs in Nebraska. Biologists collected nearly 2000 dead swallows from beneath the cliffs and captured around 1000 living ones. By measuring the body mass of the birds, they found that birds with a larger than average body mass survived the cold spell better than ones with a smaller than average body mass. State, giving your reasons, which type of selection was taking place here.

This illustrates stabilising selection because the mortality rate is greater at the two extremes. The infants with the highest and lowest birth masses are more likely to die (are being selected against) while those around the mean are less likely to die (are being selected for/favoured). The population's characteristics are being preserved rather than changed.

Stabilising selection therefore results in phenotypes around the mean of the population being selected for and those at both extremes being selected against. These events are summarised in Figure 5.

Natural selection results in species that are better adapted to the environment that they live in. These adaptations may be:

- **Anatomical**, such as shorter ears and thicker fur in arctic foxes compared to foxes in warmer climates.

- **Physiological**, for example oxidising of fat rather than carbohydrate in kangaroo rats to produce additional water in a dry desert environment.

- **Behavioural**, such as the autumn migration of swallows from the UK to Africa to avoid food shortages in the UK winter.

They must be cuckoo!

Cuckoos lay their eggs in the nests of other birds. The host birds will often raise these parasite chicks alongside their own.

In many valleys in southern Spain, great cuckoos and common magpies have lived together for hundreds of years. In some valleys, however, magpies have been around for centuries but cuckoos have only recently arrived.

▲ **Figure 6** *Cuckoos lay their eggs (bottom row) in the nests of magpies whose eggs are shown on the upper row*

Scientists placed artificial cuckoo eggs into magpie nests in both types of valley. Where cuckoos and magpies had lived together for a long period, 78% of the magpies removed the cuckoo eggs from their nests. Where cuckoos had only recently colonised the valleys, only 14% of the magpies removed the cuckoo eggs.

It would appear that, in the valleys where cuckoos are well established, selection has favoured those magpies that removed the cuckoo eggs.

1 Suggest **one** advantage to the magpies of removing cuckoo eggs from their nest.

2 Explain how removing cuckoo eggs increases the probability of the alleles for this type of behaviour being passed on to subsequent generations.

3 Suggest why this form of behaviour is not shown by magpies in those valleys where cuckoos have only recently arrived.

4 State, with your reasons, which type of selection is taking place here.

5 Explain how the different behaviour of the two groups of magpies might lead to selection that produces a change within the magpie population.

1 Phenylketonuria is a disease caused by mutations of the gene coding for the enzyme PAH. The table shows part of the DNA base sequence coding for PAH. It also shows a mutation of this sequence which leads to the production of non-functioning PAH.

DNA base sequence coding for PAH	C	A	G	T	T	C	G	C	T	A	C	G
DNA base sequence coding for non-functioning PAH	C	A	G	T	T	C	C	C	T	A	C	G

(a) (i) What is the maximum number of amino acids for which this base sequence could code? *(1 mark)*

(ii) Explain how this mutation leads to the formation of non-functioning PAH. *(3 marks)*

PAH catalyses a reaction at the start of two enzyme-controlled pathways.

The diagram shows these pathways.

(b) Use the information in the diagram to give two symptoms you might expect to be visible in a person who produces non-functioning PAH. *(2 marks)*

(c) One mutation causing phenylketonuria was originally only found in one population in central Asia. It is now found in many different populations across Asia. Suggest how the spread of this mutation may have occurred. *(1 mark)*

AQA Jan 2012

2 **Figure 3** shows a pair of chromosomes at the start of meiosis. The letters represent alleles.

▲ **Figure 3**

(a) What is an allele? *(1 mark)*
(b) Explain the appearance of one of the chromosomes in **Figure 3**. *(2 marks)*
(c) The cell containing this pair of chromosomes divided by meiosis. **Figure 4** shows the distribution of chromosomes from this pair in four of the gametes produced.

▲ **Figure 4**

(i) Some of the gametes formed during meiosis have new combinations of alleles. Explain how the gametes with the combinations of alleles Ef and eF have been produced. *(2 marks)*

(ii) Only a few gametes have the new combination of alleles Ef and eF. Most gametes have the combination of alleles EF and ef. Suggest why only a few gametes have the new combination of alleles, Ef and eF. *(1 mark)*

(d) **Figure 5** shows a cell with six chromosomes.

▲ **Figure 5**

(i) This cell produces gametes by meiosis. Draw a diagram to show the chromosomes in one of the gametes. *(2 marks)*

(ii) How many different types of gametes could be produced from this cell as a result of different combinations of maternal and paternal chromosomes? *(1 mark)*

AQA June 2010

(e) (i) Calculate the number of different types of gametes that can be produced in a species with a diploid number of 24. *(1 mark)*

(ii) Assuming random fertilisation, calculate the number of different combinations of maternal and paternal chromosomes in the zygotes of this species. *(1 mark)*

3 **(a)** Explain what is meant by genetic diversity. *(1 mark)*

(b) Apart from genetic factors what other type of factor causes variation within a species? *(1 mark)*

(c) The spotted owl is a bird. Numbers of spotted owls have decreased over the past 50 years. Explain how this decrease may affect genetic diversity. *(2 marks)*

AQA June 2011

4 The table shows some differences between three varieties of banana plant.

	Variety **A**	Variety **B**	Variety **C**
Number of chromosomes in a leaf cell	22	33	44
Growth rate of fruit / cm^3 week^{-1}	2.9	6.9	7.2
Breaking strength of leaf / arbitrary units	10.8	9.4	7.8

(a) (i) How many chromosomes are there in a male gamete from variety **C**? *(1 mark)*

(ii) Variety **B** cannot produce fertile gametes. Use information in the table to explain why. *(2 marks)*

In some countries very strong winds may occur. Banana growers in these countries choose to grow variety **B**.

(b) (i) Use the data in the table to explain why banana growers in these countries choose to grow variety **B** rather than variety **A**. *(1 mark)*

(ii) Use the data in the table to explain why banana growers in these countries choose to grow variety **B** rather than variety **C**. *(1 mark)*

(c) Banana growers can only grow new variety **B** plants from suckers. Suckers grow from cells at the base of the stem of the parent plant.

Use your knowledge of cell division to explain how growing variety **B** on a large scale will affect the genetic diversity of bananas. *(2 marks)*

AQA Jan 2011

Scientists have identified and named around 1.8 million different living organisms. No one knows how many types remain to be identified. Estimates for the total number of species on Earth vary from 10 million to 100 million. The figure is likely to be around 14 million. These represent only the species that exist today. Some scientists have estimated that 99% of the species that have existed on Earth are now extinct, and almost all of them have left no fossil record. With such a vast number of organisms it is clearly important for scientists to name them and sort them into groups.

Classification is the organisation of living organisms into groups. This process is not random but is based on a number of accepted principles. Before we examine how organisms are grouped according to these principles, consider how scientists distinguish one type of organism from another.

The concept of a species

A species is the basic unit of classification. A definition of a species is not easy, but members of a single species have one main thing in common:

- **They are capable of breeding to produce living, fertile offspring.** They are therefore able to produce more offspring. This means that, when a species reproduces sexually, any of the genes of its individuals can, in theory, be combined with any other.

Naming species – the binomial system

At one time scientists often gave new organisms a name that described their features, for example blackbird, rainbow trout. This practice resulted in the same names being used in different parts of the world for very different species. Therefore, it was difficult for scientists to be sure they were referring to the same organism. Over 200 years ago the Swedish botanist Linnaeus overcame this problem by devising a common system of naming organisms. This system is still in use today.

Organisms are identified by two names and hence the system is called the **binomial system**. Its features are as follows:

- It is a universal system based upon Latin or Greek names.
- The first name, called the **generic name**, denotes the genus to which the organism belongs. This is equivalent to the surname used to identify people and shared by their close relatives.
- The second name, called the **specific name**, denotes the species to which the organism belongs. This is equivalent to the first (or given) name used to identify people. However, unlike in humans, it is never shared by other species within the genus.

There are a number of rules that are applied to the use of the binomial system in scientific writing:

- The names are printed in italics or, if handwritten, they are underlined to indicate that they are scientific names.

Learning objectives

→ Explain the concept of a species is.

→ Outline how species are named.

→ Explain how courtship is a precursor to mating.

→ Explain the principles of classification.

→ Explain how classification is related to evolution.

Specification reference: 3.4.5

Study tip

A common error is to state that members of the same species are capable of breeding to produce viable offspring rather than fertile offspring. Viable simply means alive not fertile.

▲ **Figure 1** *(From top to bottom) the fungus* Mucor mucedo *(bread mould); the plant* Lathyrus odoratus *(sweet pea); the animal* Felix tigris *(tiger). The classification of these organisms is shown in Table 1 on the next page*

- The first letter of the generic name is in upper case (capitals), but the specific name is in lower case (small letters).
- If the specific name is not known, it can be written as '*sp.*', for example, *Felix sp.*

The naming of organisms is in a constant state of change. Current names reflect the present state of scientific knowledge and understanding. In the same way, the classification of species is regularly changing as our knowledge of their evolution, physical features, biochemistry and behaviour increases.

Courtship behaviour

Members of the same species have similar, or have the same genes and therefore resemble one another physically and biochemically. This helps them to distinguish members of their own species from those of other species. The same is true of behaviour. The behaviour of members of the same species is more alike than that of members of different species. Individuals can therefore recognise members of their own species by the way they act. Like the physical and biochemical features of a species, the ability to display a behaviour is genetically determined. It too has evolved and it influences the chances of survival. When it comes to survival of the species (as opposed to the individuals), courtship and mating are essential.

No individual lives forever. Reproduction is therefore the means by which a species can survive over time. Each individual has adaptations that help to ensure that their DNA is passed on, through the reproductive process, to the next generation. The females of most species only produce eggs at specific times, often as little as once a year. It is therefore important to ensure that mating is successful and that the offspring have the maximum chance of survival. Courtship behaviour helps to achieve this by enabling individuals to:

- **recognise members of their own species** to ensure that mating only takes place between members of the same species because only members of the same species can produce fertile offspring
- **identify a mate that is capable of breeding** because both partners need to be sexually mature, fertile and receptive to mating
- **form a pair bond** that will lead to successful mating and raising of offspring
- **synchronise mating** so that it takes place when there is the maximum probability of the sperm and egg meeting.
- **become able to breed** by bringing a member of the opposite sex into a physiological state that allows breeding to occur.

The females of many species undergo a cycle of sexual activity in which they can only conceive during a very short time. They are often only receptive to mating for a period around the time when they produce eggs. Courtship behaviour is used by males to determine whether the female is at this receptive stage. If she responds with the appropriate behavioural response, courtship continues and is likely to result in the production of offspring. If she is not receptive, she exhibits a different pattern of behaviour and the male ceases to court her, turning his attentions elsewhere.

During courtship, animals use signals to communicate with a potential mate and with members of their own sex. Typically there is a chain of actions between a male and female. The chain of actions is the same for all members of a species but differs for members of different species. In this way both individuals recognise that their partner is of the same species and that they may be prepared to mate.

Grouping species together – the principles of classification

With so many species, past and present, it makes sense to organise them into manageable groups. This allows better communication between scientists and avoids confusion. The grouping of organisms is known as **classification**, while the theory and practice of biological classification is called **taxonomy**.

There are two main forms of biological classification, each used for a different purpose.

- **Artificial classification** divides organisms according to differences that are useful at the time. Such features may include colour, size, number of legs, leaf shape, etc. These are described as analogous characteristics where they have the same function but do not have the same evolutionary origins. For example, the wings of butterflies and birds are both used for flight but they originated in different ways.

- **Phylogenetic classification:**
  a is based upon the evolutionary relationships between organisms and their ancestors
  b classifies species into groups using shared features derived from their ancestors
  c arranges the groups into a hierarchy, in which the groups are contained within larger composite groups with no overlap.

Relationships in a phylogenetic classification are partly based on homologous characteristics. Homologous characteristics have similar evolutionary origins regardless of their functions in the adult of a species. For example, the wing of a bird, the arm of a human and the front leg of a horse all have the same basic structure and evolutionary origins and are therefore homologous.

Organising the groups of species – taxonomy

Each group within a phylogenetic biological classification is called a taxon (plural taxa). Taxonomy is the study of these groups and their positions in a hierarchical order, where they are known as taxonomic ranks. These are based upon the evolutionary line of descent of the group members. A **domain** is the highest taxonomic rank and three are recognised: **Bacteria**, **Archaea** (a group of prokaryotes) and **Eukarya**.

▲ **Figure 1** *Courtship of great crested grebes – the weed presentation dance*

Study tip

When considering animal behaviour always remember that animals do not think like humans.

Hint

A useful mnemonic for remembering the order of the taxonomic ranks is 'Delicious King Prawn Curry Or Fat Greasy Sausages'.

Bacteria are a group of single-celled prokaryotes with the following features:

- the absence of membrane-bounded organelles such as nuclei or mitochondria
- unicellular, although cells may occur in chains or clusters
- ribosomes are smaller (70S) than in eukaryotic cells
- cell walls are present and made of murein (but never **chitin** or cellulose)
- single loop of naked DNA made up of nucleic acids but no **histones**

Archaea are a group of single-celled prokaryotes that were originally classified as bacteria which they resemble in appearance.

They differ from bacteria because:

- their genes and protein synthesis are more similar to eukaryotes
- their membranes contain fatty acid chains attached to glycerol by **ether** linkages
- there is no murein in their cell walls
- they have a more complex form of RNA polymerase.

Eukarya are a group of organisms made up of one or more eukaryotic cells. Their features are:

- their cells possess membrane-bounded organelles such as mitochondria and chloroplasts
- they have membranes containing fatty acid chains attached to glycerol by **ester** linkages
- not all possess cells with a cell wall, but where they do it contains no murein
- ribosomes are larger (80S) than in Bacteria and Archaea.

The Eukarya domain is divided into four kingdoms: **Protoctista**, **Fungi**, **Plantae** and **Animalia**. Within each kingdom the largest groups are known as **phyla**. Organisms in each phylum have a body plan radically different from organisms in any other phylum. Diversity within each phylum allows it to be divided into **classes**. Each class is divided into **orders** of organisms that have additional features in common. Each order is divided into **families** and at this level the differences are less obvious. Each family is divided into **genera** and each genus (singular) into **species**. As examples of how the system works (rather than names to be learnt), the classification of three organisms is given in Table 1.

▼ **Table 1** *Classification of three organisms from different kingdoms*

Rank	Pin mould	Sweet pea	Tiger
kingdom	Fungi	Plantae	Animalia
phylum	Zygomycota	Angiospermophyta	Chordata
class	Zygomycetes	Dicotyledonae	Mammalia
order	Mucorales	Rosales	Carnivora
family	Mucoraceae	Fabaceae	Felidae
genus	*Mucor*	*Lathyrus*	*Felix*
species	*mucedo*	*odoratus*	*tigris*

Phylogeny

The hierarchical order of taxonomic ranks is based upon the supposed evolutionary line of descent of the group members. This evolutionary relationship between organisms is known as **phylogeny**. The term is derived from the word 'phylum', which, in classification, is a group of related or similar organisms. The phylogeny of an organism reflects the evolutionary branch that led up to it. The phylogenetic relationships of different species are usually represented by a tree-like diagram called a phylogenetic tree. In these diagrams, the oldest species is at the base of the tree while the most recent ones are represented by the ends of the branches. An example is shown in Figure 3.

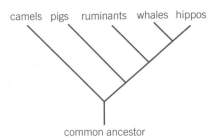

The closer the branches, the closer the evolutionary relationship. Hippos and whales are more closely related than hippos and ruminants.

▲ **Figure 3** *A phylogenetic tree showing the evolutionary relationship between certain mammals*

Summary questions

1 State the one thing that all members of a species share.

2 List the **three** features of a phylogenetic system of classification.

3 Explain why species recognition is important in courtship.

4 Suggest a way in which the courtship behaviour of one species might be used to determine which of two other species is most closely related to it.

5 *Rana temporaria* is the frog commonly found in Britain. Table 2, which is incomplete, shows part of its classification. State the most appropriate name for each of the blanks represented by the numbers 1–7.

▼ **Table 2**

Kingdom	Animalia
1	chordata
2	amphibia
3	anura
4	ranidae
genus	5
6	7

6 a State which group is the closest relative of the snakes.

b State whether dinosaurs are more closely related to crocodiles or birds.

c Suggest what C represents.

d Suggest a reason why dinosaurs are not shown along the time line like all the other groups.

▲ **Figure 5** *Turtles and crocodiles evolved from the same ancestor - but a very long time ago*

Figure 4 shows a phylogenetic tree for birds and certain reptiles.

▲ **Figure 4**

Application

The difficulties of defining species

A species may be defined in terms of observable similarities and the ability to produce fertile offspring. There are, however, certain difficulties with this definition. These include:

- Species are not fixed forever, but change and evolve over time. In time, some individuals may develop into a new species.

- Within a species there can be considerable variation among individuals. All dogs, for example, belong to the same species, but artificial selection has led to a variety of different breeds.

- Many species are extinct and most of these have left no fossil record.

- Some species rarely, if ever, reproduce sexually.

- Members of different populations of the same species may be isolated, for example by oceans, and so never meet and therefore never got the opportunity to interbreed.

- Populations of organisms that are isolated from one another may be classified as different species. These groups may turn out to be of the same species when their ability to interbreed is tested.

- Some types of organism are sterile (see below).

1 Even where groups of extinct organisms have left fossil records, it is very difficult to distinguish different species. Suggest **two** reasons why.
2 Suggest reasons why it is often difficult to classify organisms as distinct species.

A horse and a donkey (Figure 2) are capable of mating and producing offspring, which are known as mules. A horse and a donkey are, however, different species and the resulting mules are infertile, i.e. they almost never produce offspring when mated with each other. Why are mules infertile? It is all down to the number of chromosomes and the first stage of meiosis. A horse has 64 chromosomes (32 pairs) and a donkey has 62 chromosomes (31 pairs). The gametes of a horse and a donkey therefore have 32 and 31 chromosomes respectively. When the gametes of a horse and a donkey fuse, the offspring (the mule) has 63 chromosomes. Gametes are formed by meiosis but there is an odd number of chromosomes — they cannot pair up appropriately and so the gametes produced are not functional and so mules are infertile. However, mitosis can take place and therefore a mule grows and develops normally.

There have been occasional cases of a fertile female mule. This event is very rare, so much so that the Romans had a saying that meant 'when a mule foals', which was the equivalent of our modern 'once in a blue moon'.

3 Does the fact that fertile mules occasionally occur make a mule a distinct species? Give reasons for your answer.

▲ **Figure 2** *A horse (right) and a donkey (left), although different species, are capable of mating and producing offspring called mules*

Biodiversity is the general term used to describe variety in the living world. It refers to the number and variety of living organisms in a particular area and has three components:

- **Species diversity** refers to the number of different species and the number of individuals of each species within any one **community**.
- **Genetic diversity** refers to the variety of genes possessed by the individuals that make up a population of a species.
- **Ecosystem diversity** refers to the range of different **habitats**, from a small local habitat to the whole of the Earth.

One measure of species diversity is **species richness**. This is the number of different species in a particular area at a given time (community). Two communities may have the same number of species but the proportions of the community made up of each species may differ markedly. For example, a natural meadow and a field of wheat may both have 25 species. However, in the meadow, all 25 species might be equally abundant, whereas, in the wheat field, over 95% of the plants may be a single species of wheat.

Measuring the index of diversity

Consider the data shown in Table 1 about two different **habitats**. It does not tell us much about the differences between the two habitats because, in both cases, the total number of species and the total number of individuals are identical. However, if we measure the species diversity, we get a different picture.

▼ **Table 1** *Number and types of species found in two different habitats within the same ecosystem*

Species found	Numbers found in habitat X	Numbers found in habitat Y
A	10	3
B	10	5
C	10	2
D	10	36
E	10	4
no. of species	5	5
no. of individuals	50	50

One way of measuring species diversity is to use an index that is calculated as follows:

$$d = \frac{N(N-1)}{\sum n(n-1)}$$

Where:

d = **index of diversity**

N = **total number of organisms of all species**

n = **total number of organisms of each species**

$\sum$ = **the sum of**

Learning objectives

→ Describe what we understand by species diversity.

→ Explain how a diversity index is used as a measure of species diversity.

Specification reference: 3.4.6

▲ **Figure 1** *In a tropical rainforest there is high species diversity*

Maths link √x̄

MS 2.3, see Chapter 22.

▲ **Figure 2** *In the sub-arctic tundra there is low species diversity*

Worked example

Use the index to calculate the species diversity of the two habitats.

You must first calculate $n(n-1)$ for each species in each habitat. You can then calculate the sum of $n(n-1)$ for each species. These calculations are shown in Table 2.

▼ Table 2 *Calculation of $n(n-1)$ and $\sum n(n-1)$ for habitats X and Y*

Species	Numbers (n) found in habitat X	$n(n-1)$	Numbers (n) found in habitat Y	$n(n-1)$
A	10	10(9) = **90**	3	3(2) = 6
B	10	10(9) = **90**	5	5(4) = 20
C	10	10(9) = **90**	2	2(1) = 2
D	10	10(9) = **90**	36	36(35) = 1260
E	10	10(9) = **90**	4	4(3) = 12
	$\sum n(n-1)$	450	$\sum n(n-1)$	1300

You can now calculate the species diversity index for each habitat.

Habitat X: $d = \dfrac{50(49)}{450} = \dfrac{2450}{450} = 5.44$

Habitat Y: $d = \dfrac{50(49)}{1300} = \dfrac{2450}{1300} = 1.88$

The higher the value d, the greater is the species diversity. So, in this case, although the total number of species and the total number of individuals are the same in both habitats, the species diversity of habitat X is much greater.

Summary questions

1 Explain what is meant by species diversity.

2 Table 3 shows the numbers of each of six species of plant found in a salt-marsh community. Calculate the species diversity index for this salt-marsh community using the formula shown earlier. Show your working.

▼ Table 3

Species	Numbers in salt marsh
salicornia maritima	24
halimione portulacoides	20
festuca rubra	7
aster tripolium	3
limonium humile	3
suaeda maritima	1

3 Explain why it is more useful to calculate a species diversity index than just to record the number of species present.

Species diversity and ecosystems

Biodiversity reflects how well an **ecosystem** is likely to function. The higher the species diversity index, the more stable an ecosystem usually is and the less it is affected by change, for example, climate change. If there is a drought, a community with a high species diversity index is much more likely to have at least one species able to tolerate drought than a community with a low species diversity index. At least some members are therefore likely to survive the drought and maintain a community.

In extreme environments, such as hot deserts, only a few species have the necessary adaptations to survive the harsh conditions. The species diversity index is therefore normally low. This usually results in an unstable ecosystem in which communities are dominated by climatic factors rather than by the organisms within the community. In less hostile environments, the species diversity index is normally high. This usually results in a stable ecosystem in which communities are dominated by living organisms rather than climate.

1 Scientists believe that the production of greenhouse gases by human activities is contributing to climate change. Explain why an increase in greenhouse gases is more likely to result in damage to communities with a low species diversity index than communities with a high index.

2 The graph in Figure 3 shows the effect of environmental change on the stability and the functioning of ecosystems.

   a Describe the relationship between environmental change and the community with a low species diversity index.

   b Explain the different responses to environmental change in communities with a low and a high species diversity.

key
— community with low species diversity
--- community with high species diversity
— environmental change

▲ Figure 3

▲ **Figure 4** *In harsh environments, like this hot desert, only a few species are adapted to survive the extreme conditions and therefore species diversity is low*

Maths link √x̄

MS 1.3, see Chapter 22.

Learning objectives

→ Describe the impact of agriculture on species diversity.

→ Explain the balance between conservation and farming.

Specification reference: 3.4.6

In our efforts to provide enough food for the human population at a low cost, mankind has had a considerable impact on the natural world. This impact has led to a reduction in **biodiversity**. In this topic we will look at how agriculture reduces species diversity and the balance between conservation and farming.

Impact of agriculture

As natural ecosystems develop over time, they become complex **communities** with many individuals of a large number of different species. In other words, these communities have a high index of diversity. Agricultural **ecosystems** are controlled by humans and are different. Farmers often select species for particular qualities that make them more productive. As a result the number of species, and the genetic variety of **alleles** they possess, is reduced to the few that exhibit the desired features. To be economic, the number of individuals of these desirable species needs to be large. Any particular area can only support a certain amount of **biomass**. If most of the area is taken up by the one species that the farmer considers desirable, it follows that there is a smaller area available for all the other species. These many other species have to compete for what little space and resources are available. Many will not survive this competition. Even if species evolved to adapt to the changes, the population of the species would be considerably reduced. In addition, pesticides are used to exclude these species because they compete for the light, mineral ions, water and food required by the farmed species. The overall effect is a reduction in species diversity. The index of species diversity is therefore low in agricultural ecosystems.

▲ **Figure 1** *High species diversity in a hay meadow*

▲ **Figure 2** *Low species diversity in a field grown for silage*

▲ **Figure 3** *Less intensively farmed agricultural land with hedgerows.*

The balance between conservation and farming

Food is essential for life, and with an ever-expanding human population there is pressure to produce it more and more intensively. In the UK, food production has doubled over the past 40 years. This has been achieved by the use of improved genetic varieties of plant and animal species, greater use of chemical fertilisers and pesticides, greater use of biotechnology and changes in farm practices, leading to larger farms and the conversion of land supporting natural communities into farmland. These changes have had many ecological impacts, but the overriding effect of intensive food production has been to diminish the variety of **habitats** within ecosystems and consequently reduce species diversity.

Certain practices have directly removed habitats and reduced species diversity. For example:

- removal of hedgerows and grubbing out woodland
- creating monocultures, for example replacing natural meadows with cereal crops or grass for silage
- filling in ponds and draining marsh and other wetland
- over-grazing of land, for example upland areas by sheep, thereby preventing regeneration of woodland.

Other practices have had a more indirect effect:

- use of pesticides and inorganic fertilisers
- escape of effluent from silage stores and slurry tanks into water courses
- absence of crop rotation and lack of **intercropping** or undersowing.

Despite the obvious conflicts between intensive food production and conservation, there are a number of management techniques that can be applied to increase species and habitat diversity, without unduly raising food costs or lowering yields. Examples of these conservation techniques include the following:

- Maintain existing hedgerows at the most beneficial height and shape. An A-shape provides better habitats than a rectangular one.
- Plant hedges rather than erect fences as field boundaries.
- Maintain existing ponds and where possible create new ones.
- Leave wet corners of fields rather than draining them.
- Plant native trees on land with a low species diversity rather than in species-rich areas.
- Reduce the use of pesticides – use biological control where possible or genetically modified organisms that are resistant to pests.
- Use organic, rather than inorganic, fertilisers.
- Use crop rotation that includes a **nitrogen-fixing** crop, rather than fertilisers, to improve soil fertility.
- Use intercropping rather than herbicides to control weeds and other pests.
- Create natural meadows and use hay rather than grasses for silage.
- Leave the cutting of verges and field edges until after flowering and when seeds have dispersed.
- Introduce conservation headlands – areas at the edges of fields where pesticides are used restrictively so that wild flowers and insects can breed.

It is recognised that these practices will make food slightly more expensive to produce, and therefore to encourage farmers there are a number of financial incentives from the Department for Environment, Food and Rural Affairs (DEFRA) and the European Union. Maintaining biodiversity is very important. If biodiversity is reduced the global living system becomes increasingly unstable and we all rely on the global system for food and other resources.

▲ **Figure 4** *Intensively farmed agricultural land with hedgerows removed.*

Summary questions

1 Explain how agriculture has reduced species diversity.

2 Explain why there is a reduction in species diversity when a forest is replaced by grassland for grazing sheep or cattle.

3 Suggest why the draining of ponds on agricultural land might have a greater effect on biodiversity than removing a hedgerow.

 Human activity and loss of species in the UK

The present rate of species extinction is thought to be between 100 and 1000 times greater than at any time in evolutionary history. The main cause of species loss is the clearance of land in order to grow crops and meet the demand for food from an ever-increasing human population. An area of rainforest roughly the size of the UK is cleared every year. Throughout the world habitats are being lost. Most of this habitat loss has entailed the replacement of natural communities of high species diversity with agricultural ones of low species diversity. The conservation agencies in the UK have made estimates of the percentage of various habitats that have been lost in the UK since 1900. These estimates are shown in Table 1.

▲ **Figure 5** *Deforestation (left) Heathland (middle) and mixed woodland (right)*

▼ **Table 1**

Habitat	Habitat loss since 1900/%	Main reason for habitat loss
hay meadow	95	conversion to highly productive grass and silage
chalk grassland	80	conversion to highly productive grass and silage
lowland fens and wetlands	50	drainage and reclamation of land for agriculture
limestone pavements in England	45	removal for sale as rockery stone
lowland heaths on acid soils	40	conversion to grasslands and commercial forests
lowland mixed woodland	40	conversion to commercial conifer plantations and farmland
hedgerows	30	to make larger fields to accommodate farm machinery

1 There are currently approximately 350 000 km of hedgerow in the UK. Calculate how many kilometres there were in 1900.

2 Some lowland mixed woodlands have been replaced by other woodland. Explain how this change might still result in a lower species diversity.

3 Suggest **one** benefit and **one** risk associated with the conversion of hay meadows and chalk grasslands to highly productive grass and silage.

4 Suggest in what ways the information in the table might be used to inform decision-making on preserving habitats and biodiversity.

5 The European Union gives grants to farmers to replant hedges. Explain how replanting hedges might affect the species diversity found on farms.

Hedge rows!

Hedgerows typify the conflict between conservation and farming.

They were originally created to mark the boundaries of fields and to contain livestock. Over the past 50 years there has been a farming revolution with an increase in the use of large farm machinery and larger farm sizes. Small fields are not suited to the new machinery and so hedgerows are removed to make it easier to manoeuvre large equipment. Hedgerows also take up land that could produce crops and so farmers removed them, often with grants that were once available to increase productive land area.

1 Suggest **three** ways in which the removal of hedges might benefit the farmer by increasing crop yields.

Hedges do, however, have a number of uses. They increase species diversity and act as corridors along which many species move to disperse themselves. They also produce food for both animals that live in the hedgerow as well as those that do not. Overall they add diversity and interest to the countryside.

2 Suggest **two** ways in which hedges could help farmers to increase crop yields in the long term.

10.4 Investigating diversity

In Topic 10.1, we saw that classification systems were originally based on features that could easily be observed. As science has developed it has become possible to use a wider range of evidence to determine the evolutionary relationships between organisms.

When organisms evolve it is not only their visible internal and external features that change, but also the molecules of which they are made. DNA determines the proteins of an organism, including **enzymes** and proteins determine the features of an organism. It follows that changes in the features of a **species** are due to changes in its DNA. Comparing the genetic diversity within, and between, species helps scientists to determine the evolutionary relationships between them. Let us look at the different ways that these comparisons can be made.

Comparison of observable characteristics

Traditionally, genetic diversity was measured by observing the characteristics of organisms. This method is based on the fact that each observable characteristic is determined by a gene or genes (with environmental influences). The variety within a characteristic depends on the number and variety of alleles of that gene (plus environmental influences).

Using observable characteristics has its limitations because a large number of them are coded for by more than one gene. They are polygenic. This means they are not discrete from one another but rather vary continuously. It is often difficult to distinguish one from another. Characteristics can also be modified by the environment. Differences may therefore be the result of different environmental conditions rather than different alleles. Height in humans for example is determined by a number of genes. However, environmental factors like diet can influence the actual height of an individual.

For these reasons, inferring DNA differences from observable characteristics has been replaced by directly observing DNA sequences themselves. This has been made possible through the advances in gene technology made over recent years.

Comparison of DNA base sequences

With the advent of gene technology, we can now read the base sequences of the DNA of any organism. Using various techniques, we can now accurately determine the exact order of nucleotides on DNA. DNA sequencing is now routinely done by automatic machines and the data produced analysed by computers. In these computerised systems, each nucleotide base can be tagged with a different coloured fluorescent dye – adenine (green), thymine (red), cytosine (blue) and guanine (yellow). This produces a series of coloured bands, each of which represents one of the four nucleotide bases as shown in Figure 1. We can measure the genetic diversity of a species by sampling the DNA of its members and sequencing it to produce a pattern of

Learning objectives

→ Explain the use of the following techniques in comparing genetic diversity within, and between, species:

- observable characteristics
- base sequence of DNA
- base sequence of mRNA
- amino acid sequence of proteins

→ Explain how immunological comparisons are used to investigate variations in proteins.

Specification reference: 3.4.7

▲ **Figure 1** *Computer screen display of a DNA sequence. Each coloured band represents one of the four nucleotide bases*

A further study compared the number of base differences in the first 200 bases of a gene found in five species of primate. The results are shown in Table 2 below.

▼ Table 2

Human	0				
Gorilla	12	0			
Chimpanzee	15	15	0		
Orang-utan	29	33	26	0	
Lemur	48	49	49	50	0
	Human	Gorilla	Chimpanzee	Orang-utan	Lemur

5   Evaluate what evidence there is from Table 2 to show that humans are more closely related to orang-utans than to lemurs.

6   Evaluate whether these data support the evolutionary relationships of these primates suggested by the other two studies. Explain your answer.

The conflicting evidence for the relationships between different primates illustrates the need to use a variety of evidence from different sources in drawing valid scientific conclusions.

Immunological comparisons of proteins

The proteins of different species can also be compared using immunological techniques. The principle behind this method is the fact that antibodies of one species will respond to specific antigens on proteins, such as albumin, in the blood serum of another. The process is carried out as follows:

*   Serum albumin from species A is injected into species B.

*   Species B produces antibodies specific to all the antigen sites on the albumin from species A.

*   Serum is extracted from species B; this serum contains antibodies specific to the antigens on the albumin from species A.

*   Serum from species B is mixed with serum from the blood of a third species C.

*   The antibodies respond to their corresponding antigens on the albumin in the serum of species C.

*   The response is the formation of a precipitate.

*   The greater the number of similar antigens, the more precipitate is formed and the more closely the species are related.

*   The fewer the number of similar antigens, the less precipitate is formed and the more distantly the species are related.

An example of this technique is illustrated in Figure 3. In this case, species A is a human, species B is a rabbit and species C is represented by a variety of other mammals.

The results show that humans are very closely related to chimpanzees, less so to baboons and even less so to spider monkeys. They are only distantly related to dogs.

▲ Figure 4 *Immunological comparisons of human serum with that of other species*

One look around us and it is clear that living things differ. If one species differs from another this is called **interspecific variation** (Figure 1). But members of the same species also differ from each other. This is called **intraspecific variation**. Every one of the billions of organisms on planet Earth is unique. Even identical twins, who are born with the same DNA, vary as a result of their different experiences. How then do we measure the differences between these characteristics?

Making measurements

All scientists measure things, but this is a particular problem for biologists. This is because they are usually measuring some aspect of living organisms and all living organisms are different. For this reason, biologists have to take many measurements of the same thing. They cannot reliably determine the height of buttercups or the number of red cells in 1 mm³ of human blood by taking a single measurement. Equally, they cannot measure every buttercup or human being in existence. What they do is take samples.

Random sampling

Sampling involves taking measurements of individuals, selected from the population of organisms which is being investigated. In theory, if these individuals are representative of the population as a whole, then the measurements can be relied upon. But are the measurements representative? There are several reasons why they might not be, including:

- **sampling bias.** The selection process may be biased. The investigators may be making unrepresentative choices, either deliberately or unwittingly. Are they as likely to take samples of buttercups from a muddy area as a dry one? Will they avoid areas covered in cow dung or rich in nettles?

- **chance.** Even if sampling bias is avoided, the individuals chosen may, by pure chance, not be representative. The 50 buttercup plants selected might just happen to be the 50 tallest in the population.

The best way to prevent sampling bias is to eliminate, as far as possible, any human involvement in choosing the samples. This can be achieved by carrying out **random sampling**. One method is:

1. Divide the study area into a grid of numbered lines, for example by stretching two long tape measures at right angles to each other.
2. Using random numbers, from a table or generated by a computer, obtain a series of coordinates.
3. Take samples at the intersection of each pair of coordinates.

We cannot completely remove chance from the sampling process but we can minimise its effect by:

- **using a large sample size.** The more individuals that are selected the smaller is the probability that chance will influence the result,

Learning objectives

→ Describe how variation is measured.

→ Explain what sampling is and why it is used.

→ Describe the types of variation and their causes.

→ Explain what is meant by the mean and standard variation.

Specification reference: 3.4.7

▲ **Figure 1** *Variation between species (interspecific variation)*

Study tip

Remember that intraspecific variation is variation *within* a species.

and the less influence anomalies will have. If we sample only five buttercups there is a high probability that they may all be taller than average. If we sample 500 there is a much lower probabilty that they will all be taller than average. The greater the sample size the more reliable the data will be.

- **analysis of the data collected.** Accepting that chance will play a part, the data collected can be analysed using statistical tests to determine the extent to which chance may have influenced the data. These tests allow us to decide whether any variation observed is the result of chance or is more likely to have some other cause.

The normal distribution curve

Figure 2 shows a normal distribution curve: its bell-shape is typical for a feature that shows continuous variation, for example height in humans. The graph is symmetrical about a central value. Occasionally the curve is shifted slightly to one side. This is called a skewed distribution and is illustrated in Figure 3. There are three terms associated with normal distribution curves. To illustrate these terms, consider the values given in Table 1 that compares the number of children in 11 different families.

▼ Table 1

Family	Number of children
A	0
B	1
C	1
D	1
E	2
F	2
G	3
H	4
I	6
J	6
K	7

The mean (arithmetic mean)

This the sum of the sampled values divided by the number of items. In our example, total the number of children in all families and divide by the number of families:

Total children = 0 + 1 + 1 + 1 + 2 + 2 + 3 + 4 + 6 + 6 + 7
= 33

Total number of families A–K = 11

Mean $= \dfrac{33}{11} = 3$

The mode

This is the single value of a sample that occurs most often. In our example, more families have one child than any other number. The mode is therefore equal to 1.

The median

This is the central or middle value of a set of values. This requires arranging the values in ascending order. In our example, they are already arranged in ascending order of the number of children in each family. There are 11 families. The sixth family is therefore the middle family in the sample. This is family F. Family F has two children and so the median = 2.

Figure 2 shows a typical normal distribution curve in which the mean and mode (and often the median) have the same value. Figure 3 shows a skewed distribution in which the mean, mode and median all have different values.

Maths link $\sqrt{x}$

MS 1.2 and 1.6, see Chapter 22.

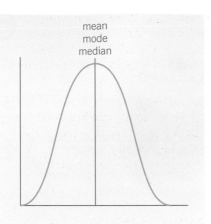

▲ **Figure 2** *A normal distribution curve where the mean, mode and median have the same value*

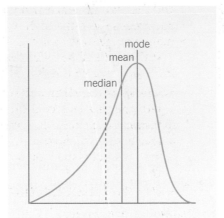

▲ **Figure 3** *A skewed distribution where the mean, mode and median have different values*

Mean and standard deviation

A normal distribution curve always has the same basic shape (Figure 2). It differs in two measurements: its maximum height and its width.

* The **mean** is the measurement at the maximum height of the curve. As we have seen, the mean of a sample of data provides an average value and is useful information when comparing one sample with another. It does not, however, provide any information about the range of values within the sample. For example, the mean number of children in a sample of eight families may be two. However, this could be made up of eight families each with two children or six families with no children and two families with eight children each.

* The **standard deviation** (*s*) is a measure of the width of the curve. It gives an indication of the range of values either side of the mean. A standard deviation is the distance from the mean to the point where the curve changes from being convex to concave (the point of inflexion) **68%** of all the measurements lie within ± 1.0 standard deviation. Increasing this width to almost ± 2.0 (actually ± 1.96) standard deviations takes in 95% of all measurements. These measurements are illustrated in Figure 4. To calculate the standard deviation with any accuracy there needs to be a minimum number of values.

Calculating standard deviation

At first sight, the formula for standard deviation can look complex:

$$\text{standard deviation} = \sqrt{\frac{\sum (x - \bar{x})^2}{n - 1}}$$

Where:

$\sum$ = **the sum of**

x = **measured value (from the sample)**

$\bar{x}$ = **mean value**

n = **total number of values in the sample.**

However, it is straightforward to calculate and less frightening if you take it step by step.

Maths link √x̄

MS 1.2, see Chapter 22.

Study tip

A large standard variation means a lot of variety, while a small standard deviation means little variety.

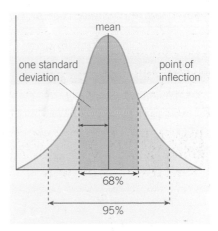

▲ **Figure 4** *The normal distribution curve showing values for standard deviation*

Hint

Remember to square all the numbers, not just the negative ones. It is also good practice not to round up figures too early in a calculation.

Maths link $\sqrt{x}$

MS 1.10, see Chapter 22.

Worked example

The following very simple example, using the six measured values (x) 4, 1, 2, 3, 5 and 0, illustrates each step in the process.

- Calculate the mean value ($\bar{x}$), i.e. $4 + 1 + 2 + 3 + 5 + 0 = 15$
$$15 \div 6 = 2.5$$

- Subtract the mean value from each of the measured values ($x - \bar{x}$). This gives: +1.5, −1.5, −0.5, +0.5, +2.5, −2.5.

- As some of these numbers are negative, you need to make them positive. To do this, square *all* the numbers $(x - \bar{x})^2$. Remember to square all the numbers and not just the negative ones. This gives: 2.25, 2.25, 0.25, 0.25, 6.25, 6.25.

- Add all these squared numbers together:
$$\sum (x - \bar{x})^2 = 17.5$$

- Divide this number by the original number of measurements less one, i.e. 5:
$$\frac{\sum (x - \bar{x})^2}{n - 1} = \frac{17.5}{5} = 3.5$$

- As all the numbers have been squared, the final step is to take the square root in order to get back to the same units as the mean:
$$\sqrt{\frac{\sum (x - \bar{x})^2}{n - 1}} = \sqrt{3.5} = 1.87$$

Study tip

You may need to calculate standard deviations as part of practical work but you will not be asked to do so in written papers.

Maths link $\sqrt{x}$

MS 0.4 and 1.1, see Chapter 22.

You will need to use a calculator to find out the value of standard deviations as it will considerably speed up your calculation. In doing so you will usually find the calculator provides a long figure running to many decimal places. In this example, the calculation $\sqrt{3.5}$ produces the answer 1.870828693. Clearly the significance of the latter digits is less than the earlier ones. It is normal to reduce these figures to a certain number of significant figures. In this case the answer has been rounded down to three significant figures, namely 1.87.

The number of significant figures to use can vary and you should always follow any instructions on how many significant figures to use in your answer. In the absence of specific instructions, it is best to use the same number of significant figures as the data you are given or are using. For example, in this case the square values were calculated to be 2.25, 0.25, 6.25, etc. As these had three significant figures, the same number was used in the final calculation. Remember, however, that where a numerical answer to one part of a question has to be used in a subsequent calculation, the answer to the first part should be carried forward without rounding it up or down.

Summary questions

1 List **two** reasons why a sample may not be representative of the population as a whole.

2 Explain how sampling bias may be prevented.

3 Distinguish between the terms 'mean', 'mode' and 'median'.

1 (a) What is a *species*?   (*2 marks*)
   (b) Scientists investigated the diversity of plants in a small area within a forest.
        The table shows their results.

Plant species	Number of individuals
Himalayan raspberry	20
Heartwing sorrel	15
Shala tree	09
Tussock grass	10
Red cedar	04
Asan tree	06
Spanish needle	8
Feverfew	8

The index of diversity can be calculated by the formula

$$d = \frac{N(N-1)}{\sum n(n-1)}$$

where
d = index of diversity
N = total number of organisms of all species
n = total number of organisms of each species

   (i)   Use the formula to calculate the index of diversity of plants in the forest.
         Show your working.   (*2 marks*)
   (ii) The forest was cleared to make more land available for agriculture.
         After the forest was cleared the species diversity of insects in the area
         decreased.
         Explain why.   (*3 marks*)
                                                                                     AQA June 2013

2 Organisms can be classified using a hierarchy of phylogenetic groups.
   (a) Explain what is meant by:
        (i)   a hierarchy   (*2 marks*)
        (ii) a phylogenetic group.   (*1 mark*)
   (b) Cytochrome c is a protein involved in respiration. Scientists determined
        the amino acid sequence of human cytochrome c. They then:
        • determined the amino acid sequences in cytochrome c from five other animals
        • compared these amino acid sequences with that of human cytochrome c
        • recorded the number of differences in the amino acid sequence compared with
          human cytochrome c.
        The table shows their results.

Animal	Number of differences in the amino acid sequence compared with human cytochrome c
A	1
B	12
C	12
D	15
E	21

(b) Give the scientific name of this otter. *(1 mark)*

Scientists investigated the effect of hunting on the genetic diversity of otters. Otters are animals that were killed in very large numbers for their fur in the past.

The scientists obtained DNA from otters alive today and otters that were alive before hunting started.

For each sample of DNA, they recorded the number of base pairs in alleles of the same gene. Mutations change the numbers of base pairs over time.

Figure 6 shows the scientists' results.

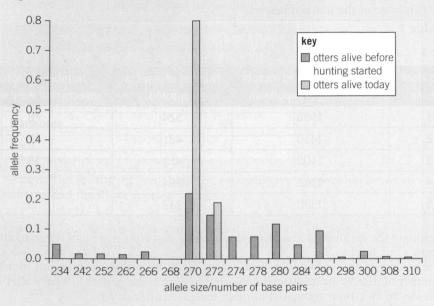

(c) The scientists obtained DNA from otters that were alive before hunting started. Suggest **one** source of this DNA. *(1 mark)*

(d) What can you conclude about the effect of hunting on genetic diversity in otters? Use data from **Figure 6** to support your answer. *(2 marks)*

(e) Some populations of animals that have never been hunted show very low levels of genetic diversity.

Other than hunting, suggest **two** reasons why populations might show very low levels of genetic diversity. *(2 marks)*

AQA SAMS AS PAPER 1

Section 5
Energy transfer in and between organisms

Introduction

Organisms require a constant input of energy to maintain their highly ordered structures and systems. Life depends on energy, usually from the Sun, being transferred continuously between organisms. Living organisms live isolated lives but form part of interdependent communities. Each community interacts with other communities and with its non-living environment within ecosystems. While ecosystems as a whole remain relatively stable, their biotic and abiotic components are constantly changing. Ecosystems are maintained by light energy from the Sun that photosynthesising organisms absorb with chlorophyll to produce ATP and carbohydrates. These carbohydrates, and other substrates, are broken down by all organisms to produce the ATP needed for survival.

ATP production in both photosynthesis and respiration is formed when protons (hydrogen ions) diffuse down an electrochemical gradient through molecules of ATP synthase. This enzyme is embedded in the membranes of chloroplasts and mitochondria. As respiration is common to all organisms, and photosynthesis to all photoautotrophic ones, they provide indirect evidence for evolution.

The Sun constantly provides energy, which flows through ecosystems. In communities, molecules produced by photosynthetic organisms (producers) are consumed by other organisms such as bacteria, fungi, and animals (consumers). While the supply of energy from the Sun will be constant for the foreseeable future, other nutrients are finite and are recycled.

Working scientifically

The study of energy transfer in and between organisms provides many opportunities to carry out practical work and to develop practical skills. Required practical activities are:

- The use of chromatography to investigate the pigments isolated from leaves of different plants, e.g., leaves from shade-tolerant and shade-intolerant plants or leaves of different colours.
- Investigation into the effect of a named factor on the rate of dehydrogenase activity in extracts of chloroplasts.
- Investigation into the effect of a named variable on the rate of respiration of cultures of single-celled organisms.

In carrying out these activities you could develop practical skills such as:

- using appropriate apparatus to record a range of quantitative measurements
- using appropriate instrumentation to record quantitative measurements
- separating biological compounds using thin layer/paper chromatography
- using microbiological aseptic techniques.

You will require a range of mathematical skills. In particular the ability to recognise and make use of appropriate units in calculations, use fractions and percentages, and solve algebraic equations.

What you already know

The material in this unit is intended to be self-explanatory, but there is certain information from GCSE that will prove helpful to the understanding of this section. This information includes:

- ◯ Photosynthesis uses light energy to combine carbon dioxide and water to form glucose and oxygen.

- ◯ Light energy is absorbed by chlorophyll, which is found in chloroplasts in some plant cells and algae.

- ◯ The rate of photosynthesis may be limited by a shortage of light, low temperature or a shortage of carbon dioxide.

- ◯ Aerobic respiration takes place continuously in plants and animals during which oxygen is used to release energy from glucose.

- ◯ Aerobic respiration is summarised as: glucose + oxygen → carbon dioxide + water (+ energy)

- ◯ Energy released during respiration is used by the organism to build larger molecules from smaller ones.

- ◯ Radiation from the Sun is the source of energy for most communities of living organisms.

- ◯ The mass of living material (biomass) at each stage in a food chain is less than it was at the previous stage.

- ◯ The amounts of material and energy contained in the biomass of organisms is reduced at each successive stage in a food chain.

- ◯ Living things remove materials from the environment for growth and other processes and these materials are returned to the environment either in waste materials or when living things die and decay.

- ◯ Materials decay because they are broken down by microorganisms. The decay process releases substances that plants need to grow.

- ◯ Anaerobic respiration is the incomplete breakdown of glucose and produces lactic acid and it releases much less energy than during aerobic respiration.

Learning objectives

→ Explain how the plant leaf is adapted to carry out photosynthesis.

→ Describe the main stages of photosynthesis.

Specification reference: 3.5.1

Synoptic link

The structure of ATP and of the chloroplast were considered in Topics 2.3, Energy and ATP, and 3.4, Eukaryotic cell structure, respectively. A review of these topics will help you to follow how both are linked in the process of photosynthesis as described here.

Humans, along with almost every other living organism, owe their continued existence to photosynthesis. The energy we rely on, whether it comes from food when we respire or from the wood, coal, oil or gas that we burn in our homes, has been captured by photosynthesis from sunlight. Photosynthesis likewise produces the oxygen we breathe by releasing it from water molecules.

Life depends on continuous transfers of energy. How this energy enters an organism depends on its type of nutrition. In plants, energy in light is absorbed by chlorophyll and then transferred into the chemical energy of the molecules formed during photosynthesis. These molecules are used by the plant to produce **ATP** during respiration. Non-photosynthetic organisms feed on the molecules produced by plants and then also use them to make ATP during respiration.

Site of photosynthesis

The leaf is the main photosynthetic structure in eukaryotic plants. Chloroplasts are the cellular organelles within the leaf where photosynthesis takes place.

Structure of the leaf

Photosynthesis takes place largely in the leaf, the structure of which is shown in Figure 2. Leaves are adapted to bring together the three raw materials of photosynthesis (water, carbon dioxide, and light) and remove its products (oxygen and glucose). These adaptations include:

- a large surface area that absorbs as much sunlight as possible
- an arrangement of leaves on the plant that minimises overlapping and so avoids the shadowing of one leaf by another
- thin, as most light is absorbed in the first few micrometres of the leaf and the diffusion distance for gases is kept short
- a transparent **cuticle** and epidermis that let light through to the photosynthetic mesophyll cells beneath
- long, narrow upper mesophyll cells packed with chloroplasts that collect sunlight
- numerous stomata for gaseous exchange so that all mesophyll cells are only a short diffusion pathway from one
- stomata that open and close in response to changes in light intensity
- many air spaces in the lower mesophyll layer to allow rapid diffusion in the gas phase of carbon dioxide and oxygen
- a network of xylem that brings water to the leaf cells, and phloem that carries away the sugars produced during photosynthesis.

An outline of photosynthesis

The overall equation for photosynthesis is:

▲ **Figure 1** *Overview of photosynthesis*

$$6CO_2 + 6H_2O \xrightarrow{\text{light}} C_6H_{12}O_6 + 6O_2$$

carbon dioxide   water   glucose   oxygen

▲ **Figure 2** *Leaf and chloroplast structure. Bottom left, false colour transmission electron micrograph (TEM) of a chloroplast*

The equation shown is highly simplified. Photosynthesis is a complex metabolic pathway involving many intermediate reactions. It is a process of energy transferral in which some of the energy in light is conserved in the form of chemical bonds. There are three main stages to photosynthesis, see Figure 1:

1 **capturing of light energy** by chloroplast pigments such as chlorophyll
2 **the light-dependent reaction**, in which some of the light energy absorbed is conserved in chemical bonds. During the process an electron flow is created by the effect of light on chlorophyll, causing water to split (**photolysis**) into **protons**, **electrons**, and oxygen. The products are reduced NADP, ATP, and oxygen.

Practical link 🧪

Required practical 7. Use of chromatography to investigate pigments isolated from leaves of different plants, for example, leaves from shade-tolerant and shade-intolerant plants or leaves of different colours.

3 **the light-independent reaction**, in which these protons (hydrogen ions) are used to produce sugars and other organic molecules.

Structure and role of chloroplasts in photosynthesis

In eukaryotic plants, photosynthesis takes place within cell organelles called chloroplasts, the structure of which is shown in Figure 2d. These vary in shape and size but are typically disc-shaped, 2–10 µm long, and 1 µm in diameter. They are surrounded by a double membrane. Inside the chloroplast membranes are two distinct regions:

- **The grana** are stacks of up to 100 disc-like structures called **thylakoids** where the light-dependent stage of photosynthesis takes place. Within the thylakoids is the photosynthetic pigment called chlorophyll. Some thylakoids have tubular extensions that join up with thylakoids in adjacent grana. These are called inter-granal lamellae.

- **The stroma** is a fluid-filled matrix where the light-independent stage of photosynthesis takes place. Within the stroma are a number of other structures such as starch grains.

▲ **Figure 3** *Photomicrograph of a moss leaf showing cells that contain chloroplasts (green) around their margins*

Summary questions

1 List **two** molecules that are the raw materials of photosynthesis.

2 List **two** molecules that are the products of photosynthesis.

3 State in which parts of the chloroplast each of the following occur:

  a the light-dependent reaction

  b the light-independent reaction.

4 Name the products of each of the following:

  a the light-dependent reaction

  b the light-independent reaction.

The light-dependent reaction of photosynthesis involves the capture of light whose energy is used for two purposes:

- to add an inorganic phosphate (P_i) molecule to ADP, thereby making **ATP**
- to split water into H^+ ions (protons) and OH^- ions. As the splitting is caused by light, it is known as **photolysis**.

Oxidation and reduction

Before we look at what happens in the light-dependent reaction, it is necessary to understand what oxidation and reduction are.

When a substance gains oxygen or loses hydrogen the process is called **oxidation**. The substance to which oxygen has been added or hydrogen has been lost is said to be oxidised. When a substance loses oxygen, or gains hydrogen, the process is called **reduction**. In practice, when a substance is oxidised it loses **electrons** and when it is reduced it gains electrons. This is the more usual way to define oxidation and reduction. Oxidation results in energy being given out, whereas reduction results in it being taken in. Oxidation and reduction always take place together.

The making of ATP

When a chlorophyll molecule absorbs light energy, it boosts the energy of a pair of electrons within this chlorophyll molecule, raising them to a higher energy level. These electrons are said to be in an excited state. In fact the electrons become so energetic that they leave the chlorophyll molecule altogether. As result the chlorophyll molecule becomes ionised and so the process is called **photoionisation**. The electrons that leave the chlorophyll are taken up by a molecule called an **electron carrier**. Having lost a pair of electrons, the chlorophyll molecule has been oxidised. The electron carrier, which has gained electrons, has been reduced.

The electrons are now passed along a number of electron carriers in a series of oxidation-reduction reactions. These electron carriers form a transfer chain that is located in the membranes of the **thylakoids**. Each new carrier is at a slightly lower energy level than the previous one in the chain, and so the electrons lose energy at each stage. Some of this energy is used to combine an inorganic phosphate molecule with an ADP molecule in order to make ATP.

The precise mechanism by which ATP is produced can be explained by the **chemiosmotic theory**. This is described here and illustrated in Figure 1.

- Each thylakoid is an enclosed chamber into which protons (H^+) are pumped from the stroma using protein carriers in the thylakoid membrane called proton pumps.
- The energy to drive this process comes from electrons released when water molecules are split by light – photolysis of water (see later).

Learning objectives

→ Explain the processes of oxidation and reduction.

→ Explain how ATP is made during the light-dependent reaction.

→ Describe the role of photolysis in the light-dependent reaction.

→ Explain how chloroplasts are adapted to carry out the light-dependent reaction.

Specification reference: 3.5.1

Hint

Oxidation and reduction can each be described in three ways:

Oxidation – loss of electrons or loss of hydrogen or gain of oxygen.

Reduction – gain of electrons or gain of hydrogen or loss of oxygen.

Study tip

Make sure you know the three ways in which something can be oxidised or reduced.

Synoptic link

An understanding of the light-independent reaction depends on knowledge of membrane structure, diffusion, and active transport. Now would be a good time to revise Topics 4.1, 4.2, and 4.4.

- The photolysis of water also produces protons which further increases their concentration inside the thylakoid space.

- Overall this creates and maintains a concentration gradient of protons across the thylakoid membrane with a high concentration inside the thylakoid space and a low concentration in the stroma.

- The protons can only cross the thylakoid membrane through ATP synthase channel proteins – the rest of the membrane is impermeable to protons. These channels form small granules on the membrane surface and so are also known as stalked granules.

- As the protons pass through these ATP synthase channels they cause changes to the structure of the enzyme which then catalyses the combination of ADP with inorganic phosphate to form ATP.

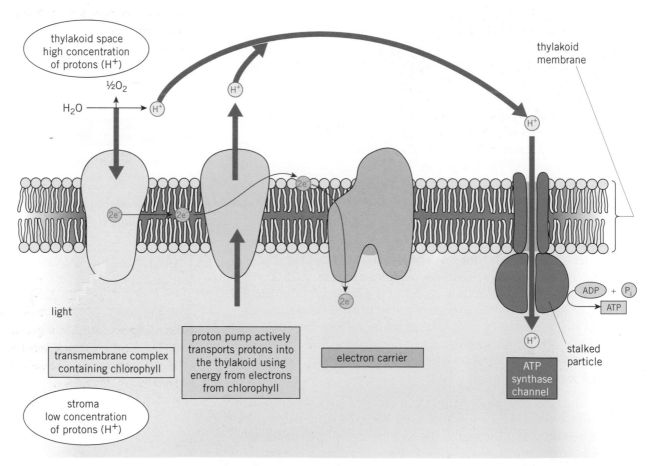

▲ **Figure 1** *Chemiosmosis in a thylakoid*

Photolysis of water

The loss of electrons when light strikes a chlorophyll molecule leaves it short of electrons. If the chlorophyll molecule is to continue absorbing light energy, these electrons must be replaced. The replacement electrons are provided from water molecules that are split using light energy. This photolysis of water also yields protons. The equation for this process is:

$$2H_2O \longrightarrow 4H^+ + 4e^- + O_2$$
$$\text{water} \qquad \text{protons} \qquad \text{electrons} \qquad \text{oxygen}$$

These protons pass out of the thylakoid space through the ATP synthase channels and are taken up by an electron carrier called NADP. On taking up the protons the NADP becomes reduced. The reduced NADP is the main product of the light-dependent stage and it enters the light-independent reaction (Topic 11.3) taking with it the electrons from the chlorophyll molecules. The reduced NADP is important because it is a further potential source of chemical energy to the plant. The oxygen by-product from the photolysis of water is either used in respiration or diffuses out of the leaf as a waste product of photosynthesis. Figure 2 summarises how ATP and reduced NADP are produced during the light-dependent stage of photosynthesis.

▲ **Figure 2** *Summary of light-dependent stage of photosynthesis*

▲ **Figure 3** *False-colour TEM of grana in a chloroplast from a leaf of maize. The grana are made up of disc-like thylakoids where the light-dependent reaction of photosynthesis takes place*

Site of the light-dependent reaction

As we have seen, the light-dependent reaction of photosynthesis takes place in the thylakoids of chloroplasts. The thylakoids are disc-like structures that are stacked together in groups called grana.

Chloroplasts are structurally adapted to their function of capturing sunlight and carrying out the light-dependent reaction of photosynthesis in the following ways:

- The thylakoid membranes provide a large surface area for the attachment of chlorophyll, electron carriers and enzymes that carry out the light-dependent reaction.
- A network of proteins in the grana hold the chlorophyll in a very precise manner that allows maximum absorption of light.
- The granal membranes have ATP synthase channels within them, which catalyse the production of ATP. They are also selectively permeable which allows establishment of a proton gradient.
- Chloroplasts contain both DNA and ribosomes so they can quickly and easily manufacture some of the proteins involved in the light-dependent reaction.

Summary questions

1 State precisely where within a plant cell the electron carriers involved in the light-dependent reaction are found.

2 Describe what happens in the photolysis of water.

3 In each of the following, state whether the process involves oxidation or reduction of the molecule named.

   a An unsaturated fat molecule gains a hydrogen atom.

   b Oxygen is lost from a carbon dioxide molecule.

   c Light causes an electron to leave a chlorophyll molecule.

Chloroplasts and the light-dependent reaction

Figure 4 shows the structure of a chloroplast.

▲ Figure 4

1 Name the parts labelled A, C and D.

2 State in which of these labelled parts the light-dependent reaction takes place?

3 Structure B is used for storage. Suggest the name of the substance likely to be stored in B.

4 ATP is produced in the light-dependent reaction of photosynthesis. Suggest **two** reasons why plants cannot use this as their only source of ATP.

5 √x̄ The actual length of X—Y in this chloroplast is 2 μm. Calculate the magnification used in Figure 4. Show your working.

The products of the light-dependent reaction of photosynthesis, namely **ATP** and reduced **NADP**, are used to reduce glycerate 3-phosphate in the second stage of photosynthesis. Unlike the first stage, this stage does not require light directly and, in theory, occurs whether or not light is available. It is therefore called the light-independent reaction. In practice, it requires the products of the light-dependent stage and so rapidly ceases when light is absent. The light-independent reaction takes place in the stroma of the chloroplasts. The details of this stage were worked out by Melvin Calvin and his co-workers and so it is often referred to as the Calvin cycle.

The Calvin cycle

▲ **Figure 1** *Summary of the light-independent reaction of photosynthesis (or Calvin cycle)*

In the following account of the Calvin cycle, the numbered stages are illustrated in Figure 1.

1   Carbon dioxide from the atmosphere diffuses into the leaf through **stomata** and dissolves in water around the walls of the mesophyll cells. It then diffuses through the cell-surface membrane, cytoplasm and chloroplast membranes into the **stroma** of the chloroplast.
2   In the stroma, the carbon dioxide reacts with the 5-carbon compound **ribulose bisphosphate (RuBP)** a reaction catalysed by an enzyme called ribulose bisphosphate carboxylase, otherwise known as **rubisco**.
3   The reaction between carbon dioxide and RuBP produces two molecules of the 3-carbon **glycerate 3-phosphate (GP)**.
4   Reduced NADP from the light-dependent reaction is used to reduce glycerate 3-phosphate to **triose phosphate (TP)** using energy supplied by ATP.

Learning objectives

→ Explain how carbon dioxide absorbed by plants is incorporated into organic molecules.

→ Describe the roles of ATP and reduced NADP in the light-independent reaction.

→ Describe the events in the Calvin cycle.

Specification reference: 3.5.1

Any substance whose name ends in 'ose' is a sugar. The ending 'ate' usually means that the substance is an acid (in solution).

5 The NADP is re-formed and goes back to the light-dependent reaction to be reduced again by accepting more protons.

6 Some triose phosphate molecules are converted to organic substances that the plant requires such as starch, cellulose, lipids, glucose, amino acids, and nucleotides..

7 Most triose phosphate molecules are used to regenerate ribulose bisphosphate using ATP from the light-dependent reaction.

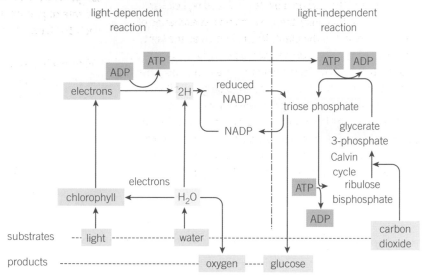

▲ **Figure 2** *Summary of photosynthesis*

Summary questions

1 Describe the role of ribulose bisphosphate (RuBP) in the Calvin cycle.

2 State how the reduced NADP from the light-dependent reaction is used in the light-independent reaction.

3 Apart from reduced NADP, which other product of the light-dependent reaction is used in the light-independent reaction?

4 State precisely where in a plant cell the enzymes involved in the Calvin cycle are found.

5 Light is not required for the Calvin cycle to take place. Explain therefore why the Calvin cycle cannot take place for long in the absence of light.

Site of the light-independent reaction

The light-independent reaction of photosynthesis takes place in the stroma of the chloroplasts.

The chloroplast is adapted to carrying out the light-independent reaction of photosynthesis in the following ways:

• The fluid of the stroma contains all the enzymes needed to carry out the light-independent reaction. Stromal fluid is membrane-bound in the chloroplast which means a chemical environment which has a high concentration of enzymes and substrates can be maintained within it – as distinct from the environment of the cytoplasm.

• The stroma fluid surrounds the grana and so the products of the light-dependent reaction in the grana can readily diffuse into the stroma.

• It contains both DNA and ribosomes so it can quickly and easily manufacture some of the proteins involved in the light-independent reaction.

Factors affecting photosynthesis √x̄

In any complex process such as photosynthesis, the factors that affect its rate all operate together. However, the rate of the process at any given moment is not affected by all the factors, but rather by the one whose level is at the least favourable value. This factor is called a **limiting factor** because it limits the rate at which the whole process can take place. Changing only the levels of the other factors will not alter the rate of the process.

The law of limiting factors can therefore be expressed as:

At any given moment, the rate of a physiological process is limited by the factor that is at its least favourable value.

When light is the limiting factor, the rate of photosynthesis is directly proportional to light intensity. As light intensity is increased, the volume of oxygen produced and carbon dioxide absorbed due to photosynthesis will increase to a point at which it is exactly balanced by the oxygen absorbed and the carbon dioxide produced by cellular respiration. At this point there will be no net exchange of gases into or out of the plant. This is known as the light **compensation point**. Further increases in light intensity will cause a proportional increase in the rate of photosynthesis and increasing volumes of oxygen will be given off and carbon dioxide taken up. A point will be reached at which further increases in light intensity will have no effect on photosynthesis. At this point some other factor, such as carbon dioxide concentration or temperature, is limiting the reaction. These events are illustrated in Figure 3.

▲ **Figure 3** *Graph showing the effect of light intensity on the rate of photosynthesis as measured by the amount of CO_2 exchange*

Carbon dioxide is present in the atmosphere at a concentration of around 0.04% and is often the factor that limits the rate of photosynthesis under normal conditions. The optimum concentration of carbon dioxide for a consistently high rate of photosynthesis is 0.1% and growers of some greenhouse crops, such as tomatoes, enrich the air in the greenhouses with more carbon dioxide to provide higher yields. Carbon dioxide concentration affects enzyme activity, in particular the enzyme that catalyses the combination of ribulose bisphosphate with carbon dioxide in the light-independent reaction.

Provided that other factors are not limiting, the rate of photosynthesis increases in direct proportion to the temperature. Between the temperatures of 0 °C and 25 °C the rate of photosynthesis is approximately doubled for each 10 °C rise in temperature.

Figure 4 illustrates the influence of light intensity, carbon dioxide and temperature on the rate of photosynthesis.

1 0.1% carbon dioxide at 25 °C
2 0.04% carbon dioxide at 35 °C
3 0.04% carbon dioxide at 25 °C
4 0.04% carbon dioxide at 15 °C

▲ **Figure 4**

1 State **one** measurement that could be taken to determine the rate of photosynthesis in this experiment.
2 Name the factor that is limiting the rate of photosynthesis over the region marked A on the graph. Explain your answer.
3 In the spring a commercial grower of tomatoes keeps his greenhouses at 25 °C and at a carbon dioxide concentration of 0.04%. The light intensity is 4 kilolux at this time of year. Using the graph, predict whether the tomato plants would grow more if the carbon dioxide level was raised to 0.1% or if the temperature was increased to 35 °C. Explain your answer.
4 Explain why there is no advantage in the grower heating his greenhouses on a dull day.
5 Using your knowledge of the light-independent reaction, explain why, at 25 °C, raising the level of carbon dioxide from 0.04% to 0.1% increases the amount of glucose produced.

▲ **Figure 5** *Student using a photosynthometer to measure the rate of photosynthesis*

 Measuring photosynthesis

The rate of photosynthesis in an aquatic plant such as Canadian pondweed (*Elodea*) can be found by measuring the volume of oxygen produced by using the apparatus (called a photosynthometer) illustrated in Figures 5 and 6.

▲ **Figure 6** *Apparatus used to measure the rate of photosynthesis under various conditions*

- The apparatus is set up as in Figure 6, taking care not to introduce any air bubbles into it and that the apparatus is completely air-tight.

- The water bath is used to maintain a constant temperature throughout the experiment and can be adjusted as necessary.

- Potassium hydrogencarbonate solution is used around the plant to provide a source of carbon dioxide.

- A source of light, whose intensity can be adjusted, is arranged close to the apparatus, which is kept in an otherwise dark room.

- The apparatus is kept in the dark for two hours before the experiment begins.

- The light source is switched on and the plant left for 30 minutes to allow the air spaces in the leaves to fill with oxygen.

- Oxygen released by the plant during photosynthesis collects in the funnel end of the capillary tube above the plant.

- After 30 minutes this oxygen is drawn up the capillary tube by gently withdrawing the syringe until its volume can be measured on the scale, which is calibrated in mm^3.

- The gas is drawn up into the syringe, which is then depressed again before the process is repeated at the same light intensity four or five times, and the mean volume of oxygen produced per hour is calculated.

- The apparatus is left in the dark for 2 hours before the procedure is repeated with the light source set at a different light intensity.

1 Explain why the apparatus needs to be airtight.
2 Explain why the temperature of the water bath needs to be kept constant.
3 Suggest an advantage of providing an additional source of carbon dioxide.
4 Suggest a reason for carrying out the experiment in a room that is dark except for the light source.
5 Suggest why the plant is kept in the dark before the experiment begins.
6 Suggest why measuring the volume of gas produced by the plant in this experiment may not be an accurate measure of photosynthesis.

Using a lollipop to work out the light-independent reaction

The details of the light-independent reaction were worked out by Melvin Calvin and his co-workers using his 'lollipop' experiment. It was so called because the apparatus, shown in Figure 7, resembled a lollipop.

In the experiment, single-celled algae are grown in the light in a thin transparent 'lollipop'. Radioactive hydrogencarbonate is injected into the 'lollipop'. This supplies radioactive carbon dioxide to the algae. At 5-second intervals, samples of the photosynthesising algae are dropped into hot methanol to stop chemical reactions instantly. The compounds in the algae are then separated out and those that are radioactive are identified. The results are given in Table 1.

▼ Table 1

Time / s	Substances found to be radioactive
0	carbon dioxide
5	glycerate 3-phosphate
10	glycerate 3-phosphate + triose phosphate
15	glycerate 3-phosphate + triose phosphate + glucose
20	glycerate 3-phosphate + triose phosphate + glucose + ribulose bisphosphate

▲ **Figure 7** *The 'lollipop' apparatus used by Melvin Calvin*

- Algae are grown under light in the thin transparent lollipop.

- Radioactive ^{14}C in the form of hydrogencarbonate is injected.

- At intervals (seconds to minutes) samples of the photosynthesising algae are dropped into the hot methanol to stop chemical reactions instantly.

- The compounds in the algae are separated by two-way chromatography.

- The radioactive compounds are identified.

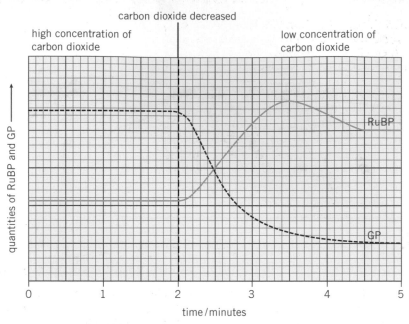

▲ **Figure 8** *Apparatus used to measure the rate of photosynthesis under various conditions*

1 Suggest why the carbon dioxide supplied to the algae was radioactively labelled.
2 Explain how information in Table 1 provides evidence that glycerate 3-phosphate is converted into triose phosphate.
3 Suggest an explanation of how the hot methanol might stop further chemical reactions taking place.

bisphosphate (RuBP) were measured. At the beginning of the experiment, the concentration of carbon dioxide supplied was high. After two minutes the concentration of carbon dioxide was reduced. The graph in Figure 8 shows the results of this experiment.

4 Describe the effects on the quantities of GP and RuBP of the decrease in carbon dioxide after two minutes.
5 Suggest explanations for these changes to the levels of GP and RuBP.

In a further experiment, samples of algae were collected at 1-minute intervals over a period of five minutes. The quantities of glycerate 3-phosphate (GP) and ribulose

1 Scientists investigated the effect of iron deficiency on the production of triose phosphate in sugar beet plants. They grew the plants under the same conditions with their roots in a liquid growth medium containing all the necessary nutrients. Ten days before the experiments, they transferred half the plants to a liquid growth medium containing no iron. The scientists measured the concentration of triose phosphate produced in these plants and in the control plants:
   • at the end of 6 hours in the dark
   • then for 16 hours in the light.
Their results are shown in the graph.

(a) (i)   The experiments were carried out at a high carbon dioxide concentration. Explain why.   *(1 mark)*

    (ii) Explain why it was important to grow the plants under the same conditions up to ten days before the experiment.   *(1 mark)*

    (iii) The plants were left in the dark for 6 hours before the experiment. Explain why.   *(1 mark)*

(b) Iron deficiency reduces electron transport. Use this information and your knowledge of photosynthesis to explain the decrease in production of triose phosphate in the iron-deficient plants.   *(4 marks)*

(c) Iron deficiency results in a decrease in the uptake of carbon dioxide. Explain why.   *(2 marks)*

AQA June 2013

2 During photosynthesis, carbon dioxide reacts with ribulose bisphosphate (RuBP) to form two molecules of glycerate 3-phosphate (GP). This reaction is catalysed by the enzyme Rubisco. Rubisco can also catalyse a reaction between RuBP and oxygen to form one molecule of GP and one molecule of phosphoglycolate. Both the reactions catalysed by Rubisco are shown in **Figure 1**.

▲ Figure 1

(a) (i)   Where exactly in a cell is the enzyme Rubisco found?   *(1 mark)*

    (ii) Use the information provided to give the number of carbon atoms in **one** molecule of phosphoglycolate.   *(1 mark)*

(b) Scientists investigated the effect of different concentrations of oxygen on the rate of absorption of carbon dioxide by leaves of soya bean plants. Their results are shown in **Figure 2**.
Use **Figure 1** to explain the results obtained in **Figure 2**.   *(2 marks)*

(c) Use the information provided and your knowledge of the light-independent reaction to explain why the yield from soya bean plants is decreased at higher concentrations of oxygen. Phosphoglycolate is not used in the light-independent reaction.

(3 marks)
AQA Jan 2013   ▲ Figure 2

3 A scientist investigated the uptake of radioactively labelled carbon dioxide in chloroplasts. She used three tubes, each containing different components of chloroplasts. She measured the uptake of carbon dioxide in each of these tubes. Her results are shown in the table.

Tube	Contents of tube	Uptake of radioactively labelled CO_2 / counts per minute
A	Stroma and grana	96 000
B	Stroma, ATP and reduced NADP	97 000
C	Stroma	4 000

(a) Name the substance which combines with carbon dioxide in a chloroplast. (*1 mark*)

(b) Explain why the results in tube **B** are similar to those in tube **A**. (*1 mark*)

(c) Use the information in the table to predict the uptake of radioactively labelled carbon dioxide if tube A was placed in the dark. Explain your answer. (*2 marks*)

(d) Use your knowledge of the light-independent reaction to explain why the uptake of carbon dioxide in tube **C** was less than the uptake in tube **B**. (*2 marks*)

(e) DCMU is used as a weed killer. It inhibits electron transfer during photosynthesis. The addition of DCMU to tube A decreased the uptake of carbon dioxide. Explain why. (*2 marks*)

AQA June 2012

4 (a) The concentrations of carbon dioxide in the air at different heights above ground in a forest changes over a period of 24 hours. Use your knowledge of photosynthesis to describe these changes and explain why they occur. (*5 marks*)

(b) In the light-independent reaction of photosynthesis, the carbon in carbon dioxide becomes carbon in triose phosphate. Describe how. (*5 marks*)

(c) Microorganisms make the carbon in polymers in a dead worm available to cells in a leaf. Describe how. (*5 marks*)

AQA June 2010

5 Scientists investigated the effects of temperature and light intensity on the rate of photosynthesis in creeping azalea. They investigated the effect of temperature on the net rate of photosynthesis at three different light intensities. They also investigated the effect of temperature on the rate of respiration. The graph shows the results.

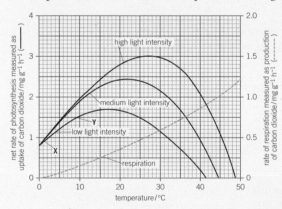

(a) (i) Name the factors that limited the rate of photosynthesis between **X** and **Y**. (*1 mark*)

  (ii) Use information from the graph to explain your answer. (*2 marks*)

(b) Use information from the graph to find the gross rate of photosynthesis at 20°C and medium light intensity.

  (*1 mark*)

(c) Creeping azalea is a plant which grows on mountains. Scientists predict that in the area where this plant grows the mean summer temperature is likely to rise from 20°C to 23°C. It is also likely to become much cloudier. Describe and explain how these changes are likely to affect the growth of creeping azalea. (*3 marks*)

AQA Jan 2011

12 Respiration
12.1 Glycolysis

We have seen in Chapter 11 that photosynthesis transfers energy in the form of sunlight into the chemical energy of carbohydrates such as glucose. We also saw, in Topic 2.3, that this glucose cannot be used directly by cells as a source of energy. Instead, cells use ATP as their immediate energy source. The formation of ATP from the break down of glucose takes place during the process of cellular respiration. There are two different forms of cellular respiration depending on whether oxygen is involved or not:

- **Aerobic respiration** requires oxygen and produces carbon dioxide, water and much ATP.
- **Anaerobic respiration** takes place in the absence of oxygen and produces lactate (in animals) or ethanol and carbon dioxide (in plants and fungi) but only a little ATP in both cases.

Aerobic respiration can be divided into four stages:

1 **glycolysis** – the splitting of the 6-carbon glucose molecule into two 3-carbon pyruvate molecules
2 **link reaction** – the 3-carbon pyruvate molecules enter into a series of reactions which lead to the formation of acetylcoenzyme A, a 2-carbon molecule.
3 **Krebs cycle** – the introduction of acetylcoenzyme A into a cycle of oxidation-reduction reactions that yield some ATP and a large quantity of reduced NAD and FAD (Topic 12.2)
4 **oxidative phosphorylation** – the use of the electrons, associated with reduced NAD and FAD, released from the Krebs cycle to synthesise ATP with water produced as a by-product.

The main respiratory pathways are summarised in Figure 1.

Glycolysis is the initial stage of both aerobic and anaerobic respiration. It occurs in the cytoplasm of all living cells and is the process by which

Learning objectives

→ Outline where glycolysis fits into the overall process of respiration.

→ Describe the main stages of glycolysis and its products.

Specification reference: 3.5.2

▲ **Figure 1** *Summary of respiratory pathways*

a hexose (6-carbon) sugar, usually glucose, is split into two molecules of the 3-carbon molecule, pyruvate. Although there are a number of smaller enzyme-controlled reactions in glycolysis, these can be conveniently grouped into four stages:

1 **phosphorylation of glucose to glucose phosphate**. Before it can be split into two, glucose must first be made more reactive by the addition of two phosphate molecules (phosphorylation). The phosphate molecules come from the **hydrolysis** of two ATP molecules to ADP. This provides the energy to activate glucose and lowers the **activation energy** for the enzyme-controlled reactions that follow (Topic 1.7).

2 **splitting of the phosphorylated glucose**. Each glucose molecule is split into two 3-carbon molecules known as triose phosphate.

3 **oxidation of triose phosphate**. Hydrogen is removed from each of the two triose phosphate molecules and transferred to a hydrogen-carrier molecule known as NAD to form reduced NAD.

4 **the production of ATP**. Enzyme-controlled reactions convert each triose phosphate into another 3-carbon molecule called pyruvate. In the process, two molecules of ATP are regenerated from ADP.

The events of glycolysis are summarised in Figure 2.

Hint

It must be remembered that for each molecule of glucose at the start of the process, two molecules of triose phosphate are produced. Therefore the yields must be doubled, that is, four molecules of ATP and two molecules of reduced NAD.

Hint

Glucose, a sugar, is oxidised to pyruvate, an acid.

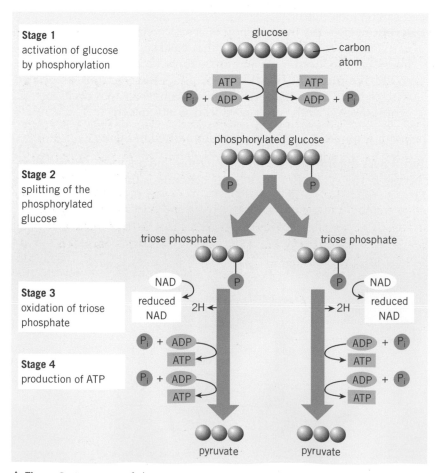

▲ **Figure 2** *Summary of glycolysis*

Energy yields from glycolysis

The overall yield from one glucose molecule undergoing glycolysis is therefore:

- two molecules of ATP (four molecules of ATP are produced, but two were used up in the initial phosphorylation of glucose and so the net increase is two molecules)

- two molecules of reduced NAD (these have the potential to provide energy to produce more ATP as we shall see in Topic 12.3)

- two molecules of pyruvate.

Glycolysis is a universal feature of every living organism and therefore provides indirect evidence for evolution. The enzymes for the glycolytic pathway are found in the cytoplasm of cells and so glycolysis does not require any organelle or membrane for it to take place. It does not require oxygen and therefore it can take place whether or not it is present. In the absence of oxygen the pyruvate produced by glycolysis can be converted into either lactate or ethanol during anaerobic respiration. This is necessary in order to re-oxidise NAD so that glycolysis can continue. This is explained, along with details of the reactions, in Topic 12.4. Anaerobic respiration, however, yields only a small fraction of the potential energy stored in the pyruvate molecule. In order to release the remainder of this energy, most organisms use oxygen to break down pyruvate further.

Summary questions

In the following passage, state the most suitable word to replace each of the numbers 1–10.

Glycolysis takes place in the (1) of cells and begins with the activation of the main respiratory substrate, namely the hexose sugar called (2). This activation involves the addition of two (3) molecules provided by two molecules of (4). The resultant activated molecule is known as (5) and in the next stage of glycolysis it is split into two molecules called (6). The third stage entails the oxidation of these molecules by the removal of (7), which is transferred to a carrier called (8). The final stage is the production of the 3-carbon molecule (9), which also results in the formation of two molecules of (10).

Learning objectives

→ Outline the nature of the link reaction.

→ Explain what happens during the Krebs cycle.

→ Describe the nature of hydrogen carrier molecules and explain their role in the Krebs cycle.

Specification reference: 3.5.2

The pyruvate molecules produced during glycolysis possess potential energy that can only be released in a process called the Krebs cycle. Before they can enter the Krebs cycle, these pyruvate molecules must first be **oxidised** in a procedure known as the **link reaction**. In **eukaryotic cells** both the Krebs cycle and the link reaction take place exclusively inside mitochondria.

The link reaction

The pyruvate molecules produced in the cytoplasm during **glycolysis** are actively transported into the matrix of mitochondria. Here pyruvate undergoes a series of reactions during which the following changes take place:

- The pyruvate is oxidised to acetate. In this reaction, the 3-carbon pyruvate loses a carbon dioxide molecule and two hydrogens. These hydrogens are accepted by NAD to form reduced NAD, which is later used to produce ATP (Topic 12.4).
- The 2-carbon acetate combines with a molecule called coenzyme A (CoA) to produce a compound called **acetylcoenzyme A**.

The overall equation can be summarised as:

$$\text{pyruvate} + \text{NAD} + \text{CoA} \rightarrow \text{acetyl CoA} + \text{reduced NAD} + CO_2$$

The Krebs cycle

The Krebs cycle was named after the British biochemist, Hans Krebs, who worked out its sequence. The Krebs cycle involves a series of oxidation-reduction reactions that take place in the matrix of mitochondria. Its events are illustrated in Figure 1 and can be summarised as follows:

- The 2-carbon acetylcoenzyme A from the link reaction combines with a 4-carbon molecule to produce a 6-carbon molecule.
- In a series of reactions this 6-carbon molecule loses carbon dioxide and hydrogen to give a 4-carbon molecule and a single molecule of ATP produced as a result of **substrate-level phosphorylation** (Topic 2.3).
- The 4-carbon molecule can now combine with a new molecule of acetylcoenzyme A to begin the cycle again.

For each molecule of pyruvate, the link reaction and the Krebs cycle therefore produce:

- **reduced coenzymes** such as NAD and FAD. These have the potential to provide energy to produce ATP molecules by oxidative phosphorylation (Topic 12.3) and are therefore the important products of Krebs cycle
- one molecule of ATP
- three molecules of carbon dioxide.

As two pyruvate molecules are produced for each original glucose molecule, the yield from a single glucose molecule is double the quantities above.

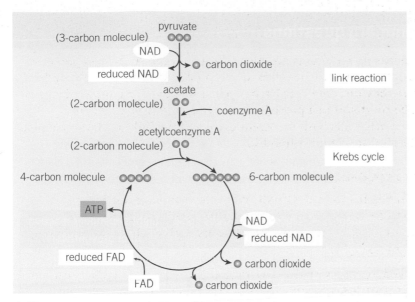

▲ **Figure 1** *Summary of the link reaction and the Krebs cycle*

A summary of the link reaction and the Krebs cycle is shown in Figure 1.

Coenzymes

Despite their name, coenzymes are not enzymes. They are molecules that some enzymes require in order to function. Coenzymes play a major role in photosynthesis and respiration where they carry hydrogen atoms from one molecule to another. Examples include:

- **NAD**, which is important throughout respiration
- **FAD**, which is important in the Krebs cycle
- **NADP**, which is important in photosynthesis (Topic 11.2).

In respiration, NAD is the most important carrier. It works with dehydrogenase enzymes that catalyse the removal of hydrogen atoms from substrates and transfer them to other molecules involved in oxidative phosphorylation (Topic 12.3).

The significance of the Krebs cycle

The Krebs cycle performs an important role in the cells of organisms for four reasons:

- It breaks down macromolecules into smaller ones – pyruvate is broken down into carbon dioxide.
- It produces hydrogen atoms that are carried by NAD to the electron transfer chain and provide energy for oxidative phosphorylation. This leads to the production of ATP that provides metabolic energy for the cell.
- It regenerates the 4-carbon molecule that combines with acetylcoenzyme A, which would otherwise accumulate.
- It is a source of intermediate compounds used by cells in the manufacture of other important substances such as fatty acids, amino acids and chlorophyll.

Summary questions

1 State how many carbon molecules there are in a single molecule of pyruvate.

2 Name the 2-carbon molecule that pyruvate is converted to during the link reaction.

3 State precisely in which part of the cell the Krebs cycle takes place.

4 Table 1 lists statements about some biochemical processes in a plant cell. State whether each of the letters a–r represents true or false.

▼ Table 1

Statement	Glycolysis	Krebs cycle	Light-dependent reaction of photosynthesis
ATP is produced	a	b	c
ATP is needed	d	e	f
NAD is reduced	g	h	i
NADP is reduced	j	k	l
CO_2 is produced	m	n	o
CO_2 is needed	p	q	r

Coenzymes in respiration

Coenzymes such as NAD are important in respiration. They help enzymes to function by carrying hydrogen atoms from one molecule to another. Scientists can model the way coenzymes work in cells using a blue dye called methylene blue. It can accept hydrogen atoms and so become reduced. Reduced methylene blue is colourless.

methylene blue + hydrogen ⟶ reduced methylene
  (blue colour)   blue (colourless)

In an investigation into respiration in yeast, three test tubes were set up as follows:

Tube A	Tube B	Tube C
2 cm³ yeast suspension	2 cm³ distilled water	2 cm³ yeast suspension
2 cm³ glucose solution	2 cm³ glucose solution	2 cm³ distilled water
1 cm³ methylene blue	1 cm³ methylene blue	1 cm³ methylene blue

All three tubes were incubated at a temperature of 30 °C. The colour of each tube was recorded at the start of the

experiment and after 5 and 15 minutes. The results are shown in the table below:

Time / min	Colour of tube contents		
	Tube A	Tube B	Tube C
0	blue	blue	blue
5	colourless	blue	blue
15	colourless	blue	pale blue

1 Tube B acts as a control. Explain why this control was necessary in this investigation.

2 Using your knowledge of respiration, suggest an explanation for the colour change after 15 minutes in:
  a tube A   b tube C.

3 How might the results in tube A after 15 minutes have been different if the experiment had been carried out at 70 °C? Explain your answer.

4 After 20 minutes the contents of tube A were mixed with air by shaking it vigorously, turning the methylene blue back to a blue colour. Suggest a reason for this colour change.

5 Suggest why conclusions made only on the basis of the results of this experiment may not be reliable.

So far in the process of **aerobic** respiration, we have seen how hexose sugars such as glucose are split (glycolysis) and how the 3-carbon pyruvate that results is fed into the Krebs cycle to yield carbon dioxide and hydrogen atoms. The carbon dioxide is a waste product and is removed during the process of gaseous exchange. The hydrogen atoms (or more particularly the electrons they possess) are valuable as a potential source of energy. These hydrogen atoms are carried by the coenzymes NAD and FAD into the next stage of the process, **oxidative phosphorylation**. This is the mechanism by which some of the energy of the **electrons** within the hydrogen atoms is conserved in the formation of **adenosine triphosphate (ATP)**.

Oxidative phosphorylation and mitochondria

Mitochondria are organelles that are found in **eukaryotic cells**. Each mitochondrion is bounded by a smooth outer membrane and an inner one that is folded into extensions called cristae. The inner space, or matrix, of the mitochondrion contains proteins lipids, and traces of DNA.

Mitochondria are the site of oxidative phosphorylation. Within the inner folded membrane (cristae) are the enzymes and other proteins involved in oxidative phosphorylation and hence ATP synthesis.

As mitochondria play a vital role in respiration it is hardly surprising that they occur in greater numbers in metabolically active cells, such as those of the muscles, liver and epithelial cells, which carry out active transport. The mitochondria in these cells also have more densely packed cristae which provide a greater surface area of membrane incorporating enzymes and other proteins involved in oxidative phosphorylation.

The electron transfer chain and the synthesis of ATP

The synthesis of ATP by oxidative phosphorylation involves the transfer of electrons down a series of electron carrier molecules which together form the **electron transfer chain**. The process takes place as follows:

- The hydrogen atoms produced during glycolysis and the Krebs cycle combine with the coenzymes NAD and FAD.

- The reduced NAD and FAD donate the electrons of the hydrogen atoms they are carrying to the first molecule in the electron transfer chain.

- The electrons pass along a chain of electron transfer carrier molecules in a series of **oxidation-reduction** reactions. As the electrons flow along the chain, the energy they release causes the active transport of protons across the inner mitochondrial membrane and into inter-membranal space.

- The protons accumulate in the inter-membranal space before they diffuse back into the mitochondrial matrix through ATP synthase channels embedded in the inner mitochondrial membrane.

Learning objectives

→ Describe where oxidative phosphorylation takes place.

→ Explain how ATP is synthesised during oxidative phosphorylation.

→ Explain the role of oxygen in aerobic respiration.

Specification reference: 3.5.2

▲ **Figure 1** *Coloured TEM of a sectioned mitochondrion (red and yellow). It has two membranes: an outer surrounding membrane and an inner membrane that forms folds called cristae, seen here as red lines. The cristae are the sites of oxidative phosphorylation*

Synoptic link

Re-reading about mitochondria in Topic 3.4 and membranes in Topic 4.1 will help you follow the processes of the electron transport chain.

Hint

Remember that a single hydrogen atom is made up of one proton (H^+) and one electron (e^-).

- At the end of the chain the electrons combine with these protons and oxygen to form water. Oxygen is therefore the final acceptor of electrons in the electron transfer chain.

The process described above is the chemiosmotic theory of oxidative phosphorylation and is summarised in Figure 2. You will notice that it involves the same types of processes as those used to explain photophosphorylation in the light-dependent stage of photosynthesis (Topic 11.2).

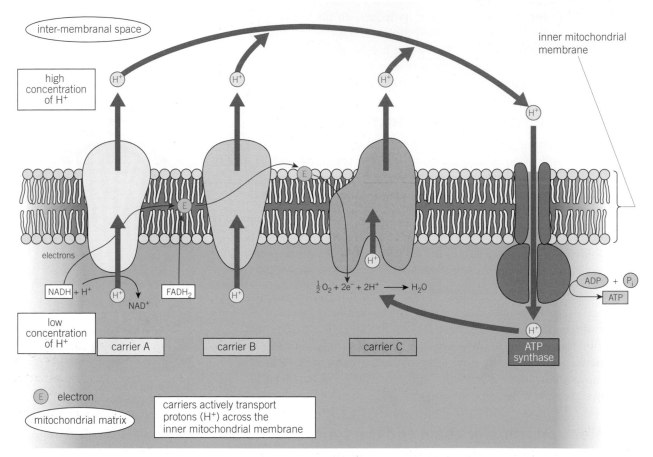

▲ **Figure 2** *Summary of the chemiosmotic theory of oxidative phosphorylation*

The importance of oxygen in respiration is to act as the final acceptor of the hydrogen atoms produced in glycolysis and the Krebs cycle. Without its role in removing hydrogen atoms at the end of the chain, the hydrogen ions (protons) and electrons would 'back up' along the chain and the process of respiration would come to a halt.

Releasing energy in stages

In general, the greater the energy that is released in a single step, the more of it is released as heat and the less there is available for more useful purposes. When energy is released a little at a time, more of it can be harvested for the benefit of the organism. For this reason, the electrons carried by NAD and FAD are not transferred in one explosive step. Instead they are passed along a series of electron transfer carrier molecules, each of which is at a slightly lower energy level (Figure 3). The electrons therefore move down an energy gradient. The transfer of electrons down this gradient allows their energy to be released gradually and therefore more usefully.

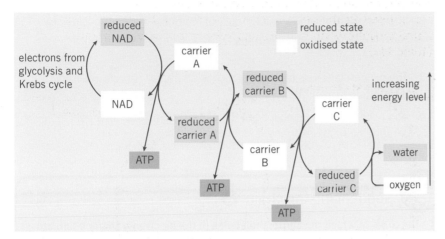

▲ **Figure 3** *Summary of electron transfer chain*

Alternative respiratory substrates

Sugars are not the only substances which can be oxidised by cells to release energy. Both lipids and protein may, in certain circumstances, be used as respiratory substrates, without first being converted to carbohydrate.

Respiration of lipids

Before being respired, lipids are first hydrolysed to glycerol and fatty acids. The glycerol is then phosphorylated and converted to triose phosphate which enters the glycolysis pathway and subsequently the Krebs cycle. The fatty acid component is broken down into 2-carbon fragments which are converted to acetyl coenzyme A. This then enters the Krebs cycle.

The oxidation of lipids produces 2-carbon fragments of carbohydrate and many hydrogen atoms. The hydrogen atoms are used to produce ATP during oxidative phosphorylation. For this reason lipids release more than double the energy of the same mass of carbohydrate.

Respiration of protein

Protein is another potential source of energy. It is first hydrolysed to its constituent amino acids. These have their amino group removed (deamination) before entering the respiratory pathway at different points depending on the number of carbon atoms they contain. 3-carbon compounds are converted to pyruvate, while 4- and 5-carbon compounds are converted to intermediates in the Krebs cycle.

Practical link

Required practical 9. Investigation into the effect of a named variable on the rate of respiration of cultures of single-celled organisms.

Study tip

Oxygen is used as the final acceptor of hydrogen atoms at the end of the electron transfer chain. It is therefore used to form water and not carbon dioxide, as you may think.

Summary questions

1 The processes that occur in the electron transfer chain are also known as oxidative phosphorylation. Suggest why this term is used.

2 The surface of the inner mitochondrial membrane is highly folded to form cristae. State **one** advantage of this arrangement to the electron transfer chain.

3 The oxygen taken up by organisms has an important role in aerobic respiration. Explain this role.

4 As part of which molecule does the oxygen taken into an organism leave after being respired?

 Sequencing the chain

The order in which the carrier molecules of the electron transfer chain are arranged can be determined experimentally. The experiments rely on the fact that each transfer of electrons between one molecule and the next is catalysed by a specific enzyme. In a series of experiments, three different inhibitors, 1, 2, and 3, are added to four electron transfer molecules, A, B, C and D. Table 1 shows whether the molecules A–D are oxidised or reduced after the inhibitor is added.

▼ Table 1

Inhibitor added	Electron transfer molecules			
	A	B	C	D
1	reduced	oxidised	reduced	oxidised
2	oxidised	oxidised	reduced	oxidised
3	reduced	oxidised	reduced	reduced

1 Using the information in the table, state the order of the electron transfer molecules in this chain. Explain your answer.

12.4 Anaerobic respiration

We saw in Topic 12.3, that oxygen is needed if the hydrogen atoms produced in **glycolysis** and the **Krebs cycle** are to be used in the production of ATP. What happens if oxygen is temporarily or permanently unavailable to a tissue or a whole organism?

In the absence of oxygen, neither the Krebs cycle nor the electron transfer chain can continue because soon all the FAD and NAD will be reduced. No FAD or NAD will be available to take up the H^+ produced during the Krebs cycle and so the enzymes stop working. This leaves only the anaerobic process of glycolysis as a potential source of ATP. For glycolysis to continue, its products of pyruvate and hydrogen must be constantly removed. In particular, the hydrogen must be released from the reduced NAD in order to regenerate **NAD**. Without this, the already tiny supply of NAD in cells will be entirely converted to reduced NAD, leaving no NAD to take up the hydrogen newly produced from glycolysis. Glycolysis will then grind to a halt. The replenishment of NAD is achieved by the pyruvate molecule from glycolysis accepting the hydrogen from reduced NAD. The oxidised NAD produced can then be used in further glycolysis.

In eukaryotic cells, only two types of anaerobic respiration occur with any regularity:

- In plants, and in microorganisms such as yeast, the pyruvate is converted to ethanol and carbon dioxide.
- In animals, the pyruvate is converted to lactate.

Production of ethanol in plants and some microorganisms

Anaerobic respiration leading to the production of ethanol occurs in organisms such as certain bacteria and fungi (e.g., yeast) as well as in some cells of higher plants, for example, root cells under waterlogged conditions.

The pyruvate molecule formed at the end of glycolysis loses a molecule of carbon dioxide and accepts hydrogen from reduced NAD to produce ethanol. The summary equation for this is:

pyruvate + reduced → ethanol + carbon + oxidised
NAD   dioxide   NAD

This form of anaerobic respiration in yeast has been exploited by humans for thousands of years in the brewing industry. In brewing, ethanol is the important product. Yeast is grown in anaerobic conditions in which it ferments natural carbohydrates in plant products, such as grapes (wine production) or barley seeds (beer production) into ethanol.

Production of lactate in animals

Anaerobic respiration leading to the production of lactate occurs in animals as a means of overcoming a temporary shortage of oxygen. Clearly, such a mechanism has considerable survival value, for example, in a baby mammal in the period immediately after birth, and in an animal living in water where the amount of oxygen may sometimes be very low.

Learning objectives

→ Explain how energy is released by respiration in the absence of oxygen.

→ Explain how ethanol is produced by anaerobic respiration.

→ Explain how lactate is produced by anaerobic respiration.

Specification reference: 3.5.2

▲ **Figure 1** *Coloured SEM of yeast cells. Yeast produces ethanol and carbon dioxide during anaerobic respiration making it useful in brewing*

Hint

During strenuous exercise, muscles carry out aerobic respiration. If this cannot supply ATP fast enough, they also carry out some anaerobic respiration as well. It is a case not of one or the other but of both together.

▲ **Figure 2** *During strenuous exercise, muscles may temporarily respire anaerobically*

Summary questions

The diagram below shows the relationship between some respiratory pathways.

1 State which of the pathways, A, B, C or D, apply to each of the following statements. There may be more than one answer in each case.

  a Only occurs in the presence of oxygen.

  b Takes place in animals.

  c Produces ATP.

  d Is carried out by yeast in the absence of oxygen.

  e Produces reduced NAD.

  f Regenerates NAD from reduced NAD.

  g Is known as glycolysis.

However, lactate production occurs most commonly in muscles as a result of strenuous exercise. In these conditions oxygen may be used up more rapidly than it can be supplied and therefore an oxygen debt occurs. It is often essential, however, that the muscles continue to work despite the shortage of oxygen, for example, if the organism is fleeing from a predator. When oxygen is in short supply, NAD from glycolysis can accumulate and must be removed. To achieve this, each pyruvate molecule produced takes up the two hydrogen atoms from the reduced NAD produced in glycolysis to form lactate as shown below:

pyruvate + reduced NAD → lactate + oxidised NAD

At some point the lactate produced is oxidised back to pyruvate. This can then be either further oxidised to release energy or converted into glycogen. This happens when oxygen is once again available. In any case, lactate will cause cramp and muscle fatigue if it is allowed to accumulate in the muscle tissue. As lactate is an acid it also causes pH changes which affects enzymes. Although muscle has a certain tolerance to lactate, it is nevertheless important that it is removed by the blood and taken to the liver to be converted to glycogen. Figure 3 shows how the NAD needed for glycolysis to continue is regenerated in both common forms of anaerobic respiration.

▲ **Figure 3** *How the NAD needed for glycolysis is regenerated in various organisms*

Energy yields from anaerobic and aerobic respiration

Energy from cellular respiration is derived in two ways:

* substrate-level phosphorylation in glycolysis and the Krebs cycle. This is the direct transfer of phosphate from a respiratory intermediate to ADP to produce ATP.

* oxidative phosphorylation in the electron transfer chain. This is the indirect linking of energy from phosphate to ADP to produce ATP involving energy from the hydrogen atoms that are carried on NAD and FAD. Cells produce most of their ATP in this way.

In anaerobic respiration, pyruvate is converted to either ethanol or lactate. Consequently it is not available for the Krebs cycle. Therefore in anaerobic respiration neither the Krebs cycle nor the electron transfer chain can take place. The only ATP that can be produced by anaerobic respiration is therefore that formed by glycolysis.

Investigating where certain respiratory pathways take place in cells

Most people are aware that cyanide is a very potent poison that causes death rapidly. It is lethal because it is a non-competitive inhibitor of the final enzyme in the electron transport chain. This enzyme is called cytochrome oxidase and it catalyses the addition of the hydrogen ions and electrons to oxygen to form water. The inhibition of cytochrome oxidase causes hydrogen ions and electrons to accumulate on their carrier molecules, bringing the electron transport chain and Krebs cycle to a halt.

To determine where in the cell some of the respiratory pathways take place, scientists carried out the following experiment involving cyanide.

- Mammalian liver cells were broken up (homogenised) and the resulting homogenate was centrifuged.
- Portions containing only nuclei, ribosomes, mitochondria, and the remaining cytoplasm were separated out.

> **Synoptic link**
>
> To help you follow the experiment described in the extension and to answer the questions, it is necessary to understand cell fractionation and enzyme inhibition. It is therefore advisable to review Topics 3.1 and 1.9.

- Samples of each portion, and of the complete homogenate, were incubated as follows:
  - with glucose
  - with glucose and cyanide
  - with pyruvate and cyanide
  - with pyruvate

After incubation the presence or absence of carbon dioxide and lactate in each sample was recorded. The results are shown in Table 1, in which ✓ = present and ✗ = absent.

▼ **Table 1** Presence (✓) or absence (✗) of CO_2 and lactate in sample shown

Incubated with	Complete homogenate		Nuclei only		Ribosomes only		Mitochondria only		Remaining cytoplasm only	
	Carbon dioxide	Lactate	Carbon dioxide	Lactate	Carbon dioxide	Lactate	Carbon dioxide	Lactate	Carbon dioxide	Lactate
Glucose	✓	✓	✗	✗	✗	✗	✗	✗	✗	✓
Pyruvate	✓	✓	✗	✗	✗	✗	✓	✗	✗	✓
Glucose + cyanide	✗	✓	✗	✗	✗	✗	✗	✗	✗	✓
Pyruvate + cyanide	✗	✓	✗	✗	✗	✗	✗	✗	✗	✓

1 Briefly describe how the different portions of the homogenate may have been separated out by centrifuging.

2 From the results of this experiment, name two organelles that appear not to be involved in respiration. Explain your answer.

3 In which cell organelle would you expect to find the enzymes of the Krebs cycle?
Explain how the results in the table support your answer.

4 Suggest which portion of the homogenate contains the enzymes that convert pyruvate into lactate.

5 Explain why lactate is produced in the presence of cyanide but carbon dioxide is not.

6 Explain why carbon dioxide can be produced by the complete homogenate when none of the separate portions can do so.

7 Suggest which two products might be formed if glucose was incubated with cytoplasm from yeast cells.

8 Giving your reason in each case, assess the relative number of mitochondria in the following: xylem vessel, liver cell, red blood cell, epithelial cell of intestine, myofibril (muscle fibre).

9 Mature red blood cells do not possess mitochondria. Suggest two advantages of this to the functioning of these cells.

1 **(a)** The table contains statements about three biological processes. Copy and complete the table with a tick if the statement in the first column is true, for each process.

	Photosynthesis	Anaerobic respiration	Aerobic respiration
ATP produced			
Occurs in organelles			
Electron transport chain involved			

(3 marks)

(b) Write a simple equation to show how ATP is synthesised from ADP. *(1 mark)*

(c) Give two ways in which the properties of ATP make it a suitable source of energy in biological processes. *(2 marks)*

(d) Humans synthesise more than their body mass of ATP each day. Explain why it is necessary for them to synthesise such a large amount of ATP. *(2 marks)*

AQA June 2011

2 **(a)** The table contains statements about three stages of respiration. Copy and complete the table with a tick if the statement in the first column is true for each stage of respiration in an animal.

	Glycolysis	Link reaction	Krebs cycle
Occurs in mitochondria			
Carbon dioxide produced			
NAD is reduced			

(3 marks)

(b) The following reaction occurs in the Krebs cycle.

$$\text{Succinate} \xrightarrow{\text{Enzyme}} \text{Fumarate}$$

A scientist investigated the effect of the enzyme inhibitor malonate on this reaction. The structure of malonate is very similar to the structure of succinate. The scientist added malonate and the respiratory substrate, pyruvate, to a suspension of isolated mitochondria. She also bubbled oxygen through the suspension.

(i) Explain why the scientist did not use glucose as the respiratory substrate for these isolated mitochondria. *(2 marks)*

(ii) Explain how malonate inhibits the formation of fumarate from succinate. *(2 marks)*

(iii) The scientist measured the uptake of oxygen by the mitochondria during the investigation. The uptake of oxygen decreased when malonate was added. Explain why. *(2 marks)*

AQA June 2013

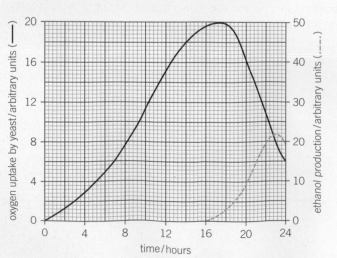

3 Yeast is a single-celled organism. A student investigated respiration in a population of
yeast growing in a sealed container. His results are shown in the graph.
   (a) Calculate the rate of oxygen uptake in arbitrary units per hour between 2 and
      4 hours.   (*1 mark*)
   (b) **(i)** Use the information provided to explain the changes in oxygen uptake during
         this investigation.   (*3 marks*)
      (ii) Use the information provided to explain the changes in production of ethanol
         during this investigation.   (*2 marks*)
   (c) Sodium azide is a substance that inhibits the electron transport chain in respiration.
      The student repeated the investigation but added sodium azide after 4 hours.
      Suggest and explain how the addition of sodium azide would affect oxygen uptake
      and the production of ethanol.   (*3 marks*)
                                                                          AQA Jan 2013

4 A student investigated the rate of gas exchange in aerobically
respiring seeds using the apparatus shown in the diagram. She
carried out two experiments.
   • In Experiment **1**, she put potassium hydroxide solution
      in the beaker. Potassium hydroxide solution absorbs
      carbon dioxide.
   • In Experiment **2**, she put water in the beaker.
   (a) Both experiments were carried out at the same
      temperature. Explain why.   (*2 marks*)
   (b) **(i)** The level of coloured liquid in the right-hand
         side of the manometer tube went down during
         Experiment 1. Explain why.

                                          (*3 marks*)

The results from both experiments are shown in the table.

Experiment	Solution in the beaker	Distance moved by the meniscus in the right hand side of the manometer (mm)
1	Potassium hydroxide	6.4
2	Water	1.2

   (ii) The diameter of the manometer tube was 1mm.
      Use these results to calculate the volume of carbon dioxide produced
      during experiment 1.   (*3 marks*)
   (c) The student repeated Experiment **1** using seeds which were respiring anaerobically.
      What would happen to the level of coloured liquid in the right-hand side of
      the manometer tube? Explain your answer.   (*2 marks*)
                                                          AQA June 2012 (apart from 4 (b) (ii))

Learning objectives

→ Explain how energy enters an ecosystem.

→ Explain how energy is transferred between the organisms in the ecosystem.

→ Define the terms: trophic level, food chain, food web, producer, consumer, and decomposer.

→ Define biomass and explain how it is measured.

Specification reference: 3.5.3

Hint

You may be familiar with the following terms:

- Herbivore – an animal that eats plants (producers) and is therefore a primary consumer.

- Carnivore – an animal that eats animals and may therefore be a secondary or a tertiary consumer.

- Omnivore – an animal that eats both plants and animals and is therefore a primary consumer and also a secondary or a tertiary consumer.

Study tip

Make sure you learn simple definitions of ecological terms and use them appropriately.

The organisms found in any ecosystem rely on a source of energy to carry out all their activities. The ultimate source of this energy for almost all organisms is sunlight, which is conserved as chemical energy by plants. Most plants use sunlight in making organic compounds from carbon dioxide in the air or water that surrounds them. These organic compounds include sugars, most of which are used by the plants as respiratory substrates. The remainder are used to make other groups of biological molecules. These biological molecules form the **biomass** of plants that is the means by which energy is passed between other organisms.

In this chapter we shall look at how this energy is transferred, how nutrients are cycled and how we use artificial fertilisers to supplement natural nutrients in order to improve productivity. Before we do, let us first recap some of the basic terminology of ecology.

Organisms can be divided into three groups according to how they obtain their energy and nutrients. These three groups are – producers, consumers, and saprobionts.

- **Producers** are photosynthetic organisms that manufacture organic substances using light energy, water, carbon dioxide, and mineral ions.

- **Consumers** are organisms that obtain their energy by feeding on (consuming) other organisms rather than using the energy of sunlight directly. Animals are consumers. Those that directly eat producers (green plants) are called **primary consumers** because they are the first in the chain of consumers. Those animals eating primary consumers are called **secondary consumers** and those eating secondary consumers are called **tertiary consumers**. Secondary and tertiary consumers are usually predators but they may also be scavengers or parasites.

- **Saprobionts** (decomposers) are a group of organisms that break down the complex materials in dead organisms into simple ones. In doing so, they release valuable minerals and elements in a form that can be absorbed by plants and so contribute to recycling. The majority of this work is carried out by fungi and bacteria.

- A **food chain** describes a feeding relationship in which the producers are eaten by primary consumers. These in turn are eaten by secondary consumers, which are then eaten by tertiary consumers. In a long food chain the tertiary consumers may in turn be eaten by further consumers called quaternary consumers. Each stage in this chain is referred to as a **trophic level**. The arrows on food chain diagrams represent the direction of energy flow.

- **Food webs** – in reality, most animals do not rely on a single food source and within a single **habitat** many food chains will be linked together to form a food web. The problem with food webs is their complexity. In practice, it is likely that all organisms within a habitat, even within an ecosystem, will be linked to others in the food web.

Biomass

Biomass is the total mass of living material in a specific area at a given time. The fresh mass is quite easy to assess, but the presence of varying amounts of water makes it unreliable. Measuring the mass of carbon or dry mass overcomes this problem but, because the organisms must be killed, it is usually only made on a small sample, and this sample may not be representative. Biomass is measured using dry mass per given area, in a given time. More specifically it is measured in grams per square metre ($g\,m^{-2}$) where an area is being sampled, for example, on grassland or a seashore. Where a volume is being sampled, for example, in a pond or an ocean, it is measured in grams per cubic metre ($g\,m^{-3}$).

▲ **Figure 1** *The snake (tertiary consumer) is swallowing an insect-eating frog (secondary consumer) on a plant leaf (producer)*

The chemical energy store in dry mass can be estimated using **calorimetry**. In bomb calorimetry, a sample of dry material is weighed and is then burnt in pure oxygen within a sealed chamber called a bomb. The bomb is surrounded by a water bath and the heat of combustion causes a small temperature rise in this water. As we know how much heat (energy) is required to raise the temperature of 1g of water by 1°C, if we know the volume of water and the temperature rise, we can calculate the energy released from the mass of burnt biomass in units such as $kJ\,kg^{-1}$.

Summary questions

The diagram below shows a simplified food web within an aquatic ecosystem.

1 State which organisms are secondary consumers.

2 State which organisms carry out photosynthesis.

3 State which organisms are at the fourth trophic level.

4 Explain what the arrows in the diagram show.

5 When the organisms in this web die they will be broken down by bacteria and fungi. Name the general term used to describe these bacteria and fungi.

13.2 Energy transfer and productivity

Learning objectives

→ Calculate the percentage of energy that is transferred from one trophic level to the next.

→ Explain how energy is lost along a food chain.

→ Explain what is meant by gross primary productivity and net primary productivity.

Specification reference: 3.5.3

The Sun is the source of energy for **ecosystems**. However, as little as one % of this light energy may be captured by green plants and so made available to organisms in the food chain. These organisms in turn pass on only a small fraction of the energy that they receive to each successive stage in the chain. How then is so much energy lost?

Plants normally convert between one % and three % of the Sun's energy available to them into organic matter. Most of the Sun's energy is not converted to organic matter by photosynthesis because:

- over 90% of the Sun's energy is reflected back into space by clouds and dust or absorbed by the atmosphere
- not all wavelengths of light can be absorbed and used for photosynthesis
- light may not fall on a chlorophyll molecule
- a factor, such as low carbon dioxide levels, may limit the rate of photosynthesis.

The total quantity of the chemical energy store in plant biomass, in a given area or volume, in a given time is called the **gross primary production** (**GPP**). However, plants use 20–50% of this energy in respiration. The chemical energy store which is left when these losses to respiration have been taken into account, is called **net primary productivity** (**NPP**).

net primary production = gross primary production – respiratory losses
NPP   *GPP*   *R*

The net primary production is available for plant growth and reproduction. It is also available to other trophic levels in the ecosystem, such as consumers and decomposers. Usually less than 10% of this net primary production in plants can be used by primary consumers for growth. Secondary and tertiary consumers are slightly more efficient, transferring up to about 20% of the energy available from their prey into their own bodies. The low percentage of energy transferred at each stage is the result of the following:

- Some of the organism is not consumed.
- Some parts are consumed but cannot be digested and are therefore lost in faeces.
- Some of the energy is lost in excretory materials, such as urine.
- Some energy losses occur as heat from respiration and lost to the environment. These losses are high in mammals and birds because of their high body temperature. Much energy is needed to maintain their body temperature when heat is constantly being lost to the environment.

The net production of consumers can therefore be calculated as:

$$N = I - (F + R)$$

where:

N represents the net production
 I represents the chemical energy store of ingested food
F represents the energy lost in faeces and urine
R represents the energy lost in respiration.

Energy flow along food chains, showing the percentage transferred at each trophic level, is summarised in Figure 1.

It is the relative inefficiency of energy transfer between trophic levels that explains why:

- most food chains have only four or five trophic levels because insufficient energy is available to support a large enough breeding population at trophic levels higher than these

- the total mass of organisms in a particular place (biomass) is less at higher trophic levels

- the total amount of energy available is less at each level as one moves up a food chain.

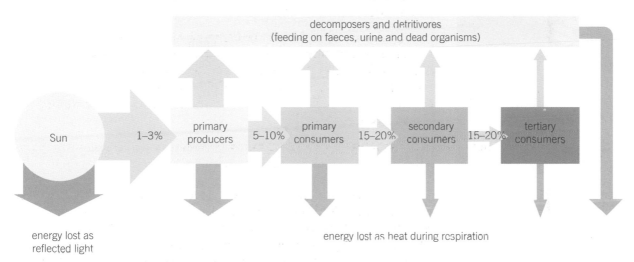

▲ **Figure 1** *Energy flow through different trophic levels of a food chain. The arrows are not to scale and give only an idea of the proportion of energy transferred at each stage. Likewise, the figures for % energy transfer between trophic levels are only a rough average as they vary considerably between different plants, animals, and habitats*

Summary questions √x̄

1 State **three** reasons for the small percentage of energy transferred at each trophic level.

2 Explain why most food chains rarely have more than four trophic levels.

3 √x̄ An area of vegetation 5 m by 5 m produces 4×10^4 kJ of potential energy in a year. Calculate the gross primary production of this area.

Hint

When making calculations involving energy transfer, always remember that energy cannot be created or destroyed. In this type of question, this means that the total amount of energy entering a box must equal the amount of energy in the box plus the amount leaving the box.

Maths link $\sqrt{x}$

MS 2.4, see Chapter 22.

▲ **Figure 2** *Only about 10% of the energy in the plant being eaten by this swallowtail butterfly larva will be used for its growth*

 Calculating the efficiency of energy transfers

Data are often presented showing the amount of energy available at each trophic level of a food chain. The energy available is usually measured in kilojoules per square metre per year ($kJ\ m^{-2}\ year^{-1}$). It is often useful to calculate the efficiency of the energy transfer between each trophic level of these food chains. This is calculated as follows:

$$\text{percentage efficiency} = \frac{\text{energy available after the transfer}}{\text{energy available before the transfer}} \times 100$$

Let us take an example. Look at Figure 3, which shows the amount of energy available at different trophic levels in a lake in the USA. Suppose we wanted to calculate the percentage efficiency of the transfer of energy from trout to humans. We would make the calculation as follows.

Energy available after the transfer $= 50\ kJ\ m^{-3}\ year^{-1}$
(i.e., energy available to humans)

Energy available before the transfer $= 250\ kJ\ m^{-3}\ year^{-1}$
(i.e., energy available to trout)

$$\text{percentage efficiency} = \frac{50}{250} \times 100 = \frac{5000}{250} = 20\%$$

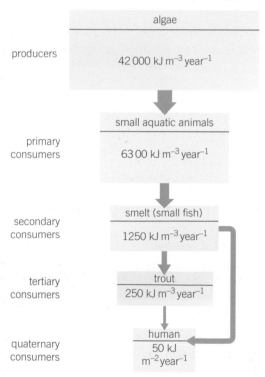

▲ **Figure 3** *Food chain in Cayuga Lake, New York State. Figures illustrate the relative amount of energy available at each trophic level in the food chain*

1 $\sqrt{x}$ Using Figure 3, calculate the percentage efficiency of energy transfer between:
   a primary consumers and secondary consumers
   b algae and humans.
Show your working in both cases.

Maths link $\sqrt{x}$

MS 0.1 and 2.4, see Chapter 22.

Adding up the totals

Figure 4 shows the flow of energy through a terrestrial ecosystem each year. All the values are in kJ m⁻² year⁻¹.

Maths link

MS 2.4 and 3.1, see Chapter 22.

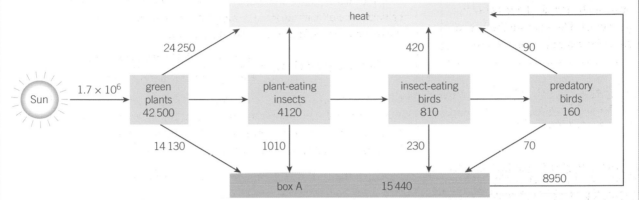

▲ Figure 4

1. Give the name of the group of organisms represented by Box A.
2. Which group of organisms are secondary consumers?
3. Calculate the percentage efficiency with which light energy is transferred to energy in green plants. Show your working.
4. State **three** reasons why so little of the solar energy is transferred to energy in green plants.
5. Calculate the amount of energy that is lost as heat from plant-eating insects. Show your working.

Productivity and farming practices

Maths link

MS 0.3 and 3.1, see Chapter 22.

Many farming practices are employed as methods of increasing yields by increasing the efficiency of energy transfer along the food chains which produce our food. As energy passes along a food chain only a small percentage passes from one organism in the chain to the next. This is because much of the energy is lost as heat during respiration. Any practice that reduces the respiratory losses in a human food chain will therefore reduce energy loss and increase the yield.

One farming practice that achieves this is the intensive rearing of domestic livestock. This is about converting the smallest possible quantity of food energy into the greatest quantity of animal mass. One way to achieve this is to minimise the energy losses from domestic animals during their lifetime. This means that more of the food energy taken in by the animals will be converted into body mass, ready to be passed on to the next link in the food chain, namely us. Energy conversion can be made more efficient by ensuring that as much energy from

respiration as possible goes into growth rather than other activities or other organisms. This is achieved by keeping animals in confined spaces, such as small enclosures, barns or cages, a practice often called factory farming. This increases the energy-conversion rate because:

- movement is restricted and so less energy is used in muscle contraction
- the environment can be kept warm in order to reduce heat loss from the body (most intensively reared species are warm-blooded)
- feeding can be controlled so that the animals receive the optimum amount and type of food for maximum growth with no wastage
- predators are excluded so that there is no loss to other organisms in the food web.

1 Suggest a reason why keeping animals in the dark for longer periods might improve the energy conversion rate.

Another farming practice that increases the efficiency of energy transfer is to reduce losses to non-human food chains by simplifying food webs. In other words, to reduce or eliminate organisms that are part of a food web and which compete with the plant or animal that is being farmed. Weeds compete with crop plants for water, mineral ions, carbon dioxide, space, and light. As these resources are often in limited supply, any amount taken by the pest means less is available for the crop plant. Insect pests may damage the leaves of crops, limiting their ability to photosynthesise and thus reducing their productivity. Alternatively, they may be in direct competition with humans, eating the crop itself. Many crops are now grown in **monoculture**, and this enables insect and fungal pests to spread rapidly. Pests of domesticated animals may cause disease. The animals may not grow as rapidly, be unfit for human consumption or die – all of which lead to reduced productivity.

2 Pesticides are used to increase productivity. Suggest how their use might sometimes reduce productivity.

The aim of pest control is to simplify the food web and so limit the effect of pests on productivity to a commercially acceptable level. In other words, to balance the cost of pest control with the benefits it brings. The problem is that at least two different interests are involved: the farmer who has to satisfy our demand for cheap food while still making a living, and the **conservation** of natural resources, which will enable us to continue to have food in the future. The trick is to balance these two, often conflicting, interests.

The graph in Figure 5 shows the effects of weeds on the productivity of two crops: wheat and soya bean.

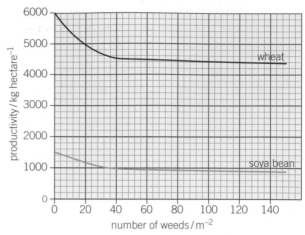

▲ Figure 5

3 Describe the effects of weeds on the productivity of wheat.
4 A herbicide that reduces the number of weeds from 40 m^{-2} to 0 m^{-2} is applied to both crops. Which crop would show the greatest percentage change in productivity?
5 ⓥ̅ It will cost a farmer £100 to treat each hectare of his wheat crop with a herbicide. The herbicide will reduce the number of weeds from 40 m^{-2} to 20 m^{-2}. He can sell his wheat at £150 a tonne (one tonne = 1000 kg). Is it economically worthwhile for the farmer to apply the herbicide to his crop? Use calculations to support your answer.

+ A mighty problem

The two-spotted spider mite, *Tetranychus urticae*, is an important pest of crops, especially those in greenhouses. Control is mostly achieved using chemicals. However, the spider mite has increasingly developed resistance to these chemicals and they are therefore less effective in controlling its populations.

Studies have been carried out to investigate the use of biological control to combat spider mites. In one such study the predatory mite *Phytoseiulus persimilis* was used to test its effectiveness against the two-spotted spider mite. This predatory mite feeds on the spider mite. Mites were introduced into two separate groups of 100 bean plants as follows:

1 Describe and explain the differences between the spider mite populations in experiment 1 and experiment 2.
2 Comment on the effectiveness of predatory mites in controlling populations of spider mites.
3 Predict what the levels of the two populations in experiment 2 might be over a period of 150–300 days if the experiment was continued. Explain the reasons for the levels you suggest.

• Experiment 1 – spider mites only

• Experiment 2 – spider mites and predatory mites

▲ **Figure 8**

▲ **Figure 7** *Biological pest control: orange predatory mite* (Phytoseiulus persimilis) *attacking the red spider mite* (Tetranychus urticae)

Learning objectives

→ Summarise the common features of all nutrient cycles.

→ Describe the features of the phosphorus cycle.

→ Describe the features of the nitrogen cycle.

→ Define the terms ammonification, nitrification, nitrogen fixation, and denitrification.

→ Explain the roles of saprobiotic organisms in nutrient recycling.

Specification reference: 3.5.4

We saw in Topic 13.1, that energy enters an **ecosystem** as sunlight and is lost as heat. This heat cannot be recycled. The flow of energy through an ecosystem is therefore in one direction, that is, it is linear. Provided the Sun continues to supply energy to Earth, this is not a problem. Nutrients, by contrast, do not have an extraterrestrial source. There is limited availability of nutrient ions in a usable form. It is important therefore that elements such as carbon, nitrogen, and phosphorus are recycled. The flow of nutrients within an ecosystem is not linear, but mostly cyclic.

All nutrient cycles have one simple sequence at their heart.

- The nutrient is taken up by **producers** (plants) as simple, inorganic molecules.
- The producer incorporates the nutrient into complex organic molecules.
- When the producer is eaten, the nutrient passes into **consumers** (animals).
- It then passes along the food chain when these animals are eaten by other consumers.
- When the producers and consumers die, their complex molecules are broken down by **saprobiontic microorganisms** (decomposers) that release the nutrient in its original simple form. The cycle is then complete. The role of these saprobionts in nutrient cycles cannot be overestimated. They are in many ways the driving forces that ensure that nutrients are released for reuse. Without them nutrients would remain locked up as part of complex molecules that cannot be taken up and used again by plants.

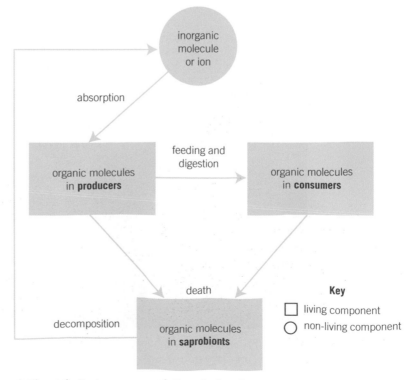

▲ **Figure 1** *Basic sequence of all nutrient cycles*

Although other processes and non-living sources are also involved, it is this sequence, illustrated in Figure 1, that forms the basis of all nutrient cycles.

The nitrogen cycle

Living organisms require a source of nitrogen from which to manufacture proteins, nucleic acids and other nitrogen-containing compounds. Although 78% of the atmosphere is nitrogen, there are very few organisms that can use nitrogen gas directly. Plants take up most of the nitrogen they require in the form of nitrate ions (NO_3^-), from the soil. These ions are absorbed, using **active transport**, by the roots. This is where nitrogen enters the living component of the ecosystem. Animals obtain nitrogen-containing compounds by eating and digesting plants.

Nitrate ions are very soluble and easily leach (wash) through the soil, beyond the reach of plant roots. In natural ecosystems, the nitrate concentrations are restored largely by the recycling of nitrogen-containing compounds. In agricultural ecosystems, the concentration of soil nitrate can be further increased by the addition of fertilisers. When plants and animals die, the process of decomposition begins, in a series of steps by which microorganisms replenish the nitrate concentrations in the soil. This release of nitrate ions by decomposition is most important because, in natural ecosystems, there are very few nitrate ions available from other sources.

There are four main stages in the nitrogen cycle (Figure 2), **ammonification**, **nitrification**, **nitrogen fixation** and **denitrification**, each of which involves saprobiontic microorganisms.

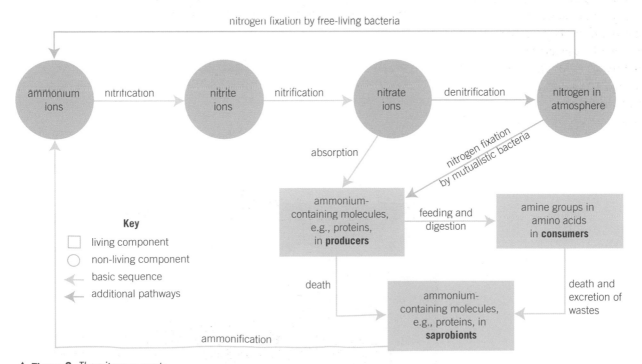

▲ **Figure 2** *The nitrogen cycle*

Ammonification

Ammonification is the production of ammonia from organic nitrogen-containing compounds. In nature, these compounds include urea (from the breakdown of excess amino acids) and proteins, nucleic acids and vitamins (found in faeces and dead organisms). Saprobiontic microorganisms, mainly fungi and bacteria, feed on faeces and dead organisms materials, releasing ammonia, which then forms ammonium ions in the soil. This is where nitrogen returns to the non-living component of the ecosystem.

Nitrification

Plants use light energy to produce organic compounds. Some bacteria, however, obtain their energy from chemical reactions involving inorganic ions. One such reaction is the conversion of ammonium ions to nitrate ions. This is an **oxidation** reaction and so releases energy. It is carried out by free-living soil microorganisms called nitrifying bacteria. This conversion occurs in two stages:

1 oxidation of ammonium ions to nitrite ions (NO_2^-)
2 oxidation of nitrite ions to nitrate ions (NO_3^-).

Nitrifying bacteria require oxygen to carry out these conversions and so they require a soil that has many air spaces. To raise productivity, it is important for farmers to keep soil structure light and well aerated by ploughing. Good drainage also prevents the air spaces from being filled with water and so prevents air being forced out of the soil.

Nitrogen fixation

This is a process by which nitrogen gas is converted into nitrogen-containing compounds. It can be carried out industrially and also occurs naturally when lightning passes through the atmosphere. By far the most important form of nitrogen fixation is carried out by microorganisms, of which there are two main types:

- **free-living nitrogen-fixing bacteria**. These bacteria reduce gaseous nitrogen to ammonia, which they then use to manufacture amino acids. Nitrogen-rich compounds are released from them when they die and decay.

- **mutualistic nitrogen-fixing bacteria**. These bacteria live in nodules on the roots of plants such as peas and beans (Figure 3). They obtain carbohydrates from the plant and the plant acquires amino acids from the bacteria.

▲ **Figure 3** *Nitrogen-fixing nodules on the roots of a pea plant allow the plant to use free nitrogen in the atmosphere and soil. Mutualistic bacteria in the nodules fix the nitrogen, transforming it into a form usable by the plant*

Denitrification

When soils become waterlogged, and have a low oxygen concentration, the type of microorganism present changes. Fewer **aerobic** nitrifying and nitrogen-fixing bacteria are found, and there is an increase in **anaerobic** **denitrifying bacteria**. These convert soil nitrates into gaseous nitrogen. This reduces the availability of nitrogen-containing compounds for plants. For land to be productive, the soils on which crops grow must therefore be kept well aerated to prevent the build-up of denitrifying bacteria.

As with any nutrient cycle, the delicate balance can be easily upset by human activities. Some of the effects of these activities are considered in Topic 13.4, Use of natural and artificial fertilisers.

▲ **Figure 4** *Ploughing helps to aerate the soil and so prevents the build-up of denitrifying bacteria that can reduce the level of soil nitrates*

The phosphorus cycle

Phosphorus is an important biological element as it is a component of ATP, phospholipids and nucleic acids. Life therefore depends on it being constantly recycled.

In the carbon and nitrogen cycles the main reservoir of each element is in the atmosphere. In the phosphorus cycle however the main reservoir is in mineral form rather than in the atmosphere – in fact the phosphorus cycle lacks a gaseous phase altogether.

Phosphorus exists mostly as phosphate ions (PO_4^{3-}) in the form of sedimentary rock deposits. These have their origins in the seas but are brought to the surface by the geological uplifting of rocks. The weathering and erosion of these rocks helps phosphate ions to become dissolved and so available for absorption by plants which incorporate them into their biomass. The phosphate ions pass into animals which feed on the plants. Excess phosphate ions are excreted by animals and may accumulate in waste material such as guano formed from the excretory products of some sea birds.

On the death of plants and animals, decomposers such as certain bacteria and fungi break them down releasing phosphate ions into the water or soil. Some phosphate ions remain in parts of animals, such as bones or shells, that are very slow to breakdown. Phosphate ions in excreta, released by decomposition and dissolved out of rocks, are transported by streams and rivers into lakes and oceans where they form sedimentary rocks thus completing the cycle.

The phosphorus cycle is illustrated in Figure 5.

> ### Synoptic link
>
> To revise the part phosphorus plays in the structure of biological molecules you will need to consult Topic 1.5, Lipids, Topic 2.1, Structure of RNA and DNA, and 2.3, Energy and ATP.

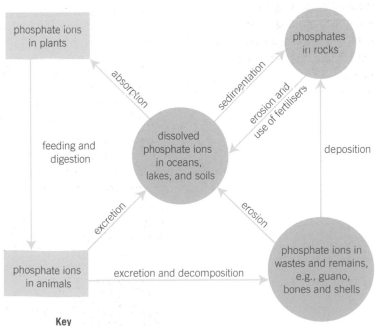

Key
☐ living component
◯ non-living component

▲ **Figure 5** *The phosphorus cycle*

The role of mycorrhizae in nutrient cycles

Mycorrhizae are associations between certain types of fungi and the roots of the vast majority of plants. The fungi act like extensions of the plant's root system and vastly increase the total surface area for the absorption of water and minerals. The mycorrhiza acts like a sponge and so holds water and minerals in the neighbourhood of the roots. This enables the plant to better resist drought and to take up inorganic ions more readily. The mycorrhiza plays a part in nutrient cycles by improving the uptake of relatively scarce ions such as phosphates ions.

The mycorrhizal relationship between plants and fungi is a **mutualistic** one. The plant benefits from improved water and inorganic ion uptake while the fungus receives organic compounds such as sugars and amino acids from the plant.

Summary questions

In the following passage, suggest the most appropriate word to replace each of the numbers in brackets.

A few organisms can convert nitrogen gas into compounds useful to other organisms in a process known as (**1**). These organisms can be free-living or live in a relationship with certain (**2**). Most plants obtain their nitrogen by absorbing (**3**) from the soil through their (**4**) by active transport. They then convert this to (**5**), which is passed to animals when they eat the plants. On death, (**6**) break down these organisms, releasing (**7**), which can then be oxidised to form nitrite ions by (**8**) bacteria. Further oxidation by the same type of bacteria forms (**9**) ions. These ions may be converted back to atmospheric nitrogen by the activities of (**10**) bacteria.

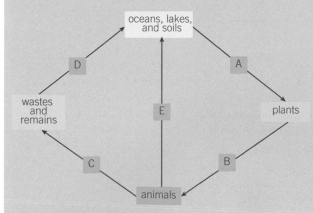

▲ Figure 6

11 Figure 6 is a simplified illustration of the phosphorus cycle. Each box represents a process. Name the process in each of the boxes A, B, C, D and E.

13.4 Use of natural and artificial fertilisers

Agricultural ecosystems increase the efficiency of energy transfer along human food chains. In doing so they improve productivity. One farming practice that contributes to this improved productivity is the use of fertilisers. Let us see how this is achieved.

The need for fertilisers

All plants need mineral ions, especially nitrates, from the soil. Much food production in the developed world is intensive, that is, it is concentrated on specific areas of land that are used repeatedly to achieve maximum yield from the crops and animals grown on them. Intensive food production makes large demands on the soil because mineral ions are continually taken up by the crops being grown on it. These crops are either used directly as food or as fodder for animals that are then eaten. Either way, the mineral ions that the crops have absorbed from the soil are removed.

In natural ecosystems the minerals that are removed from the soil by plants are returned when the plant is decomposed by microorganisms on its death. In agricultural systems the crop is harvested and then transported from its point of origin for consumption. The urine, faeces and dead remains of the consumer are rarely returned to the same area of land. Under these conditions the concentrations of the mineral ions in agricultural land will fall. It is therefore necessary to replenish these mineral ions because, otherwise, their reduced concentrations will become the main limiting factor to plant growth. Productivity will consequently be reduced. To offset this loss of mineral ions, fertilisers need to be added to the soil. These fertilisers are of two types:

- **natural (organic) fertilisers**, which consist of the dead and decaying remains of plants and animals as well as animal wastes such as manure, slurry and bone meal
- **artificial (inorganic) fertilisers**, which are mined from rocks and deposits and then converted into different forms and blended together to give the appropriate balance of minerals for a particular crop. Compounds containing the three elements, nitrogen, phosphorus, and potassium are almost always present.

Research suggests that a combination of natural and artificial fertilisers gives the greatest long-term increase in productivity. However, it is important that minerals are added in appropriate quantities as there is a point at which further increases in the quantity of fertiliser no longer results in increased productivity. This is illustrated in Figure 1.

How fertilisers increase productivity

Plants require minerals for their growth. Let us look at nitrogen as an example. Nitrogen is an essential component of amino acids, ATP, and nucleotides in DNA. Both are needed for plant growth. Where nitrate ions are readily available, plants are likely to develop earlier, grow taller and have a greater leaf area. This increases the rate of photosynthesis and improves crop productivity. There can be no doubt that nitrogen-containing fertilisers have been of considerable benefit in providing us with cheaper food. It is

Learning objectives

→ Explain why fertilisers are needed in agricultural ecosystems.

→ Distinguish between natural and artificial fertilisers.

→ Explain how fertilisers increase productivity.

Specification reference: 3.5.4

▲ **Figure 1** *The effect of different quantities of nitrogenous fertiliser on the yield of wheat*

▲ **Figure 2** *Cattle slurry, a natural fertiliser, being spread onto a crop of wheat*

Maths link

MS 1.3, see Chapter 22.

Summary questions

1 Explain why fertilisers are needed in an agricultural ecosystem.

2 Using Figure 1, determine what concentration of fertiliser you would advise a farmer to apply to a field of wheat.

3 Suggest a reason why, after a certain point, the addition of more fertiliser no longer improves the productivity of a crop.

4 Distinguish between natural and artificial fertilisers.

Maths link

MS 3.1, see Chapter 22.

estimated that the use of fertilisers has increased agricultural food production in the UK by around 100% since 1955.

Different forms of nitrogen-containing fertiliser

Nitrogen-containing fertiliser can be applied to crops in a number of different forms. These include ammonium salts, animal manure, the ground-up bones of animals (bone meal), and urea (a waste product found in the urine of mammals). An investigation was carried out in which the same crop was grown on six separate plots of land each of the same area. No nitrogen-containing fertiliser was added to the first plot. To each of the remaining five plots, a different form of nitrogen-containing fertiliser was added at the rate of 140 kg total nitrogen per hectare. The graph in Figure 3 shows the results of the investigation.

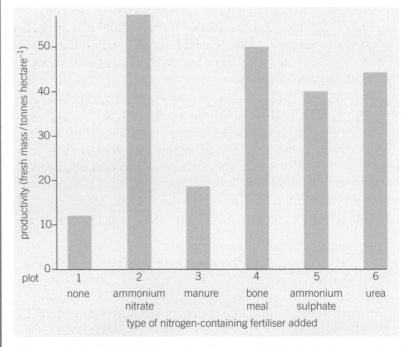

▲ Figure 3

1 State which forms of nitrogen used in the investigation are natural fertilisers.

2 Suggest why the investigation included a plot to which no nitrogen-containing fertiliser was added.

3 Suggest how the addition of nitrogen-containing fertiliser, in whatever form, increased productivity.

4 The mass of each fertiliser used was different in each case. Suggest why this was necessary.

5 It is sometimes claimed that nitrogen-containing fertilisers in the form of ammonium salts increase productivity of crops better than other forms of nitrogen-containing fertilisers. State, with your reasons, whether or not you think the results of this experiment support this view.

6 The increase in productivity when manure was applied was lower than for other forms of nitrogen-containing fertiliser. This is because the manure has to break down before its nitrogen is released and this process takes a few months. Suggest how a farmer who spreads manure on his/her crops, might use this information in order to improve productivity.

In natural ecosystems minerals such as nitrate ions, which are removed from the soil by plants, are mainly returned when the plant is decomposed. However, as we saw in Topic 13.4, in agricultural systems the crop is removed and so the nitrate is not returned and has to be replaced. This is done by the addition of natural or artificial fertilisers.

Effects of nitrogen-containing fertilisers

Nitrogen is an essential component of biological molecules such as proteins and is needed for growth and, therefore, an increase in the area of leaves. This increases the rate of photosynthesis and improves crop productivity. There can be no doubt that nitrogen-containing fertilisers have benefited us considerably by providing us with cheaper food. Most of this increase is due to additional nitrogen (Figure 1). The use of nitrogen-containing fertilisers has also had some detrimental effects. These include:

- **reduced species diversity**, because nitrogen-rich soils favour the growth of grasses, nettles and other rapidly growing species. These out-compete many other species, which die as a result. Species-rich hay meadows, such as the one in the photograph (Figure 3), only survive when soil nitrogen concentrations are low enough to allow other species to compete with the grasses
- **leaching**, which may lead to pollution of watercourses
- **eutrophication**, caused by leaching of fertiliser into watercourses.

Leaching

Leaching is the process by which nutrients are removed from the soil. Rainwater will dissolve any soluble nutrients, such as nitrate ions, and carry them deep into the soil, eventually beyond the reach of plant roots. The leached nitrate ions find their way into watercourses, such as streams and rivers, that in turn may drain into freshwater lakes. Here they may have a harmful effect on humans if the river or lake is a source of drinking water. Very high nitrate ion concentrations in drinking water can prevent efficient oxygen transport in babies and a link to stomach cancer in humans has been suggested. The leached nitrate ions are also harmful to the environment as they can cause eutrophication.

Eutrophication

Eutrophication is the process by which nutrient concentrations increase in bodies of water. It is a natural process that occurs mostly in freshwater lakes and the lower reaches of rivers. Eutrophication consists of the following sequence of events:

1 In most lakes and rivers there is naturally very low concentration of nitrate and so nitrate ions are a limiting factor for plant and algal growth.

2 As the nitrate ion concentration increases as a result of leaching, it ceases to be a limiting factor for the growth of plants and algae whose populations both grow.

3 As algae mostly grow at the surface, the upper layers of water become densely populated with algae. This is called an 'algal bloom'.

Learning objectives

→ Describe the main environmental effects of using nitrogen-containing fertilisers.

→ State the meanings of leaching and eutrophication.

→ Explain how leaching and eutrophication affect the environment.

Specification reference: 3.5.4

▲ **Figure 1** *Use of different types of fertilisers in the UK*

▲ **Figure 2** *Low species diversity in a field grown for silage that has had nitrogen-containing fertiliser added*

▲ **Figure 3** *High species diversity in a meadow grown for hay without the addition of nitrogen-containing fertiliser*

▲ **Figure 4** *Algal bloom in a canal as a result of eutrophication caused by nitrogen-containing fertiliser run-off*

4 This dense surface layer of algae absorbs light and prevents it from penetrating to lower depths.

5 Light then becomes the limiting factor for the growth of plants and algae at lower depths and so they eventually die.

6 The lack of dead plants and algae is no longer a limiting factor for the growth of saprobiontic bacteria and so these populations too grow, using the dead organisms as food.

7 The saprobiontic bacteria require oxygen for their respiration, creating an increased demand for oxygen.

8 The concentration of oxygen in the water is reduced and nitrates are released from the decaying organisms.

9 Oxygen then becomes the limiting factor for the population of **aerobic** organisms, such as fish. These organisms ultimately die as the oxygen is used up altogether.

10 Without the aerobic organisms, there is less competition for the **anaerobic** organisms, whose populations now rise.

11 The anaerobic organisms further decompose dead material, releasing more nitrates and some toxic wastes, such as hydrogen sulphide, which make the water putrid.

Organic manures, animal slurry, human sewage, ploughing old grassland and natural leaching can all contribute to eutrophication, but the leaching of artificial fertilisers is the main cause.

Summary questions

1 Explain what is meant by eutrophication.

2 Explain how an increase in algal growth at the surface can lead to the death of plants growing beneath them.

3 Explain how the death of these plants can result in the death of animals such as fish.

Maths link √x̄

MS 3.1, see Chapter 22.

Troubled waters √x̄

A farmer applied a large quantity of fertiliser to fields next to a small lake. A period of heavy rain followed. After 10 days, scientists monitoring the lake noticed changes to the algal population, the clarity of the water and the levels of dissolved oxygen. These changes are shown in the three graphs in Figure 5. Secchi depth is a measure of the clarity of water. Measurements are taken by lowering a black-and-white disc (called a Secchi disc) into the water and recording the depth at which it is no longer visible.

1 Suggest a reason why the changes in the lake do not occur until 10 days after the application of the fertiliser to the fields.

2 Explain a possible cause of the increase in the density of algae after 10 days.

3 Describe and explain the relationship between the density of algae and water clarity in the lake.

4 Describe and explain changes to the levels of dissolved oxygen over the 100-day period.

▲ **Figure 5**

1   Scientists constructed a mathematical model. They used this model to estimate the transfer of energy through consumers in a natural grassland ecosystem. The table shows their results.

	Energy transferred as percentage of energy in biomass of producers				
	Ingested Food (I)	Absorbed from gut (A)	Egested (E)	Net Production (N)	Respired (R)
Primary consumers					
Mammals	25.00	12.50	12.50	0.25	12.25
Insects	4.00	1.60	2.40	0.64	0.96
Secondary consumers					
Mammals	0.16	0.13	0.03	0.003	0.127
Insects	0.17	0.135	0.035	0.040	0.095

(a) Copy and complete the equation to show how net production is calculated from the energy in ingested food.

      P =   *(1 mark)*

(b) Describe and explain how intensive rearing of domestic livestock would affect
    (i)   the figure for **A** in the first row of the table   *(1 mark)*
    (ii) the figure for **R** in the first row of the table.   *(1 mark)*

(c) (i)   Calculate the ratio of **R** : **A** for mammalian primary consumers.   *(1 mark)*
    (ii) The **R** : **A** ratio is higher in mammalian primary consumers than in insect primary consumers. Suggest a reason for this higher value.   *(1 mark)*

(d) The scientists tested their model by comparing the values it predicted with actual measured values. The graph shows their results.

Are the values predicted by the model supported by the actual measured values? Evaluate the evidence in the graph.

                                                               (3 marks)
                                                            AQA June 2010

2 Scientists measured the mean temperature in a field each month between March and October. The table shows their results.

Month	Mean temperature / °C
March	9.0
April	11.0
May	14.0
June	17.0
July	20.0
August	18.0
September	16.0
October	14.0

(a) The gross productivity of the plants in the field was highest in July. Use the data in the table to explain why. *(2 marks)*

(b) (i) Give the equation that links gross productivity and net productivity. *(1 mark)*

  (ii) The net productivity of the plants in the field was higher in August than in July. Use the equation in part **(b)(i)** and your knowledge of photosynthesis and respiration to suggest why. *(2 marks)*

(c) A horse was kept in the field from March to October. During the summer months, the horse was able to eat more than it needed to meet its minimum daily requirements. Suggest how the horse used the extra nutrients absorbed. *(1 mark)*

(d) The horse's mean energy expenditure was higher in March than it was in August. Use information in the table to suggest why. *(2 marks)*

AQA June 2011

3 The diagram shows the nitrogen cycle.

(a) (i) Name process **P**. *(1 mark)*

  (ii) Name process **Q**. *(1 mark)*

(b) Leguminous crop plants have nitrogen-fixing bacteria in nodules on their roots. On soils with a low concentration of nitrate ions, leguminous crops often grow better than other types of crop. Explain why. *(2 marks)*

(c) Applying very high concentrations of fertiliser to the soil can reduce plant growth. Use your knowledge of water potential to explain why. *(2 marks)*

AQA Jan 2013

4   Scientists investigated the effect of a mycorrhizal fungus on the growth of pea plants with
     a nitrate fertiliser or an ammonium fertiliser. The fertilisers were identical, except for
     nitrate or ammonium.
     The scientists took pea seeds and sterilised their surfaces. They planted the seeds in soil
     that had been heated to 85 °C for 2 days before use. The soil was sand that contained no
     mineral ions useful to the plants.
     (a)   Explain why the scientists sterilised the surfaces of the seeds and grew them
           in soil that had been heated to 85 °C for 2 days.   *(2 marks)*
     (b)   Explain why it was important that the soil contained no mineral ions useful
           to the plants.   *(1 mark)*
           The pea plants were divided into four groups, A, B, C and D.
           • Group A – heat-treated mycorrhizal fungus added, nitrate fertiliser
           • Group B – mycorrhizal fungus added, nitrate fertiliser
           • Group C – heat-treated mycorrhizal fungus added, ammonium fertiliser
           • Group D – mycorrhizal fungus added, ammonium fertiliser
           The heat-treated fungus had been heated to 120 °C for 1 hour.
     (c)   Explain how groups A and C act as controls.   *(2 marks)*
           After 6 weeks, the scientists removed the plants from the soil and cut the roots from
           the shoots. They dried the plant material in an oven at 90 °C for 3 days. They then
           determined the mean dry masses of the roots and shoots of each group of pea plants.
     (d)   (i)   Suggest what the scientists should have done during the drying process to be
                 sure that all of the water had been removed from the plant samples. *(2 marks)*
                 The scientists' results are shown in **Table 3**.

     ▼ **Table 3**

Treatment	Mean dry mass / g per plant (± standard deviation)	
	Root	Shoot
A – heat-treated fungus and nitrate fertiliser	0.40 (±0.05)	1.01 (±0.12)
B – fungus and nitrate fertiliser	1.61 (±0.28)	9.81 (±0.33)
C – heat-treated fungus and ammonium fertiliser	0.34 (±0.03)	0.96 (±0.26)
D – fungus and ammonium fertiliser	0.96 (±0.18)	4.01 (± 0.47)

           (ii) the values of dry mass recorded for the shoots of the plants in group D were 3.4,
                 4.5, 4.2, 3.9, 4.1 g.
                 calculate the mean dry mass and the standard deviation, s, of this mean.

                 Use the formula $s = \sqrt{\dfrac{\Sigma(x - \bar{x})^2}{n-1}}$ *(5 marks)*

           (iii) comment on the reliability of the means shown in the table   *(2 marks)*
     (e)   What conclusions can be drawn from the data in **Table 3** about the following?
           The effects of the fungus on growth of the pea plants.
           The effects of nitrate fertiliser and ammonium fertiliser on growth of the pea plants.
                                                                                       (4 marks)
                                                     AQA Specimen 2014 (apart from 5 (d) (ii) and(iii))

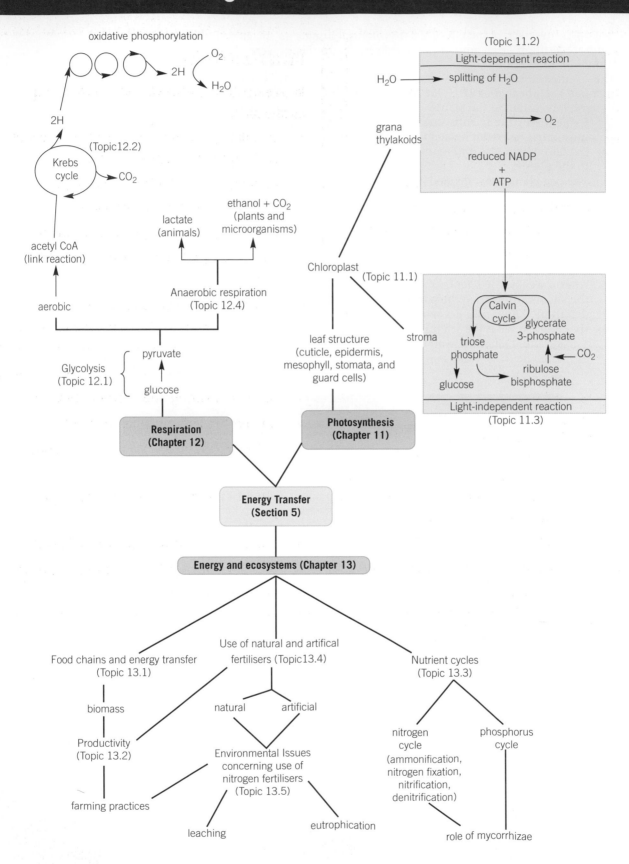

Practical skills

In this section you have met the following practical skills:

How to use appropriate apparatus to record a range of quantitative measurements in experiments such as:

- investigating the effect of named environmental variables on the rate of photosynthesis
- using a redox indicator such as methylene blue to investigate dehydrogenase activity.

Maths skills

In this section you have met the following maths skills:

- calculating gross primary productivity and deriving the appropriate units.
- calculating the efficiency of energy transfers within ecosystems.
- calculating percentage yields of crops.
- interpreting data from a variety of tables and graphs.
- translating information between graphical, numerical, and algebraic forms.

Extension

Design and carry out an experiment to determine whether there is a significant difference between the average surface area of leaves on nettle plants that are growing in a sunny position compared to leaves on nettle plants growing in a shaded position.

In planning and carrying out your experiment, consider how you will:

- select the sunny and shaded positions
- choose the individual nettle plants
- select the sample leaves from the chosen plants
- decide on the number of samples you will take
- measure the surface area of the leaves
- test whether any differences you find are statistically significant.

1 **(a)** A student measured the rate of aerobic respiration of a woodlouse using the apparatus shown in Figure 1.

▲ **Figure 1**

(i) The student closed the tap. After thirty minutes the drop of coloured liquid had moved to the left. Explain why the drop of coloured liquid moved to the left. *(3 marks)*

(ii) What measurements should the student have taken to calculate the rate of aerobic respiration in mm^3 of oxygen $g^{-1} h^{-1}$? *(3 marks)*

(b) DNP inhibits respiration by preventing a proton gradient being maintained across membranes. When DNP was added to isolated mitochondria the following changes were observed

- less ATP was produced
- more heat was produced
- the uptake of oxygen remained constant

Explain how DNP caused these changes *(3 marks)*

AQA Jan 2011

2 **(a)** Ecologists investigated photosynthesis in two species of plant found in woodland. One of the species was adapted to growing in bright sunlight (sun plant) and the other was adapted to growing in the shade (shade plant). The ecologists' results are shown in Figure 2.

▲ **Figure 2**

(i) Give two factors which could be limiting the rate of photosynthesis in the sun plant between points A and B on Figure 2. *(1 mark)*

(ii) Explain why CO_2 uptake is a measure of net productivity. *(1 mark)*

(iii) Use the information in Figure 1 to explain how the shade plant is better adapted than the sun plant to growing at low light intensities. *(2 marks)*

AQA June 2014

3 A student investigated the rate of anaerobic respiration in yeast. She put 5g of yeast into a glucose solution and placed this mixture in the apparatus shown in Figure 3. She then recorded the total volume of gas collected every ten minutes for 1 hour.

gas syringe

layer of oil

yeast in glucose solution

▲ **Figure 3**

(a) Explain why a layer of oil is required in this investigation. *(1 mark)*

(b) The student's results are shown in Table 1.

▼ **Table 1**

Time / minutes	Total volume of gas collected / cm^3
10	0.3
20	0.9
30	1.9
40	3.1
50	5.0
60	5.2

(c) (i) Calculate the rate of gas production in cm^{3}g^{-1}min^{-1} during the first 40 minutes of this investigation. Show your working. *(2 marks)*

(ii) Suggest why the rate of gas production decreased between 50 and 60 minutes. *(1 mark)*

(iii) Yeast can also respire aerobically. The student repeated the investigation with a fresh sample of yeast in glucose solution, but without the oil. All other conditions remained the same. Explain what would happen to the volume of gas in the syringe if the yeast were only respiring aerobically. *(2 marks)*

(d) Respiration produces more ATP per molecule of glucose in the presence of oxygen than it does when oxygen is absent. Explain why. *(2 marks)*

AQA June 2014

4 (a) Name the type of bacteria which convert:
(i) nitrogen in the air into ammonium compounds
(ii) nitrites into nitrates *(2 marks)*

(b) (i) Other than spreading fertilisers, describe and explain how **one** farming practice results in addition of nitrogen-containing compounds to a field. *(2 marks)*

(ii) Describe and explain how **one** farming practice results in the removal of nitrogen-containing compounds from a field. *(2 marks)*

AQA June 2006

5 Starfish feed on a variety of invertebrate animals that are attached to rocks on the seashore. The diagram shows part of a food web involving a species of starfish.

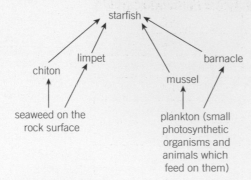

(a) Explain why a starfish can be described as both a secondary and a tertiary consumer. *(1 mark)*

(b) When starfish feed on mussels they leave behind the empty shell. Explain how quadrats could be used to determine the percentage of mussels that had been eaten by starfish on a rocky shore. *(3 marks)*

(c) Table 2 shows the composition of the diet of starfish.

▼ **Table 2**

	Prey species			
	chitons	limpets	mussels	barnacles
Percentage of total number of animals eaten	3	5	27	65
Energy provided by each species as a percentage of total energy intake	42	5	38	15

(i) The percentage of barnacles in the diet is much higher than the percentage of energy they provide. Suggest **one** explanation for this difference. *(1 mark)*

(ii) Table 2 shows that the amount of energy provided by chitons is greater than the amount of energy provided by limpets. Calculate the number of limpets a starfish would need to eat in order to obtain the same amount of energy as it would obtain from one chiton. *(1 mark)*

AQA June 2006

6 The progress of the light-dependent reaction was investigated using DCPIP indicator solution to compare the activity of chlorophyll extract with that of intact chloroplasts. The following results were obtained.

Time / s	% absorption of red light (extract)	% absorption of red light (intact chloroplasts)
60	98	96
90	92	94
120	86	88
150	74	84
180	63	76
210	55	70
240	41	64
270	35	58
300	28	55
330	26	53
360	24	52

(a) Plot a suitable graph to show the progress of the reaction in each sample. *(4 marks)*

(b) Use the graph to determine the percentage difference between the maximum rates of reaction of the two samples. *(3 marks)*

7 (a) The biochemical pathway of aerobic respiration involves a number of different steps. Name **one** step in which carbon dioxide is produced. *(1 marks)*
In an investigation, scientists transferred slices of apple from air to anaerobic conditions in pure nitrogen gas. They measured the rate of carbon dioxide production.

(b) The scientists kept the temperature constant throughout the investigation. Explain how a decrease in temperature would affect the rate of carbon dioxide production. *(2 marks)*

(c) When the apple slices were transferred to nitrogen, the following biochemical pathway took place.

CH₃   CH₃
| |
CO   ⟶   CH₂
| |
COOH   OH

pyruvic acid   ethanol

Use this pathway to explain the part played by reduced NAD when the apple slices were transferred to nitrogen. *(2 marks)*

(d) The rate of carbon dioxide production was higher when the apple slices were in nitrogen than when they were in the air. Explain why. *(3 marks)*

AQA Jan 2010

Section 6
Organisms respond to changes in their environment

Introduction

Multicellular organisms are able to respond to stimuli that originate both from outside and from within their bodies. By doing so, they can avoid harmful environments while at the same time maintaining an internal environment that provides the optimum conditions for their metabolism. These organisms control their activities through a combination of growth factors, hormones, and nerve impulses. A stimulus is a change in the internal or external environment. Stimuli are detected by receptors which are specific to one type of stimulus. A coordinator formulates a suitable response to a stimulus and an effector produces the response.

There are two forms of coordination in most multicellular animals – nervous and hormonal. The nervous system allows rapid communication between one part of an organism and another. It comprises nerve cells that pass electrical impulses along their length. The nerve impulse releases a chemical messenger on to its target cell that usually leads to a rapid, short-lived, and localised response. The hormonal system produces slower responses. Hormones are produced in endocrine glands and stimulate their target cells via the blood stream. Each hormone is specific to particular receptors that are only present on their target cells. Hormonal responses are usually slow, long-lasting, and widespread.

It is advantageous to maintain a relatively constant internal environment. Not only can chemical reactions take place at a predictable rate but also the organism has a greater degree of independence from the external environment. The maintenance of a constant internal environment is called homeostasis. Responding to changes in the internal and external environment is no less important to survival in plants. Plants lack contractile tissue and do not move from place to place. At the molecular level, some plant responses are rapid where they use hormone-like substances to control their responses to stimuli such as light and gravity.

Working scientifically

In studying how organisms respond to changes in their environment there will be opportunities to perform practical exercises and so develop practical skills. Required practical activities are:

- Investigating the effect of an environmental variable on the movement of an animal using either a choice chamber or a maze.

- Producing a dilution series of a glucose solution and using colorimetric techniques to produce a calibration curve with which to identify the concentration of glucose in an unknown 'urine' sample.

In performing these experiments you will have the chance to develop practical skills such as:

- using a colorimeter to record quantitative measurements
- using laboratory glassware apparatus to make up serial dilutions
- using a light microscope at high power and low power
- safely and ethically using organisms to measure animal responses.

You will be able to develop a range of mathematical skills. In particular the ability to use appropriate units when calculating the maximum frequency of impulse conduction.

What you already know:

The information in this unit is intended to be self-explanatory, but there is certain knowledge from GCSE that will prove beneficial to the understanding of this section. This information includes:

- ☐ Cells called receptors detect stimuli (changes in the environment). Information from receptors passes along cells (neurones) in nerves to the brain. The brain coordinates the response. Reflex actions are automatic and rapid.

- ☐ In a simple reflex action impulses from a receptor pass along a sensory neurone to the central nervous system. At a junction (synapse) between a sensory neurone and a relay neurone in the central nervous system, a chemical is released that causes an impulse to be sent along a relay neurone.

- ☐ Many processes within the body are coordinated by chemical substances called hormones. Hormones are secreted by glands and are usually transported to their target organs by the bloodstream.

- ☐ Plants are sensitive to light, moisture and gravity; their shoots grow towards light and against the force of gravity while their roots grow towards moisture and in the direction of the force of gravity.

- ☐ Plants produce hormones to coordinate and control growth. Auxin controls phototropism and gravitropism (geotropism).

- ☐ Waste products to be removed from the body include urea, produced in the liver by the breakdown of amino acids and removed by the kidneys in the urine.

- ☐ A healthy kidney produces urine by filtering the blood and then reabsorbing all the sugar, dissolved ions and water needed by the body and finally releasing urea, excess ions, and water as urine.

- ☐ The blood glucose concentration of the body is monitored and controlled by the pancreas through the production of the hormone insulin.

- ☐ A second hormone, glucagon, is produced in the pancreas when blood glucose levels fall. This causes glycogen to be converted into glucose.

- ☐ Type 1 diabetes is a disease in which a person's blood glucose concentration may rise to a high level because the pancreas does not produce enough of the hormone insulin.

Learning objectives

→ Define a stimulus and a response.

→ Examine the advantage to organisms of being able to respond to stimuli.

→ Describe taxes, kineses, and tropisms.

→ Explain how each type of response increases an organism's chances of survival.

Specification reference: 3.6.1.1

Practical link ⚗

Required practical 10. Investigation into the effect of an environmental variable on the movement of an animal using either a choice chamber or a maze.

▲ **Figure 1** *Woodlice exhibit a behaviour called kinesis, which ensures that they spend most of their time in the dark moist conditions that prevent them from drying out and hence aid their survival*

Stimulus and response

A **stimulus** is a detectable change in the internal or external environment of an organism that leads to a **response** in the organism. The ability to respond to stimuli is a characteristic of life and increases the chances of survival for an organism. For example, to be able to detect and move away from harmful stimuli, such as predators and extremes of temperature, or to detect and move towards a source of food clearly aid survival. Those organisms that survive have a greater chance of raising offspring and of passing their **alleles** to the next generation. There is always, therefore, a **selection pressure** favouring organisms with more appropriate responses.

Stimuli are detected by **receptors**. Receptors are specific to one type of stimulus. A coordinator formulates a suitable response to a stimulus. Coordination may be at the molecular level or involve a large organ such as the brain. A response is produced by an **effector**. This response may be at the molecular level or involve the behaviour of a whole organism. One means of communication in large, multicellular organisms occurs via chemicals called hormones, which is a relatively slow process found in both plants and animals (Topics 14.2 and 16.3).

In addition to hormonal communication, animals have another, more rapid, means of communication – the nervous system. Their nervous systems usually have many different receptors and control effectors. Each receptor and effector is linked to a central **coordinator** of some type. The coordinator acts like a switchboard, connecting information from each receptor with the appropriate effector. The sequence of events can therefore involve either chemical control or nerve cells and may be summarised as:

stimulus → receptor → coordinator → effector → response

Let us look first at the simplest forms of response to stimuli and how they can increase an organism's chances of survival.

Taxes

A **taxis** is a simple response whose direction is determined by the direction of the stimulus. As a result, a motile organism responds directly to environmental changes by moving its whole body either towards a favourable stimulus or away from an unfavourable one. Taxes are classified according to whether the movement is towards the stimulus (positive taxis) or away from the stimulus (negative taxis) and also by the nature of the stimulus. Some examples are given below:

• Single-celled algae will move towards light (positive phototaxis). This increases their chances of survival since, being photosynthetic, they require light to manufacture their food.

- Earthworms will move away from light (negative phototaxis). This increases their chances of survival because it takes them into the soil, where they are better able to conserve water, find food and avoid some predators.
- Some species of bacteria will move towards a region where glucose is more highly concentrated (positive chemotaxis). This increases their chances of survival because they use glucose as a source of food.

Kineses

A **kinesis** is a form of response in which the organism does not move towards or away from a stimulus. Instead, it changes the speed at which it moves and the rate at which it changes direction. If an organism crosses a sharp dividing line between a favourable and an unfavourable environment, its rate of turning increases. This raises its chances of a quick return to a favourable environment. However, if it moves a considerable distance into an unfavourable environment its rate of turning may slowly decrease so that it moves in long straight lines before it turns, often very sharply. This type of response tends to bring the organism into a new region with favourable conditions. It is important when a stimulus is less directional. Humidity and temperature, for example, do not always produce a clear gradient from one extreme to another.

An example of a kinesis occurs in woodlice. Woodlice lose water from their bodies in dry conditions. When they move from a damp area into a dry one, they move more rapidly and change direction more often. This increases their chance of moving back into the damp area. Once back in the damp area, they slow down and change direction less often. This means they are more likely to stay within the damp area. However, if after some time spent changing direction rapidly they are in the damp area, their behaviour changes. Instead they move rapidly in straight lines, which increases their chances of moving through the dry area and into a new damp one. In this way they spend more time in favourable damp conditions than in less favourable drier ones. This prevents them drying out and so increases their chances of survival.

Tropisms

A **tropism** is the growth of part of a plant in response to a directional stimulus. In almost all cases the plant part grows towards (positive response) or away from (negative response) the stimulus. Again, the type of response is named after the stimulus. Examples, and the survival value of the response, include the following:

- Plant shoots grow towards light (positive phototropism) and away from gravity (negative gravitropism) so that their leaves are in the most favourable position to capture light for photosynthesis.
- Plant roots grow away from light (negative phototropism) and towards gravity (positive gravitropism). In both cases the response increases the probability that roots will grow into the soil, where they are better able to absorb water and mineral ions.

We will learn more about tropisms in Topic 14.2.

▲ **Figure 2** *Cress seedlings exhibiting phototropism. The seedlings at the bottom have been grown with light directed at them from the left-hand side*

Summary questions

For each of the following statements, name the type of response described and the survival value of the response.

1 Some species of bacteria move away from the waste products that they produce.

2 The sperm cells of a moss plant are attracted towards a chemical produced by the female reproductive organ of another moss plant.

3 The young stems of seedlings grow away from gravity.

14.2 Plant growth factors

Learning objectives

→ Describe the stimuli that plants respond to.

→ Describe plant growth factors such as IAA.

→ Explain phototropism in flowering plants.

→ Explain gravitropism in flowering plants.

Specification reference: 3.6.1.1

Unlike animals, plants have no nervous system. Nevertheless, in order to survive, plants respond to changes in both their external and internal environments. For example, plants respond to:

- **light**. Shoots grow towards light (i.e., are positively phototropic) because light is needed for photosynthesis.
- **gravity**. Plants need to be firmly anchored in the soil. Roots are sensitive to gravity and grow in the direction of its pull (i.e., they are positively gravitropic).
- **water**. Almost all plant roots grow towards water (i.e., are positively hydrotropic) in order to absorb it for use in photosynthesis and other metabolic processes, as well as for support.

Plant responses to external stimuli involve hormone-like substances or, more correctly, **plant growth factors**. The latter term is more descriptive because:

- they exert their influence by affecting growth and, they may be made by cells located throughout the plant rather than in particular organs
- unlike animal hormones, some plant growth factors affect the tissues that release them rather than acting on a distant target organ.

Plant growth factors are produced in small quantities. An example of a plant growth factor is **indoleacetic acid** (**IAA**), which belongs to a group of substances called auxins. Among other things, IAA controls plant cell elongation.

Control of tropisms by IAA

We learnt in Topic 14.1 that a tropism is the directional growth of a plant in response to a directional stimulus. In the case of light, we can observe that a young shoot will grow towards light that is directed at it from one side (unilateral light). This is known as **positive phototropism**.

Phototropism in flowering plants

The response of the shoots of flowering plants to unilateral light is due to the following sequence of events:

1 Cells in the tip of the shoot produce IAA, which is then transported down the shoot.
2 The IAA is initially transported evenly throughout all regions as it begins to move down the shoot.
3 Light causes the movement of IAA from the light side to the shaded side of the shoot.
4 A greater concentration of IAA builds up on the shaded side of the shoot than on the light side.
5 As IAA causes elongation of shoot cells and there is a greater concentration of IAA on the shaded side of the shoot, the cells on this side elongate more.
6 The shaded side of the shoot elongates faster than the light side, causing the shoot tip to bend towards the light.

IAA also controls the bending of roots in response to light. However, whereas a high concentration of IAA increases cell elongation in shoots, it inhibits cell elongation in roots. For example, an IAA concentration of 10 parts per million increases shoot cell elongation by 200% but decreases root cell elongation by 100% (Figure 1). As a result, in roots the elongation of cells is greater on the light side than on the shaded side and so roots bend away from light, that is, they are negatively phototropic.

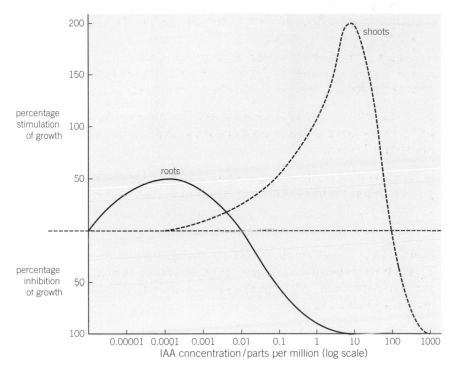

▲ **Figure 1** *Relationship between cell elongation and IAA concentration in shoots and roots*

Hint

In many plants, gravity leads to a change in the distribution of IAA carrier proteins that export IAA from cells.

Gravitropism in flowering plants

The response of a horizontally growing root to gravity is as follows:

1 Cells in the tip of the root produce IAA, which is then transported along the root.
2 The IAA is initially transported to all sides of the root.
3 Gravity influences the movement of IAA from the upper side to the lower side of the root.
4 A greater concentration of IAA builds up on the lower side of the root than on the upper side.
5 As IAA inhibits the elongation of root cells and there is a greater concentration of IAA on the lower side, the cells on this side elongate less than those on the upper side.
6 The relatively greater elongation of cells on the upper side compared to the lower side causes the root to bend downwards towards the force of gravity.

In shoots, the greater concentration of IAA on the lower side increases cell elongation and causes this side to elongate more than the upper side. As a result the shoot grows upwards away from the force of gravity.

▲ **Figure 2** *Gravitropism is a plant response to the Earth's gravitational field. This bean plant shows a turn in its stem which occurred after the pot was tipped over. The response also occurs in the dark, showing that it is not phototropism*

These events are summarised in Figure 3.

Phototropic response

cells on illuminated side are not elongated as much

greater elongation of cells on shaded side results in bending of shoot towards light.

IAA accumulates on shaded side

Light

shoot
high IAA concentration stimulates cell elongation causing a positive phototropic response

root
high IAA concentration inhibits cell elongation causing a negative phototropic response

Gravitropism

shoot
high IAA concentration stimulates cell elongation causing a negative gravitropic response

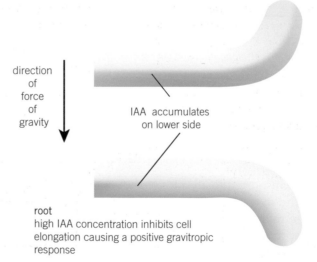

direction of force of gravity

IAA accumulates on lower side

root
high IAA concentration inhibits cell elongation causing a positive gravitropic response

▶ **Figure 3** *Mechanism of IAA action in the phototropic and gravitropic responses of shoots and roots*

Role of IAA in elongation growth

The transport of IAA is in one direction, namely away from the tip of shoots and roots where it is produced. IAA has a number of effects on plant cells including increasing the plasticity (ability to stretch) of their cell walls. The response only occurs on young cell walls where cells are able to elongate. As the cells mature they develop greater rigiditiy – therefore older parts of the shoot/root will not be able to respond. The proposed explanation of how IAA increases the plasticity of cells is called the acid growth hypothesis. It involves the active transport of hydrogen ions from the cytoplasm into spaces in the cell wall causing the cell wall to become more plastic allowing the cell to elongate by expansion.

The elongation of cells on one side only of a stem or root can lead to them bending (Figure 3). This is the means by which plants respond relatively quickly to environmental stimuli like light and gravity. These responses can be explained in terms of the stimuli causing uneven distribution of IAA, as described earlier, as it moves away from the tip of the stem or root.

Summary questions

1 Explain how the movement of IAA in shoots helps a plant to survive.

2 Suggest **two** advantages to a plant of having roots that respond to gravity by growing in the direction of its force.

3 Consider the following facts about IAA:

  i   They are easily made synthetically

  ii They are readily absorbed by plants

  iii They are not easily broken down

  iv They are lethal to some plants in low concentrations

  v   Narrow-leaved plants are less easily killed than broad-leaved plants.

  Suggest ways in which these facts might be relevant to agricultural practice.

 ### Discovering the role of IAA in tropisms

No less a person than Charles Darwin was one of the earliest scientists to investigate the response of plant shoots to light. He observed that young grass shoots grow towards the window (i.e., they were positively phototropic). Being curious, he proposed the hypothesis that the stimulus of the light was detected by the tip of the shoot, which was therefore the source of the response. He tested his hypothesis in a series of experiments in which he removed the tips of shoots or covered the tips with lightproof covers.

These experiments and the results are summarised in Figure 4 (experiments 1–3).

1 Which of Darwin's three experiments acted as a control?

Expt no	Method	Result	Explanation
1	Unilateral light — Plant shoot	Shoot bends towards light	The shoot is positively phototropic. Bending occurs behind the tip.
2	Shoot tip removed — Light — Tip discarded	No response	The tip must either detect the stimulus or produce the messenger (or both) as its removal prevents any response.
3	Lightproof cover is placed over intact tip of shoot — Light	No response	The light stimulus must be detected by the tip.

▲ **Figure 4** *Darwin's experiments to show that it is the tips of shoots that are the source of the phototropic response*

Once it had been shown that the tip is the light-sensitive region of the shoot but that the response (bending) occurs lower down the shoot, some scientists proposed another hypothesis, that a chemical substance was being produced in the tip and transported down the stem, where it caused a response. Others disagreed and put forward an alternative theory, that it was an electrical signal that was passing from the tip and causing the response. One scientist, Peter Boysen-Jensen, carried out a further set of experiments designed to prove the hypothesis that the 'messenger' was a chemical. In these experiments, he used mica, which conducts electricity but not chemicals, and gelatin, which conducts chemicals but not electricity. His experiments and the results are summarised in Figure 5 (experiments 4–6).

2   Suggest an explanation for the results in experiment 5.

EXPT NO.	METHOD	RESULT	EXPLANATION
4	Thin, impermeable barrier of mica Light	Movement of chemical down shaded side Bends towards the light	Mica on the illuminated side of the shoot allows the hormone to pass only down the shaded side where it increases growth and causes bending.
5	Mica inserted on shaded side Light	Movement of chemical down shaded side is prevented by mica No response	
6	Tip removed, gelatin block inserted and tip replaced Light Gelatin block	Movement of chemical down shaded side Bends towards the light	As gelatin allows chemicals to pass through it, but not electrical signals, the bending which occurs must be due to a chemical passing from the tip.

▲ **Figure 5** *Boysen-Jensen's experiments to show the nature of the 'messenger' in the phototropic response*

Boysen-Jenson's experiments stimulated another scientist, Arpad Paal, to investigate how the chemical messenger worked. He removed the tips of shoots and placed them on one side of the cut surface. He kept the shoots in total darkness throughout the experiment.

His experiment and its results are shown in Figure 6 (experiment 7).

3   Suggest an explanation for the results in experiment 7.

EXPT NO.	METHOD	RESULT
7	Darkness Tips removed and then replaced but displaced to one side	Shoots bend towards side where no tip is present

▲ **Figure 6** *Paal's experiment on the action of the 'messenger'*

So far, it had been established that bending was due to a chemical which was produced in the tip and caused growth on the shaded side of the shoot. This chemical was later shown to be indoleacetic acid (IAA). The next question was how did light cause the uneven distribution of IAA? Different theories were put forward, including:

- Light inhibits IAA production in the tip and so it is only produced on the shaded side.

- Light destroys the IAA as it passes down the light side of the shoot.

- IAA is transported from the light side to the shaded side of the shoot.

This prompted Winslow Briggs and his associates to test these hypotheses. They set up experiments as shown in Figure 7 (experiments 8–10).

Study experiments 8, 9 and 10.

4 Suggest reasons for using a glass plate in experiments 9 and 10.

5 State which of the three theories the results tend to support. Give reasons for your answer.

▲ **Figure 7** *Briggs's experiments to determine how IAA becomes unevenly distributed*

Learning objectives

→ Explain how a simple reflex arc works.

→ Explain the roles sensory, intermediate and motor neurones play in a reflex arc.

→ Outline the importance of reflex arcs.

Specification reference: 3.6.1.1

The simplest type of nervous response to a stimulus is a reflex arc. Before considering how a reflex arc works, it is helpful to understand how the millions of **neurones** in a mammalian body are organised.

Nervous organisation

The nervous system has two major divisions:

- the **central nervous system (CNS)**, which is made up of the brain and spinal cord
- the **peripheral nervous system (PNS)**, which is made up of pairs of nerves that originate from either the brain or the spinal cord.

The peripheral nervous system is divided into:

- **sensory neurones**, which carry nerve impulses (electrical signals) from receptors towards the central nervous system
- **motor neurones**, which carry nerve impulses away from the central nervous system to **effectors**.

The motor nervous system can be further subdivided as follows:

- the **voluntary nervous system**, which carries nerve impulses to body muscles and is under voluntary (conscious) control
- the **autonomic nervous system**, which carries nerve impulses to glands, **smooth muscle** and cardiac muscle and is not under voluntary control, that is, it is involuntary (subconscious).

A summary of nervous organisation is given in Figure 1.

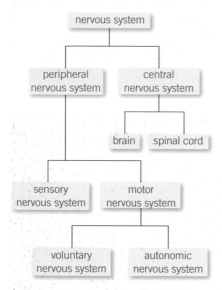

▲ **Figure 1** *Nervous organisation*

The spinal cord

The spinal cord is a column of nervous tissue that runs along the back and lies inside the vertebral column for protection. Emerging at intervals along the spinal cord are pairs of nerves as shown in Figure 2.

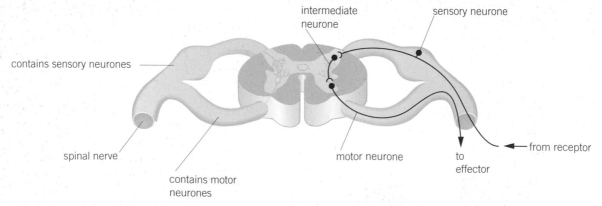

▲ **Figure 2** *Section through spinal cord showing the neurones of a reflex arc*

A reflex arc

You will have noticed that you immediately withdraw your hand if you place it on a hot or sharp object. You do not stop to consider any alternative actions. The response is rapid, short-lived, localised and totally involuntary. Indeed, by the time the brain has received nerve impulses from the receptors in the hand, the muscles in the arm have already pulled the hand clear of the danger. This type of involuntary response to a sensory stimulus is called a **reflex**. The pathway of neurones involved in a reflex is known as a **reflex arc**.

Reflex arcs, such as the withdrawal reflex described above, involve just three neurones. One of the neurones is in the spinal cord and so this type of reflex is also called a spinal reflex. The main stages of a spinal reflex arc, such as withdrawing the hand from a hot object, are described below (the numbers relate to the stages shown in Figure 3).

1 the **stimulus** – heat from the hot object
2 a **receptor** – temperature receptors in the skin on the back of the hand, which generates nerve impulses in the sensory neurone
3 a **sensory neurone** – passes nerve impulses to the spinal cord
4 a **coordinator** (intermediate neurone) – links the sensory neurone to the motor neurone in the spinal cord
5 a **motor neurone**   carries nerve impulses from the spinal cord to a muscle in the upper arm
6 an **effector** – the muscle in the upper arm, which is stimulated to contract
7 the **response** – pulling the hand away from the hot object.

Hint

Remember the sequence: stimulus, receptor, sensory neurone, intermediate neurone, motor neurone, effector, response.

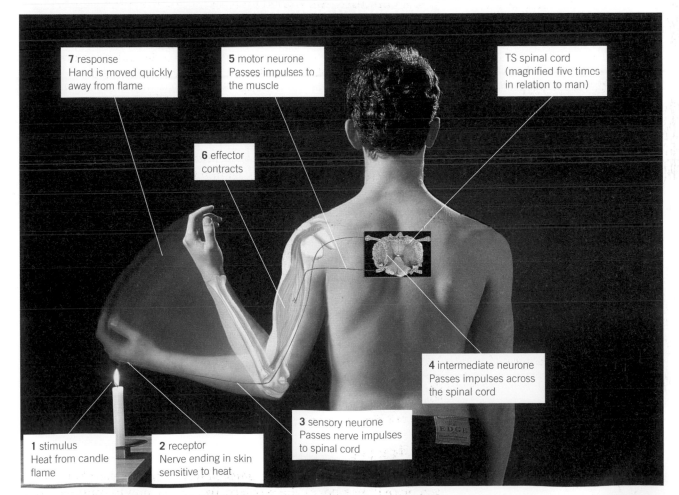

▲ **Figure 3** *Reflex arc involved in the withdrawal of the hand from a heat stimulus*

Importance of reflex arcs

Any action that makes survival more likely is clearly of value. Reflexes are involuntary – the actions they control do not need to be considered, because there is only one obvious course of action, that is, to remove the hand from the hot object. Reflex actions are important for the following reasons:

- They are involuntary and therefore do not require the decision-making powers of the brain, thus leaving it free to carry out more complex responses. In this way, the brain is not overloaded with situations in which the response is always the same. Some impulses are nevertheless sent to the brain, so that it is informed of what is happening and can sometimes override the reflex if necessary.

- They protect the body from harm. They are effective from birth and do not have to be learnt.

- They are fast, because the neurone pathway is short with very few, typically one or two, **synapses** where neurones communicate with each other (synapses are the slowest link in a neurone pathway). This is important in withdrawal reflexes.

- The absence of any decision-making process also means the action is rapid.

Summary questions

In the following passage give the word that best replaces the number in brackets.

The nervous system has two main divisions, the central nervous system (CNS), comprising the (1) and (2), and the peripheral nervous system (PNS). The peripheral nervous system is made up of the (3) nerves that carry impulses away from the CNS and (4) nerves that carry impulses towards the CNS. A spinal reflex is an (5) response that involves the spinal cord. An example is the withdrawal reflex, for example, the withdrawing of the hand from a hot object. The sequence of events begins with the heat from the hot object, which acts as the (6). This is detected by a (7) in the skin on the back of the hand, which creates nerve impulses that pass along a (8) neurone into the spinal cord. The impulse then passes to an (9) neurone, in the central region of the spinal cord. The impulse leaves the spinal cord via a (10) neurone. This neurone stimulates a muscle of the upper arm to contract and withdraw the hand from the object. Structures such as these that bring about a response to a stimulus are called (11).

The central nervous system receives sensory information from its internal and external environment through a variety of receptors, each type responding to a different and specific type of stimulus. Sensory reception is the function of these receptors, whereas sensory perception involves making sense of the information from the receptors. This is largely a function of the brain. The concepts of stimulus and response were covered in Topic 14.1. We shall now look in detail at one receptor – the **Pacinian corpuscle**.

Features of sensory reception as illustrated by the Pacinian corpuscle

Pacinian corpuscles respond to changes in mechanical pressure. As with all sensory receptors, a Pacinian corpuscle:

- **is specific to a single type of stimulus**. In this case, it responds only to mechanical pressure. It will not respond to other stimuli, such as heat, light, or sound.
- **produces a generator potential by acting as a transducer**. All stimuli involve a change in some form of energy. It is the role of the transducer to convert the change in form of energy by the stimulus into a form, namely nerve impulses, that can be understood by the body. The stimulus always involves a change in some form of energy, for example, heat, light, sound, or mechanical energy. The nerve impulse is also a form of energy. Receptors therefore convert, or transduce, one form of energy into another. Receptors in the nervous system convert the energy of the stimulus into a nervous impulse known as a **generator potential**. For example, the Pacinian corpuscle, whose action is described below, transduces the mechanical energy of the stimulus into a generator potential.

Structure and function of a Pacinian corpuscle

Pacinian corpuscles respond to mechanical stimuli such as pressure. They occur deep in the skin and are most abundant on the fingers, the soles of the feet and the external genitalia. They also occur in joints, **ligaments** and **tendons**, where they enable the organism to know which joints are changing direction. The single sensory neurone of a Pacinian corpuscle is at the centre of layers of tissue, each separated by a gel. This gives it the appearance of an onion when cut vertically (Figure 2). How does this structure transduce the mechanical energy of the stimulus into a generator potential?

In Topic 4.1 we saw that plasma membranes contain channel proteins that span them. These proteins have channels along which ions can be transported. Each channel is specific. Sodium channels, for example, carry only sodium ions.

The sensory neurone ending at the centre of the Pacinian corpuscle has a special type of sodium channel in its plasma membrane. This is called a **stretch-mediated sodium channel**. These channels are so-called because their permeability to sodium changes when they are deformed, for example, by stretching. The Pacinian corpuscle functions as follows:

- In its normal (resting) state, the stretch-mediated sodium channels of the membrane around the neurone of a Pacinian corpuscle are

▲ **Figure 1** *Pacinian corpuscle*

neurone ending

layers of connective tissue with viscous gel between

capsule

blood capillary

neurone

▲ **Figure 2** *Structure of a Pacinian corpuscle*

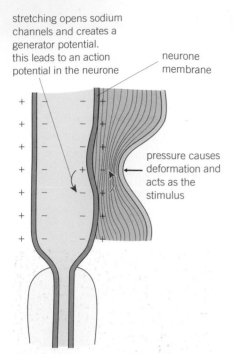

stretching opens sodium channels and creates a generator potential. this leads to an action potential in the neurone

neurone membrane

pressure causes deformation and acts as the stimulus

▲ **Figure 3** *Creation of a generator potential in a Pacinian corpuscle*

▼ **Table 1** *Differences between rod and cone cells*

Rod cells	Cone cells
Rod-shaped	Cone-shaped
Greater numbers than cone cells	Fewer numbers than rod cells
Distribution – more at the periphery of the retina, absent at the fovea	Fewer at the periphery of the retina, concentrated at the fovea
Give poor visual acuity	Give good visual acuity
Sensitive to low-intensity light	Not sensitive to low-intensity light
One type only	Three types each responding to different wavelengths of light

too narrow to allow sodium ions to pass along them. In this state, the neurone of the Pacinian corpuscle has a resting potential (Topic 15.2).

- When pressure is applied to the Pacinian corpuscle, it is deformed and the membrane around its neurone becomes stretched (Figure 3).
- This stretching widens the sodium channels in the membrane and sodium ions diffuse into the neurone.
- The influx of sodium ions changes the potential of the membrane (i.e., it becomes **depolarised**), thereby producing a generator potential.
- The generator potential in turn creates an action potential (nerve impulse) (Topic 15.2) that passes along the neurone and then, via other neurones, to the central nervous system.

These events are illustrated in Figure 3.

Receptors working together in the eye

The light receptor cells of the mammalian eye are found on its innermost layer, the retina. The millions of light receptor cells found in the retina are of two main types: rod cells and cone cells. Both rod and cone cells act as **transducers** by conserving light energy into the electrical energy of a nerve impulse.

Rod cells

Rod cells cannot distinguish different wavelengths of light and therefore lead to images being seen only in black and white. Rod cells are more numerous than cone cells – there are around 120 million in each eye.

Many rod cells are connected to a single sensory neurone in the optic nerve (Figure 4). Rod cells are used to detect light of very low intensity. A certain threshold value has to be exceeded before a **generator potential** is created in the bipolar cells to which they are connected. As a number of rod cells are connected to a single bipolar cell (= retinal convergence), there is a much greater chance that the threshold value will be exceeded than if only a single rod cell were connected to each bipolar cell. This is due to summation which is explained in Topic 15.5. As a result, rod cells allow us to see in low light intensity (i.e., at night), although only in black and white.

In order to create a generator potential, the pigment in the rod cells (rhodopsin) must be broken down. There is enough energy from low-intensity light to cause this breakdown. This explains why rod cells respond to low-intensity light.

A consequence of many rod cells linking to a single bipolar cell is that light received by rod cells sharing the same neurone will only generate a single impulse travelling to the brain regardless of how many of the neurones are stimulated. This means that, in perception, the brain cannot distinguish between the separate sources of light that stimulated them. Two dots close together cannot be resolved and so will appear as a single blob. Rod cells therefore give low **visual acuity**.

Cone cells

Cone cells are of three different types, each responding to a different range of wavelengths of light. Depending upon the proportion of each type that is stimulated, we can perceive images in full colour.

▲ **Figure 4** *Microscopic structure of the retina*

In each human eye, there are around 6 million cone cells, often with their own separate bipolar cell connected to a sensory neurone in the optic nerve (see Figure 4). This means that the stimulation of a number of cone cells cannot be combined to help exceed the threshold value and so create a generator potential. As a result, cone cells only respond to high light intensity and not to low light intensity.

In addition, cone cells contain different types of pigment from that found in rod cells. The pigment in cone cells (iodopsin) requires a higher light intensity for its breakdown. Only light of high intensity will therefore provide enough energy to break it down and create a generator potential. There are three different types of cone cell, each containing a specific type of iodopsin. As a result, each cone cell is sensitive to a different specific range of wavelengths.

Each cone cell has its own connection to a single bipolar cell, which means that, if two adjacent cone cells are stimulated, the brain receives two separate impulses. The brain can therefore distinguish between the two separate sources of light that stimulated the two cone cells. This means that two dots close together can be resolved and will appear as two dots. Therefore cone cells give very accurate vision, that is, they have good visual acuity.

The distribution of rod and cone cells on the retina is uneven. Light is focused by the lens on the part of the retina opposite the pupil. This point is known as the fovea. The fovea therefore receives the highest intensity of light. Therefore cone cells, but not rod cells, are found at the fovea. The concentration of cone cells diminishes further away from the fovea. At the peripheries of the retina, where light intensity is at its lowest, only rod cells are found.

Table 1 summarises the differences between rod and cone cells.

All this shows how the distribution of rod and cone cells, and the connections they make in the optic nerve, can explain the differences in sensitivity and visual acuity in mammals. By having different types of light receptor, each responding to different stimuli, mammals can benefit from good all-round vision both day and night.

▲ **Figure 5** *False colour SEM of rod and cone cells in the retina of the eye. Rod cells (brown) are long nerve cells responding to dim light while cone cells (green) detect colour*

Summary questions

1 Describe a stretch-mediated sodium channel.

2 Describe the sequence of events by which pressure on a Pacinian corpuscle results in the creation of a generator potential.

3 Explain why brightly coloured objects often appear grey in dim light.

4 At night, it is often easier to see a star in the sky by looking slightly to the side of it rather than directly at it. Suggest why this is so.

14.5 Control of heart rate

Learning objectives

→ Describe the autonomic nervous system.

→ Explain how the autonomic nervous system controls heart rate.

→ Explain the role chemical and pressure receptors play in the processes controlling the heart rate.

Specification reference: 3.6.1.3

Although we are not aware of it, much of the sensory information reaching our central nervous system comes from receptors within our bodies responding to internal stimuli. All the internal systems of our body need to operate efficiently and be ready to adapt to meet the changing demands made upon them. This requires the coordination of a vast amount of information. This information comes from the monitoring of all our internal environment – a process that takes place continuously. Before investigating one example, how heart rate is controlled, let us first look at the part of the nervous system responsible for this type of control – the autonomic nervous system.

The autonomic nervous system

Autonomic means self-governing. The autonomic nervous system controls the involuntary (subconscious) activities of internal muscles and glands. It has two divisions:

- **the sympathetic nervous system**. In general, this stimulates effectors and so speeds up any activity. It acts rather like an emergency controller. It controls effectors when we exercise strenuously or experience powerful emotions. In other words, it helps us to cope with stressful situations by heightening our awareness and preparing us for activity (the fight or flight response).
- **the parasympathetic nervous system**. In general, this inhibits effectors and so slows down any activity. It controls activities under normal resting conditions. It is concerned with conserving energy and replenishing the body's reserves.

The actions of the sympathetic and parasympathetic nervous systems normally oppose one another. In other words they are **antagonistic**. If one system contracts a muscle, then the other relaxes it. The activities of internal glands and muscles are therefore regulated by a balance of the two systems. Let us look at one such example, the control of heart rate.

Control of heart rate

The muscle of the heart is known as cardiac muscle. It is myogenic, that is, its contraction is initiated from within the muscle itself, rather than by nervous impulses from outside (neurogenic), as is the case with other muscles. Within the wall of the right atrium of the heart is a distinct group of cells known as the **sinoatrial node (SAN)**. It is from here that the initial stimulus for contraction originates. The sinoatrial node has a basic rhythm of stimulation that determines the beat of the heart. For this reason it is often referred to as the pacemaker. The sequence of events that controls the basic heart rate is:

- A wave of electrical excitation spreads out from the sinoatrial node across both atria, causing them to contract.
- A layer of non-conductive tissue (atrioventricular septum) prevents the wave crossing to the ventricles.

Synoptic link

The cardiac cycle is covered in Topic 7.5, and the structure of the heart in Topic 7.4. This includes information on how the sinoatrial node controls the heart beat – information that is relevant here.

- The wave of excitation enters a second group of cells called the **atrioventricular node (AVN)**, which lies between the atria.
- The atrioventricular node, after a short delay, conveys a wave of electrical excitation between the ventricles along a series of specialised muscle fibres called **Purkyne tissue** which collectively make up a structure called the **bundle of His**.
- The bundle of His conducts the wave through the atrioventricular septum to the base of the ventricles, where the bundle branches into smaller fibres of Purkyne tissue.
- The wave of excitation is released from the Purkyne tissue, causing the ventricles to contract quickly at the same time, from the bottom of the heart upwards.

These events are summarised in Figure 1.

Modifying the resting heart rate

The resting heart rate of a typical adult human is around 70 beats per minute. However, it is essential that this rate can be altered to meet varying demands for oxygen. During exercise, for example, the resting heart rate may need to more than double.

Changes to the heart rate are controlled by a region of the brain called the **medulla oblongata**. This has two centres concerned with heart rate:

- a centre that **increases heart rate**, which is linked to the **sinoatrial node** by the sympathetic nervous system
- a centre that **decreases heart rate**, which is linked to the sinoatrial node by the parasympathetic nervous system.

Which of these centres is stimulated depends upon the nerve impulses they receive from two types of receptor, which respond to stimuli of either chemical or pressure changes in the blood.

a wave of electrical activity spreads out from the sinoatrial node

b wave spreads across both atria causing them to contract and reaches the atrioventricular node

c atrioventricular node conveys wave of electrical activity between the ventricles along the bundle of His and releases it at the apex, causing the ventricles to contract.

▲ **Figure 1** *Control of the heart rate*

▲ **Figure 2** *Structure of the human heart.*

increased muscular/metabolic activity

↓

more carbon dioxide produced by tissues from increased respiration

↓

blood pH is lowered

↓

chemical receptors in the carotid arteries increase frequency of impulses to the medulla oblongata

↓

centre in medulla oblongata that speeds heart rate, increases frequency of impulses to SA node via the sympathetic nervous system

↓

SA node increases heart rate

↓

increased blood flow removes carbon dioxide faster

↓

carbon dioxide concentration returns to normal

▲ **Figure 3** *Effects of exercise on cardiac output (SA node = sinoatrial node)*

Control by chemoreceptors

Chemoreceptors are found in the wall of the carotid arteries (the arteries that serve the brain). They are sensitive to changes in the pH of the blood that result from changes in carbon dioxide concentration. In solution, carbon dioxide forms an acid and therefore lowers pH. The process of control works as follows:

- When the blood has a higher than normal concentration of carbon dioxide, its pH is lowered.
- The chemoreceptors in the wall of the carotid arteries and the aorta detect this and increase the frequency of nervous impulses to the centre in the medulla oblongata that increases heart rate.
- This centre increases the frequency of impulses via the sympathetic nervous system to the sinoatrial node. This, in turn, increases the rate of production of electrical waves by the sinoatrial node and therefore increases the heart rate.
- The increased blood flow that this causes leads to more carbon dioxide being removed by the lungs and so the carbon dioxide concentration of the blood returns to normal.
- As a consequence the pH of the blood rises to normal and the chemoreceptors in the wall of the carotid arteries and aorta reduce the frequency of nerve impulses to the medulla oblongata.
- The medulla oblongata reduces the frequency of impulses to the sinoatrial node, which therefore leads to a reduction in the heart rate.

This process is summarised in Figure 3, which shows the sequence of events that follows changes in activity levels.

Control by pressure receptors

Pressure receptors occur within the walls of the carotid arteries and the aorta. They operate as follows:

- **When blood pressure is higher than normal**, pressure receptors transmit more nervous impulses to the centre in the medulla oblongata that decreases heart rate. This centre sends impulses via the parasympathetic nervous system to the sinoatrial node of the heart, which leads to a decrease in the rate at which the heart beats.
- **When blood pressure is lower than normal**, pressure receptors transmit more nervous impulses to the centre in the medulla oblongata that increases heart rate. This centre sends impulses via the sympathetic nervous system to the sinoatrial node, which increases the rate at which the heart beats.

Figure 4 summarises the control of heart rate

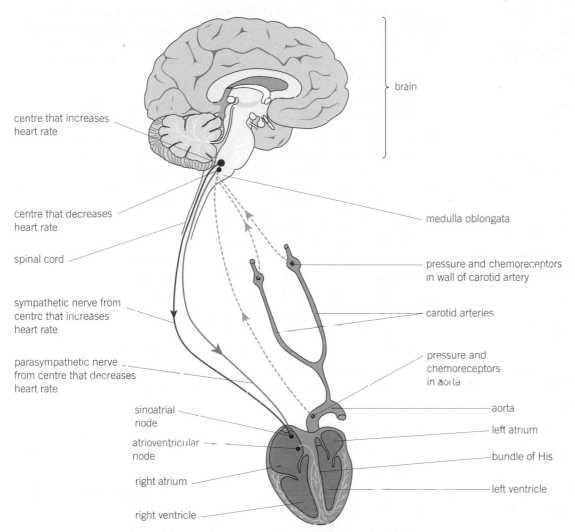

brain

centre that increases
heart rate

centre that decreases
heart rate

medulla oblongata

spinal cord

pressure and chemoreceptors
in wall of carotid artery

sympathetic nerve from
centre that increases
heart rate

carotid arteries

parasympathetic nerve
from centre that decreases
heart rate

pressure and
chemoreceptors
in aorta

sinoatrial
node

aorta

left atrium

atrioventricular
node

bundle of His

right atrium

left ventricle

right ventricle

▲ **Figure 4** *Control of heart rate*

Summary questions

1 Describe the function of the autonomic nervous system.

2 Distinguish between the functions of the sympathetic and
parasympathetic nervous systems.

3 Suppose the parasympathetic nerve connections from the medulla
oblongata to the sinoatrial node were cut. Suggest what might happen if
a person's blood pressure increases above normal.

4 The nerve connecting the carotid artery to the medulla oblongata of a
person is cut. This person then undertakes some strenuous exercise.
Suggest what might happen to the person's:

a heart rate

b blood carbon dioxide concentration.

Explain your answers.

1 IAA is a specific growth factor
   (a) Name the process by which IAA moves from the growing regions of a plant shoot to other tissues. *(1 mark)*
   (b) (i) When a young shoot is illuminated from one side, IAA stimulates growth on the shaded side. Explain why growth on the shaded side helps to maintain the leaves in a favourable environment. *(2 marks)*

Temperature °C	0	5	10	15	20	25
Rate of uptake of NAA by lower surface of leaf / arbitrary units	0	4	10	15	26	36
Rate of uptake of NAA by upper surface (arbitrary units)	0	2	4	6	9	12

   (ii) Scientists hypothesise that there is a positive correlation between temperature and the rate of uptake of NAA through the lower leaf surface.
   Use the data in the table to calculate the correlation coefficient, r, to test this, where $r = \dfrac{\Sigma(x - \bar{x}) \times (y - \bar{y})}{\sqrt{\Sigma(x - \bar{x})^2 \times \Sigma(y - \bar{y})^2}}$ *(4 marks)*

   (iii) at 10 degrees of freedom the critical value of r at the 5% level is 0.576. Comment on the support for the scientists' hypothesis. *(2 marks)*
   (iv) NAA is a similar substance to IAA. It is used to control the growth of cultivated plants. Plant physiologists investigated the effect of temperature on the uptake of NAA by leaves. They sprayed a solution containing NAA on the upper and lower surfaces of a leaf. The graph shows their results.
   (c) Explain the effect of temperature on the rate at which NAA is taken up by the lower surface of the leaf. *(2 marks)*
   (d) There are differences in the properties of the cuticle on the upper and lower surfaces of leaves.
   (i) Suggest how these differences in the cuticle might explain the differences in rates of uptake of NAA by the two surfaces. *(2 marks)*
   (ii) In this investigation, the physiologists investigated the leaves of pear trees. Explain why the results might be different for other species. *(1 mark)*
   AQA June 2011 (apart from 1 (b) (ii) and (iii))

2 Scientists investigated the response of lateral roots to gravity. Lateral roots grow from the side of main roots.
   The diagrams show four stages, **A** to **D**, in the growth of a lateral root and typical cells from the tip of the lateral root in each stage. All of the cells are drawn with the bottom of the cell.
   (a) Describe **three** changes in the root tip cells between stages **A** and **D**. *(3 marks)*
   (b) The scientists' hypothesis was that there was a relationship between the starch grains in the root tip cells and the bending and direction of growth of lateral roots. Does the information in the diagram support this hypothesis? Give reasons for your answer. *(3 marks)*

(c) The diagram shows the distribution of indoleacetic acid (IAA) in the lateral root at Stage **B**.

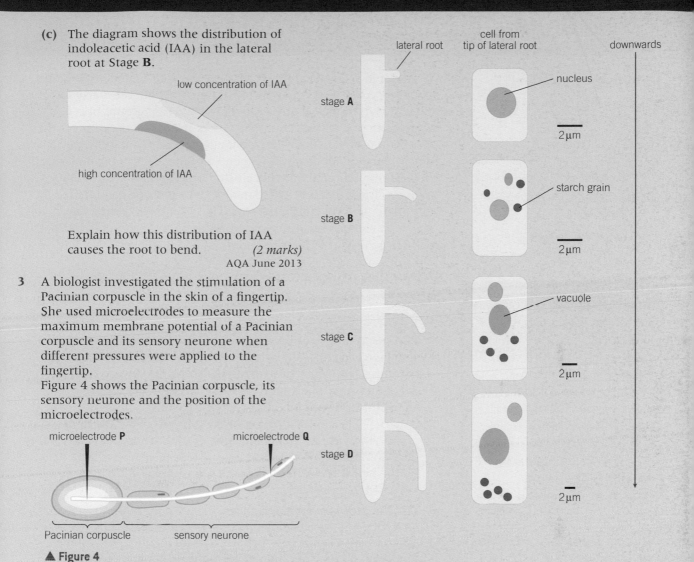

Explain how this distribution of IAA causes the root to bend. *(2 marks)*

AQA June 2013

3 A biologist investigated the stimulation of a Pacinian corpuscle in the skin of a fingertip. She used microelectrodes to measure the maximum membrane potential of a Pacinian corpuscle and its sensory neurone when different pressures were applied to the fingertip.

Figure 4 shows the Pacinian corpuscle, its sensory neurone and the position of the microelectrodes.

▲ **Figure 4**

▼ **Table 2** *shows some of the biologist's results.*

Pressure applied to the fingertip	Membrane potential at P / millivolts	Membrane potential at Q / millivolts
None	−70	−70
Light	−50	−70
Medium	+30	+40
Heavy	+40	+40

(a) Explain how the resting potential of −70 mV is maintained in the sensory neurone when no pressure is applied. *(2 marks)*

(b) Explain how applying pressure to the Pacinian corpuscle produces the changes in membrane potential recorded by microelectrode P. *(3 marks)*

(c) The membrane potential at Q was the same whether medium or heavy pressure was applied to the finger tip. Explain why. *(2 marks)*

(d) Multiple sclerosis is a disease in which parts of the myelin sheaths surrounding neurones are destroyed. Explain how this results in slower responses to stimuli. *(2 marks)*

AQA Specimen 2014

Learning objectives

→ Distinguish between how nervous and hormonal coordination.

→ Describe the structure of a myelinated motor neurone.

→ Describe the different types of neurone.

Specification reference: 3.6.2.1

Hint

The nervous system operates like a telephone system, allowing rapid communication between two specific individuals. The hormonal system can be likened to a nationwide mail shot, sending a slower, more general message to everyone, everywhere, but only those individuals who are sensitive to it respond.

As species have evolved, their cells have become adapted to perform specialist functions. By specialising in one function, cells have lost the ability to perform some other functions. Different groups of cells each carry out their own function. This makes cells dependent upon others to carry out the functions they no longer specialise in. Cells specialising in reproduction, for example, depend on other cells to obtain oxygen for their respiration, to provide glucose or to remove their waste products. These different functional systems must be coordinated if they are to perform efficiently. No body system works in isolation, all must be integrated in a coordinated fashion. In this chapter we shall look at one way in which this coordination is achieved.

Principles of coordination

There are two main forms of coordination in animals as a whole – the nervous system and the hormonal system:

- **The nervous system** uses nerve cells to pass electrical impulses along their length. They stimulate their target cells by secreting chemicals, known as **neurotransmitters**, directly on to them. This results in rapid communication between specific parts of an organism. The responses produced are often short-lived and restricted to a localised region of the body. An example of nervous coordination is a reflex action, such as the withdrawal of the hand from an unpleasant stimulus. For obvious reasons this type of action, which is covered in Topic 14.3, is rapid, short-lived and restricted to one region of the body.

- **The hormonal system** produces chemicals (hormones) that are transported in the blood plasma to their target cells. The target cells have specific receptors on their cell-surface membranes and the change in the concentration of hormones stimulates them. This results in a slower, less specific form of communication between parts of an organism. The responses are often long-lasting and widespread. An example of hormonal coordination is the control of blood glucose concentration, which produces a slower response but has a more long term and more widespread effect.

Although different, both systems work together and interact with one another. A comparison of the nervous and hormonal systems is given in Table 1.

▼ **Table 1** *Comparison of hormonal and nervous systems*

Hormonal system	Nervous system
Communication is by chemicals called hormones	Communication is by nerve impulses
Transmission is by the blood system	Transmission is by neurones
Transmission is usually relatively slow	Transmission is very rapid
Hormones travel to all parts of the body, but only target cells respond	Nerve impulses travel to specific parts of the body
Response is widespread	Response is localised
Response is slow	Response is rapid
Response is often long-lasting	Response is short-lived
Effect may be permanent and irreversible	Effect is usually temporary and reversible

Neurones

Neurones (nerve cells) are specialised cells adapted to rapidly carrying electrochemical changes called **nerve impulses** from one part of the body to another.

A mammalian motor neurone is made up of:

- a **cell body**, which contains all the usual cell organelles, including a nucleus and large amounts of rough endoplasmic reticulum. This is associated with the production of proteins and **neurotransmitters**

- **dendrons**, extensions of the cell body which subdivide into smaller branched fibres, called **dendrites**, that carry nerve impulses towards the cell body

- an **axon**, a single long fibre that carries nerve impulses away from the cell body

- **Schwann cells**, which surround the axon, protecting it and providing electrical insulation. They also carry out **phagocytosis** (the removal of cell debris) and play a part in nerve regeneration. Schwann cells wrap themselves around the axon many times, so that layers of their membranes build up around it

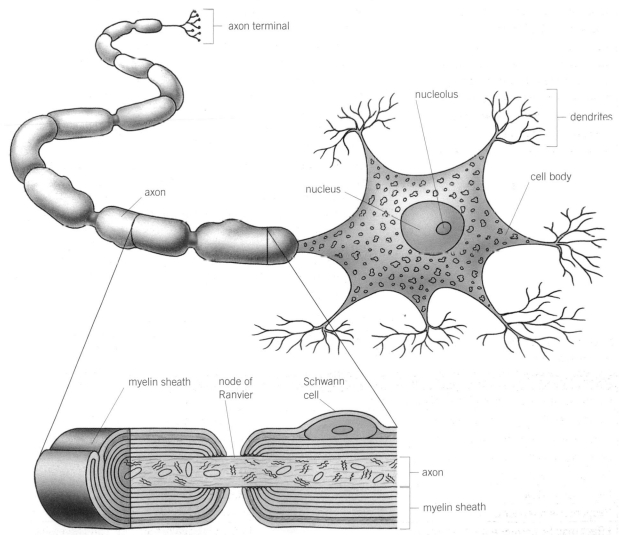

▲ **Figure 1** *Myelinated motor neurone*

LM of a node of Ranvier in a neurone. The node is the constriction in the centre. The constriction is a small area without myelin in an otherwise myelinated nerve fibre

- a **myelin sheath**, which forms a covering to the axon and is made up of the membranes of the Schwann cells. These membranes are rich in a lipid known as **myelin**. Neurones with a myelin sheath are called myelinated neurones.

- **nodes of Ranvier**, constrictions between adjacent Schwann cells where there is no myelin sheath. The constrictions are 2–3 μm long and occur every 1–3 mm in humans (Figure 2).

The structure of a myelinated motor neurone is illustrated in Figure 1.

Neurones can be classified according to their function:

- **Sensory neurones** transmit nerve impulses from a **receptor** to an intermediate or motor neurone. They have one dendron that is often very long. It carries the impulse towards the cell body and one axon that carries it away from the cell body.

- **Motor neurones** transmit nerve impulses from an intermediate or relay neurone to an **effector**, such as a gland or a muscle. Motor neurones have a long axon and many short dendrites.

- **Intermediate or relay neurones** transmit impulses between neurones, for example, from sensory to motor neurones. They have numerous short processes.

Figure 3, shows the structure of all three types of neurone.

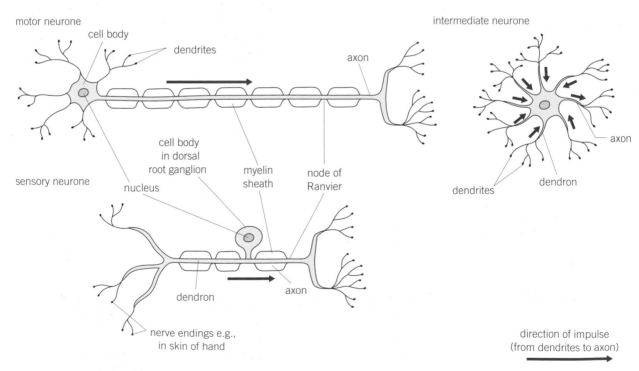

▲ **Figure 3** *Types of neurone*

Summary questions

In the following passage give the word that best replaces the numbers in brackets.

Neurones are adapted to carry electrochemical charges called (1). Each neurone comprises a cell body that contains a (2) and large amounts of (3), which is used in the production of proteins and neurotransmitters. Extending from the cell body is a single long fibre called an axon and smaller branched fibres called (4). Axons are surrounded by (5) cells, which protect and provide (6) because their membranes are rich in a lipid known as (7). There are three main types of neurone. Those that carry nerve impulses to an effector are called (8) neurones. Those that carry impulses from a receptor are called (9) neurones and those that link the other two types are called (10) neurones.

(11) List **three** ways in which a response to a hormone differs from a response to a nerve impulse.

▲ **Figure 4** *SEM of a neurone with the cell body at its centre and dendrites radiating from it*

Maths link √x̄

MS 0.3, see Chapter 22.

Ageing neurones √x̄

Neurones are found throughout the brain. In Figure 5 we see drawings of some neurones from a region of the brain of healthy humans of different ages. The last diagram in the series shows neurones from a 70-year-old with Alzheimer's disease, a condition that causes dementia.

1 Using Figure 5 describe the changes in the neurones that take place when healthy humans age.
2 Comment on the appearance of the neurone from the 70 year old who has Alzheimer's disease.
3 √x̄ After the age of 50 years, humans lose 5% of the neurones in this region of the brain every 10 years. Calculate how many neurones will be left at the age of 70 years from each 2000 neurones present at the age of 50 years.

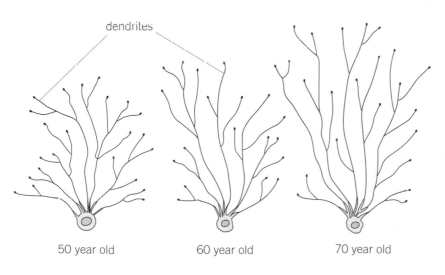

▲ **Figure 5**

15.2 The nerve impulse

Learning objectives

→ Describe the nature of the resting potential.

→ Explain how a resting potential is established in a neurone.

→ Explain what an action potential is.

Specification reference: 3.6.2.1

Synoptic link

To understand the nerve impulse requires a thorough knowledge and understanding of plasma membranes, particularly the structure of plasma membranes and the role of their carrier proteins in the sodium–potassium pump. It would be useful to revise Topics 4.1, 4.2, and 4.4 as a starting point for this section.

Study tip

Where sodium and potassium ions are actively transported through carrier proteins, it is known as the sodium-potassium pump.

Hint

As the phospholipid bilayer prevents diffusion of sodium and potassium ions, they move back across the bilayer by facilitated diffusion through channel proteins that are permanently open. These channel proteins are known as sodium or potassium 'gates', depending on which ion they transport.

A nerve impulse may be defined as a self-propagating wave of electrical activity that travels along the axon membrane. It is a temporary reversal of the electrical potential difference across the axon membrane. This reversal is between two states, called the **resting potential** and the **action potential**.

Resting potential

The movement of ions, such as sodium ions (Na^+) and potassium ions (K^+), across the axon membrane is controlled in a number of ways:

- The **phospholipid** bilayer of the axon plasma membrane prevents sodium and potassium ions diffusing across it.

- Proteins, known as **channel proteins**, span this phospholipid bilayer. These proteins have channels, called ion channels, which pass through them. Some of these channels have 'gates', which can be opened or closed so that sodium or potassium ions can move through them by facilitated diffusion at any one time, but not on other occasions. There are different gated channels for sodium and potassium ions. Some channels, however, remain open all the time, so the sodium and potassium ions move unhindered through them by facilitated diffusion.

- Some carrier proteins actively transport potassium ions into the axon and sodium ions out of the axon. This mechanism can be called a **sodium-potassium pump**.

As a result of these various controls, the inside of an axon is negatively charged relative to the outside. This is known as the **resting potential** and ranges from 50 to 90 millivolts (mV), but is usually 65 mV in humans. In this condition the axon is said to be **polarised**. The establishment of this potential difference (the difference in charge between the inside and outside of the axon) is due to the following events:

- Sodium ions are actively transported **out** of the axon by the sodium–potassium pumps.

- Potassium ions are actively transported **into** the axon by the sodium–potassium pumps.

- The active transport of sodium ions is greater than that of potassium ions, so three sodium ions move out for every two potassium ions that move in.

- Although both sodium and potassium ions are positive, the outward movement of sodium ions is greater than the inward movement of potassium ions. As a result, there are more sodium ions in the tissue fluid surrounding the axon than in the cytoplasm, and more potassium ions in the cytoplasm than in the tissue fluid, thus creating an electrochemical gradient.

- The sodium ions begin to diffuse back naturally into the axon while the potassium ions begin to diffuse back out of the axon.

- However, most of the gates in the channels that allow the potassium ions to move through are open, while most of the gates in the channels that allow the sodium ions to move through are closed.

These events are summarised in Figure 1.

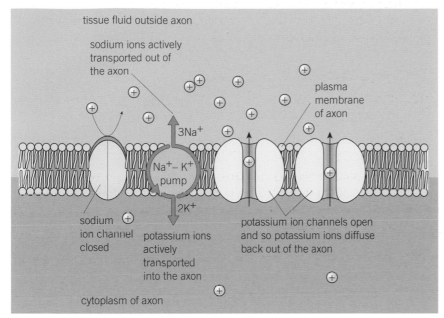

▲ **Figure 1** *Distribution of ions at resting potential*

The action potential

When a stimulus of sufficient size is detected by a receptor in the nervous system, its energy causes a temporary reversal of the charges either side of this part of the axon membrane. If the stimulus is great enough, the negative charge of $-65\,mV$ inside the membrane becomes a positive charge of around $+40\,mV$. This is known as the **action potential**, and in this condition this part of the axon membrane is said to be **depolarised**. This depolarisation occurs because the channels in the axon membrane change shape, and hence open or close, depending on the voltage across the membrane. They are therefore called voltage-gated channels. The sequence of events is described on the next page (the numbers relate to the stages illustrated in Figure 2). It is important to stress that the events described relate to a particular point on the axon membrane and not the whole of the membrane.

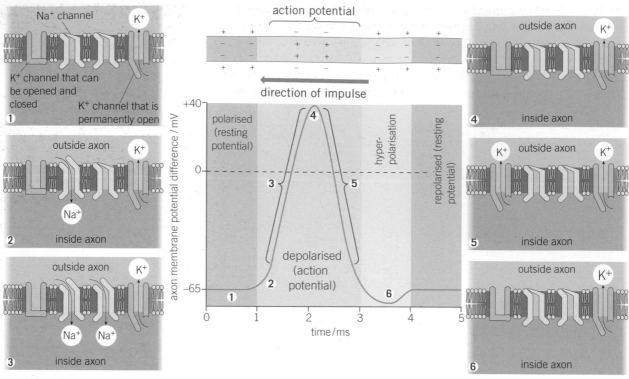

▲ Figure 2 *The action potential*

1 At resting potential some potassium voltage-gated channels are open (namely those that are permanently open) but the sodium voltage-gated channels are closed.

2 The energy of the stimulus causes some sodium voltage-gated channels in the axon membrane to open and therefore sodium ions diffuse into the axon through these channels along their electrochemical gradient. Being positively charged, they trigger a reversal in the potential difference across the membrane.

3 As the sodium ions diffuse into the axon, so more sodium channels open, causing an even greater influx of sodium ions by diffusion.

4 Once the action potential of around +40 mV has been established, the voltage gates on the sodium ion channels close (thus preventing further influx of sodium ions) and the voltage gates on the potassium ion channels begin to open.

5 With some potassium voltage-gated channels now open, the electrical gradient that was preventing further outward movement of potassium ions is now reversed, causing more potassium ion channels to open. This means that yet more potassium ions diffuse out, starting repolarisation of the axon.

6 The outward diffusion of these potassium ions causes a temporary overshoot of the electrical gradient, with the inside of the axon being more negative (relative to the outside) than usual (= hyperpolarisation) The closable gates on the potassium ion channels now close and the activities of the sodium–potassium pumps once again cause sodium ions to be pumped out and potassium ions in. The resting potential of −65 mV is re-established and the axon is said to be **repolarised**.

The terms action potential and resting potential can be misleading because the movement of sodium ions inwards during the action potential is purely due to diffusion – which is a passive process – while the resting potential is maintained by active transport – which is an active process. The term action potential simply means that the axon membrane is transmitting a nerve impulse, whereas resting potential means that it is not.

Hint

The action potential can be described as a travelling wave of depolarisation.

Summary questions

1 Describe how the movement of ions establishes the resting potential in an axon.

2 Table 1 shows the membrane potential of an axon at different stages of an action potential. The table refers to those channels that can be open and closed, not those that remain permanently open. For each of the letters A–F, indicate the state of the relevant channels, that is, open or closed.

▼ Table 1

	Resting	Beginning to depolarise	Repolarising
Membrane potential / mV	−70	−50	−20
Na⁺ channels in axon membrane	A	B	C
K⁺ channels in axon membrane	D	E	F

 Measuring action potentials

The plasma membrane of an axon will transmit an action potential when stimulated to do so. The action potential involves changes in the electrical potential across the membrane due to the movement of positive ions.

1 State which **two** positive ions are responsible for this change in electrical potential.

Figure 3 shows two action potentials that were recorded using electrodes and displayed on an instrument called an oscilloscope.

2 Between 0.5 and 2.0 ms there is a considerable change in membrane potential. Explain how this change is brought about.

3 ✓x Calculate how many action potentials will occur in 1 second if the frequency shown on the graph is maintained for this period. Show your working.

▲ Figure 3

Maths link ✓x

MS 3.1, see Chapter 22.

Learning objectives

→ Explain how an action potential passes along an unmyelinated axon.

→ Explain how an action potential passes along a myelinated axon.

Specification reference: 3.6.2.1

Once it has been created, an **action potential** moves rapidly along an **axon**. The size of the action potential remains the same from one end of the axon to the other. Strictly speaking, nothing physically moves from place to place along the axon of the neurone. As one region of the axon produces an action potential and becomes depolarised, it acts as a stimulus for the **depolarisation** of the next region of the axon. In this manner, action potentials are generated along each small region of the axon membrane. The action potential is therefore a travelling wave of depolarisation. In the meantime, the previous region of the membrane returns to its resting potential, that is, it undergoes **repolarisation**.

The process can be likened to the Mexican wave that often takes place in a crowded stadium during a sporting event. Although the wave of people standing up and raising their hands (the action potential) moves around the stadium, the people themselves do not move from seat to seat with the wave (i.e., they do not physically pass around the stadium until they reach their original seat again). Instead, their individual actions of standing and raising their hands are stimulated by the action of the person on one side of them and are reproduced by the person on the other side.

Passage of an action potential along an unmyelinated axon

It is easier to understand how a nerve impulse is propagated in a myelinated axon if we first look at how it is propagated in an unmyelinated one. This process is described and illustrated in Figure 2.

▲ **Figure 1** *False-colour TEM of the myelin sheath (orange bands at top) around the axon (bottom)*

polarised

depolarised

repolarised

1

stimulus

2

3

4

5

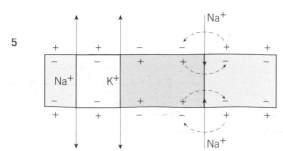

1 At resting potential the concentration of sodium ions outside the axon membrane is high relative to the inside, whereas that of the potassium ions is high inside the membrane relative to the outside. The overall concentration of positive ions is, however, greater on the outside, making this positive compared with the inside. The axon membrane is polarised. In our Mexican wave analogy, this is equivalent to the whole stadium being seated, that is, at rest.

2 A stimulus causes a sudden influx of sodium ions and hence a reversal of charge on the axon membrane. This is the action potential and the membrane is depolarised. In our analogy, a prompt leads a vertical line of people to stand and wave their arms, that is, they are stimulated into action.

3 The localised electrical currents established by the influx of sodium ions cause the opening of sodium voltage-gated channels a little further along the axon. The resulting influx of sodium ions in this region causes depolarisation. Behind this new region of depolarisation, the sodium voltage-gated channels close and the potassium ones open. Potassium ions begin to leave the axon along their electrochemical gradient. So, once initiated, the depolarisation moves along the membrane. The sight of the person next to them standing and waving prompts the person in the adjacent seat to stand and wave. A new vertical line of people stands and waves, while the original line of

4 The action potential (depolarisation) is propagated in the same way further along the axon. The outward movement of the potassium ions has continued to the extent that the axon membrane behind the action potential has returned to its original charged state (positive outside, negative inside), that is, it has been repolarised. The second line of people standing and waving prompts the third line of people to do the same. Meanwhile, the first line have now resumed their original positions, that is, they are re-seated.

5 Repolarisation of the axon allows sodium ions to be actively transported out, once again returning the axon to its resting potential in readiness for a new stimulus if it comes. The people who have just sat down settle back in their seats and readjust themselves in readiness to repeat the process should they be prompted to do so again.

▲ **Figure 2** *Passage of an impulse along the axon of an unmyelinated neurone*

Passage of an action potential along a myelinated axon

In myelinated axons, the fatty sheath of myelin around the axon acts as an electrical insulator, preventing action potentials from forming. At intervals of 1–3 mm there are breaks in this myelin insulation, called nodes of Ranvier (see Topic 15.1). Action potentials can occur at these points. The localised circuits therefore arise between adjacent nodes of Ranvier and the action potentials in effect jump from node to node in a process known as saltatory conduction (Figure 3). As a result, an action potential passes along a myelinated neurone faster than along the axon of an unmyelinated one of the same diameter. This is because in an unmyelinated neurone, the events of depolarisation have to take place all the way along an axon and this takes more time. In our Mexican wave analogy, this is equivalent to a whole block of spectators leaping up simultaneously, followed by the next block and so on. Instead of the wave passing around the stadium in hundreds of small stages, it passes around in 20 or so large ones and is consequently more rapid.

▲ **Figure 3** *Passage of an action potential along a myelinated axon. Action potentials are produced only at nodes of Ranvier. Depolarisation therefore skips from node to node (= saltatory conduction)*

Hint

The term saltatory, in saltatory conduction, comes from the Latin word saltare, meaning to jump.

Summary questions

1 In a myelinated axon, sodium and potassium ions can only be exchanged at certain points along it.

   a State the name given to these points.

   b Explain why ions can only be exchanged at these points.

   c Describe the effect this has on the way an action potential is conducted along the axon.

   d State the name that is given to this type of conduction.

   e Describe how it affects the speed with which the action potential is transmitted compared to an unmyelinated axon.

2 Describe what happens to the size of an action potential as it moves along an axon.

Once an **action potential** has been set up, it moves rapidly from one end of the axon to the other without any decrease in size. In other words, the action potential at the end of the axon is the same size as when it starts. This transmission of an action potential along the axon of a neurone is the **nerve impulse**.

Factors affecting the speed at which an action potential travels

A number of factors affect the speed at which action potentials pass along an axon. Depending upon these factors, action potentials may travel at a speed of as little as $0.5\,m\,s^{-1}$ or as much as $120\,m\,s^{-1}$. These factors include:

- **The myelin sheath**. We saw in Topic 15.3, Passage of an action potential, that the myelin sheath acts as an electrical insulator, preventing an action potential forming in the part of the axon covered in myelin. It does, however, jump from one node of Ranvier to another (**saltatory conduction**). This increases the speed of conductance from $30\,m\,s^{-1}$ in an unmyelinated neurone to $90\,m\,s^{-1}$ in a similar myelinated one.

- **The diameter of the axon**. The greater the diameter of an axon, the faster the speed of conductance. This is due to less leakage of ions from a large axon (leakage makes membrane potentials harder to maintain).

- **Temperature**. This affects the rate of diffusion of ions and therefore the higher the temperature the faster the nerve impulse. The energy for active transport comes from respiration. Respiration, like the **sodium–potassium pump**, is controlled by enzymes. Enzymes function more rapidly at higher temperatures up to a point. Above a certain temperature, enzymes and the plasma membrane proteins are denatured and impulses fail to be conducted at all. Temperature is clearly an important factor in response times in cold-blooded (ectothermic) animals, whose body temperature varies in accordance with the environment. Temperature also affects the speed and strength of muscle contractions.

All-or-nothing principle

Nerve impulses are described as **all-or-nothing** responses. There is a certain level of stimulus, called the **threshold value**, which triggers an action potential. Below the threshold value, no action potential, and therefore no impulse, is generated. Any stimulus, of whatever strength, that is below the threshold value will fail to generate an action potential – this is the nothing part. Any stimulus above the threshold value will succeed in generating an action potential and so a nerve impulse will travel. All action potentials are more or less the same size, and so the strength of a stimulus cannot be detected by the size of the action potentials. How then can an organism perceive the size of a stimulus? This is achieved in two ways:

Learning objectives

→ Describe the factors that affect the speed of conductance of an action potential.

→ Explain what is meant by the refractory period.

→ Explain the role of the refractory period in separating one impulse from the next.

→ Explain the meaning of the all-or-nothing principle.

Specification reference: 3.6.2.1

▲ **Figure 1** *False-coloured TEM of a section through a myelinated neurone and Schwann cell. Myelin (black) surrounds the axon (purple), increasing the speed at which nerve impulses travel. It is formed when Schwann cells (green) wrap around the axon, depositing layers of myelin between each coil*

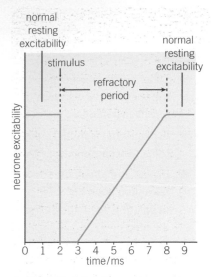

▲ **Figure 2** *Graph illustrating neurone excitability before and after a nerve impulse*

- by the number of impulses passing in a given time. The larger the stimulus, the more impulses that are generated in a given time (Figure 3)
- by having different neurones with different threshold values. The brain interprets the number and type of neurones that pass impulses as a result of a given stimulus and thereby determines its size.

The refractory period

Once an action potential has been created in any region of an axon, there is a period afterwards when inward movement of sodium ions is prevented because the sodium **voltage-gated channels** are closed. During this time it is impossible for a further action potential to be generated. This is known as the **refractory period** (Figure 2).

▲ **Figure 3** *Effect of stimulus intensity on impulse frequency*

The refractory period serves three purposes:

- **It ensures that action potentials are propagated in one direction only**. Action potentials can only pass from an active region to a resting region. This is because action potentials cannot be propagated in a region that is refractory, which means that they can only move in a forward direction. This prevents action potentials from spreading out in both directions, which they would otherwise do.

- **It produces discrete impulses**. Due to the refractory period, a new action potential cannot be formed immediately behind the first one. This ensures that action potentials are separated from one another.

- **It limits the number of action potentials**. As action potentials are separated from one another this limits the number of action potentials that can pass along an axon in a given time, and thus limits the strength of stimulus that can be detected.

Different axons, different speeds

Table 1 below shows the speeds at which different axons conduct action potentials.

▼ **Table 1**

Axon	Myelin	Axon diameter / μm	Transmission speed / m s^{-1}
Human motor axon to leg muscle	Yes	20	120
Human sensory axon from skin pressure receptor	Yes	10	50
Squid giant axon	No	500	25
Human motor axon to internal organ	No	1	2

1 Using data from the table, describe the effect of axon diameter on the speed of conductance of an action potential.

2 The data show that a myelinated axon conducts an action potential faster than an unmyelinated axon. Explain why this is so.

3 Name the cells whose membranes make up the myelin sheath around some types of axon.

4 State which has the greater effect on the speed of conductance of an action potential: the presence of myelin or the diameter of the axon. Use information from Table 1 to explain your answer.

5 The squid is an ectothermic animal. This means that its body temperature fluctuates with the temperature of the water in which it lives. Suggest how this might affect the speed at which action potentials are conducted along a squid axon.

6 Assuming it is circular in cross section, calculate the surface area of a squid giant axon. Give your answer in mm^2 to five significant figures.

Summary questions

1 Explain how the refractory period ensures that nerve impulses are kept separate from one another.

2 State the all-or-nothing principle.

3 Earthworms have unmyelinated axons and so to increase the speed of conduction of action potentials these are relatively large in diameter. Suggest **two** reasons why mammals do not require large diameter axons to achieve rapid transmission of action potentials.

A synapse is the point where one **neurone** communicates with another or with an **effector**. They are important in linking different neurones together and therefore coordinating activities.

Structure of a synapse

Synapses transmit information, but not impulses, from one neurone to another by means of chemicals known as **neurotransmitters**. Neurones are separated by a small gap, called the **synaptic cleft**, which is 20–30 nm wide. The neurone that releases the neurotransmitter is called the **presynaptic neurone**. The axon of this neurone ends in a swollen portion known as the **synaptic knob**. This possesses many mitochondria and large amounts of endoplasmic reticulum. These are required in the manufacture of the neurotransmitter which takes place in the axon. The neurotransmitter is stored in the **synaptic vesicles**. Once the neurotransmitter is released from the vesicles it diffuses across to the postsynaptic neurone, which possesses specific receptor proteins on its membrane to receive it.

The structure of a chemical synapse is illustrated in Figure 1.

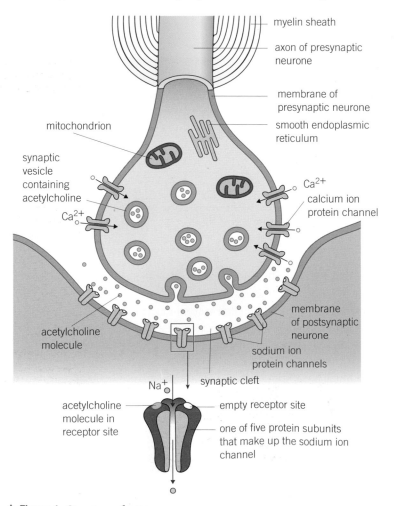

▲ **Figure 1** *Structure of a synapse*

Features of synapses

The basic way in which synapses function means they have a number of different features.

Unidirectionality

Synapses can only pass information in one direction – from the presynaptic neurone to the postsynaptic neurone. In this way, synapses act like valves.

Summation

Low-frequency action potentials often lead to the release of insufficient concentrations of neurotransmitter to trigger a new action potential in the postsynaptic neurone. They can, however, do so in a process called summation. This entails a rapid build-up of neurotransmitter in the synapse by one of two methods:

- **spatial summation**, in which a number of different presynaptic neurones together release enough neurotransmitter to exceed the threshold value of the postsynaptic neurone. Together they therefore trigger a new action potential.

- **temporal summation**, in which a single presynaptic neurone releases neurotransmitter many times over a very short period. If the concentration of neurotransmitter exceeds the threshold value of the postsynaptic neurone, then a new action potential is triggered.

▲ **Figure 2** *TEM of synapse. The synaptic cleft between the two neurones (centre) appears deep red. The cell above the cleft has many small vesicles (red–yellow spheres) containing neurotransmitter, whereas the two larger spheres above the vesicles are mitochondria*

spatial summation

presynaptic neurone A

no action potential

presynaptic neurone B

Neurone A releases neurotransmitter but concentration is below threshold to trigger action potential in postsynaptic neurone.

no action potential

Neurone B releases neurotransmitter but concentration is below threshold to trigger action potential in postsynaptic neurone.

action potential

Neurone A and B release neurotransmitter. Concentration is above threshold and so an action potential is triggered in the postsynaptic neurone.

temporal summation

low-frequency action potentials — no action potential

Low-frequency action potentials lead to neurotransmitter being broken down rapidly. Concentration is below the threshold to trigger an action potential in the postsynaptic neurone.

high-frequency action potentials — action potential

High-frequency action potentials lead to release of of neurotransmitter in a short time. Concentration is above the threshold to trigger an action potential in the postsynaptic neurone.

presynaptic neurone — postsynaptic neurone

neurotransmitter

▲ **Figure 3** *Spatial and temporal summation*

Inhibition

Some synapses make it less likely that a new action potential will be created on the postsynaptic neurone. These are known as **inhibitory synapses**. They operate as follows:

- The presynaptic neurone releases a type of neurotransmitter that binds to chloride ion protein channels on the postsynaptic neurone.
- The neurotransmitter causes the chloride ion protein channels to open.
- Chloride ions (Cl^-) move into the postsynaptic neurone by facilitated diffusion.
- The binding of the neurotransmitter causes the opening of nearby potassium (K^+) protein channels.
- Potassium ions move out of the postsynaptic neurone into the synapse.
- The combined effect of negatively charged chloride ions moving in and positively charged potassium ions moving out is to make the inside of the postsynaptic membrane more negative and the outside more positive.
- The membrane potential increases to as much as $-80\,mV$ compared with the usual $-65\,mV$ at resting potential.
- This is called hyperpolarisation and makes it less likely that a new action potential will be created because a larger influx of sodium ions is needed to produce one.

Maths link √x̄

MS 0.3, see Chapter 22.

Functions of synapses

Synapses transmit information from one neurone to another. In so doing, they act as junctions, allowing:

- a single impulse along one neurone to initiate new impulses in a number of different neurones at a synapse. This allows a single stimulus to create a number of simultaneous responses
- a number of impulses to be combined at a synapse. This allows nerve impulses from receptors reacting to different stimuli to contribute to a single response.

We shall look in more detail at how synapses transmit information in Topic 15.6, Transmission across a synapse. However, to understand the basic functioning of synapses as described here, it is sufficient to appreciate the following:

- A chemical (the neurotransmitter) is made **only** in the presynaptic neurone and not in the postsynaptic neurone.
- The neurotransmitter is stored in synaptic vesicles. When an **action potential** reaches the synaptic knob the membranes of these vesicles fuse with the pre-synaptic membrane to release the neurotransmitter.
- When released, the neurotransmitter diffuses across the synaptic cleft to bind to specific receptor proteins which are found **only** on the postsynaptic neurone.
- The neurotransmitter binds with the receptor proteins and this leads to a new action potential in the postsynaptic neurone. Synapses that produce new action potentials in this way are called **excitatory synapses**.

Summary questions √x̄

1 Explain how a presynaptic neurone is adapted for the manufacture of neurotransmitter.

2 Explain how the postsynaptic neurone is adapted to receive the neurotransmitter.

3 Outline the events in the transmission of information from one neurone to another across a synapse.

4 If a neurone is stimulated in the middle of its axon, an action potential will pass both ways along it to the synapses at each end of the neurone. However, the action potential will only pass across the synapse at one end. Explain why.

5 When walking along a street we barely notice the background noise of traffic. However, we often respond to louder traffic noises, such as the sound of a horn.

   a From your knowledge of summation, explain this difference.

   b Suggest an advantage in responding to high-level stimuli but not to low-level ones.

6 Explain why hyperpolarisation reduces the likelihood of a new action potential being created.

7 √x̄ Table 1 compares the number of synapses with the speed of transmission in three neural pathways: A, B and C.

   a Calculate the percentage increase in transmission speed when the number of synapses is reduced from 13 to 9.

   b Explain why the neural pathways of reflex arcs have very few synapses.

▼ Table 1

Neural pathway	Number of synapses	Speed of transmission /m s⁻¹
A	13	40
B	9	64
C	5	93

15.6 Transmission across a synapse

In Topic 15.5 we outlined how **neurotransmitters** transmit information from one neurone to another. Let us now consider this in more detail by looking at a cholinergic synapse.

A **cholinergic** synapse is one in which the neurotransmitter is a chemical called **acetylcholine**. Acetylcholine is made up of two parts: acetyl (more precisely ethanoic acid) and choline. Cholinergic synapses are common in vertebrates, where they occur in the central nervous system and at neuromuscular junctions (junctions between neurones and muscles). Details of the neuromuscular junction are given in Topic 15.7.

The process of transmission across a cholinergic synapse is described in the series of diagrams in Figure 2 later in this topic. To simplify matters, only the relevant structures are shown on each diagram. Each receptor is a protein that binds specifically to a neurotransmitter because they have complementary shapes.

Effects of drugs on synapses

There are many different neurotransmitters responsible for the exchange of information across a synapse. There are also many different types of receptor on the postsynaptic neurone. Each receptor is a protein that binds specifically to a neurotransmitter because they have complementary shapes. Some of these neurotransmitters and receptors are excitatory, that is, they lead to a new action potential in the postsynaptic neurone. Others are inhibitory, that is, they make it less likely that a new action potential will be created in the postsynaptic neurone. Overall, the action of a specific neurotransmitter depends on the specific receptor to which it binds.

Given that our perception of the world is through stimuli detected by receptors and information transferred to the brain as nerve impulses by neurones that connect via synapses, it is not surprising that the effects of many medicinal and recreational drugs are due to their actions on synapses. Drugs act on synapses in two main ways:

- **They stimulate the nervous system by creating more action potentials in postsynaptic neurones**. A drug may do this by mimicking a neurotransmitter, stimulating the release of more neurotransmitter, or inhibiting the enzyme that breaks down the neurotransmitter. The outcome is to enhance the body's responses to impulses passed along the postsynaptic neurone. For example, if the neurone transmits impulses from sound receptors, a person will perceive the sound as being louder.

- **They inhibit the nervous system by creating fewer action potentials in postsynaptic neurones**. A drug may do this by inhibiting the release of neurotransmitter or blocking receptors on sodium/potassium ion channels on the postsynaptic neurone. The outcome is to reduce the impulses passed along the postsynaptic neurone. In this case, if the neurone transmits impulses from sound receptors, a person will perceive the sound as being quieter.

The effects of a drug on the synapse depend on the type of transmitter. For example, a drug that inhibits the action of an excitatory neurotransmitter will reduce a particular effect, but a drug that inhibits an inhibitory neurotransmitter will enhance a particular effect. Let us look at some examples of the effects of drugs on synapses.

Endorphins are neurotransmitters used by certain sensory nerve pathways, especially pain pathways. Endorphins block the sensation of pain. Drugs such as morphine and codeine bind to specific receptors in the brain used by endorphins and so mimic the effects of endorphins.

1   Suggest the likely effect of drugs like morphine and codeine on the body.
2   Explain how the effect you suggest might be brought about.

Serotonin is a neurotransmitter involved in the regulation of sleep and certain emotional states. Reduced activity of the neurones that release serotonins is thought to be one cause of clinical depression. **Prozac** is an antidepressant drug that affects serotonin within synaptic clefts.

3   Suggest a way that the drug Prozac might affect serotonin within synaptic clefts.
4   Explain how the effect you suggest makes Prozac an effective antidepressant.

GABA is a neurotransmitter that inhibits the formation of action potentials when it binds to postsynaptic neurones. **Valium** is a drug that enhances the binding of GABA to its receptors.

5   Suggest the likely effect of Valium on the nerve pathways that cause muscle contractions.
6   Explain the reasoning for your answer.
7   Epilepsy can be the result of an increase in the activity of neurones in the brain due to insufficient GABA. An enzyme breaks down GABA on the postsynaptic membrane. A drug called Vigabatrin has a molecular structure similar to GABA and is used to treat epilepsy. Suggest a way in which Vigabatrin might be effective in treating epilepsy.

Summary questions

1   For each of the following, state as accurately as possible the name of the substance described.

   a   They diffuse into the postsynaptic neurone where they generate an action potential.

   b   A neurotransmitter found in a cholinergic synapse.

   c   It is released by mitochondria to enable the neurotransmitter to be reformed.

   d   Their influx into the presynaptic neurone causes synaptic vesicles to release their neurotransmitter.

2   State why it is necessary for acetylcholine to be hydrolysed by acetylcholinesterase.

▲ **Figure 1** *Many drugs function by acting on synapses*

1 The arrival of an action potential at the end of the presynaptic neurone causes calcium ion protein channels to open and calcium ions (Ca^{2+}) enter the synaptic knob by facilitated diffusion.

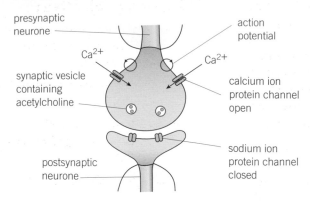

4 The influx of sodium ions generates a new action potential in the postsynaptic neurone.

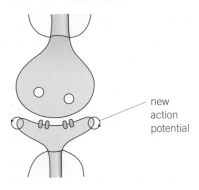

2 The influx of calcium ions into the presynaptic neurone causes synaptic vesicles to fuse with the presynaptic membrane, releasing acetylcholine into the synaptic cleft.

5 Acetylcholinesterase hydrolyses acetylcholine into choline and ethanoic acid (acetyl), which diffuse back across the synaptic cleft into the presynaptic neurone (= recycling). In addition to recycling the choline and ethanoic acid, the rapid breakdown of acetylcholine also prevents it from continuously generating a new action potential in the postsynaptic neurone, and so leads to discrete transfer of information across synapses

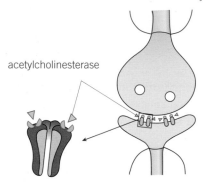

3 Acetylcholine molecules diffuse across the narrow synaptic cleft very quickly because the diffusion pathway is short. Acetylcholine then binds to receptor sites on sodium ion protein channels in the membrane of the postsynaptic neurone.
This causes the sodium ion protein channels to open, allowing sodium ions (Na^+) to diffuse in rapidly along a concentration gradient.

6 ATP released by mitochondria is used to recombine choline and ethanoic acid into acetycholine. This is stored in synaptic vesicles for future use. Sodium ion protein channels close in the absence of acetylcholine in the receptor sites.

▲ **Figure 2** *Mechanism of transmission across a cholinergic synapse*

Muscles are effector organs that respond to nervous stimulation by contracting and so bring about movement. There are three types of muscle in the body. **Cardiac muscle** is found exclusively in the heart while **smooth muscle** is found in the walls of blood vessels and the gut. Neither of these types of muscle is under conscious control and we remain largely unaware of their contractions. The third type, **skeletal muscle**, makes up the bulk of body muscle in vertebrates. It is attached to bone and acts under voluntary, conscious control.

A rope is made up of millions of separate threads. Each thread has very little individual strength and can easily be snapped. Yet grouped together in a rope, these threads can support a mass running into hundreds of tonnes. In the same way, individual muscles are made up of millions of tiny muscle fibres called **myofibrils**. In themselves, they produce almost no force while collectively they can be extremely powerful. Just as the threads in a rope are lined up parallel to each other in order to maximise its strength, so the myofibrils are arranged in order to give maximum force. And just as the threads of a rope are grouped into strings, the strings are grouped into small ropes and small ropes are grouped into bigger ropes, so muscle is composed of smaller units bundled into progressively larger ones (Figure 1).

If muscle was made up of individual cells joined end to end it would not be able to perform the function of contraction very efficiently. This is partly because the junction between adjacent cells would be a point of weakness that would reduce the overall strength of the muscle. To overcome this, muscles have a different structure. The separate cells have become fused together into muscle fibres. These muscle fibres share nuclei and also cytoplasm, called **sarcoplasm**, which is mostly found around the circumference of the fibre (Figure 1). Within the sarcoplasm is a large concentration of mitochondria and endoplasmic reticulum.

Learning objectives

→ Describe the gross and microscopic structure of a skeletal muscle.

→ Describe the ultrastructure of a myofibril.

→ Explain how actin and myosin are arranged within a myofibril.

Specification reference: 3.6.3

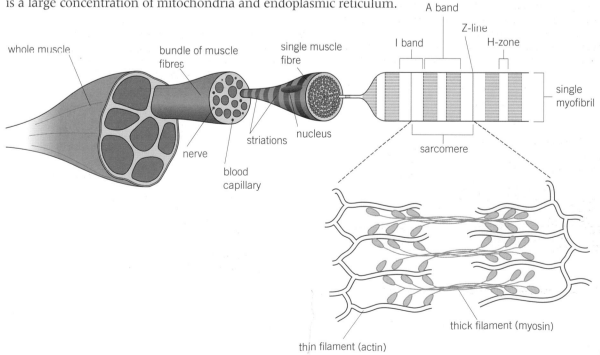

▲ **Figure 1** *The gross and microscopic structure of skeletal muscle*

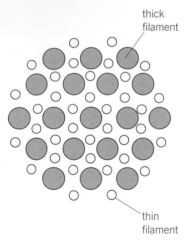

▲ **Figure 3** *Transverse section through part of a myofibril showing the arrangement of thick and thin filaments*

▶ **Figure 2** *Structure of actin and myosin molecules and their arrangement into thick and thin filaments*

Microscopic structure of skeletal muscle

We can see from Figure 1 that each muscle fibre is made up of myofibrils. Myofibrils are made up mainly of two types of protein filament:

- **actin**, which is thinner and consists of two strands twisted around one another
- **myosin**, which is thicker and consists of long rod-shaped tails with bulbous heads that project to the side.

The structure of these filaments and their constituent molecules shown in Figure 2.

Myofibrils appear striped due to their alternating light-coloured and dark-coloured bands. The light bands are called **I bands** (isotropic bands). They appear lighter because the thick and thin filaments do not overlap in this region. The dark bands are called **A bands** (anisotropic bands). They appear darker because the thick and thin filaments overlap in this region.

At the centre of each A band is a lighter-coloured region called the **H-zone**. At the centre of each I band is a line called the **Z-line**. The distance between adjacent Z-lines is called a **sarcomere** (Figure 1). When a muscle contracts, these sarcomeres shorten and the pattern of light and dark bands changes (Topic 15.8).

Another important protein found in muscle is **tropomyosin**, which forms a fibrous strand around the actin filament.

Hint

To help you remember which band is the dark band and which is the light band, look at the vowels in the words light and dark. This vowel is the first letter of the relevant band. Therefore the d<u>a</u>rk band is the **A**-band and the l<u>i</u>ght band is the **I**-band.

Hint

The arrangement of sarcomeres into a long line means that, when one sarcomere contracts a little, the line as a whole contracts a lot! In addition, having the lines of sarcomeres running parallel to each other means that all the force is generated in one direction.

Types of muscle fibre

There are two types of muscle fibre, the proportions of which vary from muscle to muscle and person to person. The two types are:

- **slow-twitch fibres**. These contract more slowly than fast-twitch fibres and provide less powerful contractions but over a longer period. They are therefore adapted to endurance work, such as running a marathon. In humans they are more common in muscles like the calf muscle, which must contract constantly to maintain the body in an upright position. They are suited to this role by being adapted for **aerobic** respiration in order to avoid a build-up of lactic acid, which would cause them to function less effectively and prevent long-duration contraction. These adaptations include having:
  - a large store of myoglobin (a bright red molecule that stores oxygen, which accounts for the red colour of slow-twitch fibres)
  - a rich supply of blood vessels to deliver oxygen and glucose for aerobic respiration
  - numerous mitochondria to produce ATP.

- **fast-twitch fibres**. These contract more rapidly and produce powerful contractions but only for a short period. They are therefore adapted to intense exercise, such as weight-lifting. As a result they are more common in muscles which need to do short bursts of intense activity, like the biceps muscle of the upper arm. Fast-twitch fibres are adapted to their role by having:
  - thicker and more numerous myosin filaments
  - a high concentration of glycogen
  - a high concentration of enzymes involved in **anaerobic** respiration which provides ATP rapidly
  - a store of phosphocreatine, a molecule that can rapidly generate ATP from ADP in anaerobic conditions and so provide energy for muscle contraction.

Before we look at how muscle contracts in the next topic, let us explore how the muscle is stimulated. To do so, we must first look at where neurones meet muscle – the neuromuscular junction.

Neuromuscular Junctions

A neuromuscular junction is the point where a motor neurone meets a skeletal muscle fibre. There are many such junctions along the muscle. If there were only one junction of this type it would take time for a wave of contraction to travel across the muscle, in which case not all the fibres would contract simultaneously and the movement would be slow. As rapid and coordinated muscle contraction is frequently essential for survival there are many neuromuscular junctions spread throughout the muscle. This ensures that contraction of a muscle is rapid and powerful when it is simultaneously stimulated by action potentials. All muscle fibres supplied by a single motor neurone act together as a single functional unit and are known as a motor unit. This arrangement gives control over the force that the muscle exerts. If only slight force is needed, only a few units are stimulated. If a greater force is required, a larger number of units are stimulated.

▲ **Figure 4** Light micrograph of a neuromuscular junction

axon of motor neurone

myelin sheath

neurone mitochondria

synaptic vesicle containing acetylcholine

presynaptic membrane

cytoplasm of muscle

myofibril membrane

endoplasmic reticulum

isotropic band of myofibril · anisotropic band of myofibril · mitochondrion of muscle · H-zone · Z-line · nucleus of muscle

▲ **Figure 5** *The neuromuscular junction*

▲ **Figure 6** *TEM of skeletal muscle*

When a nerve impulse is received at the neuromuscular junction, the synaptic vesicles fuse with the presynaptic membrane and release their acetylcholine. The acetylcholine diffuses to the postsynaptic membrane (which is the membrane of the muscle fibre), altering its permeability to sodium ions (Na^+), which enter rapidly, depolarising the membrane. A description of how this leads to the contraction of the muscle is given in Topic 15.8.

The acetylcholine is broken down by acetylcholinesterase to ensure that the muscle is not over-stimulated. The resulting choline and ethanoic acid (acetyl) diffuse back into the neurone, where they are recombined to form acetylcholine using energy provided by the mitochondria found there.

The structure of a neuromuscular junction is shown in Figure 5 and an account of how it functions is provided in Topic 15.8.

Comparison of the neuromuscular junction and a synapse

The neuromuscular junction has some similarities with a cholinergic synapse but also differs in some respects. Their similarities include that both:

- have neurotransmitters that are transported by diffusion
- have receptors, that on binding with the neurotransmitter, cause an influx of sodium ions
- use a sodium–potassium pump to repolarise the axon
- use enzymes to breakdown the neurotransmitter.

Some of the differences are shown in Table 1.

▼ **Table 1** *Differences between a neuromuscular junction and cholinergic synapse*

Neuromuscular junction	Cholinergic synapse
Only excitatory (Topic 15.5)	May be excitatory or inhibitory (Topic 15.5)
Only links neurones to muscles	Links neurones to neurones, or neurones to other effector organs
Only motor neurones are involved	Motor, sensory and intermediate neurones may be involved
The action potential ends here (it is the end of a neural pathway)	A new action potential may be produced along another neurone (the postsynaptic neurone)
Acetylcholine binds to receptors on membrane of muscle fibre	Acetylcholine binds to receptors on membrane of post-synaptic neurone

Summary questions

1 Suggest a reason why there are numerous mitochondria in the sarcoplasm.

2 Study Figure 6 and name the structures labelled A–D.

3 If we cut across a myofibril at certain points, we see only thick myosin filaments. Cut at a different point we see only thin actin filaments. At yet other points we see both types of filament. Explain why.

4 Explain how slow-twitch fibres differ from fast-twitch fibres in the way they function.

5 Describe how each type of fibre is adapted to its functions.

To understand how skeletal muscles bring about movement, it is important to appreciate that these muscles are attached to the skeleton. In humans, this skeleton is made up of bone which is incompressible. Therefore, if muscle exerts a force, via tendons the bone moves rather than the muscle changing shape. The different parts of the skeleton can be moved relative to one another around a series of points called joints.

The contraction of a skeletal muscle will move a part of the skeleton, for example, a limb, in one direction but the same muscle cannot move it in the opposite direction. Muscles cannot push they can only pull. To move the limb in the opposite direct requires a second muscle that works antagonistically to the first one, i.e. in the opposite direction. In doing so it stretches its partner muscle (which has relaxed) returning it to its original state ready to contract again.

Skeletal muscles therefore occur, and act, in antagonistic pairs. These pairs pull in opposite directions and when one is contracted the other is relaxed.

We have looked at the structure of skeletal muscle in Topic 15.7, now let us turn our attention to how exactly the arrangement of the various proteins brings about contraction of the muscle fibre. The process involves the actin and myosin filaments sliding past one another and is therefore called the **sliding filament mechanism**.

Learning objectives

→ Explain what is meant by antagonistic muscles and how they operate.

→ Summarise the evidence that supports the sliding filament mechanism of muscle contraction.

→ Explain how the sliding filament mechanism causes a muscle to contract and relax.

→ State where the energy for muscle contraction comes from.

Specification reference: 3.6.3

Synoptic link

The functioning of muscle depends on the molecular shapes of the four main proteins involved. The importance of shape on the functioning of proteins is covered in Topic 1.6.

This Figure is simplified - see Figure 2 in Topic 15.7 for the detailed arrangement of myosin filaments

▲ **Figure 1** *Comparison of two sarcomeres in a relaxed and a contracted muscle*

Evidence for the sliding filament mechanism

In Topic 15.7, we saw that myofibrils appear darker in colour where the actin and myosin filaments overlap and lighter where they do not. If the sliding filament mechanism is correct, then there will be more overlap of actin and myosin in a contracted muscle than in a relaxed one. If you look at Figure 1, you will see that, when a muscle contracts, the following changes occur to a **sarcomere**:

* The I-band becomes narrower.
* The Z-lines move closer together or, in other words, the sarcomere shortens.
* The H-zone becomes narrower.

The A-band remains the same width. As the width of this band is determined by the length of the myosin filaments, it follows that the myosin filaments have not become shorter. This discounts the theory that muscle contraction is due to the filaments themselves shortening.

Before we look at how the sliding filament mechanism works, let us take a closer look at the three main proteins involved in the process:

* Myosin is made up of two types of protein:
    * a fibrous protein arranged into a filament made up of several hundred molecules (the tail)
    * a globular protein formed into two bulbous structures at one end (the head).
* Actin is a globular protein whose molecules are arranged into long chains that are twisted around one another to form a helical strand.
* Tropomyosin forms long thin threads that are wound around actin filaments.

The arrangement of the molecules of actin and tropomyosin are shown in Figure 2.

The sliding filament mechanism of muscle contraction

The theory that actin and myosin filaments slide past one another during muscle contraction is supported by the changes seen in the band pattern on myofibrils. The next question for the scientists was, by what mechanism do the filaments slide past one another? Clues to the answer lie in the shape of the various proteins involved.

The bulbous heads of the myosin filaments form cross-bridges with the actin filaments. They do this by attaching themselves to binding sites on the actin filaments, and then flexing in unison, pulling the actin filaments along the myosin filaments. They then become detached and, using ATP as a source of energy, return to their original angle and re-attach themselves further along the actin filaments. This process is repeated up to 100 times per second. The action is similar to the way a ratchet operates. This process is illustrated in Figure 4.

actin filaments spiralled around one another

tropomyosin molecule

▲ **Figure 2** *The relationship of tropomyosin to an actin filament*

Hint

The action of the myosin heads is similar to the rowing action of oarsmen in a boat. The oars (myosin heads) are dipped into the water, flexed as the oarsmen pull on them, removed from the water and then dipped back into the water further along. The oarsmen work in unison and the boat and water move relative to one another.

The following account describes the sliding filament mechanism of muscle contraction in detail. The process is continuous but, for ease of understanding, has been divided into stimulation, contraction, and relaxation.

Muscle stimulation

- An **action potential** reaches many **neuromuscular junctions** simultaneously, causing calcium ion protein channels to open and calcium ions to diffuse into the synaptic knob.

- The calcium ions cause the synaptic vesicles to fuse with the presynaptic membrane and release their **acetylcholine** into the synaptic cleft.

- Acetylcholine diffuses across the synaptic cleft and binds with receptors on the muscle cell-surface membrane, causing it to depolarise.

Muscle contraction

- The action potential travels deep into the fibre through a system of tubules (T-tubules) that are extensions of the cell-surface membrane and branch throughout the cytoplasm of the muscle (sarcoplasm).

- The tubules are in contact with the endoplasmic reticulum of the muscle (sarcoplasmic reticulum) which has actively transported calcium ions from the cytoplasm of the muscle leading to very low Ca^{2+} concentration in cytoplasm.

- The action potential opens the calcium ion protein channels on the endoplasmic reticulum and calcium ions diffuse into the muscle cytoplasm down a concentration gradient.

- The calcium ions cause the tropomyosin molecules that were blocking the binding sites on the actin filament to pull away (Figure 4, stages 1 and 2).

- ADP molecules attached to the myosin heads mean they are in a state to bind to the actin filament and form a cross-bridge (Figure 4, stage 3).

- Once attached to the actin filament, the myosin heads change their angle, pulling the actin filament along as they do so and releasing a molecule of ADP (Figure 4, stage 4).

- An ATP molecule attaches to each myosin head, causing it to become detached from the actin filament (Figure 4, stage 5).

- The calcium ions then activate the enzyme ATPase, which hydrolyses the ATP to ADP. The hydrolysis of ATP to ADP provides the energy for the myosin head to return to its original position (Figure 4, stage 6).

- The myosin head, once more with an attached ADP molecule, then reattaches itself further along the actin filament and the cycle is repeated as long as the concentration of calcium ions in the myofibril remains high (Figure 4, stage 7).

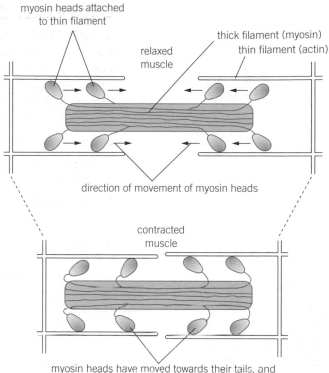

myosin heads attached to thin filament

thick filament (myosin)

thin filament (actin)

relaxed muscle

direction of movement of myosin heads

contracted muscle

myosin heads have moved towards their tails, and the centre of the sarcomere, pulling the attached thin filaments (actin) towards one another so shortening the sarcomere.

▲ **Figure 3** *The action of myosin in shortening a sarcomere and causing muscle contraction*

Hint

The contraction of muscle is the result of a wave of excitation that spreads across it.

Hint

The hydrolysis of ATP releases energy and produces ADP and inorganic phosphate (P_i). For simplicity, this account just refers to ADP as the product.

1 Tropomyosin molecule prevents myosin head from attaching to the binding site on the actin molecule.

2 Calcium ions released from the endoplasmic reticulum cause the tropomyosin molecule to change shape and so pull away from the binding sites on the actin molecule.

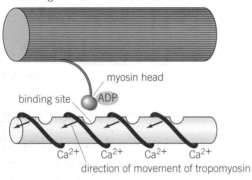

3 Myosin head now attaches to the binding site on the actin filament.

4 Head of myosin changes angle, moving the actin filament along as it does so. The ADP molecule is released.

5 ATP molecule fixes to myosin head, causing it to detach from the actin filament.

6 Hydrolysis of ATP to ADP by ATPase provides the energy for the myosin head to resume its normal position.

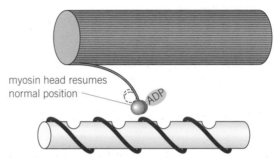

7 Head of myosin reattaches to a binding site further along the actin filament and the cycle is repeated.

Synoptic link

The presence of calcium ions changes the environment of the protein tropomyosin leading to a change in its tertiary structure. We first met this idea in Topic 1.8.

▲ **Figure 4** *Sliding filament mechanism of muscle contraction (showing only one myosin head throughout)*

- As the myosin molecules are joined tail to tail in two oppositely facing sets, the movement of one set of myosin heads is in the opposite direction to the other set. This means that actin filaments to which they are attached also move in opposite directions.

- The movement of actin filaments in opposite directions pulls them towards each other, shortening the distance between the two adjacent Z-lines. The process is illustrated in Figure 4. The overall effect of this process taking place repeatedly and simultaneously throughout a muscle is to shorten it and so bring about movement of a part of the body.

Muscle relaxation

- When nervous stimulation ceases, calcium ions are actively transported back into the endoplasmic reticulum using energy from the **hydrolysis** of ATP.

- This reabsorption of the calcium ions allows tropomyosin to block the actin filament again.

- Myosin heads are now unable to bind to actin filaments and contraction ceases, that is, the muscle relaxes.

- In this state force from antagonistic muscles can pull actin filaments out from between myosin (to a point).

Energy supply during muscle contraction

Muscle contraction requires considerable energy. This is supplied by the hydrolysis of ATP to ADP and inorganic phosphate (P_i). The energy released is needed for:

- the movement of the myosin heads
- the reabsorption of calcium ions into the endoplasmic reticulum by active transport.

In an active muscle, there is clearly a great demand for ATP. In some circumstances, for example, escaping from danger, the ability of muscles to work intensely can be life-saving. Most ATP is regenerated from ADP during the respiration of pyruvate in the mitochondria, which are particularly plentiful in the muscle. However, this process requires oxygen. In a very active muscle the demand for ATP, and therefore oxygen, is greater than the rate at which the blood can supply oxygen. Therefore a means of rapidly generating ATP **anaerobically** is also required. This is partly achieved using a chemical called **phosphocreatine** and partly by more glycolysis.

Phosphocreatine cannot supply energy directly to the muscle, so instead it regenerates ATP, which can. Phosphocreatine is stored in muscle and acts as a reserve supply of phosphate, which is available immediately to combine with ADP and so re-form ATP. The phosphocreatine store is replenished using phosphate from ATP when the muscle is relaxed.

▲ **Figure 5** *Marathon runners undergoing strenuous exercise*

Maths link $\sqrt{x}$

MS 0.1 and 2.4, see Chapter 22.

Summary questions $\sqrt{x}$

1 Explain how the shape of the myosin molecule is adapted to its role in muscle contraction.

2 Trained sprinters have high levels of phosphocreatine in the muscles. Explain the advantage of this.

3 $\sqrt{x}$ During the contraction of a muscle sarcomere, a single actin filament moves 0.8 μm. If the hydrolysis of a single ATP molecule provides enough energy to move an actin filament 40 nm, calculate how many ATP molecules are needed to move the actin filament 0.8 μm. Show your working.

4 Dead cells can no longer produce ATP. Soon after death, muscles contract, making the body stiff — a state known as rigor mortis. From your knowledge of muscle contraction, explain the reasons why rigor mortis occurs after death.

1 During an action potential, the permeability of the cell-surface membrane of an axon changes. The graph shows changes in permeability of the membrane to sodium ions (Na⁺) and to potassium ions (K⁺) during a single action potential.

(a) Explain the shape of the curve for sodium ions between 0.5 ms and 0.7 ms. *(3 marks)*

(b) (i) During an action potential, the membrane potential rises to +40 mV and then falls. Use information from the graph to explain the fall in membrane potential. *(3 marks)*

    (ii) The refractory period of a neurone has two components, absolute and relative. Calculate the maximum number of impulses that can be generated in a neurone when the total refractory period is 5 ms. *(2 marks)*

    (iii) Calculate the percentage increase in nerve impulses if the stimulus intensity is raised high enough to overcome the relative refractory period of 4 ms. *(3 marks)*

(c) After exercise, some ATP is used to re-establish the resting potential in axons. Explain how the resting potential is re-established. *(2 marks)*

AQA June 2010 (apart from 1 (b) (ii) and (iii))

2 Serotonin is a neurotransmitter released in some synapses in the brain. It is transported back out of the synaptic gap by a transport protein in the pre-synaptic membrane.

(a) Serotonin diffuses across the synaptic gap and binds to a receptor on the post-synaptic membrane. Describe how this causes depolarisation of the post-synaptic membrane. *(2 marks)*

(b) It is important that a neurotransmitter such as serotonin is transported back out of synapses. Explain why. *(2 marks)*

(c) Scientists investigated the effect of a drug called MDMA on movement of mice. They measured the amount of movement of three groups of mice, **K**, **L** and **M**.
   • Group **K**, mice not given MDMA.
   • Group **L**, mice given MDMA.
   • Group **M**, mutant mice that did not produce a serotonin receptor on their post-synaptic membranes and were given MDMA.

The graph shows their results.
The scientists concluded that MDMA affects movement by binding to serotonin receptors.
How do these results support this conclusion? *(3 marks)*

AQA June 2012

Feature	Fast muscle fibre	Slow muscle fibre
Type of respiration	Mainly anaerobic	Mainly aerobic
Glycogen	High concentration	Low concentration
Capillaries	Few	Many

3 **(a)** Describe the part played by each of the following in myofibril contraction.
   (i) Tropomyosin *(2 marks)*
   (ii) Myosin *(2 marks)*
(b) The table shows features of fast and slow muscle fibres.
   Use information from the table to suggest and explain **one** advantage of:
   (i) the high glycogen content of fast muscle fibres *(2 marks)*
   (ii) the number of capillaries supplying slow muscle fibres. *(2 marks)*
   AQA June 2013

4 The diagram shows two relaxed sarcomeres from skeletal muscle.

length of sarcomere = 2.2 µm

(a) When the sarcomeres contract, what happens to the length of
   (i) the I-band *(1 mark)*
   (ii) the A-band? *(1 mark)*
(b) The length of each sarcomere in the diagram is 2.2 µm. Use this information to calculate the magnification of the diagram. Show your working.
   (2 marks)
(c) People who have McArdle's disease produce less ATP than healthy people. As a result, they are not able to maintain strong muscle contraction during exercise. Use your knowledge of the sliding filament theory to suggest why. *(3 marks)*
   AQA June 2012

5 The drawing is a tracing of a cross-section through skeletal muscle tissue. This muscle contains fast muscle fibres and slow muscle fibres. The section has been stained to show the distribution of the enzyme succinate dehydrogenase. This enzyme is found in mitochondria.
(a) (i) Succinate dehydrogenase catalyses one of the reactions in the Krebs cycle. What is the evidence from the drawing that muscle fibre S is a slow muscle fibre? Explain your answer. *(2 marks)*
   (ii) Use evidence from the diagram to describe the distribution of mitochondria inside the slow muscle fibres. Explain the importance of this distribution. *(3 marks)*
(b) (i) You could use an optical microscope and a slide of stained muscle tissue to find the diameter of one of the muscle fibres. Explain how. *(2 marks)*

muscle fibre **S**

   (ii) A student found the mean diameter for the slow muscle fibres in a section. Give two precautions that she should have taken when sampling the fibres. Give a reason for each precaution. *(2 marks)*
   AQA June 2010

In the previous chapter we looked at how complex organisms control and coordinate their activities. In particular we considered the way in which such organisms respond rapidly to environmental changes using their nervous system. A feature of an increase in complexity is the ability of organisms to control their internal environment. By maintaining a relatively constant internal environment for their cells, organisms can limit the external changes these cells experience. This maintenance of a constant internal environment is called **homeostasis**. In this chapter we shall learn about homeostasis and the role of the other coordination system, hormonal coordination, in an organism's physiological control.

The internal environment is made up of **tissue fluids** that bathe each cell, supplying nutrients and removing wastes. Maintaining the features of this fluid at the optimum levels protects the cells from changes in the external environment, thereby giving the organism a degree of independence.

What is homeostasis?

Homeostasis is the maintenance of an internal environment within restricted limits in organisms. It involves trying to maintain the chemical make-up, volume and other features of blood and tissue fluid within restricted limits. Homeostasis ensures that the cells of the body are in an environment that meets their requirements and allows them to function normally despite external changes. This does not mean that there are no changes. On the contrary, there are continuous fluctuations brought about by variations in internal and external conditions, such as changes in temperature, pH and water potential. These changes, however, occur around an optimum point. Homeostasis is the ability to return to that optimum point and so maintain organisms in a balanced equilibrium.

The importance of homeostasis

Homeostasis is essential for the proper functioning of organisms for the following reasons amongst others:

- The enzymes that control the biochemical reactions within cells, and other proteins, such as channel proteins, are sensitive to changes in pH and temperature. Any change to these factors reduces the rate of reaction of enzymes or may even prevent them working altogether, for example, by denaturing them. Even small fluctuations in temperature or pH can impair the ability of enzymes to carry out their roles effectively. Maintaining a fairly constant internal environment means that reactions take place at a suitable rate.

- Changes to the **water potential** of the blood and tissue fluids may cause cells to shrink and expand (even to bursting point) as a result of water leaving or entering by osmosis. In both instances the cells cannot operate normally. The maintenance of a constant blood glucose concentration is essential in ensuring a constant water potential. A constant blood glucose concentration also ensures a reliable source of glucose for respiration by cells.

- Organisms with the ability to maintain a constant internal environment are more independent of changes in the external environment. They may have a wider geographical range and therefore have a greater chance of finding food, shelter, etc. Mammals, for example, with their ability to maintain a constant temperature, are found in most habitats, ranging from hot arid deserts to cold, frozen polar regions.

Control mechanisms

The control of any self-regulating system involves a series of stages that feature:

- the **optimum point**, the point at which the system operates best. This is monitored by a …
- **receptor**, which detects any deviation from the optimum point (ie., a stimulus) and informs the …
- **coordinator**, which coordinates information from receptors and sends instructions to an appropriate …
- **effector**, often a muscle or gland, which brings about the changes needed to return the system to the optimum point. This return to normality creates a …
- **feedback mechanism**, by which a receptor responds to a stimulus created by the change to the system brought about by the effector.

▲ **Figure 1** *Homeostasis allows animals such as these camels in the desert (top) and these penguins in the Antarctic (bottom) to survive in extreme environments*

Figure 2 illustrates the relationship between these stages using the everyday example of controlling a central heating system.

Input Change to the system	Receptor Change detected in a living system	Coordinator Operational information is stored here and used to coordinate effectors	Effector Brings about changes to the system in order to return it to the optimum point	Output System returned to optimum point
Room temperature drops from 20°C to 18°C	Room thermostat signals that the temperature is below the optimum point	Programmer checks that heating should be on at this time. If so, it starts boiler and circulation pump	Boiler fires up, pump circulates water, radiators become hot	Room temperature is raised to 20°C

Feedback loop – in this case = negative feedback as it turns system off

Circulation of air in room takes air at 20°C from radiator to thermostat

▲ **Figure 2** *Components of a typical control system*

Coordination of control mechanisms

Most systems, including biological ones, use **negative feedback** Negative feedback is when the change produced by the control system leads to a change in the stimulus detected by the receptor and turns the system off. We shall meet an example of negative feedback when we look at the regulation of blood glucose in Topic 16.3.

Positive feedback occurs when a deviation from an optimum causes changes that result in an even greater deviation from the normal. One example occurs in neurones where a stimulus leads to a small influx of sodium ions. This influx increases the permeability of the neurone membrane to sodium ions, more ions enter, causing a further increase

Summary questions

1 Describe homeostasis.

2 Explain why maintaining a constant temperature is important in mammals.

3 Suggest why maintaining a constant blood glucose concentration might be important in mammals.

Maths link $\sqrt{x}$

MS 3.1, see Chapter 22.

in permeability and even more rapid entry of ions. In this way, a small stimulus can bring about a large and rapid response.

Control systems normally have many receptors and effectors. This allows them to have separate mechanisms taht each produce a positive movement towards an optimum. This allows a greater degree of control of the particular factor being regulated. Having separate mechanisms that controls departures in different directions from the original state is a general feature of homeostasis. It is important to ensure that the information provided by receptors is analysed by the coordinator before action is taken. For example, temperature receptors in the skin may signal that the skin itself is cold and that the body temperature should be raised. However, information from regions in the hypothalamus in the brain may indicate that blood temperature is already above normal. This situation might arise during strenuous exercise when blood temperature rises but sweating cools the skin. By analysing the information from all detectors, the brain can decide the best course of action – in this case not to raise the body temperature further. In the same way, the control centre must coordinate the action of the effectors so that they operate harmoniously. For example, sweating would be less effective in cooling the body if it were not accompanied by **vasodilation**.

+ Comparing thermoregulation in ectotherms and endotherms $\sqrt{x}$

As with all extension boxes, the material here is to broaden understanding of material beyond the specification.

Animals such as birds and mammals derive most of their heat from the metabolic activities that take place inside their bodies. They are therefore known as **endotherms** (meaning inside heat). Some animals obtain a proportion of their heat from sources outside their bodies, namely the environment. They are therefore known as **ectotherms** (meaning outside heat).

Regulation of body temperature in ectotherms

Many ectotherms gain heat from the environment, so their body temperature fluctuates with that of the environment. They therefore control their body temperature by adapting their behaviour to changes in the external temperature. Reptiles, such as lizards, are ectotherms. They control their body temperature by:

- **exposing themselves to the Sun**. In order to gain heat lizards orientate themselves so that the maximum surface area of their body is exposed to the warming rays of the Sun.

- **taking shelter**. Lizards will shelter in the shade to prevent over-heating when the Sun's radiation is at its peak. At night they retreat into burrows in order to reduce heat loss when the external temperature is low.

- **gaining warmth from the ground**. Lizards will press their bodies against areas of hot ground to warm themselves up. When the required temperature is reached, they raise themselves off the ground on their legs.

▲ **Figure 3** *A lizard showing thermoregulatory behaviour by gaining heat both from the sun and the warm rock*

Regulation of body temperature in endotherms

Endotherms gain most of their heat from internal metabolic activities. Their body temperature remains relatively constant despite fluctuations in the external temperature. Like ectotherms, endothermic animals use behaviour to maintain a constant body temperature. Unlike ectotherms, however, they also use a wide range of physiological mechanisms to regulate their temperature.

Conserving and gaining heat in response to a cold environment

Mammals and birds that live in cold climates have evolved a number of adaptations in order to survive in these environments. One of the most important is having a body with a small surface area to volume ratio. It is from within the volume that heat is produced and from the surface area that heat is lost. Mammals and birds in cold climates therefore tend to be relatively large, for example, the polar bear and penguin. Compared with animals in warmer climates they also have smaller extremities, such as ears, and thick fur, feathers, or fat layers to insulate the body.

To make more rapid body temperature changes, mammals use one or more of the following mechanisms:

- **vasoconstriction**. The diameter of the arterioles near the surface of the skin is made smaller. This reduces the volume of blood reaching the skin surface through the capillaries. Most of the blood entering the skin passes beneath the insulating layer of fat and so loses little heat to the environment (Figure 5).

▲ **Figure 4** *The penguin and the polar bear both have large compact bodies with a small surface area to volume ratio. This helps them to conserve heat in the cold environments of the South and North Poles where they live*

▲ **Figure 5** *Vasoconstriction*

- **shivering**. The muscles of the body undergo involuntary rhythmic contractions that produce metabolic heat.

- **raising of hair**. The hair erector muscles in the skin contract, raising the hairs on the body. This enables a thicker layer of still air, which is a good insulator, to be trapped next to the skin, insulation and conserving heat in mammals with thick fur.

- **increased metabolic rate**. In cold conditions more of the hormones that increase metabolic rate are produced. As a result metabolic activity, including respiration, is increased and so more heat is generated.

- **decrease in sweating**. Sweating is reduced, or ceases altogether, in cold conditions.

- **behavioural mechanisms**. Sheltering from the wind, basking in the sun and huddling together all help animals to maintain their core body temperature.

Losing heat in response to a warm environment

Long-term adaptations to life in a warm climate include having a large surface area to volume ratio and lighter coloured fur to reflect heat. Rapid responses that enable heat to be lost when the environmental temperature is high include:

- **vasodilation**. The diameter of the arterioles near the surface of the skin becomes larger. This allows warm blood to pass close to the skin surface through the capillaries. The heat from this blood is then radiated away from the body (Figure 6).

- **increased sweating**. To evaporate water from the skin surface requires energy in the form of heat. In relatively hairless mammals, such as humans, sweating is a highly effective means of losing heat. In mammals with fur, cooling is achieved by the evaporation of water from the mouth and tongue, during panting. The high latent heat of vaporisation of water makes sweating an efficient way of losing heat.

- **lowering of body hair**. The hair erector muscles in the skin relax and the elasticity of the skin causes them to flatten against the body. This reduces the thickness of the insulating layer and allows more heat to be lost to the environment when the internal temperature is higher than the external temperature.

- **behavioural mechanisms**. Avoiding the heat of the day by sheltering in burrows and seeking out shade help to prevent the body temperature from rising.

The graphs shown in Figure 7 compare the rates of metabolic heat generation and evaporative heat loss in a mammal and a reptile as the environmental temperature changes.

1 Give a reason why the values for heat generation and heat loss are measured per gram of body mass.

2 a Describe the relationship between metabolic heat generation and evaporative heat loss in a reptile.

  b Explain how this relationship differs in a mammal.

3 Reptiles frequently seek shade when the environmental temperature rises above 25°C. Use the graphs to explain this type of behaviour.

4 Suggest a reason for the change in the evaporative heat loss in the mammal at point A on the graph.

▲ **Figure 6** *Vasodilation*

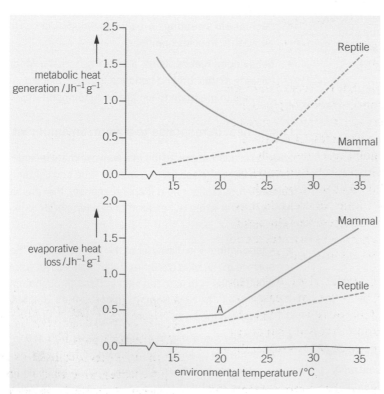

▲ **Figure 7**

16.2 Feedback mechanisms

We saw in Topic 16.1, that the **homeostatic** control of any system involves a series of stages featuring:

- **the optimum point**, or desired level (norm), at which the system operates
- **a receptor**, which detects the stimulus of any deviation from the set point (norm)
- **a coordinator**, which coordinates information from various sources
- **an effector**, which brings about the corrective measures needed to return the system to the optimum point (norm)
- **a feedback mechanism**, by which a receptor detects a stimulus created by the change to the system and the effector brings about the appropriate response.

Let us now look in more detail at the last stage in the list – the feedback mechanism. When an effector has corrected any deviation and returned the system to the optimum point, it is important that this information is fed back to the receptor. If the information is not fed back, the receptor will continue to stimulate the effector, leading to an over-correction and causing a deviation in the opposite direction. There are two types of feedback – negative feedback and positive feedback.

Negative feedback

Negative feedback occurs when the stimulus causes the corrective measures to be turned off. In doing so this tends to return the system to its original (optimum) level (and prevents any overshoot).

There are separate negative feedback mechanisms to regulate departures from the norm in each direction.

An example is in the control of blood glucose that is covered in more detail in Topic 16.3. If there is a fall in the concentration of glucose in the blood this stimulus is detected by receptors on the cell-surface membrane of the α cells (coordinator) in the pancreas. These α cells secrete the hormone **glucagon**. Glucagon causes liver cells (effectors) to convert glycogen to glucose which is released into the blood raising the blood glucose concentration. As this blood with a raised glucose concentration circulates back to the pancreas there is reduced stimulation of α cells which therefore secrete less glucagon. So the secretion of glucagon leads to a reduction in its own secretion (=negative feedback). These events are illustrated in Figure 1.

In the same way, if the blood glucose concentration rises, rather than falls, insulin will be produced from the β cells in the pancreas. **Insulin** increases the uptake of glucose by cells and its conversion to glycogen and fat. The fall in blood glucose concentration that results reduces insulin production once blood glucose concentrations return to their optimum (= negative feedback).

Having separate negative feedback mechanisms that control departures from the norm in either direction gives a greater degree of homeostatic control. This is because there are positive actions in both directions.

Learning objectives
→ Explain what negative feedback is.
→ Explain how negative feedback helps to control homeostatic processes.
→ Distinguish between negative feedback and positive feedback.

Specification reference: 3.6.4.1

Synoptic link

The concept of stimulus → receptor → coordinator → effector → response is a recurring theme in biology. For example, we met it throughout Chapter 14.

fall in blood glucose concentration → α cells in the pancreas → blood containing glucagon → liver glycogen → glucose → normal blood glucose concentration

blood at optimum glucose concentration

reduces stimulation of α cells

▲ **Figure 1** *Negative feedback in the control of blood glucose levels*

For example, if glucagon raised the blood sugar concentration above the optimum, it would take some time for it to fall again if the only way of lowering it was through metabolic activity. However, by having a second hormone, insulin, that lowers blood sugar concentration, its secretion brings about a return to optimum blood sugar concentration far more rapidly.

Study tip

If you are writing about negative feedback, for example, control of blood glucose, make certain that you focus on negative feedback rather than just giving a description of how insulin and glucagon work.

Hint

Negative feedback in the context of hormones means that the secretion of a hormone (e.g., glucagon) leads to a reduction in the secretion of that hormone.

Summary questions

1   Explain why negative feedback is important in maintaining a system at a set point.

2   Explain the advantage of having separate negative feedback mechanisms to control deviations away from normal.

 Positive feedback

Positive feedback occurs when the feedback causes the corrective measures to remain turned on. In doing so it causes the system to deviate even more from the original (normal) level. Examples are less common, but one occurs in neurones when a stimulus causes a small influx of sodium ions. This influx increases the permeability of the neurone to sodium ions so more ions enter, causing a further increase in permeability and even more rapid entry of ions. This results in a very rapid build-up of an action potential that allows an equally rapid response to a stimulus.

Positive feedback occurs more often when there is a breakdown of control systems. In certain diseases, for example typhoid fever, there is a breakdown of temperature regulation resulting in a rise in body temperature leading to hyperthermia. In the same way, when the body gets too cold (hypothermia) the temperature control system tends to break down, leading to positive feedback resulting in the body temperature dropping even lower.

1   Oxytocin is a hormone that causes contractions of the uterus at childbirth. The contractions produce a positive feedback loop that results in the release of more oxytocin. Explain the advantage of positive feedback rather than negative feedback in this situation.

Negative feedback in temperature control

If the temperature of the blood increases, thermoreceptors in a region of the brain called the **hypothalamus** send more nerve impulses to the heat loss centre, which is also in the hypothalamus. This in turn sends impulses to the skin (effector organ). **Vasodilation**, sweating and lowering of body hairs all lead to a reduction in blood temperature. If the fact that blood temperature has returned to normal is not fed back to the hypothalamus, it will continue to stimulate the skin to lose body heat. Blood temperature will then fall below normal and may continue to do so causing **hypothermia** and the death of the organism.

What happens in practice is that the cooler blood returning from the skin passes through the hypothalamus. As a result thermoreceptors send fewer impulses to the heat loss centre. This in turn stops sending impulses to the skin and so vasodilation, sweating, etc. cease, and blood temperature remains at its normal level rather than continuing to fall. The blood, having been cooled to its normal temperature, has resulted in turning *off* the effector (the skin) that was correcting the rise in temperature. This is therefore negative feedback and is illustrated in Figure 2.

1 State what would happen to the temperature of the blood if the feedback was positive rather than negative.

2 Cutting the nerves connecting the thermoreceptors to the heat loss centre in the hypothalamus might cause the death of the individual. In terms of the information in Figure 2, explain precisely why this action might cause death.

3 Cutting the nerves connecting the heat loss centre to the skin would be less likely to cause death than if those between the thermoreceptors and the heat loss centre were cut. Suggest why.

4 Negative feedback in temperature regulation occurs as a result of blood passing from the skin to the brain. In doing so, this blood passes through the heart. List, in sequence, the major vessels joined to the heart that this blood would pass through on its journey.

▲ **Figure 3** *Control of body temperature involves negative feedback mechanisms*

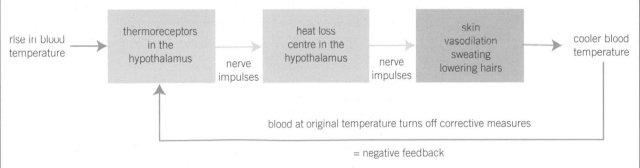

▲ **Figure 2** *Negative feedback in the control of body temperature*

Learning objectives

→ Explain how hormones work.

→ Explain the roles of the pancreas and liver in regulating blood glucose.

→ Outline the factors which influence blood glucose concentration.

→ Explain the roles of insulin, glucagon and adrenaline in regulating blood glucose.

Specification reference: 3.6.4.2

We saw in Topic 15.1 that animals possess two principal coordinating systems: the nervous system, which communicates rapidly, and the hormonal system, which usually communicates more slowly. Both systems interact in order to maintain the constancy of the internal environment while also being responsive to changes in the external environment. Both systems also use chemical messengers – the hormonal system exclusively so, and the nervous system through the use of **neurotransmitters** in chemical **synapses**.

The regulation of blood glucose is an example of how different hormones interact in achieving **homeostasis**. However, let us first look at what hormones are and how they work.

Hormones and their mode of action

Hormones differ from one another chemically but they all have certain characteristics in common. Hormones are:

- produced in glands, which secrete the hormone directly into the blood (endocrine glands)
- carried in the blood plasma to the cells on which they act – known as **target cells** – which have specific receptors on their cell-surface membranes that are complementary to a specific hormone
- are effective in very low concentrations, but often have widespread and long-lasting effects.

One mechanism of hormone action is known as the **second messenger model**. This mechanism is used by two hormones involved in the regulation of blood glucose concentration, namely adrenaline and glucagon.

The mechanism involving adrenaline is detailed below and illustrated in Figure 1.

- Adrenaline binds to a transmembrane protein receptor within the cell-surface membrane of a liver cell.
- The binding of adrenaline causes the protein to change shape on the inside of the membrane.
- This change of protein shape leads to the activation of an enzyme called adenyl cyclase. The activated adenyl cyclase converts ATP to cyclic AMP (cAMP).
- The cAMP acts as a second messenger that binds to protein kinase enzyme, changing its shape and therefore activating it.
- The active protein kinase enzyme catalyses the conversion of glycogen to glucose which moves out of the liver cell by facilitated diffusion and into the blood, through channel proteins.

The role of the pancreas in regulating blood glucose

The pancreas is a large, pale-coloured gland that is situated in the upper abdomen, behind the stomach. It produces enzymes (protease, amylase and lipase) for digestion and hormones (insulin and glucagon) for regulating blood glucose concentration.

Synoptic link

The activation of an enzyme (in this case protein kinase enzyme) as a result of it changing shape when it binds with another substance (in this case cAMP) is a feature of many biological processes. In fact it occurs twice in this process as adenyl cyclase is activated by adrenaline.

Study tip

In the second messenger model of hormone action, the hormone has its effect inside a cell even though it never enters the cell.

adrenaline (= first messenger)
receptor site
transmembrane protein
adenyl cyclase (inactive)
cell-surface membrane of liver cell

adrenaline–receptor complex
adenyl cyclase (active)
change in shape of transmembrane protein

adrenaline–receptor complex

ATP → cAMP (etc)

protein kinase enzyme (inactive) → protein kinase enzyme (active)

glycogen → glucose

1 The hormone adrenaline approaches transmembrane protein.

2 Adrenaline fuses to the receptor causing it to change shape on the inside of the membrane activating an enzyme called adenyl cyclase inside the membrane.

3 The activated adenyl cyclase converts ATP to cyclic AMP, which acts as a second messenger.

4 The cAMP, in turn changes the shape of and, activates protein kinase enzyme.

5 The active protein kinase enzyme catalyses the conversion of glycogen to glucose.

◀ **Figure 1** *Second messenger model of hormone action as illustrated by the action of adrenaline in regulating blood sugar*

▲ **Figure 2** *LM of the pancreas showing an islet of Langerhans (centre) containing α cells and β cells. The meshworks of blue and white in the islet are blood capillaries. Around the islet are the enzyme-producing pancreatic cells*

When examined microscopically, the pancreas is made up largely of the cells that produce its digestive enzymes. Scattered throughout these cells are groups of hormone-producing cells known as **islets of Langerhans**. The cells of the islets of Langerhans include:

- **α cells**, which are the larger and produce the hormone **glucagon**
- **β cells**, which are smaller and produce the hormone **insulin**.

The role of the liver in regulating blood sugar

The liver is located immediately below the diaphragm, has a mass of up to 1.5 kg and is made up of cells called hepatocytes. It serves a large variety of roles including regulating blood glucose concentration. While the pancreas produces the hormones insulin and glucagon, it is in the liver where they have their effects. There are three important processes associated with regulating blood sugar which take place in the liver.

- **Glycogenesis** is the conversion of glucose into glycogen. When blood glucose concentration is higher than normal the liver removes glucose from the blood and converts it to glycogen. It can store 75–100 g of glycogen, which is sufficient to maintain a human's blood glucose concentration for about 12 hours when at rest, in the absence of other sources.

- **Glycogenolysis** is the breakdown of glycogen to glucose. When blood glucose concentration is lower than normal, the liver can convert stored glycogen back into glucose which diffuses into the blood to restore the normal blood glucose concentration.

- **Gluconeogenesis** is the production of glucose from sources other than carbohydrate. When its supply of glycogen is exhausted, the liver can produce glucose from non-carbohydrate sources such as glycerol and amino acids.

Practical link

Required practical 11. Production of a dilution series of a glucose solution and use of colorimetric technique to produce a calibration curve with which to identify the concentration of glucose in an unknown 'urine' sample.

Study tip

Hormones only affect their target cells and not other cells because only target cells have the specific protein receptors that are complementary to the shape of that specific hormone.

Regulation of blood glucose concentration

Glucose is a substrate for respiration, providing the source of energy for almost all organisms. It is therefore essential that the blood of mammals contains a relatively constant concentration of glucose for respiration. If the concentration falls too low, cells will be deprived of energy and die – brain cells are especially sensitive in this respect because they can only respire glucose. If the concentration rises too high, it lowers the **water potential** of the blood and creates osmotic problems that can cause dehydration and be equally dangerous. Homeostatic control (Topic 16.1) of blood glucose is therefore essential.

Factors that influence blood glucose concentration

The normal concentration of blood glucose is $5\,\text{mmol}\,\text{dm}^{-3}$. Blood glucose comes from three sources:

- **directly from the diet** in the form of glucose absorbed following hydrolysis of other carbohydrates such as starch, maltose, lactose, and sucrose
- **from the hydrolysis in the small intestine of glycogen = glycogenolysis** stored in the liver and muscle cells
- **from gluconeogenesis**, which is the production of glucose from sources other than carbohydrate.

As animals do not eat continuously and their diet varies, their intake of glucose fluctuates. Likewise, glucose is used during respiration at different rates depending on the level of mental and physical activity. It is against these changes in supply and demand that the three main hormones, **insulin**, **glucagon** and **adrenaline**, operate to maintain a constant blood glucose concentration.

Insulin and the β cells of the pancreas

The β cells of the islets of Langerhans in the pancreas have receptors that detect the stimulus of a rise in blood glucose concentration and respond by secreting the hormone insulin directly into the blood plasma. Insulin is a globular protein made up of 51 amino acids.

Almost all body cells (red blood cells being a notable exception) have glycoprotein receptors on their cell-surface membranes that bind specifically with insulin molecules. When it combines with the receptors, insulin brings about:

- a change in the tertiary structure of the glucose transport carrier proteins, causing them to change shape and open, allowing more glucose into the cells by facilitated diffusion
- an increase in the number of the carrier proteins responsible for glucose transport in the cell-surface membrane. At low insulin concentrations, the protein from which these channels are made is part of the membrane of vesicles. A rise in insulin concentration results in these vesicles fusing with the cell-surface membrane so increasing the number of glucose transport channels
- activation of the enzymes that convert glucose to glycogen and fat.

As a result, the blood glucose concentration is lowered in one or more of the following ways:

- by increasing the rate of absorption of glucose into the cells, especially in muscle cells
- by increasing the respiratory rate of the cells, which therefore use up more glucose, thus increasing their uptake of glucose from the blood
- by increasing the rate of conversion of glucose into glycogen (glycogenesis) in the cells of the liver and muscles
- by increasing the rate of conversion of glucose to fat.

The effect of these processes is to remove glucose from the blood and so return its concentration to the optimum. This lowering of the blood glucose concentration causes the β cells to reduce their secretion of insulin (= negative feedback).

Glucagon and the α cells of the pancreas

The α cells of the islets of Langerhans detect a fall in blood glucose concentration and respond by secreting the hormone glucagon directly into the blood plasma. Glucagon's actions include:

- attaching to specific protein receptors on the cell-surface membrane of liver cells
- activating enzymes that convert glycogen to glucose
- activating enzymes involved in the conversion of amino acids and glycerol into glucose (= gluconeogenesis).

The overall effect is therefore to increase the concentration of glucose in the blood and return it to its optimum concentration. This raising of the blood glucose concentration causes the α cells to reduce the secretion of glucagon (= negative feedback).

Role of adrenaline in regulating the blood glucose level

There are at least four other hormones apart from glucagon that can increase blood glucose concentration. The best known of these is adrenaline. At times of excitement or stress, adrenaline is produced by the adrenal glands that lie above the kidneys. Adrenaline raises the blood glucose concentration by:

- attaching to protein receptors on the cell-surface membrane of target cells
- activating enzymes that causes the breakdown of glycogen to glucose in the liver.

Hormone interaction in regulating blood glucose

The two hormones insulin and glucagon act in opposite directions. Insulin lowers the blood glucose concentration, whereas glucagon increases it. The two hormones are said to act antagonistically. The system is self-regulating through negative feedback in that it is the concentration of glucose in the blood that determines the quantity of insulin and glucagon produced. In this way the interaction of

▲ **Figure 3** *Molecular graphic of an insulin molecule. Insulin is made up of 51 amino acids arranged in two chains (shown here as yellow and green ribbons)*

Hint

There is almost always a time lag between a hormone being produced and the response to it. This is because it takes time to produce it, transport it in the blood and for it to affect the enzyme or transport protein of the target cell.

these two hormones allows highly sensitive control of the blood glucose concentration. The concentration of glucose is not, however, constant, but fluctuates around an optimum point. This is because of the way negative feedback mechanisms work. Only when the blood glucose concentration falls below the set point is insulin secretion reduced (negative feedback), leading to a rise in blood glucose concentration. In the same way, only when the concentration exceeds the set point is glucagon secretion reduced (negative feedback), causing a fall in the blood glucose concentration.

The control of blood glucose concentration is summarised in Figure 4.

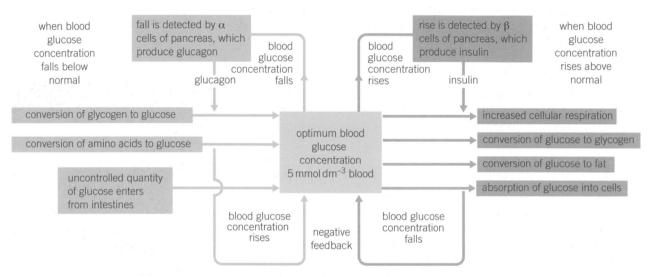

▲ **Figure 4** *Summary of regulation of blood glucose concentration*

Summary questions

In the following passage, state the most suitable word to replace the numbers in brackets.

The chemical energy in glucose is released by cells during the process known as (1). It is therefore important that the blood glucose concentration is maintained at a constant level because if it falls too low cells are deprived of energy, and (2) cells are especially sensitive in this respect. If it gets too high (3) problems occur that may cause dehydration. Blood glucose is formed directly from (4) in the diet or from the breakdown of (5), which is stored in the cells of the liver and (6). The liver can also increase the blood glucose concentration by making glucose from other sources, such as glycerol and (7), in a process known as (8). Blood glucose is used up when it is absorbed into cells, converted into fat or (9) for storage, or used up during (10) by cells. In order to maintain a constant concentration of blood glucose the pancreas produces two hormones from clusters of cells within it called (11). The β cells produce the hormone (12), which causes the blood glucose concentration to fall. The α cells produce the hormone (13), which has the opposite effect. Another hormone, called (14), can also raise blood glucose concentration.

Diabetes is a disease in which a person is unable to metabolise carbohydrate, especially glucose, properly. There are around 350 million people worldwide with diabetes, 3.2 million of whom are in the UK. In addition, a further 1 million people in the UK are thought to have the disease but are currently unaware of it. One form of diabetes is diabetes mellitus, or 'sugar diabetes'.

Types of sugar diabetes

Diabetes is a metabolic disorder caused by an inability to control blood glucose concentration due to a lack of the hormone insulin or a loss of responsiveness to insulin.

There are two forms of diabetes:

- **Type I (insulin dependent)** is due to the body being unable to produce insulin. It normally begins in childhood. It may be the result of an autoimmune response whereby the body's immune system attacks its own cells, in this case the β cells of the islets of Langerhans. Type I diabetes develops quickly, usually over a few weeks, and the signs and symptoms (see Hint) are normally obvious.

- **Type II (insulin independent)** is normally due to **glycoprotein** receptors on body cells being lost or losing their reponsiveness to insulin. However, it may also be due to an inadequate supply of insulin from the pancreas. Type II diabetes usually develops in people over the age of 40 years. There is, however, an increasing number of cases of obesity and poor diet leading to type II diabetes in adolescents. It develops slowly, and the symptoms are normally less severe and may go unnoticed. People who are overweight are particularly likely to develop type II diabetes. About 90% of people with diabetes have type II.

Figure 2 illustrates the differences in blood glucose concentration between people with and without diabetes who have swallowed a glucose solution.

Control of diabetes

Although diabetes cannot be cured, recent trials in transplanting insulin-producing cells have shown promise. Diabetes can also be successfully treated. Treatment varies depending on the type of diabetes.

- **Type I diabetes** is controlled by injections of insulin. This cannot be taken by mouth because, being a protein, it would be digested in the alimentary canal. It is therefore injected, typically either two or four times a day. The dose of insulin must be matched exactly to the glucose intake. If a person with diabetes takes too much insulin, he or she will experience a low blood glucose concentration that can result in unconsciousness. To ensure the correct dose, blood glucose concentration is monitored using **biosensors**. By injecting insulin and managing their carbohydrate intake and exercise carefully, people with diabetes can lead normal lives.

- **Type II diabetes** is usually controlled by regulating the intake of carbohydrate in the diet and matching this to the amount

Learning objectives

→ Describe the two main types of diabetes and how they differ.

→ Explain how each type of diabetes can be controlled.

Specification reference: 3.6.4.2

Hint

Signs of diabetes:

- high blood glucose concentration
- presence of glucose in urine
- need to urinate excessively
- genital itching or regular episodes of thrush
- weight loss
- blurred vision

Symptoms of diabetes:

- tiredness
- increased thirst and hunger

▲ **Figure 1** *A person with diabetes injecting insulin*

Hint

Blood glucose concentration can be controlled by changing the uptake of glucose from the gut (diet) and by changing the rate at which glucose is removed from the blood (exercise and insulin).

of exercise taken. In some cases, this may be supplemented by injections of insulin or by the use of drugs that stimulate insulin production. Other drugs can slow down the rate at which the body absorbs glucose from the intestine.

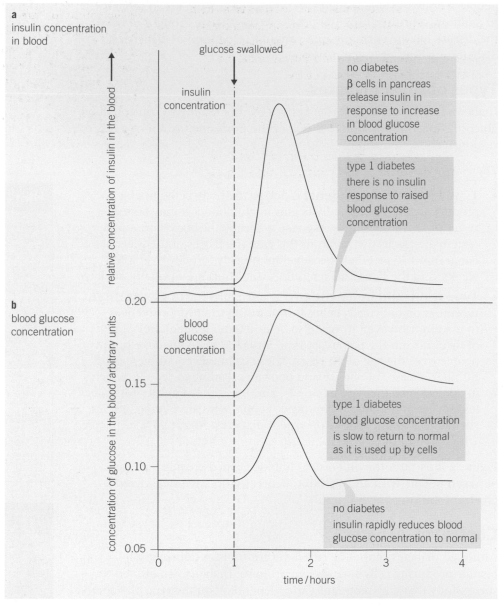

▲ **Figure 2** *Comparison of blood glucose and insulin concentrations in a person with type I diabetes and a person without diabetes after each has swallowed a glucose solution*

Summary questions

1 State **one** difference between the causes of type I and type II diabetes.

2 State **one** difference between the main ways of controlling type I and type II diabetes.

3 Suggest an explanation for why tiredness is a symptom of diabetes.

4 Suggest what lifestyle advice you might give someone in order to help them avoid developing type II diabetes.

Effects of diabetes on substance concentrations in the blood

An experiment was carried out with two groups of people. Group X had type 1 diabetes while group Y did not (control group). Every 15 minutes blood samples were taken from all members of both groups and the mean concentrations of insulin, glucagon and glucose were determined. After an hour, each person was given a glucose drink. The results are shown in the graphs below.

1 Name a hormone other than insulin and glucagon that is involved in regulating blood glucose concentration.

2 State *two* differences between groups X and Y in the way insulin secretion responds to the drinking of glucose.

3 Suggest a reason why the glucose concentration falls in both groups during the first hour.

4 Using information from the graphs, explain the changes in the blood glucose concentration in group Y after drinking the glucose.

5 Explain the difference in blood glucose concentration of group X compared with group Y.

6 Suggest what might happen to the blood glucose concentration of group X if they have no food over the next 24 hours.

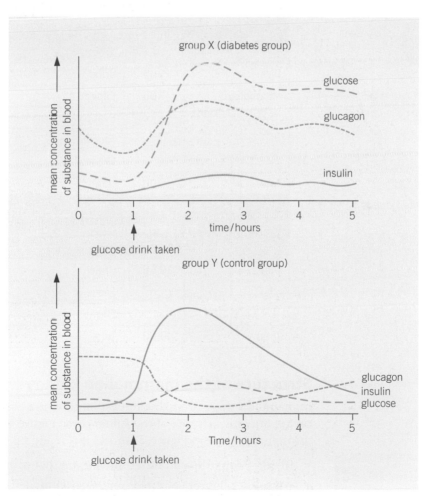

▲ Figure 3

16.5 Control of blood water potential – structure of the nephron

Learning objectives

→ Describe the structure of the mammalian kidney.

→ Describe the structure of the nephron.

Specification reference: 3.6.4.3

The amount of water and mineral ions we take in varies from day to day, as does the quantity we lose. Table 1 shows the daily balance between loss and gain of salts and water for a typical human. In the blood, however, an optimum concentration of water and salts is maintained to ensure a fairly constant water potential of blood plasma and tissue fluid. The **homeostatic** control of the water potential of the blood is called **osmoregulation**. To understand osmoregulation, we must first understand the structure of the organ that carries it out, the kidney, and in particular its functional unit – the **nephron**.

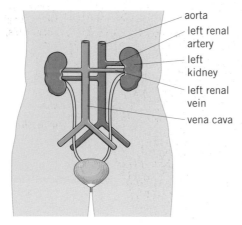

▲ **Figure 1** *Position of the kidneys in humans*

▼ **Table 1** *Daily water and sodium chloride balance in a typical human*

Water			
Volume of water / cm^3 day^{-1}			
Water gain		Water loss	
Diet	2300	Urine	1500
Metabolism (e.g., respiration)	200	Expired air	400
		Evaporation from skin	350
		Faeces	150
		Sweat	100
Total	2500	Total	2500
Sodium chloride			
Mass of sodium chloride / g day^{-1}			
Salt gain		Salt loss	
Diet	10.50	Urine	10.00
		Faeces	0.25
		Sweat	0.25
Total	10.50	Total	10.50

Structure of the mammalian kidney

In mammals there are two kidneys found at the back of the abdominal cavity, one on each side of the spinal cord (Figure 1). A section through the kidney (Figure 2) shows it is made up of the:

- **fibrous capsule** – an outer membrane that protects the kidney
- **cortex** – a lighter coloured outer region made up of **renal (Bowman's) capsules**, convoluted tubules and blood vessels
- **medulla** – a darker coloured inner region made up of **loops of Henle**, collecting ducts and blood vessels

Study tip

The large intestine also plays a part in maintaining the water content of the body and blood through its role in reabsorbing water from the contents of the large intestine.

- **renal pelvis** – a funnel-shaped cavity that collects urine into the ureter
- **ureter** – a tube that carries urine to the bladder
- **renal artery** – supplies the kidney with blood from the heart via the aorta
- **renal vein** – returns blood to the heart via the vena cava.

A microscopic examination of the cortex and medulla reveals around one million tiny tubular structures in each kidney. These are the basic structural and functional units of the kidney – the **nephrons**.

Study tip

The structure of the kidney is included to show how it relates to nephrons. Kidney structure does not need to be learned for examinations.

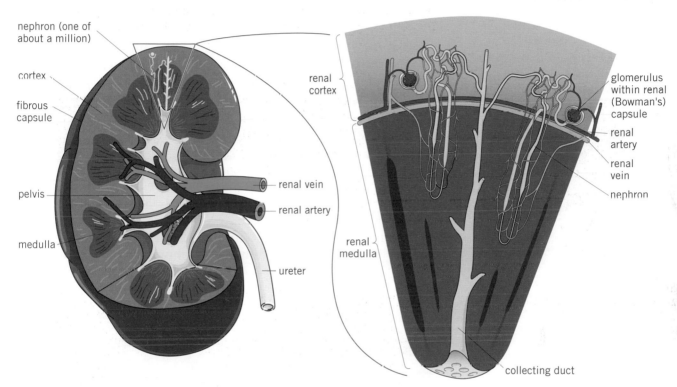

▲ **Figure 2** *Detailed structure of mammalian kidney (LS) showing the position of two of the million or more nephrons in each kidney*

The structure of the nephron

The nephron is the functional unit of the kidney. It is a narrow tube up to 14 mm long, closed at one end, with two twisted regions separated by a long hairpin loop. Each nephron is made up of a:

- **Renal (Bowman's) capsule** – the closed end at the start of the nephron. It is cup-shaped and surrounds a mass of blood capillaries known as the glomerulus. The inner layer of the renal capsule is made up of specialised cells called **podocytes**.
- **Proximal convoluted tubule** – a series of loops surrounded by blood capillaries. Its walls are made of epithelial cells which have microvilli.

▲ **Figure 3** *LM of cortex of human kidney (TS). Three glomeruli are seen as regions of small cells surrounded by a clear space – the lumen of the renal capsule. The background shows convoluted tubules*

▲ **Figure 4** *LM of medulla of human kidney showing loops of Henle (white tubes). Around them are blood capillaries containing red blood cells (red)*

- **Loop of Henle** – a long, hairpin loop that extends from the cortex into the medulla of the kidney and back again. It is surrounded by blood capillaries.

- **Distal convoluted tubule** – a series of loops surrounded by blood capillaries. Its walls are made of epithelial cells, but it is surrounded by fewer capillaries than the proximal tubule.

- **Collecting duct** – a tube into which a number of distal convoluted tubules from a number of nephrons empty. It is lined by epithelial cells and becomes increasingly wide as it empties into the pelvis of the kidney.

Associated with each nephron are a number of blood vessels (Figure 6), namely:

- **afferent arteriole** – a tiny vessel that ultimately arises from the renal artery and supplies the nephron with blood. The afferent arteriole enters the renal capsule of the nephron where it forms the

- **glomerulus** – a many-branched knot of capillaries from which fluid is forced out of the blood. The glomerular capillaries recombine to form the

- **efferent arteriole** – a tiny vessel that leaves the renal capsule. It has a smaller diameter than the afferent arteriole and so causes an increase in blood pressure within the glomerulus. The efferent arteriole carries blood away from the renal capsule and later branches to form the

- **blood capillaries** – a concentrated network of capillaries that surrounds the proximal convoluted tubule, the loop of Henle and the distal convoluted tubule and from where they reabsorb mineral salts, glucose and water. These capillaries merge together into venules (tiny veins) that in turn merge together to form the renal vein.

▲ **Figure 5** *Colourised SEM of a glomerulus (centre) surrounded by the renal capsule, seen as a white-brown membrane at centre right. Part of the proximal convoluted tubule is seen, coloured blue*

▲ **Figure 6** *Regions of the nephron and associated blood vessels*

Summary questions

Complete the passage below by stating the word or words that best replace the numbers in brackets.

The nephron is the structural unit of the kidney. It comprises a cup-shaped structure called the (1) that contains a knot of blood vessels called the (2) which receives its blood from a vessel called the (3) arteriole. The inner wall of this cup-shaped structure is lined with specialised cells called (4) and from it extends the first, or (5), convoluted tubule whose walls are lined with (6) that have (7) to increase their surface area. The next region of the nephron is a hairpin loop called the (8) which then leads onto the second, or (9), convoluted tubule. This in turn leads onto the (10) which empties into the renal pelvis. Around much of the nephron is a dense network of blood vessels called the (11) capillaries.

As a result, the hydrostatic pressure of the blood in the glomerulus is sufficient to overcome the resistance and so filtrate passes from the blood into the renal capsule. The filtrate, which contains urea, does not contain cells or plasma proteins which are too large to pass across the connective tissue. Many of the substances in the 125 cm³ of filtrate passing out of blood each minute are extremely useful to the body and are reabsorbed.

Reabsorption of glucose and water by the proximal convoluted tubule

In the proximal convoluted tubule nearly 85% of the filtrate is reabsorbed back into the blood. Ultrafiltration operates on the basis of size of molecule – small ones are removed. Some, such as urea, are wastes, but most are useful and so are reabsorbed.

The proximal convoluted tubules are adapted to reabsorb substances into the blood by having epithelial cells that have:

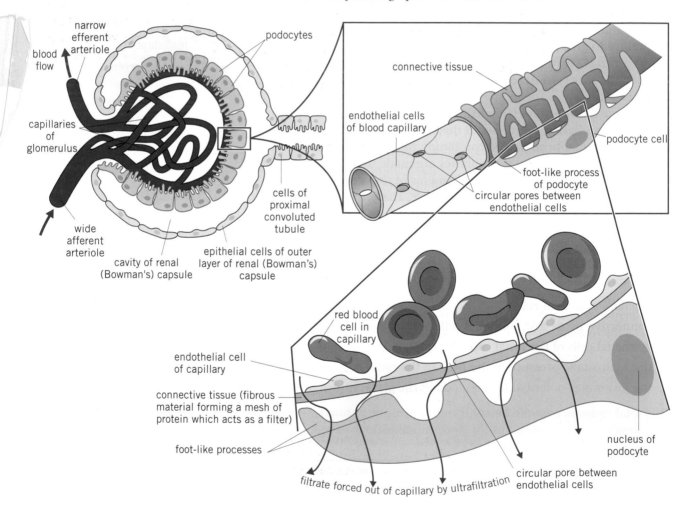

▲ **Figure 1** *Podocyte and ultrafiltration*

- microvilli to provide a large surface area to reabsorb substances from the filtrate

- infoldings at their bases to give a large surface area to transfer reabsorbed substances into blood capillaries

- a high density of mitochondria to provide ATP for active transport.

The process is as follows:

- Sodium ions are actively transported out of the cells lining the proximal convoluted tubule into blood capillaries which carry them away. The sodium ion concentration of these cells is therefore lowered.

- Sodium ions now diffuse down a concentration gradient from the lumen of the proximal convoluted tubule into the epithelial lining cells but only through special carrier proteins by facilitated diffusion.

- These carrier proteins are of specific types, each of which carries another molecule (glucose or amino acids or chloride ions, etc.) along with the sodium ions. This is known as co-transport.

- The molecules which have been co-transported into the cells of the proximal convoluted tubule then diffuse into the blood. As a result, all the glucose and most other valuable molecules are reabsorbed as well as water.

▲ **Figure 2** *Colourised SEM of podocyte cells around a glomerulus in a human kidney*

Synoptic link

Your understanding of reabsorption will be improved if you first revise Topics 4.1, 4.2, 4.4, and 4.5.

lumen of proximal convoluted tubule
reabsorption of material by facilitated diffusion, active transport and pinocytosis

- microvillus
- pinocytic vesicle
- nuclear envelope
- nucleoplasm
- intercellular space
- mitochondrion
- infolding of basal membrane
- basement membrane
- endothelial cell

lumen of blood capillary

▲ **Figure 3** *Details of cells from the wall of the proximal convoluted tubule*

About $180\,dm^3$ of water enters the nephrons each day. Of this volume, only about $1\,dm^3$ leaves the body as urine. 85% of the reabsorption of water occurs in the proximal convoluted tubule. The remainder is reabsorbed from the collecting duct as a result of the functioning of the loop of Henle.

Maintenance of a gradient of sodium ions by the loop of Henle

The loop of Henle is a hairpin-shaped tubule that extends into the medulla of the kidney. It is responsible for water being reabsorbed from the collecting duct, thereby concentrating the urine so that it has a lower **water potential** than the blood. The concentration of the urine produced is directly related to the length of the loop of Henle.

The loop of Henle has two regions:

- The descending limb, which is narrow, with thin walls that are highly permeable to water.
- The ascending limb, which is wider, with thick walls that are impermeable to water.

The loop of Henle acts as a counter-current multiplier. To understand how this works it is necessary to consider the following sequence of events in conjunction with Figure 4, to which the numbers refer.

1 Sodium ions are actively transported out of the ascending limb of the loop of Henle using ATP provided by the many mitochondria in the cells of its wall.

2 This creates a low water potential (high ion concentration) in the region of the medulla between the two limbs (called the interstitial region). In normal circumstances water would pass out of the ascending limb by osmosis. However, the thick walls are almost impermeable to water and so very little, if any, escapes.

3 The walls of the descending limb are, however, very permeable to water and so it passes out of the filtrate, by osmosis, into the

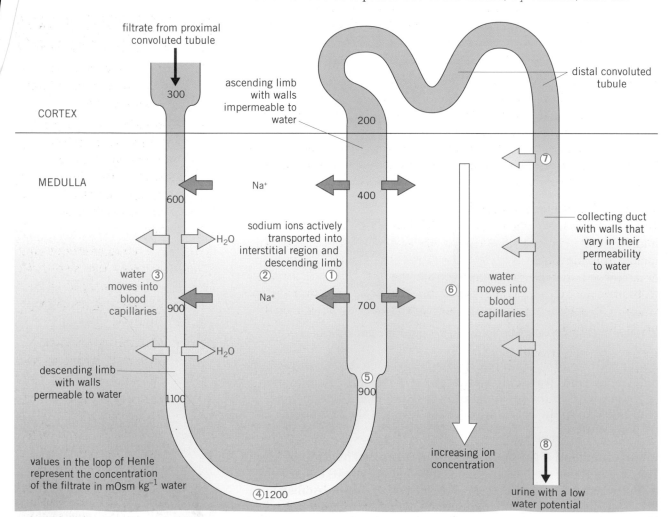

▲ **Figure 4** *Counter-current multiplier of the loop of Henle*

interstitial space. This water enters the blood capillaries in this region by osmosis and is carried away.

4 The filtrate progressively loses water in this way as it moves down the descending limb lowering its water potential. It reaches its lowest water potential at the tip of the hairpin.

5 At the base of the ascending limb, sodium ions diffuse out of the filtrate and as it moves up the ascending limb these ions are also actively pumped out (see point 1) and therefore the filtrate develops a progressively higher water potential.

6 In the interstitial space between the ascending limb and the collecting duct there is a gradient of water potential with the highest water potential (lowest concentration of ions) in the cortex and an increasingly lower water potential (higher concentration of ions) the further into the medulla one goes.

7 The collecting duct is permeable to water and so, as the filtrate moves down it, water passes out of it by osmosis. This water passes by osmosis into the blood vessels that occupy this space, and is carried away.

8 As water passes out of the filtrate its water potential is lowered. However, the water potential is also lowered in the interstitial space and so water continues to move out by osmosis down the whole length of the collecting duct. The counter-current multiplier ensures that there is always a water potential gradient drawing water out of the tubule.

The water that passes out of the collecting duct by osmosis does so through channel proteins that are specific to water (aquaporins). Antidiuretic hormone (ADH) (Topic 16.7) can alter the number of these channels and so control water loss. By the time the filtrate, now called urine, leaves the collecting duct on its way to the bladder, it has lost most of its water and so it has a lower water potential (is more concentrated) than the blood.

The distal convoluted tubule

The cells that make up the walls of the distal convoluted tubule have microvilli and many mitochondria that allow them to reabsorb material rapidly from the filtrate, by **active transport**. The main role of the distal tubule is to make final adjustments to the water and salts that are reabsorbed and to control the pH of the blood by selecting which ions to reabsorb. To achieve this, the permeability of its walls becomes altered under the influence of various hormones (Topic 16.7).

Counter-current multiplier

You may remember that when two liquids flow in opposite directions past one another, the exchange of substances (or heat) between them is greater than if they flowed in the same direction next to each other. In the case of the loop of Henle, the counter-current flow means that the filtrate in the collecting duct with a lower water potential meets interstitial fluid that has an even lower water potential. This means that, although the water potential gradient between the collecting duct and interstitial fluid is small, it exists for the whole length of the collecting duct. There is therefore a steady flow of water into the interstitial fluid, so that around 80% of the water enters the interstitial fluid and hence the blood. If the two flows were in the same direction (parallel) less of the water would enter the blood.

Synoptic link

To help you understand how the loop of Henle concentrates urine, you should look back at the counter-current exchange principle in Topic 6.3 Gas exchange in fish.

Summary questions

1 Name the structure in the nephron where the majority of water is reabsorbed.

2 The following is a list of the various parts of a nephron: distal convoluted tubule, glomerulus, loop of Henle, collecting duct, distal convoluted tubule; renal capsule.

List the sequence of structures that a molecule of water which is excreted from the body passes through on its journey to the bladder.

3 Describe how the proximal convoluted tubule is adapted to its function.

4 The length of the loop of Henle in animals living in dry environments is different from the length in those living in environments where water is abundant. Suggest if the length is longer or shorter and explain how it helps animals in dry areas to survive.

16.7 The role of hormones in osmoregulation

16.7 The role of hormones in osmoregulation

Learning objectives

→ Explain how the water potential of the blood is regulated.

→ Describe the roles of the hypothalamus, posterior pituitary and antidiuretic hormone (ADH) in osmoregulation.

Specification reference: 3.6.4.3

Hint

The name antidiuretic hormone (ADH) may, at first, seem unusual. However, it describes its function precisely. **Diuresis** is the production of large volumes of dilute urine. As the effect of ADH is to increase the permeability of collecting ducts so that more water is reabsorbed into the blood, it causes the production of small volumes of concentrated urine.

This is the opposite of diuresis – hence the name **anti**diuretic hormone.

Hint

The pituitary gland has two parts – the anterior and the posterior part.

Synoptic link

The regulation of water potential of the blood is another example of the stimulus → receptor → coordinator → effector → response pathway that we have seen many times before.

The **homeostatic** control of osmoregulation in the blood is achieved by a hormone that acts on the distal convoluted tubule and the collecting duct.

Regulation of the water potential of the blood

The **water potential** of the blood depends on the concentration of solutes like glucose, proteins, sodium chloride, and other mineral ions as well as the volume of water in the body. A rise in solute concentration lowers its water potential. This may be caused by:

- too little water being consumed
- much sweating occurring
- large amounts of ions, for example, sodium chloride, being taken in.

The body responds to this fall in water potential as follows:

- Cells called **osmoreceptors** in the **hypothalamus** of the brain detect the fall in water potential.
- It is thought that, when the water potential of the blood is low, water is lost from these osmoreceptor cells by osmosis.
- Due to this water loss the osmoreceptor cells shrink, a change that causes the hypothalamus to produce a hormone called **antidiuretic hormone (ADH)**.
- ADH passes to the posterior **pituitary gland**, from where it is secreted into the capillaries.
- ADH passes in the blood to the kidney, where it increases the permeability to water of the cell-surface membrane of the cells that make up the walls of the distal convoluted tubule and the collecting duct.
- Specific protein receptors on the cell-surface membrane of these cells bind to ADH molecules, leading to activation of an enzyme called phosphorylase within the cell.
- The activation of phosphorylase causes vesicles within the cell to move to, and fuse with, its cell-surface membrane.
- These vesicles contain pieces of plasma membrane that have numerous water channel proteins (aquaporins) and so when they fuse with the membrane the number of water channels is considerably increased, making the cell-surface membrane much more permeable to water.
- ADH increases the permeability of the collecting duct to urea, which therefore passes out, further lowering the water potential of the fluid around the duct.
- The combined effect is that more water leaves the collecting duct by osmosis, down a water potential gradient, and re-enters the blood.
- As the reabsorbed water came from the blood in the first place, this will not, in itself, increase the water potential of the blood, but merely prevent it getting lower. The osmoreceptors also send nerve impulses to the thirst centre of the brain, to encourage the individual to seek out and drink more water.

- The osmoreceptors in the hypothalamus detect the rise in water potential and send fewer impulses to the pituitary gland.
- The pituitary gland reduces the release of ADH and the permeability of the collecting ducts to water and urea reverts to its former state. This is an example of homeostasis and the principle of negative feedback (Topic 16.1).

A fall in the solute concentration of the blood raises its water potential. This may be caused by:

- large volumes of water being consumed
- salts used in metabolism or excreted not being replaced in the diet.

The body responds to this rise in water potential as follows:

- The osmoreceptors in the hypothalamus detect the rise in water potential and increase the frequency of nerve impulses to the pituitary gland to reduce its release of ADH.
- Less ADH, via the blood, leads to a decrease in the permeability of the collecting ducts to water and urea.
- Less water is reabsorbed into the blood from the collecting duct.
- More dilute urine is produced and the water potential of the blood falls.
- When the water potential of the blood has returned to normal, the osmoreceptors in the hypothalamus cause the pituitary to raise its ADH release back to normal levels (= negative feedback).

These events are summarised in Figure 1.

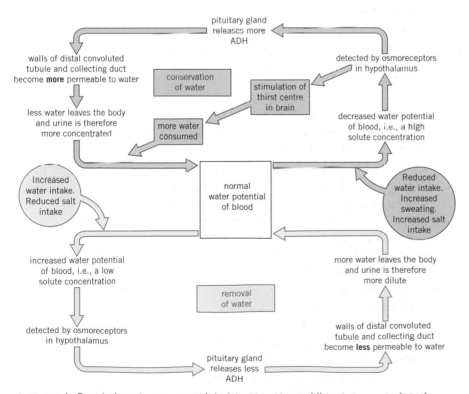

▲ **Figure 1** *Regulation of water potential of the blood by antidiuretic hormone (ADH)*

Summary questions $\sqrt{x}$

1   State where the cells which monitor the water potential of the blood are located.

2   In each of the following situations deduce whether more or less ADH would be produced by the body:

   a   drinking a large volume of water in a short time.

   b   exercising intensely for 30 minutes.

3   Explain how ADH causes the collecting ducts to reabsorb more water.

4   $\sqrt{x}$ The concentration of proteins in a sample of glomerular filtrate is 0.0625% of their concentration in the blood plasma. If the concentration of proteins in the blood plasma is 80 g dm^{-3}, calculate their concentration in the glomerular filtrate.

+ The significance of glucose in the urine $\sqrt{x}$

The presence of glucose in urine may indicate a clinical disorder. The kidney should reabsorb all glucose from the filtrate leaving the urine free of it. The presence of glucose in urine (glucosuria) suggests:

*   There may be so much glucose in the filtrate that the kidney is overwhelmed and cannot reabsorb it all.

This is often an indication of a high blood glucose concentration due to diabetes mellitus.

*   More rarely, it may be that the kidney is not functioning properly and not reabsorbing glucose. This could be due to a disease of the kidneys.

1   Name three hormones involved in controlling the level of glucose in the blood and state the effect of each on the level of glucose.

   Figure 5 shows the relative concentrations of three substances in the filtrate as it passes along the nephron. (NB the scale is not linear.)

2   Urea is a waste product of the body that is removed by the kidneys. The quantity of urea in the filtrate entering the nephron does not change significantly as the filtrate passes along the nephron. Explain why the level of urea shown in Figure 5 rises considerably as the filtrate passes along the nephron.

3   Suggest the name of substance A and explain the reasons for your answer.

4   Suggest the name of substance B and explain the reasons for your answer.

5   One symptom of diabetes is dehydration. From your knowledge of how water is reabsorbed in the collecting ducts, explain why diabetes might cause dehydration.

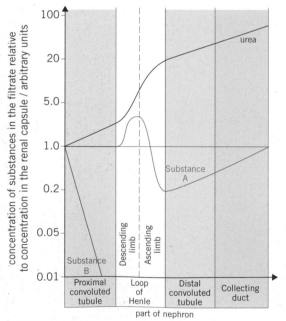

▲ **Figure 5** *Relative concentrations of three substances in the filtrate as it passes along a nephron. NB Scale is not linear*

1   The release of a substance called dopamine in some
    areas of the brain increases the desire to eat. Scientists
    measured increases in the release of dopamine in the
    brains of rats given different concentrations of sucrose
    solution to drink.
    Sucrose stimulates taste receptors on the tongue.
    The graph shows their results. Each point is the result
    for one rat.

    (a) The scientists concluded that drinking a sucrose
         solution had a positive feedback effect on the
         rats' desire to eat.
         How do these data support this conclusion?
                                                    (3 marks)
    (b) In this investigation, the higher the
         concentration of sucrose in a rat's mouth, the
         higher the frequency of nerve impulses from
         each taste receptor to the brain. If rats are given very high concentrations of
         sucrose solution to drink, the refractory period makes it impossible for information
         about the differences in concentration to reach the brain. Explain why.   (2 marks)
    (c) In humans, when the stomach starts to become full of food, receptors in the wall
         of the stomach are stimulated. This leads to negative feedback on the desire to eat.
         Suggest why this negative feedback is important.   (3 marks)
                                                              AQA June 2013

2   Different substances are involved in coordinating responses in animals.
    (a) Hormones are different from local chemical mediators such as histamine in the
         cells they affect.
         (i)   Describe how hormones are different in the cells they affect.   (1 mark)
         (ii) Describe how hormones and local chemical mediators reach the
               cells they affect.   (2 marks)
    (b) Synapses are unidirectional. Explain how acetylcholine contributes to a
         synapse being unidirectional.   (2 marks)
    (c) Cells in the stomach wall release gastric juice after a meal. The graph shows how
         the volumes of gastric juice produced by nervous stimulation and by hormonal
         stimulation change after a meal.

(i) Describe the evidence from the graph that curve **A** represents the volume of gastric juice produced by nervous stimulation. *(2 marks)*

(ii) Copy and complete the table to show the percentage of gastric juice produced by nervous stimulation at the times shown.

	Time after meal / hours		
	1	2	3
Percentage of gastric juice produced by nervous stimulation			

(1 mark)
AQA June 2011

3 **(a)** Adrenaline binds to receptors in the plasma membranes of liver cells. Explain how this causes the blood glucose concentration to increase. *(2 marks)*

   (b) Scientists made an artificial gene which codes for insulin. They put the gene into a virus which was then injected into rats with type I diabetes. The virus was harmless to the rats but carried the gene into the cells of the rats. The treated rats produced insulin for up to 8 months and showed no side-effects. The scientists measured the blood glucose concentrations of the rats at regular intervals. While the rats were producing the insulin, their blood glucose concentrations were normal.

   (i) The rats were not fed for at least 6 hours before their blood glucose concentration was measured. Explain why. *(1 mark)*

   (ii) The rats used in the investigation had type I diabetes. This form of gene therapy may be less effective in treating rats that have type II diabetes. Explain why. *(1 mark)*

   (iii) Research workers have suggested that treating diabetes in humans by this method of gene therapy would be better than injecting insulin. Evaluate this suggestion. *(4 marks)*
AQA June 2012

4 **(a)** Technicians in a hospital laboratory tested urine and blood samples from a girl with diabetes at intervals over a one-year period. Each time the technicians tested her urine, they also measured her blood glucose concentration. Their results are shown in the graph.

(i) The girl who took part in this investigation was being successfully treated with insulin. The graph shows that on some occasions, the concentration of glucose in her blood was very high. Suggest why. *(2 marks)*

(ii) Use the graph to evaluate the use of the urine test as a measure of blood glucose concentration. *(3 marks)*

(b) Diabetic people who do not control their blood glucose concentration may become unconscious and go into a coma. A doctor may inject a diabetic person who is in a coma with glucagon. Explain how the glucagon would affect the person's blood glucose concentration. *(2 marks)*

AQA June 2010

5 Osmoreceptors are specialised cells that respond to changes in the water potential of the blood.

(a) Give the location of osmoreceptors in the body of a mammal. *(1 mark)*

(b) When a person is dehydrated, the cell volume of an osmoreceptor decreases. Explain why. *(2 marks)*

(c) Stimulation of osmoreceptors can lead to secretion of the hormone ADH. Describe and explain how the secretion of ADH affects urine produced by the kidneys. *(4 marks)*

The efficiency with which the kidneys filter the blood can be measured by the rate at which they remove a substance called creatinine from the blood. The rate at which they filter the blood is called the glomerular filtration rate (GFR).

In 24 hours, a person excreted 1660 mg of creatinine in his urine. The concentration of creatinine in the blood entering his kidneys was constant at 0.01 mg cm^{-3}.

(d) Calculate the GFR in cm^3 minute^{-1} *(1 mark)*

(e) Creatinine is a breakdown product of creatine found in muscle tissues. Apart from age and gender, give **two** factors that could affect the concentration of creatinine in the blood. *(1 mark)*

AQA Specimen 2014

6 In an investigation of blood glucose levels, colorimetry was used to find the absorbance of blue light by plasma samples that had been reacted with Benedict's Reagent. The % absorbance was converted to glucose concentration using a standard calibration curve, drawn using the following data.

Glucose concentration (mMol dm^{-3})	Absorbance of blue light (%)
0.001	10
0.01	22
0.1	40
1	58
10	80

(a) Produce recipes for each of the five glucose concentrations, assuming a supply of 10mMol dm^{-3} glucose and distilled water. *(3 marks)*

(b) Plot a calibration curve of % absorbance against log$_{10}$ glucose concentration. *(3 marks)*

(c) Why is a log scale appropriate in this case? *(1 mark)*

(d) Use your graph to estimate the concentration of glucose that would correspond with a % absorbance of 45%? *(2 marks)*

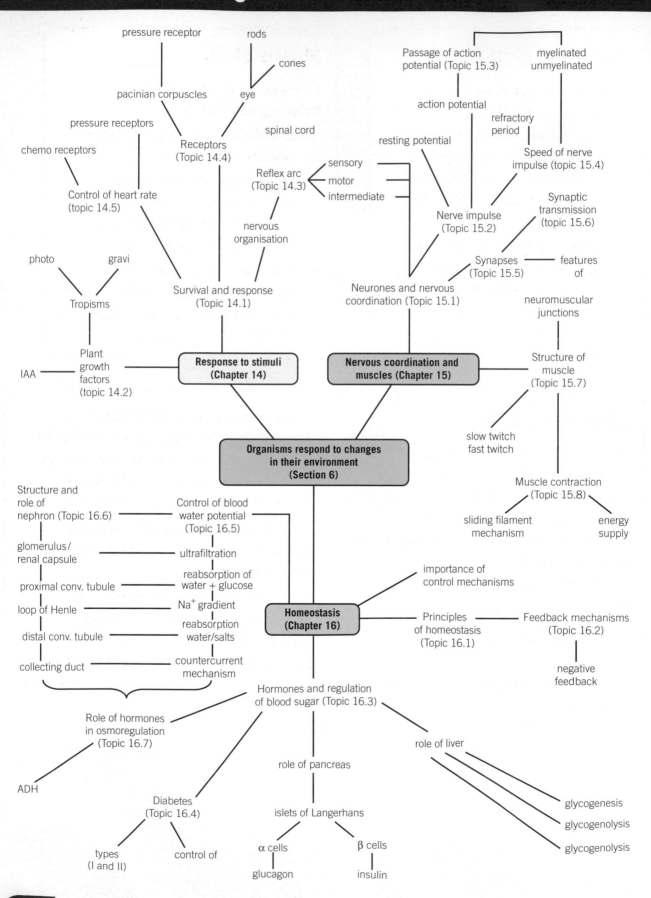

Practical skills

In this section you have met the following practical skills:

- How to carry out experiments to determine how plant growth factors such as auxins like IAA have their effects on cell growth and elongation.

- How to carry out an experiment to investigate the effects on blood sugar levels of consuming a glucose drink by diabetics and non-diabetics.

Maths skills

In this section you have met the following maths skills:

- Calculating percentage change in transmission speeds.

- Solving algebraic equations to determine the number of ATP molecules needed to contract a muscle fibre a specified distance.

- Translating information between graphical and numerical forms in calculating the number of action potentials in a given time.

Extension task

Using only the technique of a person catching a 30 cm ruler between his/her thumb and forefinger when the ruler is dropped, design and carry out a series of experiments to determine the distance travelled by the ruler before it is caught using:

1. the stimulus of sight only,

2. the stimulus of sound only,

3. the stimulus of touch only.

Using textbooks or the internet, research how to convert the distance travelled by the ruler before it is caught to the time taken for it to fall that distance. Calculate the average reaction time for each type of stimulus.

Suggest reasons for any differences you found between the reaction times for the three different stimuli.

You could also devise and carry out an experiment using the same technique to compare the reaction times between a person's dominant and non-dominant hand.

1 Scientists investigated the response of the roots of pea seedlings to gravity.
They took three samples of seedlings, **A**, **B**, and **C**, and placed them so that their roots were growing horizontally. The root tips of each sample had been given different treatments. After a set time, the scientists recorded whether the roots of the seedlings had grown upwards or downwards and the amount of curvature. Table 1 shows the treatment they gave to each sample and their results.

▼ **Table 1**

Treatment	Results	
	Direction of growth	Mean amount of curvature / degrees
A None	Downwards	60
B Root tip removed	Continues to grow horizontally	0
C Upper half of root tip removed	Downwards	30

(a) The pea seedlings were kept in the dark after each treatment. Explain why this was necessary. *(1 mark)*

(b) What conclusion can be made from the results for treatment **B**? *(1 mark)*

(c) Suggest how indoleacetic acid (IAA) could have caused the results for
    (i)   treatment **A** *(2 marks)*
    (ii) treatment **C** *(2 marks)*
AQA June 2012

2 Figure 2 shows a diagram of part of a muscle myofibril.

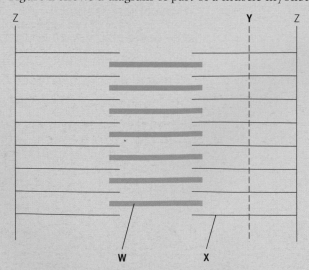

▲ **Figure 2**

(a) Name the protein present in the filaments labelled **W** and **X**. *(1 mark)*

(b) Figure 3 shows the cut ends of the protein filaments when the myofibril was cut at position Y. Figure 4 shows the protein filaments when the myofibril was cut at the same distance from a Z line at a different stage of contraction.

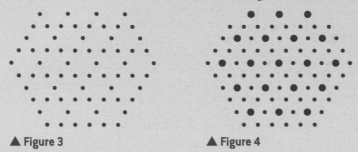

▲ **Figure 3**   ▲ **Figure 4**

Explain why the pattern of protein filaments differs in Figure 3 and 4. *(2 marks)*

(c) Describe the role of calcium ions in the concentration of a sarcomere. *(4 marks)*

AQA Jan 2004

3 (a) Describe how insulin reduces the concentration of glucose in the blood. *(3 marks)*

Some people produce no insulin. As a result they have a condition called diabetes. In an investigation, a man with diabetes drank a glucose solution. The concentration of glucose in his blood was measured at regular intervals. The results are shown in Figure 5.

▲ **Figure 5**

(b) Suggest **two** reasons why the concentration of glucose decreased after 1 hour even though this man's blood contained no insulin. *(2 marks)*

(c) The investigation was repeated on a man who did not have diabetes. The concentration of glucose in his blood before drinking the glucose solution was $80 \, mg$ per $100 \, cm^3$. Copy the graph roughly and sketch a curve on it to show the results you would expect. *(1 mark)*

(d) The diabetic man adopted a daily routine to stabilise his blood glucose concentration within narrow limits. He ate three meals a day: breakfast, a midday meal, and an evening meal. He injected insulin once before breakfast and once before the evening meal.
The injection he used before breakfast was a mixture of two types of insulin.
The mixture contained slow-acting insulin and fast-acting insulin.

(i) Explain the advantage of injecting both types of insulin before breakfast.
(2 marks)

(ii) One day, the man did not eat a midday meal. Suggest one reason why his blood glucose concentration did not fall dangerously low even though he had injected himself with the mixture of insulin before breakfast. *(1 mark)*

AQA Jan 2004

4 **(a)** Explain how a resting potential is maintained in a neurone. *(4 marks)*

(b) In an investigation, an impulse was generated in a neurone using electrodes. During transmission along the neurone, an action potential was recorded at one point on the neurone. When the impulse reached the neuromuscular junction, it stimulated a muscle cell to contract. The force generated by the contraction was measured. The results are shown in the graph.
The distance between the point on the neurone where the action potential was measured and the neuromuscular junction was exactly 18 mm.

(i) Use the graph to estimate the time between the maximum depolarisation and the start of contraction by the muscle cell. *(1 mark)*

(ii) Use the answer to part (i) to calculate the speed of transmission along this neurone to the muscle cell. Give your answer in mm per second $(mm\,s^{-1})$. Show your working. *(2 marks)*

(iii) Give one reason why the value calculated in part (ii) would be an underestimate of the speed of transmission of an impulse along a neurone. *(1 mark)*

Acetylcholine is the neurotransmitter at neuromuscular junctions.

(c) Describe how the release of acetylcholine into a neuromuscular junction causes the cell membrane of a muscle fibre to depolarise. (*3 marks*)

(d) Use your knowledge of the processes occurring at a neuromuscular junction to explain each of the following.

   (i)   The cobra is a very poisonous snake. The molecular structure of the cobra toxin is similar to the molecular structure of acetylcholine. The toxin permanently prevents muscle contraction. (*2 marks*)

   (ii) The insecticide DFP combines with the active site of the enzyme acetylcholinesterase. The muscles stay contracted until the insecticide is lost from the neuromuscular junction. (*2 marks*)

AQA Jan 2004

Section 7

Genetics, populations, evolution, and ecosystems

Introduction

The individuals of a species share the same genes but usually have different combinations of alleles of these genes. An individual inherits alleles from their parent or parents. While this process is universal, the way in which the alleles interact to produce the characteristics of the new individual depends on the type of inheritance involved. Sometimes one allele is dominant to another and so expresses itself, at other times the two alleles are equally dominant and the offspring have intermediate features. A characteristic is sometimes inherited along with the sex of an individual.

Populations of different species live in communities. Competition occurs within and between these populations for the means of survival. Populations within communities are also affected by, and in turn affect, the abiotic factors in an ecosystem. A species exists as one or more populations. The phenotypes of organisms in a population vary due to both genetic and environmental factors. Two forces affect genetic variety within a population – genetic drift and natural selection. Genetic drift can cause changes in allele frequency in small populations. Natural selection occurs when alleles that enhance the survival chances of the individuals that carry them rise in frequency. This change in the allele frequency of a population is known as evolution.

Different populations of the same species can sometimes become isolated from one another. This can be because they are geographically separated and therefore cannot interbreed. When this happens there is no flow of genes between the isolated populations. This may lead to the accumulation of genetic differences between each of these populations. These differences may ultimately lead to organisms in one population becoming unable to breed and produce fertile offspring with organisms from the other populations. This reproductive isolation means that a new species has evolved.

The theory of evolution is fundamental to biology. It states that all new species arise from existing ones by the process of natural selection. This means species, however different, have a common ancestry. This ancestry is represented in the phylogenetic classification of species. Common ancestry explains the similarities between all living organisms. These similarities include common chemistry such as all proteins having the same 20 or so amino acids, the same physiological pathways, e.g. anaerobic respiration, similar cell structure, DNA, and genetic material as well as a 'universal' genetic code.

Directional

es originally started off with sho

utation resulted in the birth c

ecks. These giraffes were be

trees and get food compa

neck. Those with longe

and breed. Over tim

longer necks incr

with shorter n

difference to modern day ferns.

showed that fossilised fern showed very little

were compared to modern day ferns. Studies

fern were studied carefully and they

ller

survive

s with

tion of giraffes

Working scientifically

In studying this unit there will be opportunities to perform practical exercises and so develop practical skills. A required practical activity is to carry out an investigation into the effect of a named environmental factor on the distribution of a given species. In performing this activity you will have the chance to develop practical skills such as: safely and ethically use organisms in investigations, using microbiological aseptic techniques such as using agar broth, using sampling techniques in fieldwork, using ICT such as computer modelling.

You will be able to develop a range of mathematical skills. In particular the ability to use ratios, fractions, logarithms, and percentages, find arithmetical means, understand simple probability, understand the principles of sampling when applied to scientific data, select and use a statistical test and solve algebraic problems,

What you already know

The material in this unit is intended to be self-explanatory, but there is certain knowledge from GCSE that will be useful to the understanding of this section. This information includes:

○ When a cell divides by meiosis to form gametes copies of the genetic information are made and then the cell divides twice to form four gametes, each with a single set of chromosomes.

○ When gametes join at fertilisation, a single body cell with new pairs of chromosomes is formed.

○ In human body cells, one of the 23 pairs of chromosomes carries the genes that determine sex. In females the sex chromosomes are the same (XX); in males the sex chromosomes are different (XY).

○ Some characteristics are controlled by a single gene. Each gene may have different forms called alleles.

○ An allele that controls the development of characteristics only if the dominant allele is not present, is a recessive allele.

○ A gene is a small section of DNA and each gene codes for a particular combination of amino acids which make a specific protein.

○ How to interpret genetic diagrams, including family trees and how to construct genetic diagrams of monohybrid crosses and predict the outcomes of monohybrid crosses.

○ Understanding the terms homozygous, heterozygous, phenotype, and genotype.

○ Individuals with characteristics most suited to the environment are more likely to survive to breed. The genes that have enabled these individuals to survive are then passed on to the next generation.

○ New species arise as a result of isolation, genetic variation, natural selection, and speciation.

○ Quantitative data on the distribution of organisms can be obtained by random sampling with quadrats and sampling along a transect.

Learning objectives

→ Define the meaning of the terms genotype and phenotype.

→ Define the terms dominant, recessive and codominant alleles.

→ Explain the nature of multiple alleles.

Specification reference: 3.7.1

Synoptic link

An understanding of inheritance depends on an understanding of the way chromosomes behave during meiosis and mutations. It would therefore be beneficial to study Topic 9.1 and Topic 9.2, again before starting this chapter.

Study tip

Not all genes code for a polypeptide, some code for ribosomal RNA or transfer RNAs

Hint

All individuals of the same species have the same genes, but not necessarily the same alleles of these genes.

The fact that children resemble both their parents to a greater or lesser degree and yet are identical to neither has long been recognised. However, it took the re-discovery, at the beginning of the last century, of the work of a scientist and monk, called Gregor Mendel, to establish the basic laws by which characteristics are inherited. In this chapter we shall look at the way in which characteristics are inherited from one generation to the next and how this can produce genetic variety within a population. Let us begin by looking at some of the terms and conventions that are used in studying inheritance.

Genotype and phenotype

Genotype is the genetic constitution (make-up) of an organism. It describes all the **alleles** that an organism has. The genotype determines the limits within which the characteristics of an individual may vary. It may determine that a human baby could grow to be 1.8 m tall, but the actual height that this individual reaches is affected by other factors, such as diet. A lack of an element like calcium (for the growth of bone) at a particular stage of development could mean that the individual never reaches his/her potential maximum height.

Phenotype is the observable or biochemical characteristics of an organism. It is the result of the interaction between the expression of the genotype and the environment. The environment can alter an organism's phenotype.

Genes and alleles

A **gene** is a length of DNA, that is, a sequence of **nucleotide** bases, that normally code for a particular polypeptide. A gene does this by coding for a particular polypeptide. This polypeptide may be an enzyme that is needed in the biochemical pathway that leads to the production of the characteristic (for example, a gene could code for a brown pigment in the iris of the eye). Genes exist in two, or more, different forms called alleles. The position of a gene on a particular DNA molecule is known as the **locus**.

An allele is one of the different forms of a gene. In pea plants, for example, there is a gene for the colour of the seed pod. This gene has two different forms, or alleles, an allele for a green pod and another allele for a yellow pod.

Only one allele of a gene can occur at the locus of any one chromosome. However, in diploid organisms the chromosomes occur in pairs called **homologous chromosomes** (Topic 8.2). There are therefore two loci that each carry one allele of a gene. If the allele on each of the chromosomes is the same (for example, both alleles for green pods are present) then the organism is said to be **homozygous** for the character. If the two alleles are different (for example, one chromosome has an allele for green pods and the other chromosome has an allele for yellow pods) then the organism is said to be **heterozygous** for the characteristic.

In most cases where two different alleles are present in the genotype (heterozygous state) only one of them shows itself in the phenotype. For instance, in our example where the alleles for green pods and yellow pods are present in the genotype, the phenotype is always green pods. The allele of the heterozygote that expresses itself in the phenotype is said to be **dominant**, while the one that is not expressed is said to be **recessive**. A homozygous organism with two dominant alleles is called **homozygous dominant**, whereas one with two recessive alleles is called **homozygous recessive**. The effect of a recessive allele is apparent in the phenotype of a **diploid** organism only when it occurs in the presence of another identical allele, that is, when it is in the homozygous state.

These different genetic types are shown in Figure 1.

Study tip

Many students do not know the difference between an allele and a gene. Make certain you are not one of them.

Study tip

Remember that, in diploid cells or organisms, there are **two** copies of each allele—one copy inherited from the mother and the other copy from the father.

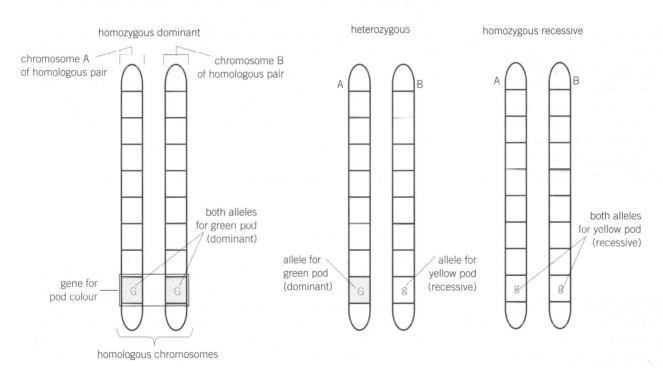

▲ **Figure 1** *Pair of homologous chromosomes showing different possible pairings of dominant and recessive alleles*

In some cases, two alleles both contribute to the phenotype, in which case they are referred to as **codominant**. In this situation when both alleles occur together, the phenotype is either a blend of both features (for example, snapdragons with pink flowers resulting from an allele for red-coloured flowers and an allele for white-coloured flowers) or both features are represented (for example, the presence of both A and B antigens in blood group AB). We will learn more about codominance in Topic 17.5.

Sometimes a gene has more than two allelic forms. In this case, the organism is said to have **multiple alleles** for the character. However,

▲ **Figure 2** *Attached earlobe. As with other inherited physical body characteristics, the earlobe is subject to genetic differences. Here, the earlobe is firmly attached to the facial skin rather than hanging freely*

as there are always only two chromosomes in a homologous pair, it follows that only two of the three or more alleles in existence can be present in a single organism. Multiple alleles occur in the human ABO blood grouping system. Again we shall learn more about multiple alleles in Topic 17.5.

Figure 3 summarises the different terms used in genetics.

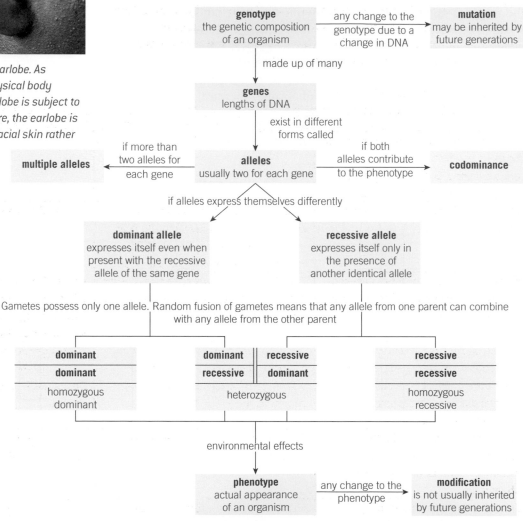

▲ **Figure 3** *Summary of genetic terms*

Summary questions

In the following passage, give the word that best replaces the number in brackets.

The genetic composition of an organism is called the (1) and any change to it is called a (2) and may be inherited by future generations. The actual appearance of an organism is called the (3). A gene is a sequence of (4) along a section of DNA that determines a single characteristic of an organism. It does this by coding for particular (5) that make up the enzymes needed in a biochemical pathway. The position of a gene on the DNA of a chromosome is called the (6). Each gene has two or more different forms called alleles. If the two alleles on a homologous pair of chromosomes are the same they are said to be (7), but if they are different, they are said to be (8). An allele that is not apparent in the phenotype when paired with a dominant allele is said to be (9). Two alleles are called (10) where they contribute equally to the appearance of a characteristic.

Representing genetic crosses

Genetic crosses are usually represented in a standard form of shorthand. This shorthand form is described in Table 1. Although you may occasionally come across variations to this scheme, that outlined in Table 1 is the one normally used. Once you have practised a number of crosses, you may be tempted to miss out stages or explanations. Not only is this likely to lead to errors, it often makes your explanations difficult for others to follow.

Learning objectives

→ Explain how to make labelled genetic diagrams.

→ Explain how a single gene is inherited.

Specification reference: 3.7.1

▼ **Table 1** *Representing genetic crosses*

Instruction	Reason/notes	Example [green pod and yellow pod]
Questions usually give the symbols to be used, in which case always use the ones provided. Choose a single letter to represent each characteristic.	An easy form of shorthand.	–
Choose the first letter of one of the contrasting features.	When more than one character is considered at one time such a logical choice means it is easy to identify which letter refers to which character.	Choose **G** (green) or **Y** (yellow).
If possible, choose the letter in which the higher and lower case forms differ in shape as well as size.	If the higher and lower case forms differ it is almost impossible to confuse them, regardless of their size.	Choose **G** because the higher case form (**G**) differs in shape from the lower case from (**g**) whereas **Y** and **y** are very similar and are likely to be confused.
Let the higher case letter represent the dominant feature and the lower case letter the recessive one. Never use two different letters where one character is dominant.	The dominant and recessive feature can easily be identified. Do not use two different letters as this indicates codominance.	Let **G** = green and **g** = yellow. Do not use **G** for green and **Y** for yellow.
Represent the parents with the appropriate pairs of letters. Label them clearly as 'parents' and state their phenotypes.	This makes it clear to any reader what the symbols refer to.	green pod × yellow pod parents **GG** **gg**
State the gametes produced by each parent. Label them clearly, and encircle them.	Encircling them reinforces the idea that they are separate.	gametes Ⓖ Ⓖ ⓖ ⓖ
Use a type of chequerboard or matrix, called a Punnett square, to show the results of the random crossing of the gametes. Label male and female gametes even though this may not affect the results.	This method is less liable to error than drawing lines between the gametes and the offspring. Labelling the sexes is a good habit to acquire – it has considerable relevance in certain types of crosses, e.g., sex-linked crosses.	♂ Gametes: Ⓖ Ⓖ; ♀ Gametes: ⓖ → Gg Gg; ⓖ → Gg Gg
State the phenotypes of each different genotype and indicate the numbers of each type. Always put the higher case (dominant) letter first when writing out the genotype.	Always putting the dominant feature first can reduce errors in cases where it is not possible to avoid using symbols with the higher and lower case letters of the same shape.	All offspring are plants producing green pods (**Gg**).

Inheritance of pod colour in peas

Monohybrid inheritance is the inheritance of a single gene. To take a simple example we will look at one of the features Gregor Mendel studied–the colour of the pods of pea plants. Pea pods come in two basic colours–green and yellow.

If pea plants with green pods are bred repeatedly with each other so that they consistently give rise to plants with green pods, they are said to be **pure-breeding** for the character of green pods. Pure-breeding strains can be bred for almost any character. This means that the organisms are homozygous (that is, they have two **alleles** that are the same) for that particular gene.

If these pure-breeding green-pod plants are then crossed with pure-breeding yellow-pod plants, all the offspring, known as the **first filial**, or **F_1, generation**, produce green pods. This means that the allele for green pods is **dominant** to the allele for yellow pods, which is therefore **recessive**. This cross is shown in Figure 1.

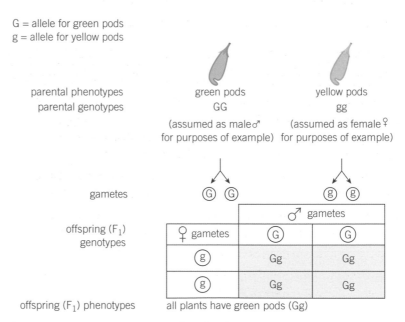

▲ **Figure 1** *Cross between a pea plant that is pure breeding for green pods and one that is pure breeding for yellow pods*

When the heterozygous plants (**Gg**) of the F_1 generation are crossed with one another (= F_1 intercross), the offspring (known as the second filial, or F_2, generation) are always in an approximate ratio of three plants with green pods to each one plant with yellow pods. This cross is shown in Figure 2.

These observed facts led to the formation of a basic law of genetics (the law of segregation). This states,

In diploid organisms, characteristics are determined by alleles that occur in pairs. Only one of each pair of alleles can be present in a single gamete.

Study tip

The larger the number of offspring the more likely that the ratio will be 3:1. If the sample is very small it is much less likely that the ratio will be 3:1.

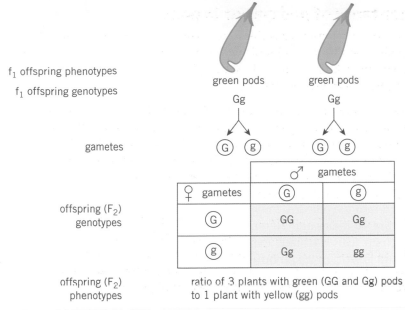

f₁ offspring phenotypes

f₁ offspring genotypes

gametes

offspring (F₂) genotypes

	♂ gametes	
♀ gametes	Ⓖ	ⓖ
Ⓖ	GG	Gg
ⓖ	Gg	gg

offspring (F₂) phenotypes

ratio of 3 plants with green (GG and Gg) pods to 1 plant with yellow (gg) pods

▲ **Figure 2** *F₁ intercross between pea plants that are heterozygous for green pods*

Summary questions

1 In humans, Huntington's disease is caused by a dominant, mutant allele of a gene. Draw a genetic diagram to show the possible genotypes and phenotypes of the offspring produced by a man with one allele for the disease and a woman who does not suffer from the disease.

2 In cocker spaniels, black coat colour is the result of a dominant allele and red coat colour is the result of a corresponding recessive allele.

  a Draw a genetic diagram to show a cross between a pure-breeding bitch with a black coat and a pure-breeding dog with a red coat.

  b If the offspring of this first cross are interbred, calculate the probability that any one of the offspring will have a red coat. Use a genetic diagram to show your working.

Learning objectives

→ Explain why results of genetic crosses often differ from predicted results.

Specification reference: 3.7.1

In Topic 17.2 we looked at monohybrid crosses and we saw that the result of any such cross theoretically produced offspring in the ratio of three offspring with one or more dominant allele to one offspring with only recessive alleles. In practice this ratio is rarely achieved exactly. Before we look at why, let us first be clear what is meant by a ratio.

Ratios

A ratio is a measure of the relative size of two classes (groups) that is expressed as a proportion. For example, any group of humans can be divided into two classes, male and female. If in a group of 60 humans there are 40 males and 20 females, then the ratio of males to females is 40 to 20. This is usually expressed as a ratio which is simplified to 2 to 1 and is written 2:1. If our group of 60 humans comprised 35 males and 25 females the ratio of males to females would be 7:5. For easy comparisons, ratios are often obtained by dividing the value of the smallest group into the value of each larger group. In which case, all ratios have their smallest value as one. For example, our ratio of 7:5 would be $7 \div 5 = 1.4$ which is written as 1.4:1.

Why actual results of genetic crosses are rarely the same as the predicted results

If you look at Table 1, you will see the results that Gregor Mendel actually obtained in his experiments. Our knowledge of genetics tells us that for each cross we would expect that, in the F_2 generation, there would be three offspring showing the dominant feature to every one showing the recessive feature. However, in no case did Mendel obtain an exact 3:1 ratio. The same is true of almost any genetic cross. These discrepancies are due to statistical error.

Imagine tossing a coin 20 times. In theory you would expect it to come down heads on 10 occasions and tails on 10 occasions. In practice it rarely does – try it. This is because each toss of the coin is an independent event that is not affected by what went before. If the coin has come down heads nine times out of 19 tosses, there is still a 50% chance it will come down tails, rather than the head needed to complete the 1:1 ratio. The coin does not know it is expected to come down heads.

The same is true of gametes. It is chance that determines which ones fuse with which. In our cross between the heterozygote (**Gg**) and the homozygous recessive (**gg**), all the gametes of the homozygous parent are recessive (**g**), whereas the heterozygote parent produces gametes of which half are dominant (**G**) and half are recessive (**g**). If it is the dominant gamete that combines with the recessive one, plants with green pods are produced (**Gg**). If it is the recessive gamete, the plants have yellow pods. The larger the sample, the more likely the actual results are to come near to matching the theoretical ones. It is therefore important to use large numbers of organisms in genetic crosses if representative results are to be obtained. It is no coincidence that the two ratios nearest to the theoretical value of 3:1 in Mendel's

Hint

Take care that you express a ratio the correct way round.

In the example opposite the ratio of males to females is 7:5. However, if you are asked for the ratio of females to males you should give the answer as 5:7.

experiments were those with the largest sample size, whereas the ratio furthest from the theoretical value had the smallest sample size (Table 1).

▼ **Table 1** *Actual results of Mendel's crosses in pea plants*

Character	F$_2$ results		Ratio
Cotyledon colour	6020 yellow	2001 green	3.01:1
Seed type	5474 smooth	1850 wrinkled	2.96:1
Pod type	882 inflated	299 constricted	2.95:1
Flower position	651 axial	207 terminal	3.14:1
Petal colour	705 purple	224 white	3.15:1
Stem height	787 long	277 short	2.84:1
Pod colour	428 green	152 yellow	2.82:1

Maths link $\sqrt{x}$

MS 0.3, see Chapter 22.

Summary questions $\sqrt{x}$

1 A cross was carried out between a pea plant producing green pods and one producing yellow pods. The seeds from this cross were germinated and, of the 63 plants grown, all produced green pods.

   a State the probable genotype of the parent plant with green pods.

   b Explain why we cannot be absolutely certain of the parent plant's genotype.

2 In a cross between a different pea plant with green pods and a pea plant with yellow pods, 96 plants were produced. 89 of these had green pods and 7 had yellow pods.

   a State the probable genotype of the parent plant with green pods.

   b Evaluate how certain we can be of the genotype of the parent plant with green pods.

   c In the cross described, state the chance of any one of the offspring plants having yellow pods.

   d $\sqrt{x}$ Calculate, to three significant figures, the percentage of offspring plants with yellow pods that were actually produced in the cross described.

In Topic 17.2 we saw how a single character is passed on from one generation to the next (monohybrid inheritance). In practice, many thousands of characters are inherited together. In this topic we shall look at how two characters, determined by two different **genes** located on different chromosomes, are inherited. This is referred to as **dihybrid inheritance**.

An example of dihybrid inheritance

In one of his experiments, Gregor Mendel investigated the inheritance of two characters of a pea plant at the same time. These were:

- **seed shape** – where round shape is dominant to wrinkled shape
- **seed colour** – where yellow-coloured seeds are dominant to green-coloured ones.

He carried out a cross between the following two pure breeding types of plants:

- one always producing round-shaped, yellow-coloured seeds (both dominant features)
- one always producing wrinkled-shaped, green-coloured seeds (both recessive features).

In the F_1 generation he obtained plants all of which produced round-shaped, yellow-coloured seeds, that is, dominant features.

He then raised the plants from these seeds and crossed them with one another to obtain the results shown in Table 1.

The explanation for these results is given in Figures 1 and 2.

▼ **Table 1** *Results obtained by Gregor Mendel when he crossed F_1 generation plants with round shaped, yellow coloured seeds*

Appearance of seeds	Condition	Number produced
Round Yellow	Dominant Dominant	315
Round Green	Dominant Recessive	108
Wrinkled Yellow	Recessive Dominant	101
Wrinkled Green	Recessive Recessive	32

R = allele for round shaped seeds

r = allele for wrinkled shaped seeds

G = allele for yellow coloured seeds

g = allele for green coloured seeds

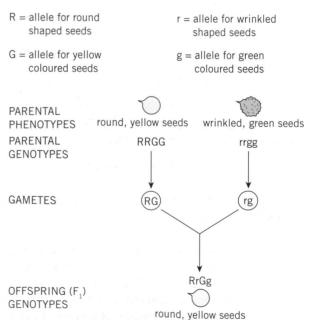

▲ **Figure 1** *Genetic explanation of Mendel's cross between a pure breeding plant for round, yellow seeds and a pure breeding one for wrinkled, green seeds*

From Figure 2 it can be seen that the plants of the F_1 generation produce four types of gamete (**RG, Rg, rG, rg**). This is because the gene for seed colour and the gene for seed shape are on separate chromosomes. As the chromosomes arrange themselves at random on the equator during meiosis, any one of the two **alleles** of the gene for seed colour (**G** and **g**) can combine with any one of the alleles for seed shape (**R** and **r**). Fertilisation is also random, so that any of the four types of gamete (with respect to seed colour and seed shape) of one plant can combine with any of the four types from the other plant.

Synoptic link

To understand much of genetics you need to appreciate what exactly is happening to alleles during a genetic cross. These alleles are attached to chromosomes and so you need to be clear about the behaviour of these chromosomes during meiosis. You can do this by recapping Topic 9.2, Meiosis and genetic variation.

R = allele for round shaped seeds

r = allele for wrinkled shaped seeds

G = allele for yellow coloured seeds

g = allele for green coloured seeds

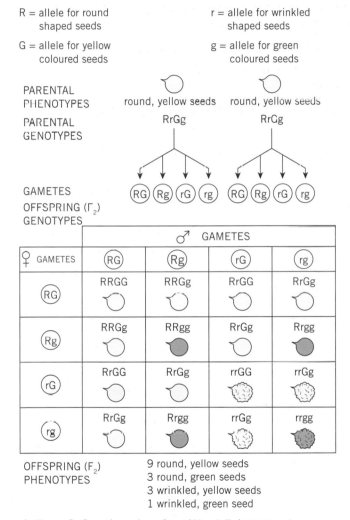

▲ **Figure 2** *Genetic explanation of Mendel's intercross between plants of the F_1 generation*

The theoretical ratio produced of 9:3:3:1 is close enough, allowing for statistical error (Topic 17.3), to Mendel's observed results of 315:108:101:32. Mendel's observations led him to formulate his **law of independent assortment** which, written in today's biological language states–**Each member of a pair of alleles may combine randomly with either of another pair.**

Summary questions

In fruit flies, a pure breeding variety with red eyes and vestigial (tiny) wings was crossed with a pure breeding variety with pink eyes and normal wings. All the F_1 flies had red eyes and normal wings. When these F_1 flies were bred with one another, the F_2 generation produced the following types and numbers:

red eyes with vestigial wings	125
red eyes with normal wings	376
pink eyes with vestigial wings	41
pink eyes with normal wings	117

1 Which characteristics are dominant and which are recessive? Explain your answer.

2 Suggest suitable symbols to represent the alleles of the genes involved.

3 Draw two suitable genetic diagrams to explain the results of this experiment.

Better late than never

Gregor Mendel was a monk with some scientific training. This training proved invaluable and led to a scientific approach to his genetic experiments. In particular he:

- chose pea plants to experiment on because they were easy to grow and possessed many contrasting features that could easily be observed

- carefully controlled pollination and hence fertilisation, by accurately transferring pollen from one plant to another with a paint brush

- ensured the plants he used were pure breeding for each feature by self-pollinating them for many generations

- produced quantitative and not just qualitative results

- counted many offspring to ensure his results were reliable and relatively free from statistical error.

In just a nine-year period, Mendel planned a well-organised programme of research, performed it accurately, painstakingly recorded masses of data and analysed his results with precision and insight. He had worked out the basic principles of inheritance by 1865, when the nature of genetics was unknown. He effectively predicted the existence of genes and meiosis long before they were discovered.

Although he circulated his work to libraries and scholars of his day, his theories were not accepted – indeed they were ignored. Even Darwin, whose theory of evolution is based on genetic variation, failed to understand the significance of Mendel's research. It took 35 years before other geneticists rediscovered his work and the significance of his remarkable experiments was at last appreciated.

Some genetic crosses may, at first glance, appear not to follow Mendel's laws because they provide unfamiliar ratios. Consider the following example of a plant that has two different varieties. The crosses made and the results they produced are shown here:

- **Cross 1** – each variety is self-fertilised and in both cases the offspring occur in the ratio 3 green-leaved plants to 1 white-leaved plant.

- **Cross 2** – the two varieties are cross-fertilised and the offspring are all green-leaved.

- **Cross 3** – some plants of the F_1 offspring of the cross between the two varieties are self-fertilised. The F_2 generation produced is in the ratio 9 green-leaved to 7 white-leaved.

Green colour is due to the presence of chlorophyll. In its absence, the plant is white.

1 The production of chlorophyll is controlled in a normal Mendelian way. Justify this statement from the information provided.

If you are having problems answering the rest of this question – try reading topic 17.8 before having another attempt.

The formation of chlorophyll in this plant is controlled by two separate genes A and B. The dominant allele of both genes is required for chlorophyll synthesis.

2 Using this information and the results of cross 1, deduce the genotypes of each variety. Show your reasoning using genetic diagrams.

3 Draw a genetic diagram to explain cross 2.

4 Draw a genetic diagram to explain cross 3.

5 Explain how the presence of both allele **A** and allele **B** might be required for the biochemical synthesis of chlorophyll.

In Topic 17.2 we dealt with straightforward situations in which there were two possible alleles at each locus on a chromosome, one of which was dominant and the other recessive. We shall now look at two different situations.

- **codominance**, in which both alleles are expressed in the **phenotype**
- **multiple alleles**, where there are more than two alleles, of which only two may be present at the loci of an individual's homologous chromosomes.

Codominance

Codominance occurs where instead of one allele being dominant and the other recessive, both alleles are equally dominant. This means that both alleles of a gene are expressed in the phenotype.

▲ **Figure 1** *Snapdragons*

One example occurs in the snapdragon plant, in which one allele codes for an enzyme that catalyses the formation of a red pigment in flowers. The other allele codes for an altered enzyme that lacks this catalytic activity and so does not produce the pigment. If these alleles showed the usual pattern of one dominant and one recessive, the flowers would have just two colours–red and white. As they are codominant, however, three colours of flower are found:

- In plants that are **homozygous** for the first allele, both alleles code for the enzyme, and hence pigment, production. These plants have red flowers.
- In plants that are **homozygous** for the other allele, no enzyme and hence no pigment is produced. These plants have white flowers.
- **Heterozygous** plants, with their single allele for the functional enzyme, produce just sufficient red pigment to produce pink flowers.

Learning objectives

→ Explain how codominance affects the inheritance of characteristics.

→ Explain how multiple alleles affect inheritance.

→ Explain how blood groups in humans are inherited.

Specification reference: 3.7.1

Hint

Remember to use different letters, such as R and W, when writing about codominance and to put these as superscripts attached to the letter of the gene (C), for example, C^R and C^W.

If a snapdragon with red flowers is crossed with one with white flowers, the resulting seeds give rise to plants with pink flowers. Note that we cannot use upper and lower case letters for the alleles, as this would imply that one (the upper case) was dominant to the other (the lower case). We therefore use different letters – in this case **R** for red and **W** for white – and put them as superscripts on a letter that represents the gene, in this case **C** for colour. Hence the allele for red pigment is written as C^R and the allele for no pigment as C^W. Figure 2 shows a cross between a red and a white snapdragon while Figure 3 shows a cross between the resultant pink-flowered plants.

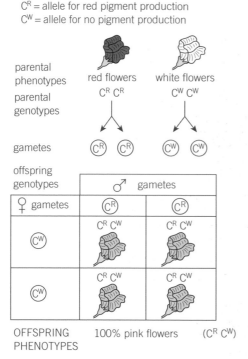

C^R = allele for red pigment production
C^W = allele for no pigment production

▲ **Figure 2** *Cross between a snapdragon with red flowers and one with white flowers*

▲ **Figure 3** *Cross between two snapdragons with pink flowers*

Synoptic link

We studied antigens in Topic 5.3 and their importance in immunity in Topics 5.4 and 5.5. Now would be a good time to look through these topics again.

▼ **Table 1** *Possible genotypes of blood groups in the ABO system*

Blood group	Possible genotypes
A	$I^A I^A$ or $I^A I^O$
B	$I^B I^B$ or $I^B I^O$
AB	$I^A I^B$
O	$I^O I^O$

Multiple alleles

A gene may have more than two alleles, that is, it has multiple alleles. The inheritance of the human ABO blood groups is an example. There are three alleles associated with the gene I (immunoglobulin gene), which lead to the presence of different **antigens** on the cell-surface membrane of red blood cells:

- allele I^A, which leads to the production of antigen A
- allele I^B which leads to the production of antigen B
- allele I^O, which does not lead to the production of either antigen.

Although there are three alleles, only two can be present in an individual at any one time, as there are only two homologous chromosomes and therefore only two gene loci. The alleles I^A and I^B are codominant, whereas the allele I^O is recessive to both. The possible genotypes for the four blood groups are shown in Table 1. There are obviously many different possible crosses between different blood groups, but two of the most interesting are:

1   A cross between an individual of blood group O and one of blood group AB, rather than producing individuals of either of the parental blood groups, produces only individuals of the other two groups, A and B (Figure 4).
2   When certain individuals of blood group A are crossed with certain individuals of blood group B, their children may have any of the four blood groups (Figure 5).

Study tip

If in a genetics question you are given a ratio of offspring that is different to the 3.1 (monohybrid) or 9.3.3.1 (dihybrid) ratio, you should begin by considering the possibility that codominance might be the explanation.

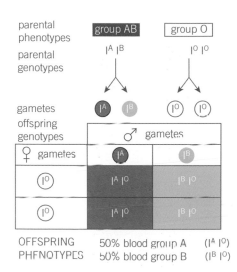

▲ **Figure 4** *Cross between an individual of blood group AB and one of blood group O*

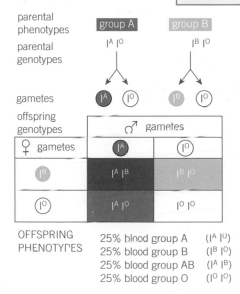

▲ **Figure 5** *Cross between an individual of blood group A and one of blood group B*

Summary questions

1   A man claims not to be the father of a child. The man is blood group O while the mother of the child is blood group A and the child is blood group AB. State, with your reasons, whether you think the man could be the father of the child.

2   In some breeds of domestic fowl, the gene controlling feather shape has two alleles that are codominant. When homozygous, the allele A^S produces straight feathers while the allele A^F produces frizzled feathers. The heterozygote for feather shape gives mildly frizzled feathers. Draw a genetic diagram to show the genotypes and phenotypes resulting from a cross between a mildly frizzled cockerel and a frizzled hen. The gene for feather shape is **not** sex-linked.

Maths link $\sqrt{x}$

MS 0.3, see Chapter 22.

Coats of many colours

In shorthorn cattle there is a gene **C** that determines coat colour. The gene has two alleles:

the allele **C^W** produces a white coat when homozygous

the allele **C^R** produces a red coat when homozygous.

In the heterozygous state the coat is light red, a colour also known as roan. The roan coat is a mixture of all white hairs and all red hairs. As each hair is either all red or all white, the **C^W** and **C^R** alleles are codominant.

▲ Figure 6

1 Draw a genetic diagram to show the possible genotypes and phenotypes of a cross between a bull with a white coat and a cow with a roan coat.

2 In each of the following crosses between shorthorn cattle what is the percentage of offspring with a roan coat?
   a red coat x white coat
   b red coat x roan coat
   c white coat x roan coat
   d roan coat x roan coat.

The phenotype of an organism often results from the influence of the environment on its genotype.

The fur on the ears, face, feet and tail of Siamese cats is darker in colour than the rest of the coat. This is due to the presence of a pigment. The production of this pigment is controlled by the action of an enzyme called tyrosinase. The action of tyrosinase is temperature-dependent.

3 Suggest why Siamese kittens are born with a completely light-coloured coat and only develop their characteristic markings some days later.

Humans have 23 pairs of chromosomes. 22 of these pairs have homologous partners that are identical in appearance, whether in a male or a female. The remaining pair are the sex chromosomes. In human females, the two sex chromosomes appear the same and are called the **X chromosomes**. In the human male there is a single **X** chromosome like that in the female, but the second one of the pair is smaller in size and shaped differently. This is the **Y chromosome**.

Sex inheritance in humans

In humans the sex-chromosomes are **X** and **Y**. This means:

- as females have two **X** chromosomes, all the gametes are the same in that they contain a single **X** chromosome
- as males have one **X** chromosome and one **Y** chromosome, they produce two different types of gamete – half have an **X** chromosome and half have a **Y** chromosome.

The inheritance of sex is shown in Figure 1.

Sex-linkage – haemophilia

Any gene that is carried on either the **X** or **Y** chromosome is said to be sex-linked. However, the **X** chromosome is much longer than the **Y** chromosome. This means that, for most of the length of the **X** chromosome, there is no equivalent homologous portion of the **Y** chromosome. Those characteristics that are controlled by **recessive alleles** on this non-homologous portion of the **X** chromosome will appear more frequently in the male. This is because there is no homologous portion on the **Y** chromosome that might have the **dominant allele**, in the presence of which the recessive allele does not express itself.

The **X** chromosome carries many genes. An **X**-linked genetic disorder is a disorder caused by a defective gene on the **X** chromosome. One example in humans is the condition called haemophilia, in which the blood clots only slowly and there may be slow and persistent internal bleeding, especially in the joints. As such it is potentially lethal if not treated. This has resulted in some selective removal of the gene from the population, making its occurrence relatively rare (about one person in 20 000 in Europe). Although haemophiliac females are known, the condition is almost entirely confined to males, in part because haemophiliac females usually died with the onset of menstruation at puberty.

One of a number of causes of haemophilia is a recessive allele with an altered sequence of DNA nucleotide bases that therefore codes for a faulty protein which does not function. This results in an individual being unable to produce a functional protein that is required in the clotting process. The production of this functional protein by genetically modified organisms means that it can now be given to haemophiliacs, allowing them to lead near-normal lives. Figure 3 shows the usual way in which a male inherits haemophilia. Note that the alleles are shown

Learning objectives

→ Explain how sex is determined genetically.

→ State what is meant by sex-linkage.

→ Explain how sex-linked diseases such as haemophilia are inherited.

Specification reference: 3.7.1

		♂ gametes	
		X	Y
♀ gametes	X	XX	XY
	X	XX	XY

offspring phenotypes

50% male (XY)
50% female (XX)

▲ **Figure 1** *Sex inheritance in humans*

▲ **Figure 2** *Scanning electron micrograph (SEM) of human X (left) and Y chromosomes as found in a male*

H = allele for production of clotting protein (rapid blood clotting)
h = allele for non-production of clotting protein (slow blood clotting)

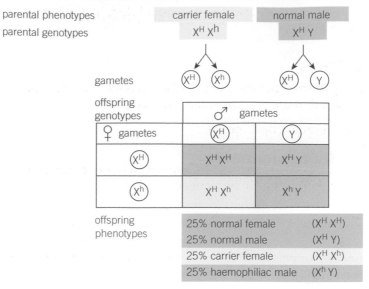

▲ **Figure 3** *Inheritance of haemophilia from a carrier female*

in the usual way (**H** = dominant allele for production of the clotting protein, and **h** = recessive allele for the non-production of clotting protein). However, as they are linked to the **X** chromosome, they are not shown separately, but always attached to the **X** chromosome, that is, as X^H and X^h respectively. There is no equivalent allele on the **Y** chromosome as it does not carry the gene for producing clotting protein.

As males can *only* obtain their **Y** chromosome from their father, it follows that their **X** chromosome comes from their mother. As the defective allele that does not code for the clotting protein is linked to the **X** chromosome, males always inherit the disease from their mother. If their mother does not suffer from the disease, she may be **heterozygous** for the character ($X^H X^h$). Such females are called carriers because they carry the allele without showing any signs of the disease in their phenotype. This is because these carriers possess one dominant **H** allele and this leads to the production of enough functional clotting protein.

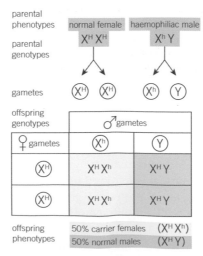

◄ **Figure 4** *Inheritance of the haemophiliac allele from a haemophiliac male*

As males pass the **Y** chromosome on to their sons, they cannot pass haemophilia to them. However, they can pass the allele to their daughters, via their **X** chromosome, who would then become carriers of the disease (Figure 4).

Pedigree charts

One useful way to trace the inheritance of sex-linked characters such as haemophilia is to use a pedigree chart. In these:

* a male is represented by a square
* a female is represented by a circle
* shading within either shape indicates the presence of a character, such as haemophilia, in the phenotype

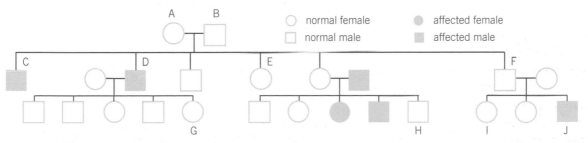

▲ **Figure 5**

Summary questions √x̄

Red-green colour blindness is linked to the **X** chromosome. The allele (**r**) for red-green colour blindness is recessive to the normal allele (**R**). Figure 5 shows the inheritance of this characteristic in a family.

1 State what sex chromosomes are present in individuals labelled E and F?

2 In terms of colour blindness, identify the phenotypes of each of the individuals labelled A, B and D.

3 In terms of colour blindness, identify the genotypes of each of the individuals labelled G, H, I and J.

4 √x̄ If individual C was to have children with a normal female (one who does not have any r alleles), determine the probability of any sons having colour blindness.

5 Individual J is colour blind. Assuming no history of colour blindness in either parent's family tree, suggest how this might have occurred.

Maths link √x̄

MS 1.4, see Chapter 22.

Maths link √x̄

MS 3.0, see Chapter 22.

A right royal disease

The royal families of Europe have been affected by haemophilia for many years. The origins of the disease stretch back to Queen Victoria. A pedigree chart showing the inheritance of haemophilia from Queen Victoria in members of various European royal families is shown in Figure 6.

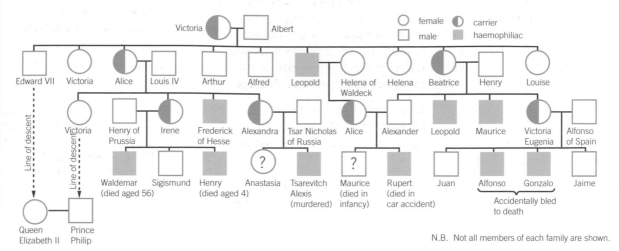

▲ **Figure 6** *Pedigree chart showing the transmission of haemophilia from Queen Victoria*

1 Suggest why haemophilia is not present in the current British royal family of Queen Elizabeth II and Prince Philip, and their children.

2 Give evidence from the chart which shows that haemophilia is:
   a sex-linked
   b recessive.

3 Using the symbols X^H for the chromosome carrying an allele that produces a clotting protein and X^h for a chromosome carrying an allele that does not produce a clotting protein, list the possible genotypes of the following people:
   a Queen Elizabeth II
   b Gonzalo
   c Irene

4 Suppose Waldemar and Anastasia had married and produced children. Using the same symbols, list all the possible genotypes of their:
   a sons
   b daughters.
   Explain your answers.

In humans, just 23 pairs of chromosomes carry the genes that determine many thousands of different characteristics. It follows that each chromosome must possess many different genes. Any two genes that occur on the same chromosome are said to be **linked**. All the genes on a single chromosome form a linkage group. We saw in Topic 17.6 that genes carried on the sex chromosomes are said to be sex linked. The remaining 22 chromosomes, other than the sex chromosomes, are called **autosomes**. The name given to the situation where two or more genes are carried on the same autosome is called **autosomal linkage**.

Assuming there is no crossing over, all the linked genes remain together during meiosis and so pass into gametes, and hence the offspring, together. They do not segregate in accordance with Mendel's Law of Independent Assortment.

Figure 1 shows the different gametes that are produced if a pair of genes **A** and **B** are linked compared with if they are on separate chromosomes.

Learning objectives

→ Describe autosomal linkage.
→ Explain how autosomal linkage affects the combinations of alleles in gametes.

Specification reference: 3.7.1

Synoptic link

You learnt about crossing over in Topic 9.2, Meiosis and genetic variation.

If genes A and B occur on separate chromosomes, that is, they are not linked.
If genes A and B occur on the same chromosome, that is, they are linked

Two homologous pairs are needed if all four alleles are to be present

According to Mendel's Law of Independent Assortment, any one of a pair of characters may combine with any of another pair. There are therefore four different possible types of gamete.

If genes A and B occur on the same chromosome, that is, they are linked

Only one homologous pair is needed if all four alleles are to be present.

Possible types of gamete

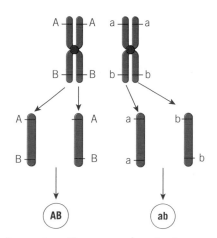

▲ **Figure 1** *Comparison of gametes produced by an organism that is heterozygous for two genes **A** and **B** when they are linked and not linked*

We see from Figure 1 that where the two genes **A** and **B** with heterozygous alleles are on different chromosomes, there are four possible combinations of the alleles in the gametes: **AB**, **Ab**, **aB** and **ab**. However, if the two genes are linked and provided there is no crossing over, there are only two possible combinations of the alleles in the gametes: **AB** and **ab**. This makes things simpler when dealing with genetic crosses.

For example, consider two linked genes of the fruit fly *Drosophila melanogaster* – one that determines body colour and the other that determines wing size. There are two alleles for body colour. One produces a grey body and is dominant to the other which produces a black body. There are also two alleles for wing size. One that produces normal sized wings and is dominant to the other that produces vestigial wings (tiny wings that do not function). One possible cross is shown in Figure 2.

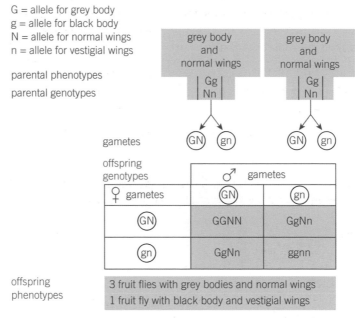

▲ **Figure 2** *The offspring of a cross between two flies that are heterozygous for both characteristics*

If the alleles for body colour and wing size were not linked but on separate chromosomes, then an individual that is heterozygous for both characters would produce four different gametes (**GN**, **Gn**, **gN**, and **gn**) rather that just two (**GN**, **gn**) as shown in our cross. We saw in Topic 17.4 that when there is random fertilisation of gametes, the offspring of this dihybrid cross would be fruit flies with the following characters:

9 grey body and normal wings (**G-N-**)

3 grey body and vestigial wings (**G-nn**)

3 black body and normal wings (**ggN-**)

1 black body and vestigial wings (**ggnn**)

Summary questions

1 Distinguish between sex-linkage and autosomal linkage.

2 A variety of rabbit has an allele (**H**) for long hairs and another allele (**h**) short hairs. It also has an allele (**G**) for grey hairs and another allele (**g**) for white hairs. The gene for hair length and the gene for hair colour are on the same chromosome.

Draw a genetic diagram to show the results of a cross between a rabbit that has short, white fur and one that is heterozygous for both characters.

Tales of the unexpected

In tomato plants, flower colour is determined by a gene that has two alleles, one producing yellow flowers and the other white flowers. Another gene determines the colour of the fruit. This gene also has two alleles, one producing red fruit and the other yellow fruit.

A tomato plant with yellow flowers and red fruit is crossed with a tomato plant with white flowers and yellow fruit. All the F_1 offspring had yellow flowers and red fruit.

A plant of this F_1 generation was self-pollinated (self-fertilised by its own gametes).

1 Suggest suitable symbols for each of the alleles involved in these crosses and annotate them to show which allele each symbol represents.

2 Using the symbols you have chosen, state the genotype of a tomato plant in the F_1 generation.

3 The two genes are linked. Predict what the ratio of plants would be in the F_2 generation.

4 Deduce what the ratio of plants would be in the F_2 generation if the two genes were not linked.

5 When a plant from the F_1 generation was self-pollinated, the actual results were:

yellow flowers and red fruit   68
yellow flowers and yellow fruit   7
white flowers and red fruit   7
white flowers and yellow fruit   18

a These results included some unexpected varieties of tomato plant. Suggest which ones these are.

b Using your knowledge of meiosis, suggest what might have happened that would explain the appearance of these unexpected varieties.

We have looked at examples of how conditions such as linkage modify Mendelian ratios but without compromising the basic ideas. Let us consider a similar situation, epistasis.

Epistasis arises when the allele of one gene affects or masks the expression of another in the phenotype.

An example occurs in mice where several genes determine coat colour. Let us look at two such genes.

- Gene A controls the distribution of a black pigment called melanin in hairs and therefore whether they are banded or not. The dominant allele **A** of this gene leads to hairs that have black bands while the recessive allele **a** produces uniform black hairs when it is present with another recessive allele **a** (homozygous = **aa**).

- Gene B controls the colour of the coat by determining or otherwise, the expression of gene A. The dominant allele **B** leads to the production of melanin while the recessive allele **b** leads to no pigment and any hair will therefore be white when it is present with another recessive allele = **bb** – the homozygous state.

The usual (wild type) mouse has a grey-brown coat known as agouti. This is the result of having hairs with black bands. If a mouse has uniform black hairs its coat is black and if the hairs lack melanin altogether its coat is albino (white).

▲ **Figure 1** *Black, agouti, and albino mice*

If an agouti mouse with the genotype **AABB** is crossed with an albino mouse with the genotype aabb, then the offspring are all agouti. If individuals from the F_1 generation are crossed to produce the F_2 generation the following ratio is produced:

    9 agouti mice
    4 albino mice
    3 black mice.

The crosses are shown in the genetic diagram in Figure 2.

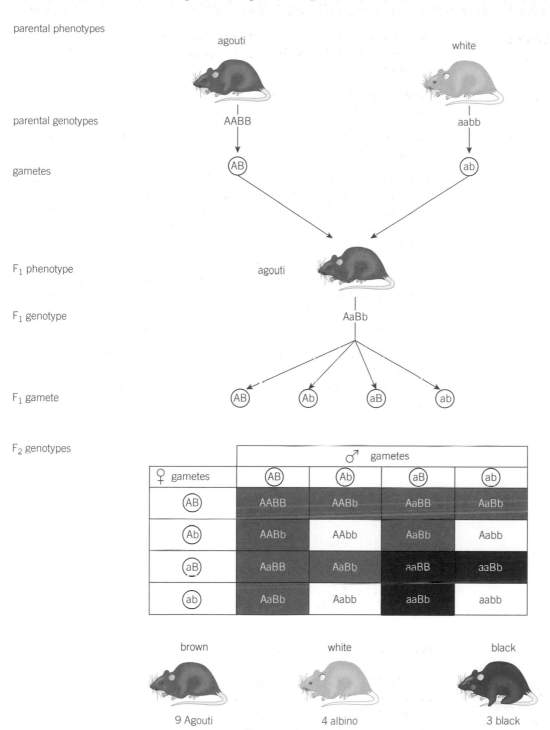

▲ **Figure 2** *Epistasis in mice*

The explanation of the results is as follows:

- The expression of gene A (black bands) is affected by the expression of gene B (production of melanin).
- If gene B is in the homozygous recessive state (**bb**) then no melanin is produced and the coat is albino.

- In the absence of melanin, gene A cannot be expressed.
- Therefore regardless of which alleles (**AA**, **Aa** or **aa**) are present, as there is no pigment, the hairs can be neither coloured nor banded.
- Where a dominant allele **B** is present, melanin is produced.
- If this allele is present with a dominant allele **A**, then banding occurs and an agouti coat results.
- Where allele **B** is present with two recessive alleles (**aa**), the hairs, and hence the coat, are uniform black.

There are other forms of epistasis. For example where genes act in sequence by determining the enzymes in a biochemical pathway. Suppose that a plant produces a red pigment in its petals using the following biochemical pathway.

$$\text{starting molecule} \xrightarrow{\text{enzyme A}} \text{intermediate molecule} \xrightarrow{\text{enzyme B}} \text{red pigment}$$

The production of enzymes A and B is coded for by genes A and B, respectively. Dominant alleles of each gene code for a functional enzyme, while recessive alleles code for a non-functional enzyme. It follows that if the alleles of either gene are both recessive, then that enzyme will be non-functional and the pathway cannot be completed. This affects the other gene in that, even if it is functional and produces its enzyme, its effects cannot be expressed because no pigment can be manufactured.

Summary questions

1 Using the example of epistasis in mice in this topic, consider a cross between mouse 1 that is heterozygous for gene A (colour distribution) and homozygous recessive for gene B (melanin production) with mouse 2 that is heterozygous for both genes.

   a State the colour of mouse 1 and the colour of mouse 2.

   b Calculate, and list, the genotypes of the offspring resulting from this cross.

   c State the ratio of different phenotypes produced by this cross.

2 Some varieties of corn, *Zea mays*, have purple seeds due to the presence of a pigment called anthocyanin in their seed coats. In the absence of the pigment, the seeds are white. The production of anthocyanin is controlled by two genes, A and B.

   One pure-breeding variety of white-seeded corn with the genotype **AAbb** was crossed with another pure-breeding variety of white-seeded corn with the genotype **aaBB**. All the offspring had purple seeds. A cross between two of the F_1 generation produced a ratio of nine purple-seeded plants to seven white-seeded plants.

   a Deduce the genotype of the F_1 generation.

   b Draw a genetic diagram showing the cross between two F_1 individuals.

   c i Analyse your table of results and deduce which nine genotypes you think might represent purple-seeded plants.

      ii State what these nine genotypes have in common.

   d Suggest how epistasis can explain the production of anthocyanin in some of the plants but not in others.

If you toss a coin 100 times it would be reasonable to expect it to land heads on 50 occasions and tails on 50 occasions. In practice, it would be unusual if these exact results were obtained. If it lands heads 55 times and tails only 45 times, does this mean that the coin is weighted or biased in some way, or is it purely a chance deviation from the expected result? How can we test which of these two options is correct?

What is the chi-squared test?

The chi-squared (χ^2) test is used to test the null hypothesis. The null hypothesis is used to examine the results of scientific investigations and is based on the assumption that there will be no statistically significant difference between sets of observations, any difference being due to chance alone. In our coin tossing example, the null hypothesis would be that there is no difference between the number of times it lands heads and the number of times it lands tails. The chi-squared test is a means of testing whether any deviation between the observed and the expected numbers in an investigation is significant or not. It is a simple test that can be used only if certain criteria are met:

- the sample size must be relatively large, that is, over 20
- the data must fall into discrete categories
- only raw counts and not percentages, rates, etc., can be used
- it is used to compare experimental results with theoretical ones, for example, in genetic crosses with expected Mendelian ratios

The formula is:

chi squared = sum of $\dfrac{[\text{observed numbers (O)} - \text{expected numbers (E)}]^2}{\text{expected numbers (E)}}$

summarised as:

$$\chi^2 = \Sigma \frac{(O - E)^2}{E}$$

The value obtained is then read off on a chi-squared distribution table (Table 1) to determine whether any deviation from the expected results is significant or not. To do this we need to know the number of **degrees of freedom**. This is simply the number of classes (categories) minus one, that is, if a human can have blood group A or B or AB or O, there are four classes and three degrees of freedom in this case.

Calculating chi-squared

Using our example of the coin tossed 100 times, we can calculate the chi-squared value:

Class (category)	Observed (O)	Expected (E)	O − E	(O − E)^2	$\dfrac{(O − E)^2}{E}$
Heads	55	50	+5	25	0.5
Tails	45	50	−5	25	0.5
					$\Sigma = 1.0$

Learning objectives

→ Explain what the chi-squared test is.

→ Calculate values for chi-squared.

→ Demonstrate how the chi-squared test is used in genetics.

Specification reference 3.7.1

Hint

The reason for squaring the value for (O−E) in the chi-squared test is to remove negative numbers.

Maths link $\sqrt{x}$

MS 1.9, see Chapter 22.

Therefore the value of $\chi^2 = 1.0$.

▼ **Table 1** *Part of a χ^2 table (based on Fisher)*

Degrees of freedom	Number of classes	χ^2							
1	2	0.00	0.10	0.45	1.32	2.71	3.84	5.41	6.64
2	3	0.02	0.58	1.39	2.77	4.61	5.99	7.82	9.21
3	4	0.12	1.21	2.37	4.11	6.25	7.82	9.84	11.34
4	5	0.30	1.92	3.36	5.39	7.78	9.49	11.67	13.28
5	6	0.55	2.67	4.35	6.63	9.24	11.07	13.39	15.09
Probability that deviation is due to chance alone		0.99 (99%)	0.75 (75%)	0.50 (50%)	0.25 (25%)	0.10 (10%)	0.05 (5%)	0.02 (2%)	0.01 (1%)

← Accept null hypothesis
(Any difference is due to chance and not significant)

Critical value

Reject null hypothesis and therefore accept experimental hypothesis. The difference is significant

of χ^2 at 0.05 p level as this is the smallest value accepted by statisticians for results being due to chance

Using the chi-squared table

To find out whether this value of 1.0 is significant or not we use a chi-squared table, part of which is given in Table 1. Before trying to read this table it is necessary to decide how many **classes of results** there are. In our case there are two classes of results, heads and tails. This corresponds to one **degree of freedom**, as the degrees of freedom are the number of classes minus one. We now look along the row showing two classes (i.e., one degree of freedom) for our calculated value of 1.0. This lies between the values 0.45 and 1.32. Reading along the probability row we see that our value lies between of 0.50 (50%) and 0.25 (25%). This means that the probability that chance alone could have produced the deviation is between 0.50 (50%) and 0.25 (25%). In the chi-squared test the critical value is p = 0.05. This is the attribution to chance accepted by statisticians, that is, 5% due to chance. If the probability that the deviation is due to chance is equal to or greater than 0.05 (5%), the deviation is said to be **not significant** and the null hypothesis would be accepted. If the deviation is less than 0.05 (5%), the deviation is said to be **significant**. In other words, some factor other than chance is affecting the results and the null hypothesis must be rejected. In our example the value is greater than 0.05 (5%) and so we assume the deviation is due to chance and accept the null hypothesis. Had we obtained 60 heads and 40 tails, a chi-squared value of slightly less than 0.05 (5%) would be obtained, in which case the null hypothesis would be rejected and we would assume that the coin might be weighted or biased in some way.

Chi-squared test in genetics

The chi-squared test is especially useful in genetics. To take the example of the genetic cross described in Topic 17.4 (also refer back to remind yourself how the expected results were obtained for the test), if we cross F_1 plants producing round, yellow seeds that we know are **heterozygous** we could expect a typical dihybrid F_2 ratio of:

9 round, yellow seeds   3 wrinkled, yellow seeds

3 round, green seeds   1 wrinkled, green seeds

Suppose we obtained 320 plants in the ratio 186:48:72:14. Could this variation be due to statistical chance or could some other factor be the reason for the differences? Our null hypothesis states that there is no significant difference between the observed and the expected results. Applying the chi-squared test:

Maths link $\sqrt{x}$

MS 1.9, see Chapter 22.

Class (category)	Observed (O)	Expected (E)	O – E	$(O - E)^2$	$\dfrac{(O - E)^2}{E}$
Round, yellow seeds	186	180	+6	36	0.2
Round, green seeds	48	60	– 12	144	2.4
Wrinkled, yellow seeds	72	60	+12	144	2.4
Wrinkled, green seeds	14	20	– 6	36	1.8
					$\Sigma = 6.8$

In this example there are four classes and therefore three degrees of freedom. Using the chi-squared table (Table 1) we see that our value falls between 6.25 and 7.82 shown on the row for three degrees of freedom and that these values correspond to between 0.1 (10%) and 0.05 (5%) probability that the deviation is due to chance alone. In this case there is between a 5% and 10% probability that the deviation is due to chance. Therefore we accept the null hypothesis and accept that the results are not significantly different from a 9:3:3:1 ratio.

Summary questions $\sqrt{x}$

In an experiment, domestic fowl with walnut combs were crossed with each other. The expected offspring ratio of comb types was 9 walnut, 3 rose, 3 pea, and 1 single. In the event, the 160 offspring produced 103 walnut combs, 20 rose combs, 33 pea combs, and 4 single combs (Figure 1).

1 Devise an appropriate null hypothesis.

2 Determine the number of degrees of freedom.

3 Calculate the value of chi-squared.

4 Determine the probability that the deviation is due to chance alone.

5 Assess whether the null hypothesis should be accepted or rejected.

▲ **Figure 1** *The different types of comb in domestic fowl are the result of dihybrid inheritance*

1 (a) In fruit flies, the genes for body colour and wing length are linked.
        Explain what this means. (*1 mark*)
        A scientist investigated linkage between the genes for body colour and wing length.
        He carried out crosses between fruit flies with grey bodies and long wings and fruit
        flies with black bodies and short wings.
        Figure 2 shows his crosses and the results.
        • **G** represents the dominant allele for grey body and **g** represents the recessive allele
          for black body.
        • **N** represents the dominant allele for long wings and **n** represents the recessive
          allele for short wings.

▼ **Figure 2**

Phenotype of parents	grey body, long wings	×	black body, short wings
Genotype of parents	**GGNN**		**ggnn**
Genotype of offspring		**GgNn**	
Phenotype of offspring		all grey body, long wings	

These offspring were crossed with flies homozygous for black body and short wings.
The scientist's results are shown in **Figure 3**.

▼ **Figure 3**

	GgNn	crossed with	ggnn	
	Grey body, long wings	Black body, short wings	Grey body, short wings	Black body, long wings
Number of offspring	975	963	963	194

   (b) Use your knowledge of gene linkage to explain these results. (*4 marks*)
   (c) If these genes were not linked, what ratio of phenotypes would the scientist
        have expected to obtain in the offspring? (*1 mark*)
   (d) Use the chi-squared test to determine whether there is a significant difference
        between the observations recorded in Figure 3 and those expected if the
        genes are not linked. (*2 marks*)

AQA Specimen 2014 (apart from 1 (d))

Key

hairless male

hairless female

male with hair

female with hair

2 A single gene controls the presence of hair
   on the skin of cattle. The gene is carried
   on the X chromosome. Its dominant allele
   causes hair to be present on the skin and
   its recessive allele causes hairlessness. The
   diagram shows the pattern of inheritance of
   these alleles in a group of cattle.
   (a) Use evidence from the diagram to explain
        (i)   that hairlessness is caused by a
              recessive allele (*2 marks*)
        (ii) that hairlessness is caused
              by a gene on the X chromosome.
                                          (*1 mark*)
   (b) What is the probability of the next calf born to animals 5 and 6 being hairless?
        Copy and complete the genetic diagram to show how you arrived at your answer.
        Phenotypes of parents   Female with hair   Male with hair
        Genotypes of parents
        Gametes
        Genotypes of offspring
        Phenotypes of offspring
        Probability of next calf being hairless
                                                                  (*4 marks*)
                                                                  AQA Jan 2012

3 The fruit fly is a useful organism for studying genetic crosses. Female fruit flies are approximately 2.5 mm long. Males are smaller and possess a distinct black patch on their bodies. Females lay up to 400 eggs which develop into adults in 7 to 14 days. Fruit flies will survive and breed in small flasks containing a simple nutrient medium consisting mainly of sugars.

(a) Use this information to explain two reasons why the fruit fly is a useful organism for studying genetic crosses. *(2 marks)*

(b) Male fruit flies have the sex chromosomes XY and the females have XX. In the fruit fly, a gene for eye colour is carried on the X chromosome. The allele for red eyes, **R**, is dominant to the allele for white eyes, **r**. The genetic diagram shows a cross between two fruit flies.

(i) Copy and complete the genetic diagram for this cross.

Phenotypes of parents	red-eyed female	×	white-eyed male
Genotype of parents			
Gametes		and	and
Phenotypes of offspring	red-eyed females	and	red-eyed males
Genotype of offspring			

(3 marks)

(ii) The number of red-eyed females and red-eyed males in the offspring was counted. The observed ratio of red-eyed females to red-eyed males was similar to, but not the same as, the expected ratio. Suggest **one** reason why observed ratios are often **not** the same as expected ratios. *(1 mark)*

(c) Male fruit flies are more likely than female fruit flies to show a phenotype produced by a recessive allele carried on the X chromosome. Explain why. *(2 marks)*

AQA Jan 2013

4 A breeder crossed a black male cat with a black female cat on a number of occasions. The female cat produced 8 black kittens and 4 white kittens.

(a) (i) Explain the evidence that the allele for white fur is recessive. *(1 mark)*

(ii) Predict the likely ratio of colours of kittens born to a cross between this black male and a white female. *(1 mark)*

(b) The gene controlling coat colour has three alleles. The allele **B** gives black fur, the allele **b** gives chocolate fur and the allele **b^i** gives cinnamon fur.
- Allele **B** is dominant to both allele **b** and **b^i**.
- Allele **b** is dominant to allele **b^i**.

(i) Copy and complete the table to show the phenotypes of cats with each of the genotypes shown.

Genotype	Phenotype
Bb^i	
bb^i	
Bb	

(1 mark)

A chocolate male was crossed several times with a black female. They produced
- 11 black kittens
- 2 chocolate kittens
- 5 cinnamon kittens.

(ii) Using the symbols given on the previous page, copy and complete the genetic diagram to show the results of this cross.

Parental phenotypes	Chocolate male		Black female
Parental genotypes			
Gametes			
Offspring genotypes			
Offspring phenotypes	Black	Chocolate	Cinnamon

(3 marks)

(iii) The breeder had expected equal numbers of chocolate and cinnamon kittens from the cross between the chocolate male and black female. Explain why the actual numbers were different from those expected. *(1 mark)*

(iv) The breeder wanted to produce a population of cats that would all have chocolate fur. Is this possible? Explain your answer. *(2 marks)*

AQA June 2011

Learning objectives

→ Define the terms gene pool and allelic frequency.

→ Define the Hardy–Weinberg principle.

→ Using the Hardy-Weinberg principle calculate allele, genotype and phenotype frequencies.

Specification reference: 3.7.2

We have so far looked at how genes and their alleles are passed between individuals in a population. Let us now consider the genes and alleles of an entire population.

A population is a group of organisms of the same species that occupies a particular space at a particular time and that can potentially interbreed. Any species exists as one or more populations. All the alleles of all the genes of all the individuals in a population at a given time time are known as the **gene pool**. The number of times an allele occurs within the gene pool is referred to as the **allelic frequency**.

Let us look at this more closely by considering just one gene that has two alleles, one of which is dominant and the other recessive. An example is the gene responsible for cystic fibrosis, a disease of humans in which the mucus produced by affected individuals is thicker than normal. The gene has a dominant allele (**F**) that leads to normal mucus production and a recessive allele (**f**) that leads to the production of thicker mucus and hence cystic fibrosis. Any individual human has two of these alleles in every one of their cells, one on each of the pair of homologous chromosomes on which the gene is found. As these alleles are the same in every cell of a single person, we only count one pair of alleles per gene per individual when considering a gene pool. If there are 10 000 people in a population, there will be twice as many (20 000) alleles in the gene pool *of this gene*.

The pair of alleles of the cystic fibrosis gene has three different possible combinations, namely homozygous dominant (**FF**), homozygous recessive (**ff**) and heterozygous (**Ff**). When we look at allele frequencies, however, it is important to appreciate that the heterozygous combination can be written as **Ff** or **fF**. (It is just conventional to put the dominant allele first in all cases.)

In our population of 10 000 people, if all 10 000 had the genotype **FF**, then:

- The probability of anyone being **FF** would be 1.0 and the probability of anyone being **ff** would be 0.0
- The frequency of the **F** allele would be 100% and the frequency of the **f** allele would be 0%.

If everyone in our population was heterozygous (**Ff**), then:

- The probability of anyone being **Ff** would be 1.0 and the frequency of the **F** allele would be 50%, and the frequency of the **f** allele would also be 50%.

Of course, in practice, the population is not made up of one genotype but of a mixture of all three, the proportions of which vary from population to population. How then can we work out the allele frequency of these mixed populations?

Hint

Whether an allele is recessive or dominant has nothing to do with it being harmful or beneficial. People with type O blood group have two recessive alleles for the gene but as it is the most common blood group it can hardly be harmful. Also, Huntingdon's disease is a fatal condition due to a dominant allele.

Maths link √x

MS 2.4, see Chapter 22.

The Hardy–Weinberg principle

The Hardy–Weinberg principle provides a mathematical equation that can be used to calculate the frequencies of the alleles of a particular gene in a population. The principle makes the assumption that the proportion of dominant and recessive alleles of any gene in a population remains the same from one generation to the next this can be the case provided that five conditions are met:

- No mutations arise.
- The population is isolated, that is, there is no flow of alleles into or out of the population.
- There is no selection, that is, all alleles are equally likely to be passed to the next generation.
- The population is large.
- Mating within the population is random.

Although these conditions are probably never totally met in a natural population, the Hardy–Weinberg principle is still useful when studying gene frequencies.

To help us understand the principle let us consider a gene that has two alleles: a dominant allele (**A**) and a recessive allele (**a**).

Let the probability of allele A = p

and the probability of allele a = q

The first equation we can write is:

$$p + q = 1.0$$

because there are only two alleles and so the probability of one plus the other must be 1.0 (100%).

As there are only four possible arrangements of the two alleles, it follows that the probability of all four added together must equal 1.0. Therefore we can state that:

AA + Aa + aA + aa = 1.0   or, expressing this as a probability:

$$p^2 + 2pq + q^2 = 1.0$$

We can now use these equations to determine the probability of any allele in a population. For example, suppose that a particular characteristic is the result of the recessive allele **a**, and we know that one person in 25 000 displays the character.

- The character, being recessive, will only be observed in individuals who have two recessive alleles **aa**.
- The probability of **aa** must be $\dfrac{1}{25\,000}$ or 0.00004.
- The probability of **aa** is q^2.
- If $q^2 = 0.00004$, then $q = \sqrt{0.00004}$ or 0.00063 approximately.
- We know that the probability of both alleles **A** and **a** is $p + q$ and is equal to 1.0.
- If $p + q = 1.0$, and $q = 0.00063$ then:
- $p = 1.0 - 0.00063 = 0.9937$, that is, the probability of allele A = 0.9937.
- We can now calculate the probability of heterozygous individuals (and therefore the probability of genotypes and phenotypes) in the population.
- From the Hardy–Weinberg equation we know that the probability of the heterozygotes is $2pq$.

Hint

A probability of 0.0125 is the equivalent of 125 in 10 000 or 313 in our population of 25 000.

- In this case, $2pq = (2 \times 0.9937 \times 0.0063) = 0.0125$.
- Heterozygous individuals act as a reservoir of recessive alleles in the population, although they themselves do not express the allele in their phenotype.

Maths link √x

MS 2.2, see Chapter 22.

Summary questions √x

1 Define the term allelic frequency.

2 State what the Hardy–Weinberg principle predicts.

3 State the **five** conditions that need to be met for this prediction to hold true.

4 √x The frequency p of a dominant allele is 0.942. Calculate the frequency of the heterozygous genotype in the population. Show your working and express your answer as a percentage of the population.

5 √x The ability to roll the tongue is determined by a dominant allele. In a sample of 416 people, 26 were unable to roll their tongues. Using the Hardy–Weinberg equation, calculate how many people in the sample were homozygous dominant for this allele.

Hint

Unless an allele leads to a phenotype with an advantage or a disadvantage compared with other phenotypes, its allele frequency in a population will probably stay the same from one generation to the next.

Maths link √x

MS 2.4, see Chapter 22.

Not as black and white as it seems √x

A gene that controls wing colour in the peppered moth has two alleles. The expression of the dominant allele produces moths with light-coloured wings while the expression of the recessive allele produces moths with dark-coloured wings. Scientists sampled a population of moths by catching them in a trap and recording their sex and wing colour. The numbers in Table 1 show how many of the 2215 moths caught were of each type.

▲ **Figure 1** Dark- and light-coloured wing forms of the peppered moth

▼ **Table 1**

	Light-coloured wings	Dark-coloured wings
Male	836	269
Female	817	293
Total	**1653**	**562**

1 State, with your reasons, whether you think the gene for wing colour is sex-linked.

2 √x Determine what proportion of the total sample has two recessive alleles.

3 √x In the Hardy–Weinberg equation $(p^2 + 2pq + q^2 = 1.0)$, $p =$ the frequency of the dominant allele and $q =$ the frequency of the recessive allele. For the population of moths that were caught, use this equation to calculate:
   a the frequency (q) of the recessive allele;
   b the frequency (p) of the dominant allele;
   c the percentage of heterozygotes.

4 Some scientists wanted to estimate the size of the total moth population. Describe how they might do this.

Individuals within a population of a species often show a wide range of variation in their phenotypes. This variation is due to both genetic and environmental factors. The primary source of genetic variation is mutation, a subject that we explored in Topic 9.1. Further genetic variation results from meiosis and the random fertilisation of gametes during sexual reproduction, subjects that were covered in Topic 9.2. Although variation is the result of genetic factors and environmental influences, it is rarely entirely due to one or the other but rather a combination of both.

Variation due to genetic factors

Within a population, all members have the same genes. Genetic differences, however, occur as members of this population will have different alleles of these genes. These differences not only occur in living individuals but also change from generation to generation. Genetic variation arises as a result of:

- **mutations**. These sudden changes to genes and chromosomes may, or may not, be passed on to the next generation. Mutations are a main source of variation.

- **meiosis**. This special form of nuclear division produces new combinations of alleles before they are passed into the gametes, all of which are therefore different.

- **random fertilisation of gametes**. In sexual reproduction this produces new combinations of alleles and the offspring are therefore different from parents. Which gamete fuses with which at fertilisation is a random process further adding to the variety of offspring two parents can produce.

Sexually reproducing organisms, increase variation by all three methods.

Where variation is very largely the result of genetic factors organisms fit into a few distinct forms and there are no intermediate types. In the ABO blood grouping system, for example, there are four distinct groups – A, B, AB and O (Figure 1). A character displaying this type of variation is usually controlled by a single gene. This variation can be represented on a bar chart or pie graph. Environmental factors have little influence on this type of variation.

Learning objectives

→ Describe variation due to genetic factors.

→ Describe variation due to environmental influences.

Specification reference: 3.7.3

Synoptic link

An understanding of mutations (Topic 9.1) and genetic variation (Topic 9.2) will help you follow this topic and so should be revised before you read on.

▲ **Figure 1** *Variation due to genetic factors illustrated by the percentage of the UK population with blood groups A, B, AB and O*

Study tip

Variation where organisms fit into a few distinct forms is known as discontinuous variation. Make certain you recognise it and are able to plot a bar chart from any data provided.

▲ **Figure 2** *Variation within a species (intraspecific variation) – despite the immense variety that they show, all dogs, including this Pug and Great Dane, belong to the same species – Canis familiaris*

Synoptic link

The normal distribution curve was covered in more depth in Topic 10.5, Quantitative investigations of variation and a recall of the information there would be helpful.

Summary questions

1 State **three** ways in which genetic variation can be increased in sexually reproducing organisms.

2 State how genetic variation is increased in asexually reproducing organisms

3 In the following list of statements, determine whether each refers to variation due to genetic or environmental factors.

   a It can be represented by a line chart.

   b It is usually controlled by a single gene.

   c It can be represented as a bar chart.

   d A mean can be calculated.

   e An example is the length of the body in rats.

Variation due largely to environmental influences

The environment exerts an influence on all organisms. These influences affect the way the organism's genes are expressed. The genes set limits, but it is largely the environment that determines where, within those limits, an organism lies. In buttercups, for example, one plant may be determined by its genes to grow much taller than other plants. If, however, the seed germinated in an environment of poor light or low soil nitrate, the plant may not grow properly and it will be short. Environmental influences include climatic conditions (e.g., temperature, rainfall, and sunlight), soil conditions, pH, and food availability.

Some characteristics of organisms grade into one another, forming a continuum. In humans, two examples are height and mass. Characters that display this type of variation are not controlled by a single gene, but by many genes (polygenes). Environmental factors play a major role in determining where on the continuum an organism actually lies. For example, individuals who are genetically predetermined to be the same height actually grow to different heights due to variations in environmental factors, such as diet. This type of variation is the product of polygenes and the environment. If we measure the heights of a large population of people and plot the number of individuals against heights on a graph we will most probably obtain a bell-shaped curve known as a **normal distribution curve** (Figure 3).

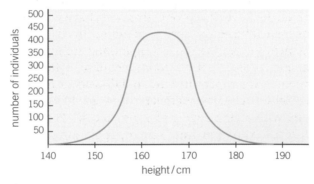

▲ **Figure 3** *Graph of frequency against height for a sample of humans*

In most cases variation is due to the combined effects of genetic differences and environmental influences. It is very hard to distinguish between the effects of the many genetic and environmental influences that combine to produce differences between individuals. As a result, it is very difficult to draw conclusions about the causes of variation in any particular case. Any conclusions that are drawn are usually tentative and should be treated with caution.

Every organism is subjected to a process of selection, based on its suitability for survival under the conditions that exist at the time. The environmental factors that limit the population of a species are called **selection pressures**. These selection pressures include predation, disease and competition. Selection pressures vary from time to time and place to place. These selection pressures determine the frequency of all alleles within the gene pool. A **gene pool** is the total number of all the alleles of all the genes of all the individuals within a particular population at a given time.

The process of evolution by means of natural selection depends upon a number of factors. These include:

- organisms produce more offspring than can be supported by the available supply of food, light, space, etc.
- there is genetic variety within the **populations** of all species.
- a variety of phenotypes that selection operates against.

The role of over-production of offspring in natural selection

Charles Darwin appreciated that all species have the potential to increase their numbers exponentially (Figure 1). He realised that, in nature, populations rarely, if ever, increased in size at such a rate. He rightly concluded that the death rate of even the most slow-breeding species must be extremely high. High reproductive rates have evolved in many species to ensure a sufficiently large population survives to breed and produce the next generation. This compensates for high death rates from predation, competition for food (including light in plants) and water, extremes of temperature, natural disasters such as earthquake, and fire, disease etc. Some species have evolved lower reproductive rates along with a high degree of parental care. The lower death rates that result help to maintain their population size.

Learning objectives

→ Define a gene pool.

→ Explain the role of overproduction of offspring in natural selection.

→ Explain the role of variation in natural selection.

Specification reference: 3.7.3

▲ **Figure 1** *The exponential rise in the population of a species whose growth is left unchecked by environmental factors such as predation, climate, disease, and competition*

The link between over-production and natural selection lies in the fact that, where there are too many offspring for the available resources, there is competition amongst individuals (**intraspecific competition**) for the limited resources available. The greater the numbers, the greater this competition and the more individuals will die in the struggle to survive. These deaths, however, are not totally random. Those individuals in a population best suited to prevailing conditions (e.g., better able to hide from or escape predators, better able to obtain light or catch prey, better able to resist disease or find a mate) will be more likely to survive than those less well adapted. These individuals will be more likely to breed and so pass on their more favourable allele combinations to the next generation, which will therefore have a different allele frequency from the previous one. The population will have evolved a combination of alleles that is better adapted to the prevailing conditions. This selection process, however, depends on individuals of a population being genetically different from one another.

The role of variation in natural selection

If an organism can survive in the conditions in which it lives, you may wonder why it doesn't always produce offspring that are identical to itself. These will, after all, be equally likely to survive in these conditions, whereas variation may produce individuals that are less suited. However, conditions change over time and having a wide range of genetically different (and therefore phenotypes) in the population means that some will have the combination of genes needed to survive in almost any new set of circumstances. Populations showing little individual genetic variation are often more vulnerable to new diseases and climate changes. It is also important that a species is capable of adapting to changes resulting from the evolution of other species.

The larger a population is, and the more genetically varied the individuals within it, the greater the chance that one or more individuals will have the combination of alleles that lead to a phenotype which is advantageous in the struggle for survival. These individuals will therefore be more likely to breed and pass their allele combinations on to future generations. Variation therefore provides the potential for a population to evolve and adapt to new circumstances.

The influence of variation on natural selection is best summarised by Darwin himself who, nearly a hundred and fifty years ago, wrote:

> How can it be doubted, from the struggle each individual has to obtain subsistence, that any minute variation in structure, habits or instinct, adapting that individual better to the new conditions, would tell upon its vigour and health? In the struggle it would have a better chance of surviving, and those of its offspring which inherited the variation, be it ever so slight, would have a better chance.

Summary questions

1 Define a gene pool.

2 State four factors that lead to differential survival and reproduction.

3 Sickle cell anaemia is a debilitating genetic disease that causes premature death but provides some resistance to the malarial parasite. Explain how selection might affect the distribution of the gene causing sickle cell anaemia in both malarial and non-malarial regions.

Maths link $\sqrt{x}$

MS 1.3 and 2.4, see Chapter 22.

How genetic variation leads to natural selection – copper tolerance in grasses

Species of grasses such as *Agrostis tenuis* and *Agrostis capillaris* have a variety of different forms with respect to copper tolerance (Figure 2). Some varieties are very tolerant and grow readily in soils with a high concentration of copper **ions**. Other varieties have a very low tolerance and do not survive, even when soil copper concentrations are very low. The majority of the varieties lie somewhere between these two extremes.

The soil around old copper workings is heavily contaminated with copper. If seeds of *Agrostis* spp. are planted in this soil, only those plants with a high tolerance to copper ions will survive. These plants are therefore the only ones to breed and so pass on their alleles to the next generation. Amongst these alleles will be those giving resistance to copper ions. Over time, the frequency of these alleles in the gene pool of the population of *Agrostis spp* increases as a result of the selection pressure from copper ions in the soil. The frequency of certain alleles in the population has changed and so the gene pool has changed and the population of *Agrostis spp* has evolved. On soils with low levels of copper, there is no selective advantage in being copper-tolerant and so the proportion

of these varieties remains low. They nevertheless survive where soil copper concentrations are low because there is no disadvantage in having tolerance to copper but there is no particular advantage either.

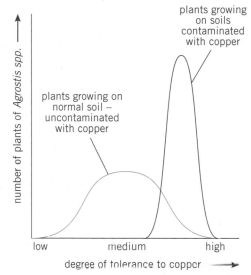

▲ **Figure 2** *Copper tolerance in* Agrostis *spp.*

1 Identify the types of curve shown in Figure 2.
2 Name the type of inheritance these curves suggest.
3 Suggest why the population of copper-tolerant varieties where the soil is heavily contaminated with copper is much larger than the population of non-tolerant varieties where the soil is less contaminated.
4 Describe how you might determine whether a plant was a variety of *Agrostis capillaris* rather than a separate species.
5 A gene associated with copper tolerance has two alleles, **T** and **t**. Varieties of *Agrostis tenuis* that have both recessive alleles are copper tolerant, while varieties of *Agrostis tenuis* possessing one or more dominant alleles are not. A sample of plants was tested for copper tolerance and 72 plants were found to be copper tolerant while 378 were not.
   a Assuming the Hardy–Weinberg principle applies, calculate the percentage of the sample that have the genotypes **TT**, **Tt**, and **tt**.
   b Justify, giving **two** reasons, why the Hardy–Weinberg principle does not apply in the situation described here.

Environmental factors help create variation within a population. These environmental factors may be an agent for constancy or an agent for change according to the type of selection pressure they exert. Earlier we looked at two different forms of selection, stabilising and directional, and touched on the effects these forms of selection have on evolution. Let us now explore these effects more closely and investigate a third form of selection – disruptive selection.

The three main types of selection affect the characteristics of a population in the following ways:

- **stabilising selection** preserves the average phenotype (phenotypes around the mean) of a population by favouring average individuals, in other words, selection against the extreme phenotypes
- **directional selection** changes the phenotypes of a population by favouring phenotypes that vary in one direction from the mean of the population, in other words, selection for one extreme phenotype
- **disruptive selection** favours individuals with extreme phenotypes rather than those with phenotypes around the mean of the population.

Stabilising selection

Stabilising selection tends to eliminate the extremes of the **phenotype** range within a population and with it the capacity for evolutionary change. It tends to occur where the environmental conditions are constant over long periods of time. One example is fur length in a particular mammalian species. In years when the environmental temperatures are hotter than usual, the individuals with shorter fur length will be at an advantage because they can lose body heat more rapidly. In colder years the opposite is true and those with longer fur length will survive better as they are better insulated.

Therefore, if the environment fluctuates from year to year, both extremes will survive because each will have some years when it can thrive at the expense of the other. If, however, the environmental temperature is constantly 10 °C, individuals at the extremes will never be at an advantage. They will therefore be selected against in favour of those with average fur length. The mean will remain the same, but there will be fewer individuals at either extreme (Figure 1). An example of stabilising selection is the body mass of human children at birth. We saw in Topic 9.4 that babies born with a body mass greater or less than the optimum of 3.2 kg have a higher risk of dying in the few months after birth.

Directional selection

Within a population there will be a range of genetically different individuals in respect of any one phenotype. The continuous variation amongst these individuals forms a normal distribution curve. This curve has a mean that represents the optimum value for the phenotypic character under the existing conditions. If the environmental conditions

1. Initially there is a wide range of fur length about the mean of 1.5 cm. The fur lengths of less than 1.0 cm or greater than 2.0 cm in individuals are maintained by rapid breeding in years when the average temperature is much warmer or colder than normal.

2. When the average environmental temperature is consistently around 10 °C with little annual variation, individuals with very long or very short hair are eliminated from the population over a number of generations.

▲ **Figure 1** *Stabilising selection*

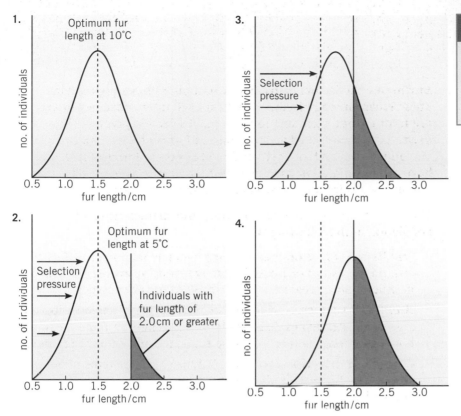

Figure 2 *Directional selection*

Synoptic link

Revising Topic 9.4, Types of selection, will be invaluable in helping you understand the ideas in this topic.

In a population of a particular mammal, fur length shows continuous variation.

1. When the average environmental temperature is 10°C, the optimum fur length is 1.5cm. This then represents the mean fur length of the population.

2. A few individuals in the population already have a fur length of 2.0 cm or greater. If the average environmental temperature falls to 5°C, these individuals are better insulated and so are more likely to survive to breed. There is a selection pressure favouring individuals with longer fur.

3. The selection pressure causes a shift in the mean fur length towards longer fur over a number of generations. The selection pressure continues.

4. Over further generations the shift in the mean fur length continues until it reaches 2.0 cm – the optimum length for the prevailing average environmental temperature of 5°C. The selection pressure now ceases.

change, so will the optimum value for survival. Some individuals, either to the left or the right of the mean, will possess a combination of alleles with the new optimum for the phenotypic character. As a result there will be a selection pressure favouring the combination of alleles that results in the mean moving to either the left or the right of its original position. Directional selection therefore results in one extreme of a range of variation being selected against in favour of the other extreme or even the average. Figure 2 illustrates a theoretical example of directional selection. A specific example is antibiotic resistance in bacteria.

Disruptive selection

Disruptive selection is the opposite of stabilising selection. It favours extreme phenotypes at the expense of the intermediate phenotypes. Although the least common form of selection, it is the most important in bringing about evolutionary change. Disruptive selection occurs when an environmental factor, such as temperature, takes two or more distinct forms. In our example this might arise if the

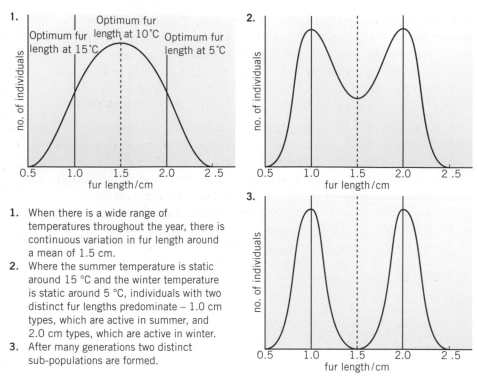

1. When there is a wide range of temperatures throughout the year, there is continuous variation in fur length around a mean of 1.5 cm.
2. Where the summer temperature is static around 15 °C and the winter temperature is static around 5 °C, individuals with two distinct fur lengths predominate – 1.0 cm types, which are active in summer, and 2.0 cm types, which are active in winter.
3. After many generations two distinct sub-populations are formed.

▲ **Figure 3** *Disruptive selection*

temperature alternated between 5 °C in winter (favouring long fur length) and 15 °C in summer (favouring short fur length). This could ultimately lead to two separate species of the mammal – one with long fur and active in winter, the other with short fur and active in summer (Figure 3). An example of disruptive selection is the coho salmon where large males and small males have a selective advantage over intermediate-sized males in passing on their alleles to the next generation. The small males are able to sneak up to the females in the spawning grounds. The large males are fierce competitors. This leaves intermediate-sized males at a disadvantage.

Selection in the peppered moth

Some species of organisms have two or more distinct forms. These different forms are genetically distinct but exist within the same interbreeding population. This situation is called polymorphism (poly = many; morph = form). One example is the peppered moth (*Biston betularia*). It existed almost entirely in its natural light form until the middle of the nineteenth century. Around this time a melanic (black) variety arose as the result of a mutation. These mutant moths had undoubtedly occurred before (one existed in a collection made before 1819) but they were highly conspicuous against the light background of the lichen-covered trees and rocks on which they normally rest. As a result, the black mutants were subjected to greater predation from insect-eating birds, for example, robins and hedge sparrows, than were the better camouflaged, light forms.

When, in 1848, a melanic form of the peppered moth was captured in Manchester, most buildings, walls and trees were blackened by

Synoptic link

You previously learnt about the peppered moth in the context of natural selection in Topic 9.3. This is explored in more detail here.

the soot of 50 years of industrial development. The sulfur dioxide in smoke emissions killed the lichens that formerly covered trees and walls. Against this black background the melanic form was less, not more, conspicuous than the light form. As a result, the light form was eaten by birds more frequently than the melanic form and, by 1895, 98% of Manchester's population of the moth was of the melanic type.

This is an example of selective predation by birds favouring individuals that lie at one extreme or the other of a range of different colour types. It illustrates directional selection of different types in different populations. The melanic form is is selected for in industrial areas while the natural form is selected for in rural areas. It also shows evolution in action whereby there has been a change in the allelic frequency in populations of moths in industrial areas. However, as the two populations overlap and interbreed they are still one species. As we shall see in Topic 18.5, to become two distinct species, the two populations would need to become reproductively separated from one another.

▲ **Figure 4** *Melanism in the peppered moth (Biston betularia). Against a natural background the dark melanic form is far more visible and more readily predated on by birds. This natural selection leads to a predominance of the light form in rural areas. In polluted areas with blackened buildings, however, the melanic form is better camouflaged and this selective advantage leads to this form predominating (Topic 18.1)*

Summary questions

1 Consider each of the following statements and suggest which form of selection it best relates to.

   a A baby with a birth weight greater than 4.0 kg or less than 2.5 kg has an increased risk of dying.

   b Some species of insects have changed very little over millions of years.

   c Elephants have evolved longer trunks enabling them to reach leaves higher up in trees.

   d Small mammals can escape from predators by hiding in small spaces while large ones can resist attack by predators.

   e It is the most important type of selection in bringing about evolutionary change.

   f The mean ear length in arctic foxes has reduced over time.

   g It preserves the characteristics of a population.

2 Suggest which form of the peppered moth, *Biston betularia*, is now most common in cities like Manchester and explain why in terms of selection pressure.

Specification reference: 3.7.3

Learning objectives

→ Explain how selection affects allelic frequencies.

→ Explain how new species are formed.

→ Explain how populations can become geographically isolated.

→ Describe allopatric and sympatric speciation.

Study tip

It should be remembered that some environmental factors may influence the overall **mutation** rate (Topic 9.1) but that this is a general and random process rather than one that affects a specific allele in a specific way.

In topic 18.4, we examined the different forms of selection. We learnt that those organisms with phenotypes that gave them a selective advantage were more likely to produce offspring and so pass on their favourable alleles to the next generation. Now we will turn our attention to the effect that this differential reproductive success has on the allelic frequencies within a **gene pool** of a population.

Allelic frequencies and how selection affects them

In theory, any sexually mature individual in a **population** is capable of breeding with any other. This means that the **alleles** of any individual organism may be combined with the alleles of any other. We saw in Topic 18.3 that all the alleles of all the genes of all the individuals in a population at a given time is known as the gene pool. The number of times an allele occurs within the gene pool is referred to as the **allelic frequency**. The allelic frequency is affected by **selection** and, as we saw in Topic 18.4, selection is due to environmental factors. Environmental changes therefore affect the probability of an allele being passed on in a population and hence the number of times it occurs within the gene pool. It must be emphasised that environmental factors do not affect the probability of a particular mutant allele arising, they simply affect the frequency of a mutant allele that is already present in the gene pool.

Evolution by natural selection is a change in the allelic frequencies within a population.

Speciation

Speciation is the evolution of new species from existing ones. A **species** is a group of individuals that have a common ancestry and so share the same genes but different alleles and are capable of breeding with one another to produce fertile offspring. In other words, members of a species are **reproductively separated** from other species.

It is through the process of speciation that evolutionary change has taken place over millions of years. This has resulted in great diversity of forms amongst organisms, past and present.

How new species are formed

By far the most important way in which new species are formed is through reproductive separation followed by genetic change due to natural selection. Within a species there are one or more populations. Although individuals tend to breed only with others in the same population, they are nevertheless capable of breeding with individuals in other populations.

Suppose that a population becomes separated in some way from other populations and undergoes different mutations – it will become genetically different from the other populations. Each of the populations will experience different selection pressures because the environment of each will be slightly different. Natural selection will then lead to changes in the allelic frequencies of each population. The different phenotypes each combination of alleles produces will be subject to selection pressure that will lead to each population becoming adapted to its local environment. This is known as **adaptive radiation** and results in changes to the allele frequencies (evolution) of each population, in other words, each population evolves. As a result of these genetic differences it may be that, even if the populations were no longer physically separated from one another, they would be unable to interbreed successfully. Each population would now be a different species, each with its own gene pool. An example is shown in Figure 1.

Genetic drift is something that can take place in small populations. This is because the relatively few members of a small population possess a smaller variety of alleles than the members of a large population. In other words, their genetic diversity is less. As these few individuals breed, the genetic diversity of the population is restricted to those few alleles in the original population. As there are only a small number of different alleles there is not an equal chance of each being passed on. Those that are passed on will quickly affect the whole population as their frequency is high. Any mutation to one of these alleles that is selectively favoured will also more quickly affect the whole population because its frequency will be high. The effects of genetic drift will be greater and the population will change relatively rapidly, making it more likely to develop into a separate species. In large populations the effect of a mutant allele will be diluted because its frequency is far less in the much larger gene pool. The effects of genetic drift are likely to be less, and development into a new species is likely to be slower.

Two forms of speciation are **allopatric speciation** and **sympatric speciation**.

Allopatric speciation

Allopatric means different countries and describes the form of speciation where two populations become **geographically separated**. Geographical separation may be the result of any physical barrier between two populations which prevents them interbreeding. These barriers include oceans, rivers, mountain ranges and deserts. What proves a barrier to one species may be no problem to another. While an ocean may separate populations of hedgehogs, it can be crossed by many birds and for marine fish it is their very means of getting from place to place. A tiny stream may be a barrier to snails, whereas the whole of the Pacific Ocean fails to separate populations of certain birds.

If environmental conditions either side of the barrier vary, then natural selection will influence the two populations differently and each will evolve leading to adaptations to their local conditions. These changes may take many hundreds or even thousands of generations, but ultimately may lead to reproductive separation and the formation of separate species. Figure 1 shows how speciation might occur when two populations of a forest-living species become geographically separated by a region of arid grassland.

▲ **Figure 1** *Speciation as a result of geographical separation*

An example of allopatric speciation is the Galapagos finch. A single ancestral species is thought to have colonised one of the Galapagos islands. In the absence of competition its population increased and populations became established in other habitats on the same and different islands. Each population evolved adaptations to suit its new environment, including available food resources. These adaptations included different shapes and sizes of beak to deal with different seed types. Being geographically separated from its mainland population, these changes have led to the various populations being so different that they can no longer interbreed and now form separate species.

Sympatric speciation

Sympatric means same country and describes the form of speciation that results within a population in the same area leading to them becoming reproductively separated.

A likely example of sympatric speciation taking place is the apple maggot fly. Originally this insect only laid its eggs inside the fruit of hawthorns, which are native to North America. When apples trees were introduced, the fly started to lay its eggs in apples also. Females tend to lay their eggs on the type of fruit in which they developed and males tend to look for mates on the type of fruit in which they developed. So flies raised in hawthorns usually mate with each other and flies raised in apples tend to mate with each other. While the two types of flies are not yet separate species, mutations in each population have led to the evolution of genetic differences. In time this could result in them being incapable of successfully breeding with one another and therefore being separate species.

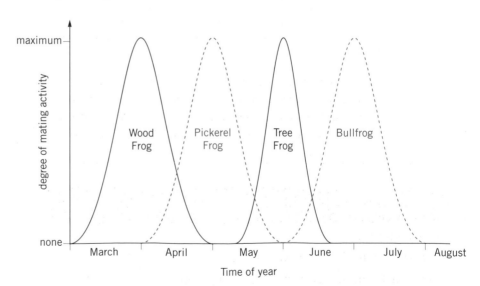

▲ **Figure 2** *Seasonal reproductive separation as illustrated by four varieties of frog*

Table 1 summarises forms of isolating mechanisms.

▼ **Table 1** *Summary of the forms of isolating mechanisms*

Type of variation	Number produced
Geographical	Populations are isolated by physical barriers such as oceans, mountain ranges, rivers, etc.
Ecological	Populations inhabit different habitats within the same area and so individuals rarely meet
Temporal	The breeding seasons of each population do not coincide and so they do not interbreed. Figure 2 illustrates this in relation to four types of frog
Behavioural	Mating is often preceded by courtship, which is stimulated by the colour or markings of the opposite sex, the call or particular actions of a mate. Any mutations which cause variations in these patterns may prevent mating, for example, if a female stickleback does not respond appropriately to the actions of the male, he ceases to court her
Mechanical	Anatomical differences may prevent mating occurring, for example, it may be physically impossible for the penis to enter the vagina in mammals
Gametic	The gametes may be prevented from meeting due to genetic or biochemical incompatibility. For instance, some pollen grains fail to germinate or grow when they land on a stigma of different genetic makeup. Some sperm are destroyed by chemicals in the female reproductive tract
Hybrid sterility	Hybrids formed from the fusion of gametes from different species are often sterile because they cannot produce viable gametes. For example, in a cross between a horse ($2n = 64$) and a donkey ($2n = 62$) the resultant mule has 63 chromosomes. It is impossible for these chromosomes to pair up appropriately during meiosis and so the gametes formed are not viable and the mule is sterile

Study tip

The examples of isolating mechanisms in Table 1 are for illustration only. You do not need to learn them.

Summary questions

1 Define a species.
2 Explain the meaning of the term speciation.
3 Describe the process of geographical separation.
4 Explain how geographical separation of two populations of a species can result in the accumulation of the differences in their gene pools.
5 Distinguish between allopatric and sympatric speciation.

Practice questions: Chapter 18

City	Frequency of allele			
	White	Non-agouti	Blotched	Long-haired
Athens	0.001	0.72	0.25	0.50
Paris	0.011	0.71	0.78	0.24
London	0.004	0.76	0.81	0.33

1 (a) What does the Hardy–Weinberg principle predict? *(3 marks)*

The table shows the frequencies of some alleles in the population of cats in three cities.

(b) White cats are deaf. Would the Hardy–Weinberg principle hold true for white cats? Explain your answer. *(2 marks)*

(c) What is the evidence from the table that non-agouti and blotched are alleles of different genes? *(1 mark)*

(d) Hair length in cats is determined by a single gene with two alleles. The allele for long hair (h) is recessive. The allele for short hair (H) is dominant. Use the information in the table and the Hardy–Weinberg equation to estimate the percentage of cats in London that are heterozygous for hair length. Show your working.

(2 marks)

AQA June 2010

	United Kingdom	Sudan
Life expectancy males/years	76.5	50.5
Life expectancy females/years	81.6	52.4

2 (a) Explain what is meant by birth rate. *(1 mark)*

(b) The table shows life expectancies for babies born in the United Kingdom and in the Sudan in 2009.

  (i) Describe the patterns shown by these data. *(2 marks)*

  (ii) Suggest reasons for the differences in the life expectancy shown by these data.

(2 marks)

AQA June 2011

▼ Table 1

Days after release	Number of marked insects remaining in population	Number of insects captured	Number of captured insects that were marked
1	1508	524	78
2	1430	421	30
3	1400	418	18
4	1382	284	2
5	1380	232	9

3 Ecologists investigated the size of an insect population on a small island. They used a mark-release-recapture method. To mark the insects they used a fluorescent powder. This powder glows bright red when exposed to ultraviolet (UV) light.

(a) The ecologists captured insects from a number of sites on the island. Suggest how they decided where to take their samples. *(2 marks)*

(b) Give **two** assumptions made when using the mark-release-recapture method.

(2 marks)

(c) Suggest the advantage of using the fluorescent powder in this experiment. *(2 marks)*

The ecologists did **not** release any of the insects they captured 1–5 days after release of the marked insects.

Table 1 shows the ecologists' results.

Calculate the number of insects on this island 1 day after release of the marked insects. Show your working.

(2 marks)

(d) The ecologists expected to obtain the same result from their calculations of the number of insects on this island on each day during the period 1–5 days after release. In fact, their estimated number increased after day 1.

During the same period, the number of insects they caught decreased.

The method used by the ecologists might have caused these changes. Use the information provided to suggest **one** way in which the method used by the ecologists might have caused the increase in their estimates of the size of the insect population.

(2 marks)

AQA Specimen 2014

4 Snow geese fly north to the Arctic in the spring and form breeding colonies. Different colonies form at different latitudes. The greater the latitude, the further north is the colony. The further north a breeding colony forms, the colder the temperature and the greater the risk of snow.

Colony	Latitude in degrees north	Percentage of white snow geese each year			
		1930	1950	1960	1970
A	72	100		100	100
B	71		>99	>99	>99
C	66	95	85	76	
D	63	86	75	67	65
E	55		62		28

(a) There is a positive correlation between the size of snow geese and how far north they breed. A large size results in snow geese being adapted for breeding in colder conditions. Explain how. *(2 marks)*

Snow geese are either white or blue in colour. The table shows the percentage of white snow geese in colonies at different latitudes at different times over a 40-year period. The blank cells in the table are years for which no figures are available.

(b) (i) Describe how the percentage of white snow geese varies with distance north.
   (1 mark)

   (ii) The further north, the greater the risk of snow. Use this information to explain how natural selection might have accounted for the effect of latitude on the percentage of white snow geese. *(3 marks)*

(c) The percentage of white snow geese in these colonies changed over the period shown in the table. Use your knowledge of climate change to suggest an explanation.
   (2 marks)

(d) Snow geese breed in large colonies. Scientists studied the nests in one colony. For each nest, they recorded the day on which the first egg hatched. They also recorded the number of young that survived from the nest. They used the data to plot a graph.
   (i) What type of natural selection is shown in the graph? *(1 mark)*
   (ii) Describe the evidence for your answer. *(1 mark)*
   AQA Jan 2010

5 The Amazonian forest today contains a very high diversity of bird species.
   • Over the last 2 000 000 years, long periods of dry climate caused this forest to separate into a number of smaller forests.
   • Different plant communities developed in each of these smaller forests.
   • Each time the climate became wetter again, the smaller forests grew in size and merged to reform the Amazonian forest.

(a) Use the information provided to explain how a very high diversity of bird species has developed in the Amazonian forest. *(5 marks)*

(b) Speciation is far less frequent in the reformed Amazonian forest. Suggest **one** reason for this. *(1 mark)*
   AQA June 2013

Specification reference: 3.7.4

Learning objectives

→ Define the terms environment, biotic, abiotic, and biosphere.

→ Explain what is meant by an ecosystem.

→ Explain what is meant by the terms population, community, and habitat.

→ Explain what a niche is.

In this chapter we shall look at how living organisms form communities within ecosystems through which energy is transferred and elements are recycled.

We shall learn how populations of different species live in communities and how competition for survival arises both within and between these populations. We shall also see how populations within a single community are affected by living and non-living factors in an ecosystem.

Ecology is the study of the inter-relationships between organisms and their environment. The environment includes both non-living (**abiotic**) factors, such as temperature and rainfall, and living (**biotic**) factors, such as competition and predation.

Ecosystems

Ecosystems are dynamic systems made up of a community and all the non-living factors of its environment. Ecosystems can range in size from very small to very large. Within an ecosystem there are two major processes to consider:

- the flow of energy through the system
- the cycling of elements within the system.

An example of an ecosystem is a freshwater pond or lake. It has its own community of plants to collect the necessary sunlight energy to supply the organisms within it. Nutrients such as nitrate ions and phosphate ions are recycled within the pond or lake. There is little or no loss or gain between it and other ecosystems. Another example of an ecosystem is an oak woodland (Figure 2). Within each ecosystem, there are a number of species. Each species is made up of a group of individuals that make up a **population**.

Populations

A **population** is a group of individuals of one **species** that occupy the same habitat at the same time and are potentially able to interbreed. An ecosystem supports a certain size of population of a species called the **carrying capacity**. The size of a population can vary as a result of:

- the effect of abiotic factors
- interactions between organisms, for example, intraspecific and interspecific competition and predation.

In the different habitats of an oak woodland there are populations of nettles, worms, green woodpeckers, beetles, etc. The boundaries of a population are often difficult to define. In our oak woodland, for example, all the mature green woodpeckers can breed with one another and so form a single population. Populations of different species form a **community**.

Hint

Many students confuse the terms biotic and abiotic. Biotic refers to living things, for example, predation or competition between different species or individuals of one species. Abiotic refers to non-living factors, such as temperature, rainfall, or light intensity.

▲ **Figure 1** *Woodland ecosystem*

Study tip

Population is a very important concept that needs to be understood. It comes up in many areas of biology, including genetics and evolution.

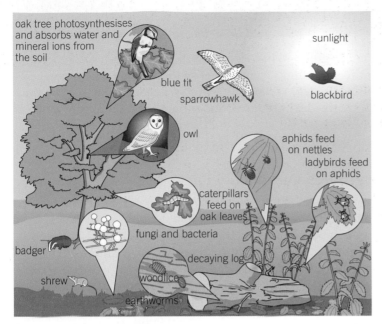

▲ **Figure 2** *Part of an oak woodland ecosystem*

Community

A **community** is defined as all the populations of different species living and interacting in a particular place at the same time. Within an oak woodland, a community may include a large range of organisms, such as oak trees, hazel shrubs, bluebells, nettles, sparrowhawks, blue tits, ladybirds, aphids, woodlice, earthworms, fungi, and bacteria (Figure 2).

Habitat

A **habitat** is the place where an organism normally lives and is characterised by physical conditions and the other types of organisms present. Within an ecosystem there are many habitats. For example, in an oak woodland, the leaf canopy of the trees may be a habitat for blue tits while a decaying log is the habitat for woodlice. A stream flowing through the woodland provides a very different habitat, within which aquatic plants and water beetles live. For a water vole, the stream and its banks are its habitat. Within each habitat there are smaller units, each with their own microclimate. These are called **microhabitats**. The mud at the bottom of the stream may be the microhabitat for a bloodworm while a crevice on the bark of an oak tree may be the microhabitat for a lichen.

Ecological niche

A **niche** describes how an organism fits into the environment. A niche refers to where an organism lives and what it does there. It includes all the biotic and abiotic conditions to which an organism is adapted in order to survive, reproduce and maintain a viable population. Some species may appear very similar, but their nesting habits or other aspects of their behaviour will be different, or they may show different levels of tolerance to environmental factors, such as a pollutant or a shortage of oxygen or nitrates. No two species occupy exactly the same niche. This is known as the competitive exclusion principle.

▲ **Figure 3** *This lake is an example of a habitat*

Summary questions

In the following passage, state the word that best replaces each of the numbers in brackets.

The study of the interrelationships between organisms and their environment is called (1). An ecosystem is a more or less self-contained functional unit made up of all the living or (2) features and non-living or (3) features in a specific area. Within each ecosystem are groups of different organisms, called a (4), which live and interact in a particular place at the same time. A group of organisms occupying the same place at the same time is called a (5), and the place where they live is known as a (6). The population size of a species that an ecosystem can support is known as (7).

Learning objectives

→ Describe the factors that determine the size of a population.

→ Describe the abiotic factors that affect the size of a population.

→ Explain how each of these factors influence population size.

Specification reference: 3.7.4

▲ **Figure 1** *A population of lesser flamingos*

▲ **Figure 1** *The collared dove only arrived in Britain in the 1950s but its population has increased rapidly since then*

A population is a group of individuals of the same species that occupy a habitat at the same time. The number of individuals in a population is the **population size**. We saw in Topic 19.1 that all the populations of the different organisms that live and interact together are known as a community. Populations are dynamic in that they vary in size and composition over time.

Plotting growth curves

Where a population grows in size slowly over a period of time it is possible to plot a graph of numbers in a population against time. Where the population grows rapidly over a short period of time this may not be possible. This is often the case when measuring the growth of microorganisms. Consider Table 1 which shows the increase in population size of a bacterium that initially doubles its numbers each hour. If we try to plot a graph of numbers against time using a time scale that allows us to differentiate each point, the curve runs off the graph after the point plotted at 4 hours (Figure 2).

In these cases it is necessary to use a logarithmic scale to represent the number of bacteria. The logarithms of bacterial numbers are shown in Table 1. When the graph of log bacterial numbers is plotted against time all points can be represented on the graph (Figure 3) and we can see that the rate of growth starts to slow after 8 hours.

Population size

Imagine a situation in which a single photosynthetic bacterial cell, capable of asexual reproduction, is placed in a newly created pond. It is summer and so there is plenty of light and the temperature of the water is around 12 °C – mineral nutrients have been added to the water. In these circumstances the bacterial cell divides rapidly because

▼ **Table 1**

Time/hours	Number of bacteria	Log number of bacteria
0	50	1.7
1	100	2.0
2	200	2.3
3	400	2.6
4	800	2.9
5	1 600	3.2
6	3 200	3.5
7	6 400	3.8
8	12 800	4.1
9	20 300	4.3
10	31 500	4.5

▲ **Figure 2** *Graph of number of bacteria against time*

▲ **Figure 3** *Graph of log number of bacteria against time*

all the factors needed for the growth of the population are present. There are no **limiting factors**. In time, however, things change. For example:

- Mineral ions are consumed as the population becomes larger.
- The population becomes so large that the bacteria at the surface prevent light reaching those at deeper levels.
- Other species are introduced into the pond, carried by animals or the wind, and some of these species may use the bacteria as food or compete for light or minerals.
- Winter brings much lower temperatures and lower light intensity of shorter duration.

In short, the good life ends and the going gets tough. As a result the growth of the population slows, and possibly ceases altogether, and the population size may even diminish. Over the winter the population is likely to reach a relatively constant size. There are many factors, living (biotic) and non-living (abiotic), which affect this population size. Changes in these factors will influence the rate of growth and the size of the population.

In summary, no population continues to grow indefinitely because certain factors limit growth, for example, the availability of food, light, water, oxygen and shelter, and the accumulation of toxic waste, disease and predators. Each population has a certain size, the **carrying capacity**, that can be sustained over a relatively long period and this is determined by these limiting factors.

Abiotic factors

The abiotic conditions that influence the size of a population include:

- **temperature**. Each species has a different optimum temperature at which it is best able to survive. The further away from this

▲ **Figure 6** *A population of migrating birds, like these terns, fluctuates seasonally*

Hint

Remember that the size of any population is eventually determined by a limiting factor.

Hint

A species can only live within a certain range of abiotic factors and this range differs from species to species.

Synoptic link

To remind yourself of the effects of temperature and pH on enzyme action revisit Topic 1.8 , Factors affecting enzyme action.

Practical link

Required practical 12. Investigation into the effect of a named environmental factor on the distribution of a given species.

optimum, the fewer individuals in a population are able to survive and the smaller is the population that can be supported. In plants and cold-blooded animals, as temperatures fall below the optimum, the enzymes work more slowly and so their metabolic rate is reduced. Populations therefore have a smaller carrying capacity. At temperatures above the optimum, the enzymes work less efficiently because they gradually undergo **denaturation**. Again the population's carrying capacity is reduced.

The warm-blooded animals, that is, birds and mammals, can maintain a relatively constant body temperature regardless of the external temperature. Therefore you might think that their carrying capacity would be unaffected by temperature. However, the further the temperature of the external environment gets from their optimum temperature, the more energy these organisms expend in trying to maintain their normal body temperature. This leaves less energy for individual growth and so they mature more slowly and their reproductive rate slows. The carrying capacity of the population is therefore reduced.

- **Light**. As the ultimate source of energy for most **ecosystems**, light is a basic necessity of life. The rate of photosynthesis increases as light intensity increases. The greater the rate of photosynthesis, the faster plants grow and the more spores or seeds they produce. Their carrying capacity is therefore potentially greater. In turn, the carrying capacity of animals that feed on plants is potentially larger.

- **pH**. This affects the action of enzymes. Each enzyme has an optimum pH at which it operates most effectively. A population of organisms is larger where the appropriate pH exists and smaller, or non-existent, where the pH is different from the optimum.

- **Water and humidity**. Where water is scarce, populations are small and consist only of species that are well adapted to living in dry conditions. Humidity affects the **transpiration** rates in plants and the evaporation of water from the bodies of animals. Again, in dry air conditions, the populations of species adapted to tolerate low humidity will be larger than those with no such adaptations.

In general terms, when any abiotic factor is below the optimum for a population, fewer individuals are able to survive because their adaptations are not suited to the conditions. If no individuals have adaptations that allow survival, the population becomes extinct.

▲ **Figure 7** *This cactus is adapted to survive in conditions where water is scarce. Its population in dry regions is therefore relatively large as there is little competition from other species, most of which are not adapted to survive in such conditions*

Maths link $\sqrt{x}$

MS 0.5 and 3.2, see Chapter 22.

The influence of abiotic factors on plant populations

Species X and species Y are two species of flowering plants. Each is able to tolerate different temperatures and different pHs. In this way they avoid direct competition by occupying different niches. This is called niche separation. The chart (Figure 8) below illustrates the way each species is able to tolerate each of these two abiotic factors.

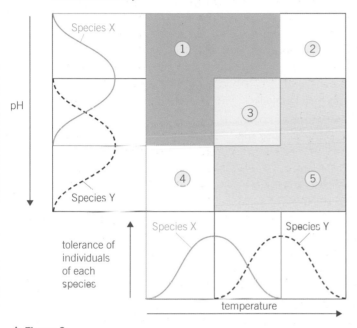

▲ Figure 8

1 State the numbered box that best fits each of the descriptions below.
  a Only a population of species X is found.
  b Both temperature and pH allow a population of both species to exist.
  c The temperature is too high for a population of species X and the pH is too low for a population of species Y.
  d There is competition between species X and species Y.

2 Explain why there is no population of either species in box 4.

Summary questions √x̄

1 Explain why populations never grow indefinitely.

2 Distinguish between biotic and abiotic factors.

3 Suggest the level and type of abiotic factor that is most likely to limit the population size of the organisms and their habitats given below.
  a Ground plants on a forest floor
  b Hares in a sandy desert
  c Bacteria on the summit of a high mountain.

4 √x̄ Table 2 shows the estimated world population over the past 12 000 years.

▼ Table 2

Time / years before present (BP)	Estimated human population / billions
0	600
2 000	200
4 000	35
6 000	20
8 000	10
10 000	5
12 000	1

  a Explain the benefits of using a logarithmic scale for population numbers when plotting a graph of these data.
  b Calculate to three significant places the log values for the human population in each case.
  c Plot a suitable graph to show the growth of the human population over the past 12 000 years.

Maths link √x̄

MS 0.3 and 3.1, see Chapter 22.

The growth and size of human populations $\sqrt{x}$

The human population has doubled in less than 50 years and now totals over 7 billion. The basic factors that affect the growth and size of human populations are the **birth rate** and the **death rate**. It is the balance between these two factors that determines whether a human population increases, decreases or remains the same.

Individual populations are further affected by **migration**, which occurs when individuals move from one population to another. There are two types of migration:

- **immigration**, where individuals join a population from outside
- **emigration**, where individuals leave a population.

Again it is a balance between these two components that affects population size.

population growth = (births + immigration) − (deaths + emigration)

$$\text{percentage population growth rate (in a given period)} = \frac{\text{population change during the period}}{\text{population at the start of the period}} \times 100$$

1 $\sqrt{x}$ The figures below show some population statistics for a country.

Total population at the start of 2007 = 1 000 000

Birth rate in 2007 = 25 per 1000 of population

Death rate in 2007 = 20 per 1000 of population

Calculate the percentage population growth for this country in 2007. Show your working.

As the future size of a human population depends upon the number of females of child-bearing age, it is useful to know the age and gender profile of a population. This is displayed graphically by a series of stacked bars representing the percentages of males and females in each age group. These graphs, called age **population pyramids**, give useful information on the future trends of different populations. Three typical types of population are represented in the age population pyramids in Figure 10. These are:

- **stable population** (Figure 10a), where the birth rate and death rate are in balance and so there is no increase or decrease in the population size.
- **increasing population** (Figure 10b), where there is a high birth rate, giving a wider base to the population pyramid (compared to a stable population) and fewer older people, giving a narrower apex to the pyramid. This type of population is typical of economically less developed countries.
- **decreasing population** (Figure 10c), where there is a lower birth rate (narrower base of the population pyramid) and a lower mortality rate leading to more elderly people (wider apex to pyramid). This type of population occurs in certain economically more developed countries, such as Japan.

As countries have developed economically their human populations have, so far, displayed a pattern of growth known as demographic transition. This pattern can be divided into four stages depending on the birth rate, death rate and total population size. The relationship between these four stages and the birth rates, death rates and total population are illustrated in Figure 11.

▲ **Figure 9** *The human population now exceeds 7 billion*

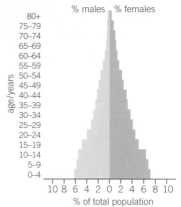

a *Population pyramid for a stable population*

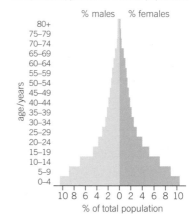

b *Population pyramid for an increasing population*

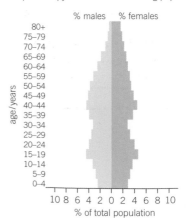

c *Population pyramid for a decreasing population*

▲ **Figure 10** *Age population pyramids*

2 Using Figure 11, suggest which of the four stages (1, 2, 3 or 4) best applies to each of the descriptions below.

a A country that has a rapidly falling birth rate and a relatively low death rate.

b A country in which there is a high birth rate but much starvation and periodic epidemic disease.

c A country where there have been many years of improved nutrition, far less infectious disease and a large number of children.

d Britain 20 000 years ago when famine and disease led to regular population crashes.

e Britain today.

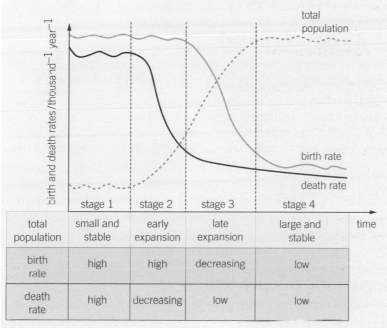

total population	small and stable	early expansion	late expansion	large and stable
birth rate	high	high	decreasing	low
death rate	high	decreasing	low	low

▲ **Figure 11** *Demographic transition*

▲ **Figure 12** *Age population pyramids A and B*

3 Figure 12 shows two age population pyramids: A and B. Suggest which stage of the demographic transition model shown in Figure 11 each pyramid represents. Give reasons for your answer in each case.

Learning objectives

→ Describe what is meant by intraspecific competition.

→ Summarise the factors that different species compete for.

→ Describe interspecific competition.

→ Explain how interspecific competition influences population size.

Specification reference: 3.7.4

Synoptic link

To help you understand competition and predation you should first revise the information on energy and ecosystems in Topics 13.1 and 13.2.

Hint

Which of two species in competition has the competitive advantage depends on the conditions at any point in time. If one species can tolerate a higher temperature than another, a rise in environmental temperature will favour it. If however there is a fall in environmental temperature, the other species is more likely to become dominant.

Where two or more individuals share any resource (e.g., light, food, space, oxygen) that is insufficient to satisfy all their requirements fully, then competition results. Where such competition arises between members of the same species it is called **intraspecific competition**. Where it arises between members of different species it is termed **interspecific competition**.

Intraspecific competition

Intraspecific competition occurs when individuals of the *same* species compete with one another for resources such as food, water, breeding sites, etc. It is the availability of such resources that determines the size of a **population**. The greater the availability, the larger the population. The lower the availability, the smaller the population. Availability of resources also affects the degree of competition between individuals which results in a smaller population. Examples of intraspecific competition include:

*   limpets competing for algae, which is their main food. The more algae available, the larger the limpet population becomes.
*   oak trees competing for resources. In a large population of small oak trees some will grow larger and restrict the availability of light, water and minerals to the rest, which then die. In time the population will be reduced to relatively few large dominant oaks.
*   robins competing for breeding territory. Female birds are normally only attracted to males who have established territories. Each territory provides adequate food for one family of birds. When food is scarce, territories become larger to provide enough food. There are therefore fewer territories in a given area and fewer breeding pairs, leading to a smaller population size.

Interspecific competition

Interspecific competition occurs when individuals of *different* species compete for resources such as food, light, water, etc. When populations of two species are in competition one will normally have a competitive advantage over the other. The population of this species will gradually increase in size while the population of the other will diminish. If conditions remain the same, this will lead to the complete removal of one species. This is known as the competitive exclusion principle.

This principle states that where two species are competing for limited resources, the one that uses these resources most effectively will ultimately eliminate the other. In other words, no two species can occupy the same niche indefinitely when resources are limiting. Two species of sea birds, shags and cormorants, appear to occupy the same niche, living and nesting on the same type of cliff face and eating fish from the sea. Analysis of their food, however, shows that shags feed largely on sand eels and herring, whereas cormorants eat mostly flat fish, gobies, and shrimps. They therefore occupy different niches.

To show how a factor influences the size of a population it is necessary to link it to the birth rate and death rate of individuals in a population. For example, an increase in food supply does not necessarily mean there will be more individuals - it could just result in bigger individuals. It is therefore important to show how a factor, such as a change in food supply, affects the number of individuals in a population. For example, a decrease in food supply could lead to individuals dying of starvation and directly reduce the size of a population. An increase in food supply means that more individuals are likely to survive and so there is an increased probability that they will produce offspring and the population will increase. This effect therefore takes longer to influence population size.

The effects of interspecific competition on population size

The red squirrel is native to the British Isles and exclusively occupied a particular niche until around 130 years ago, when the grey squirrel was introduced from North America. Since then the two species have been competing for food and territory. There are now an estimated 2.5 million grey squirrels and just 160 000 red squirrels in the British Isles. The red squirrel population lives mostly in Wales and Scotland, with smaller groups in north eastern England and on islands such as Anglesey and the Isle of Wight. Figure 1 illustrates the changes in red and grey squirrel populations in Wales and Scotland between 1970 and 1990.

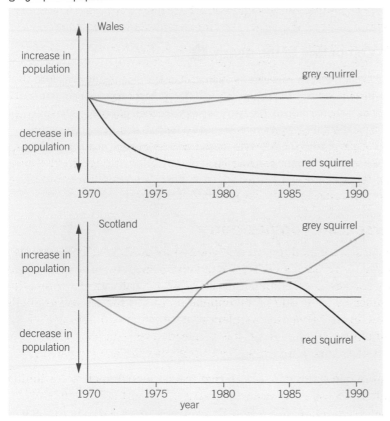

▲ **Figure 1** *Changes in red and grey squirrel populations in Wales and Scotland between 1970 and 1990. The lines show changes in comparison with the 1970 population*

In many cases we suspect that competition is the reason for variations in population. In practice it is difficult to prove for a number of reasons:

- There are many other factors that influence population size, such as abiotic factors.
- A causal link has to be established to show that competition is the cause of an observed correlation.

▲ **Figure 2** *Red squirrel*

▲ **Figure 3** *Grey squirrel*

Maths link √x̄

MS 3.1, see Chapter 22.

- There is a time lag in many cases of competition and so a population change may be due to competition that took place many years earlier.
- Data on natural population sizes are hard to obtain and not always reliable.

Study Figure 1 and answer the following questions.

1 State one piece of evidence from the graph for Scotland which indicates that changes in the red squirrel population are due to competition from the grey squirrel.

2 In Wales the populations of both grey and red squirrels declined between 1970 and 1975. Suggest a possible reason for this.

3 Both types of squirrels eat nuts, seeds and fruit as part of their diet. Grey squirrels spend more time foraging on the forest floor than red squirrels. Suggest how this behaviour might give the grey squirrel a competitive advantage over the red squirrel.

4 Suggest an explanation why islands such as Anglesey and the Isle of Wight still have significant red squirrel populations while they have disappeared from much of the rest of England and Wales.

 Competing to the death

In an experiment, two species of a genus of unicellular organism called *Paramecium* were grown separately in different test tubes that contained yeast as a source of food. The two species were then grown together in the same test tube – again with yeast as a food source. In each case the populations of both species were measured over a period of 20 days. The results are shown in the graph in Figure 4.

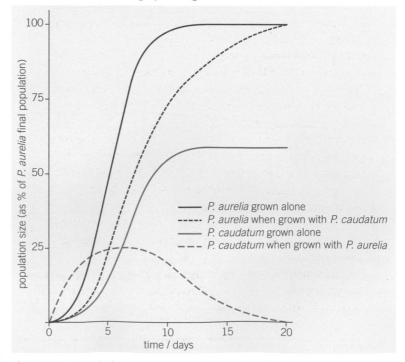

▲ **Figure 4** *Population growth of* Paramecium aurelia *and* P. caudatum *grown separately and together*

1 Describe the population growth curve of *P. caudatum* when grown alone over the 20-day period.
2 Compare the population growth curve of *P. caudatum* when grown with *P. aurelia* to the curve when *P. caudatum* is grown alone.
3 Suggest an explanation for the difference in the final population size of *P. caudatum* when grown with *P. aurelia* compared with when it is grown alone.
4 Suggest why the growth rate of *P. aurelia* is slower in the presence of *P. caudatum* than when grown alone.
5 Suggest why, after 20 days, the population size of *P. aurelia* grown with *P. caudatum* is the same as that when *P. aurelia* is grown alone.

Hint

Although the population of one species may increase as another decreases, this does not prove that this is due to direct competition between them. To be certain, it is necessary to establish a causal link for the observed correlation.

Effects of abiotic and biotic factors on population size $\sqrt{x}$

Oak trees produce acorns in the autumn. Deer mice feed on acorns. Table 1 shows the dry mass of acorns produced per hectare (ha) from 1992 to 1997 in an area of woodland. It also shows the estimated population size of deer mice per hectare of the same area of woodland in the spring of each year from 1993 to 1998.

Maths link $\sqrt{x}$

MS 1.2 and 1.3, see Chapter 22.

1 Suggest a method by which the population of deer mice might be estimated.
2 $\sqrt{x}$ Calculate the mean annual growth rate in deer mice population over the period 1993 to 1995. Show your working.
3 With reference to the data in the table, describe the relationship between acorn production in autumn and the deer mice population the following spring.
4 Acorn seeds begin to form in spring. It has been suggested that the higher the temperature in spring, the more acorns are produced the following autumn. From the table, state which year probably had the coldest spring.
5 The caterpillars of the gypsy moth feed on oak leaves. When the population of gypsy moth caterpillars is large, the damage they cause to oak trees reduces acorn production. Suggest how and why a rise in the population of gypsy moth caterpillars might affect the population of deer mice.
6 As well as acorns, deer mice also eat the pupae of gypsy moths.
   a Explain how a warm spring might result in a fall in the gypsy moth population the following year.
   b Owls are natural predators of deer mice. Suggest the possible effect of an increase in the owl population on the production of acorns. Explain your answer.

▼ Table 1

Year	Dry mass of acorns/kg ha^{-1} produced in autumn	Estimated deer mice population/ number ha^{-1} in spring
1992	28	–
1993	131	260
1994	318	550
1995	211	1320
1996	726	990
1997	39	3440
1998	–	340

In Topic 19.3 we looked at interspecific competition. We shall now turn our attention to one type of interspecific relationship, the predator–prey relationship. A **predator** is an organism that feeds on another organism, known as their **prey**.

As predators have evolved they have become better adapted for capturing their prey - faster movement, more effective camouflage, better means of detecting prey. Prey have equally become more adept at avoiding predators - better camouflage, more protective features such as spines, concealment behaviour. In other words the predator and the prey have evolved alongside each other. If either of them had not matched the adaptations of the other, it would most probably have become extinct.

Predation

Predation occurs when one organism is consumed by another. When a population of a predator and a population of its prey are brought together in a laboratory, the prey is usually exterminated by the predator. This is largely because the range and variety of the habitat provided is normally limited to the confines of the laboratory. In nature the situation is different. The area over which the population can travel is far greater and the variety of the environment is much more diverse. In particular, there are many more potential refuges. In these circumstances some of the prey can escape predation because the fewer there are the harder they are to find and catch. Therefore, although the prey population falls to a low level, it rarely becomes extinct.

Evidence collected on predator and prey populations in a laboratory does not necessarily reflect what happens in the wild. At the same time, it is difficult to obtain reliable data on natural populations because it is not possible to count all the individuals in a natural population. Its size can only be estimated from sampling and surveys. These are only as good as the techniques used, none of which guarantee complete accuracy. We must therefore treat all data produced in this way with caution.

Effect of predator–prey relationship on population size

The relationship between predators and their prey and its effect on population size can be summarised as follows:

- Predators eat their prey, thereby reducing the population of prey.
- With fewer prey available the predators are in greater competition with each other for the prey that are left.
- The predator population is reduced as some individuals are unable to obtain enough prey for their survival or to reproduce.
- With fewer predators left, fewer prey are eaten and so more survive and are able to reproduce.
- The prey population therefore increases.
- With more prey now available as food, the predator population in turn increases.

This general predator–prey relationship is illustrated in Figure 1. In natural **ecosystems**, however, organisms eat a range of foods and therefore the fluctuations in population size shown in the graph are often less severe.

▲ **Figure 1** *Relationships between prey and predator populations*

Although predator–prey relationships are significant reasons for cyclic fluctuations in populations, they are not the only reasons, disease and climatic factors also play a part. These periodic population crashes are important in evolution as there is a **selection pressure** which means that those individuals who are able to escape predators, or withstand disease or an adverse climate, are more likely to survive to reproduce. The population therefore evolves to be better adapted to the prevailing conditions.

 The Canadian lynx and the snowshoe hare

The long-term study of the predator–prey relationship of the Canadian lynx and the snowshoe hare was made possible because records exist of the number of furs traded by companies such as the Hudson Bay Company in Canada over 200 years. By analysing these records the relative population size of the Canadian lynx and the snowshoe hare can be determined. The data collected are shown as a graph in Figure 3.

1 State what assumption is being made if we use the number of each type of fur traded as a measure of the population size of each species.
2 Describe the changes that occur in the populations of Canadian lynx and snowshoe hare.
3 Explain the changes that you have described.

It has long been observed that the population of snowshoe hares fluctuates in cycles. The question is whether these fluctuations are due mostly to predation by the lynx, mostly to changes in the food supply or mostly to a combination of both. To find out, ecologists fenced off 1 km² areas of coniferous forest in Canada where the hares lived. Separate areas were treated in four different ways:

* In the first set of areas, the hares were given extra food.
* In the second set of areas, lynx were excluded.
* In the third set of areas, the hares were given extra food and lynx were excluded.
* In the fourth set of areas, conditions were left unaltered as a control.

The results of the experiment are shown in Figure 4.

Summary questions

1 Explain why a predator population often exterminates its prey population in a laboratory but rarely does so in natural habitats.

2 Explain how a fall in the population of a predator can lead to a rise in its prey population.

3 A species of mite (A) is fed on oranges in a laboratory tank until its population is stable. A second mite species (B), that preys on species A, is introduced into the tank. Sketch a graph of the likely cycle of population change that the two species will undergo. Explain the changes that the graph illustrates.

▲ **Figure 2** *Canadian lynx catching a snowshoe hare*

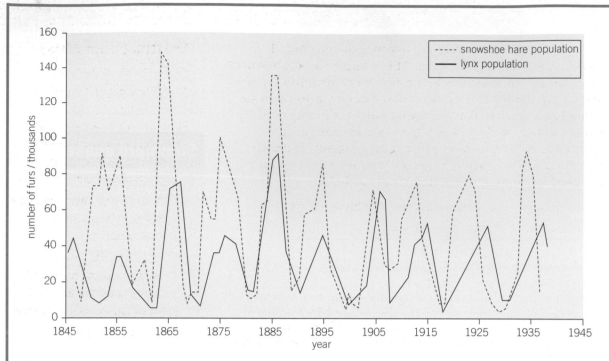

▲ **Figure 3** *The predator–prey relationship illustrated by the number of snowshoe hare and lynx trapped for the Hudson Bay Company between 1845 and 1940*

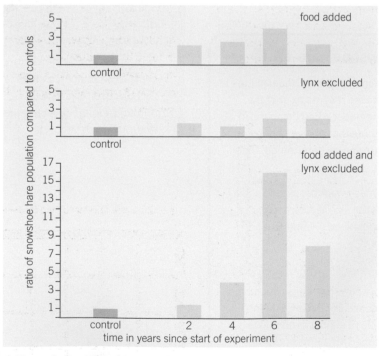

▲ **Figure 4** *Snowshoe hare population experiment*

4 √x̄ Calculate by how many times the addition of food increased the population after six years compared with the control.

5 Deduce which had the greater influence on the population of hares— the addition of food or the exclusion of the lynx. Explain your answer.

6 Deduce what conclusions can be drawn from this experiment.

19.5 Investigating populations

To study a **habitat**, it is often necessary to count the number of individuals of a species in a given space. This is known as **abundance**. It is virtually impossible to identify and count every organism. To do so would be time-consuming and would almost certainly cause damage to the habitat being studied. For this reason only small samples of the habitat are usually studied in detail. As long as these samples are representative of the habitat as a whole, any conclusion drawn from the findings will be reliable. There are a number of sampling techniques used in the study of habitats. These include:

- random sampling using frame quadrats or point quadrats
- systematic sampling along a belt transect.

Quadrats

Two types of quadrat frequently used are:

A point quadrat which consists of a horizontal bar supported by two legs. At set intervals along the horizontal bar are ten holes, through each of which a long pin may be dropped (Figure 1). Each species that the pin touches is then recorded.

A frame quadrat which is a square frame divided by string or wire into equally sized subdivisions (Figure 2). It is often designed so that it can be folded to make it more compact for storage and transport. The quadrat is placed in different locations within the area being studied. The abundance of each species within the quadrat is then recorded.

Learning objectives

→ Name the factors to be considered when using a quadrat.

→ Explain how a transect is used to obtain quantitative data about changes in communities along a line.

→ Describe how the abundance of different species is measured.

→ Explain how the mark-release-recapture method can be used to measure the abundance of motile species.

Specification reference: 3.7.4

Synoptic link

Random sampling was considered in Topic 10.5, Quantitative investigations of variation, and provides further information on the subject.

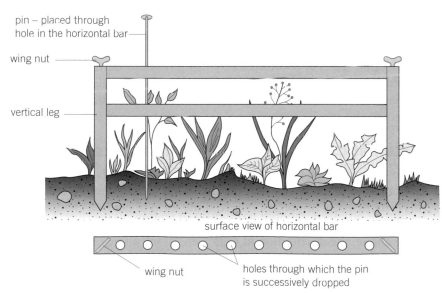

▲ **Figure 1** *A point quadrat*

There are three factors to consider when using quadrats:

- **The size of quadrat to use**. This will depend on the size of the plants or animals being counted and how they are distributed within the area. Larger species require larger quadrats. Where a population of species is not evenly distributed throughout the area, a large number of small quadrats will give more representative results than a small number of large ones.

▲ **Figure 2** *A frame quadrat*

to its climax community is the same as described above, except that it normally occurs more rapidly. This is because soil already exists in which spores and seeds often remain alive in the soil, and there is an influx of animals and plants through dispersal and migration from the surrounding area. This type of succession is called **secondary succession**. Because the land has been altered in some way, for example, by fire, some of the species in the climax community will be different.

Figure 3 summarises the events of ecological succession on land.

> **Maths link** $\sqrt{x}$
>
> MS 3.1, see Chapter 22.

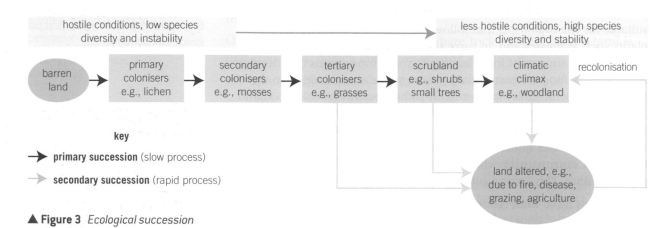

▲ **Figure 3** *Ecological succession*

Summary questions

1 State the general name given to the first organisms to colonise bare land.

2 Describe how changes in the environment lead to increased biodiversity during succession.

3 State the name that is given to the stable, final stage of any succession.

▲ **Figure 4** *The grassland in the foreground is grazed by sheep and so is prevented from reaching its natural climax. The land behind the fence has not been grazed for many years and has reverted to the climax community of woodland. This is therefore an example of secondary succession*

Warming to succession

Many glaciers in the northern hemisphere have been melting over the past 200 years. This retreat is, in part, the result of the additional global warming that has taken place since the industrial revolution and the burning of fossil fuels that has accompanied it. When glaciers melt and retreat they leave behind gravel deposits known as moraines. The retreat of the glaciers in Glacier Bay, Alaska, has been measured since 1794 and so the age of the moraines in this region is recorded.

Although no ecologist has been present to watch the succession that has taken place on these moraines, they can infer the changes that have occurred by examining the plant and animal communities on the moraines of different ages. The youngest moraines (those nearest the retreating glacier) have the earliest colonisers (pioneer species), whereas those successively further away from the glacier show a time sequence of later communities.

Each stage of a succession has its own distinctive community of plants and animals that alters the environment in a way that allows the next stage and its community to develop. The stages that follow the retreat of an Arctic glacier are:

*   **pioneer stage**. In the early years after the ice has retreated, photosynthetic bacteria and lichens colonise patches of land. Both of these pioneers fix nitrogen This is essential because nitrogen is virtually absent from glacial moraines. They also form tough mats that help to stabilise the loose surface of the moraines. When these pioneer species die, they decompose to form humus. Humus provides the nutrients that enable mosses to colonise. The pioneer stage occurs when the land has been ice free for 10–20 years.

- *Dryas* stage. Some 30 years after the ice has retreated, the ground is an almost continuous mat of the herbaceous plant *Dryas*. Its roots stabilise the thin and fragile soil layer formed from the erosion of the rocks that make up the moraine. Bacteria in root nodules of *Dryas* also fixes nitrogen, further adding nitrogenous nutrients to this poor-quality soil. Other plants found at this stage are the Arctic poppy and moss campion.

- alder stage. This arises about 60 years after the ice has retreated. Alder is a tree that has nitrogen-fixing nodules on its roots, enabling it to grow on nitrogen-poor soil. Alder sheds its leaves, which decompose into nitrogen-rich humus that further enriches the soil. The alder stage occurs some 50–70 years after the retreat of the glacial ice.

- spruce stage. About 100 years after the ice has retreated, spruce trees develop amongst the alder. A period of transition takes place and during the next 50 years or so the taller spruce out-competes the alder and ultimately displaces it altogether.

Figure 5 summarises changes in soil nitrogen, plant diversity and biomass following the retreat of an Arctic glacier.

▲ **Figure 6** Dryas (mountain avens) is the most common pioneer species in Glacier Bay, Alaska. It is able to fix nitrogen and forms dense mats and therefore enriches and stabilises the thin fragile soil

▲ Figure 5

▲ **Figure 7** Arctic poppy (yellow flower) and moss campion (pink flowers) are early flowering pioneer species on Arctic moraines

1 Using the information on the graphs, describe and explain the changes in above-ground biomass over the 200-year period.

2 a Using your knowledge of the nitrogen cycle, explain how nitrogen from the atmosphere becomes incorporated into the soil, causing its level to increase during the first 100 years after the glacier retreats.

  b Suggest two reasons for the fall in soil nitrogen levels after 150 years.

3 Suggest a reason for:

  a the rapid increase in plant species during the first 30 years after the retreat of the glacier.

  b the fall in the number of plant species 100 years after the retreat of the glacier.

4 Explain why it would be more appropriate to use a transect rather than quadrats placed at random when investigating this succession.

▲ **Figure 8** Spruce trees are the final succession stage following the retreat of glacial ice in the Arctic. They begin to grow around 100 years after the ice has retreated and persist as the dominant vegetation for centuries

Learning objectives

→ Describe what is meant by conservation.

→ Explain how managing succession can help to conserve habitats.

Specification reference: 3.7.4

▲ **Figure 1** *Moorland is an example of the conservation of a habitat by managing succession. Burning of heather and grazing by sheep has prevented shrubs and trees from developing*

What is conservation?

Conservation is the management of the Earth's natural resources by humans in such a way that maximum use of them can be made in the future. This involves active intervention by humans to maintain **ecosystems** and **biodiversity**. It is therefore a dynamic process that entails careful management of existing resources and reclamation of those already damaged by human activities. The main reasons for conservation are:

- **personal** to maintain our planet and therefore our life support system.
- **ethical**. Other species have occupied the Earth far longer than we have and should be allowed to coexist with us. Respect for living things is preferable to disregard for them.
- **economic**. Living organisms contain a gigantic pool of genes with the capacity to make millions of substances, many of which may prove valuable in the future. Long-term productivity is greater if ecosystems are maintained in their natural balanced state.
- **cultural and aesthetic**. Habitats and organisms enrich our lives. Their variety adds interest to everyday life and inspires writers, poets, artists, composers, and others who entertain and fulfill us.

Conserving habitats by managing succession

We saw in Topic 19.6 that any **climax community** has undergone a series of successional changes to reach its current state. Many of the species that existed in the earlier stages are no longer present as part of the climax community. This is because their habitats have disappeared as a result of succession, or species have been out-competed by other species or they have been taken over for human activities. One way of conserving these habitats, and hence the species they contain, is by managing succession in a way that prevents a change to the next stage.

One example is the moorland that exists over much of the higher ground in the UK. The burning of heather and grazing by sheep has prevented this land from reaching its climax community. The burning and grazing destroy the young tree saplings and so prevent the natural succession into deciduous woodland.

Around 4000 years ago, much of lowland UK was a climax community of oak woodland, but most of this forest was cleared to allow grazing and cultivation. The many heaths and grasslands that we now refer to as natural are the result of this clearance and subsequent grazing by animals. An example is chalk downland which was cleared of forest and where sheep and rabbits now eat any new saplings preventing these saplings from developing into full grown trees.

If the factor that is preventing further succession is removed, then the ecosystem develops naturally into its climatic climax (secondary succession). For example, if grasslands are no longer grazed or mowed, or if farmland is abandoned, shrubs initially take over, followed by deciduous woodland. Sand dunes can be managed to prevent succession to woodland leaving wet areas where species like natterjack toads can thrive.

Summary questions

Fenland is an area of waterlogged marsh and peat land. It supports a rich and unique community of plants and animals. If left alone, reeds initially dominate and the area gradually dries out as dead vegetation accumulates. Grasses, shrubs and trees in turn replace the fenland species.

1 Identify reasons for conserving habitats such as fenland.

2 Suggest practical measures that may be taken to prevent succession by grasses, shrubs, and trees in fenland.

Conflicting interests

One challenging conservation issue in the UK is the conflict between the conservation of hen harriers and the commercial hunting of red grouse. One scientific survey investigated the effect of predation by hen harriers on the breeding success of red grouse on managed moorland in Scotland. Some of the results included:

- On moorland where hen harriers were present there were, on average, 17% fewer young grouse than on moorlands without hen harriers.

- Over a three-year period grouse nests were intensively observed during the six weeks following the hatching of chicks. In this period, predation by harriers accounted for 91% of grouse chick losses.

- Prey remains found around harrier nests were examined. Of the 300 items identified, 32% were grouse chicks.

1 √x̄ Calculate how many of the items of prey identified around harrier nests were grouse chicks.

2 Harriers also feed on voles and meadow pipits. Explain how a rise in the population of these organisms might affect the population of grouse.

Moorland is considered one of the most attractive landscapes in the UK. Many of the national parks are made up of moorland and are visited by millions of people each year. To rear grouse, moorland has to be carefully managed. Controlled grazing by sheep and the periodic burning of vegetation are used to maintain low-growing plant populations of heather, bilberry, and crowberry that grouse feed on and nest within. The money to support this management comes largely from charges made to those who shoot grouse.

3 Explain what might happen to moorland if sheep-grazing and burning of the vegetation ceased.

The population of grouse in the UK is in decline due mainly to disease. Currently there are around 250 000 breeding pairs. The hen harrier was persecuted to such an extent that, by 1900, it was only found on a few Scottish islands. It recolonised the UK mainland in the 1970s and there are now around 750 breeding pairs. Both harriers and grouse normally produce one clutch of eggs each year. Hen harriers are protected by law and it is illegal to kill them, collect their eggs or destroy their nests. Conservationists want to retain this protection so that the population of hen harriers can increase. Grouse managers want to be allowed to control hen harrier populations to prevent them threatening the declining grouse populations.

4 Outline the arguments for and against continued protection of hen harrier populations.

To try to help resolve this conflict, scientists are currently conducting experiments to test whether hen harrier populations can be increased at the same time as reducing their negative impact on grouse populations. The information can then be used to inform decisions about how best to conserve grouse, harriers and moorland habitats.

The experiment will be carried out in two large areas where harriers are currently rare. Within these areas, the results of two strategies on the size of harrier and grouse populations will be measured:

- Killing hen harrier chicks, or moving them to a different location, when the harrier population size reaches an agreed ceiling.

- Providing alternative sources of food for hen harriers.

▲ **Figure 2** *Hen harrier*   ▲ **Figure 3** *Red grouse*

5 In each of the following, suggest a reason why:
  a The experiment will take at least five years to produce any findings.
  b An independent body, acceptable to both conservation groups and grouse managers will be needed to oversee the experiment.
  c The sites chosen for the experiment are ones where harriers can be expected to colonise relatively quickly.
  d Each experimental area will contain a number of different moorland sites managed by different individuals.
  e Some people are concerned about the long-term implications of a suspension, however temporary, to the legal protection of harriers that would be required during the experiment.
6 Explain how scientific experiments such as this one help to inform decision-making.

1   The young of frogs and toads are called tadpoles. Ecologists investigated the effect of predation on three species of tadpole. They set up four artificial pond communities. Each community contained
   • 200 spadefoot toad tadpoles
   • 300 spring peeper frog tadpoles
   • 300 southern toad tadpoles.
   The ecologists then added a different number of newts to each pond. Newts are predators.
   Figure 1 shows the effect of increasing the number of newts on the percentage survival of the tadpoles of each species.

   (a) (i)   Describe the effect of an increase in the number of newts on the percentage survival of the tadpoles of each of the toad species.   *(2 marks)*
   (ii) Suggest an explanation for the effect of an increase in the number of newts on the percentage survival of the tadpoles of spring peeper frogs.   *(2 marks)*

   Figure 2 shows how the masses of the tadpoles were affected in each pond during the investigation.

   (b) Using the information provided in **Figure 1** explain the results obtained in **Figure 2**.

                                                                                                        (2 marks)
                                                                                                    AQA Jan 2011

▲ Figure 1

▲ Figure 2

2 Algae are photosynthesising organisms. Some algae grow on rocky shores. A scientist investigated succession involving different species of algae. He placed concrete blocks on a rocky shore. At regular intervals over 2 years, he recorded the percentage cover of algal species on the blocks. His results are shown in the graph.

(a) Name the pioneer species. *(1 mark)*

(b) **(i)** The scientist used percentage cover rather than frequency to record the abundance of algae present. Suggest why. *(1 mark)*

  (ii) Some scientists reviewing this investigation were concerned about the validity of the results because of the use of concrete blocks. Suggest one reason why these scientists were concerned about using concrete blocks for the growth of algae.

(1 mark)

(c) Use the results of this investigation to describe and explain the process of succession.

(4 marks)

AQA June 2013

3 The diagram shows the dominant plants in communities formed during a succession from bare soil to pine forest.

Time in years	1	2–4	5–24	25–100
	Bare field	Grassland	Shrub	Forest

Key: Crabgrass   Aster   Broomsedge   Dogwood   Pine

(a) Name the pioneer species shown in the diagram. *(1 mark)*

(b) The species that are present change during succession. Explain why. *(2 marks)*

(c) The pine trees in the forest have leaves all year. Explain how this results in a low species diversity of plants in the forest.

(1 mark)

AQA June 2012

Section 7 Summary

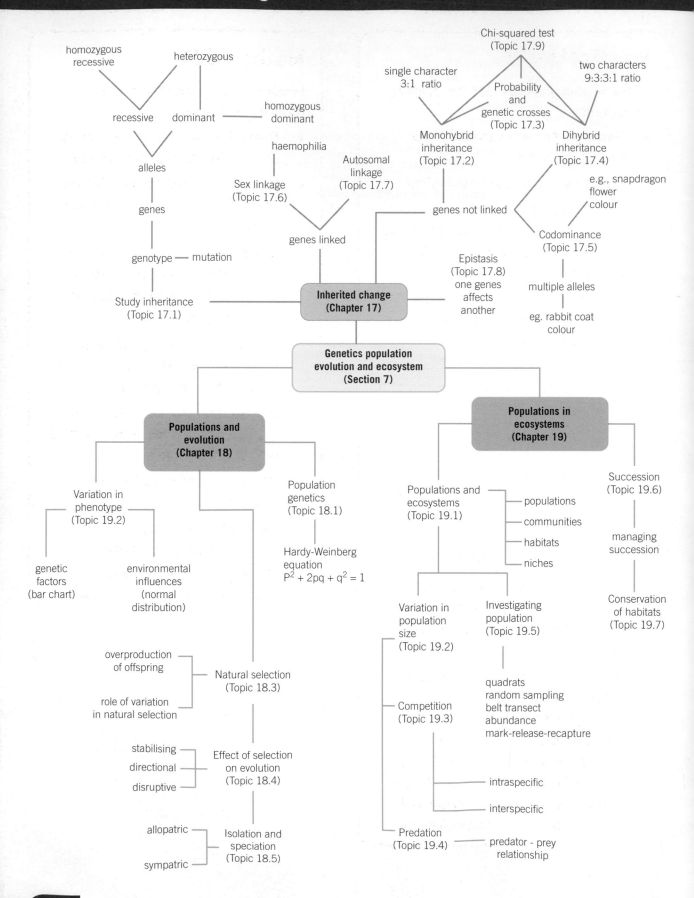

homozygous recessive

heterozygous

recessive — dominant — homozygous dominant

alleles

genes

genotype — mutation

Study inheritance (Topic 17.1)

haemophilia

Sex linkage (Topic 17.6)

Autosomal linkage (Topic 17.7)

genes linked

Chi-squared test (Topic 17.9)

single character 3:1 ratio

Probability and genetic crosses (Topic 17.3)

two characters 9:3:3:1 ratio

Monohybrid inheritance (Topic 17.2)

Dihybrid inheritance (Topic 17.4)

e.g., snapdragon flower colour

genes not linked

Codominance (Topic 17.5)

multiple alleles

eg. rabbit coat colour

Inherited change (Chapter 17)

Epistasis (Topic 17.8) one genes affects another

Genetics population evolution and ecosystem (Section 7)

Populations and evolution (Chapter 18)

Populations in ecosystems (Chapter 19)

Variation in phenotype (Topic 19.2)

Population genetics (Topic 18.1)

Populations and ecosystems (Topic 19.1)

populations

communities

habitats

niches

Succession (Topic 19.6)

managing succession

genetic factors (bar chart)

environmental influences (normal distribution)

Hardy-Weinberg equation
$P^2 + 2pq + q^2 = 1$

Variation in population size (Topic 19.2)

Investigating population (Topic 19.5)

Conservation of habitats (Topic 19.7)

overproduction of offspring

role of variation in natural selection

Natural selection (Topic 18.3)

Competition (Topic 19.3)

quadrats
random sampling
belt transect
abundance
mark-release-recapture

stabilising

directional

disruptive

Effect of selection on evolution (Topic 18.4)

intraspecific

interspecific

allopatric

sympatric

Isolation and speciation (Topic 18.5)

Predation (Topic 19.4)

predator - prey relationship

Practical skills

In this section you have met the following practical skills:

- How to plot growth curves using a logarithmic scale.
- How to carry out random sampling.
- Investigate the distribution of organisms in a habitat using randomly placed framed quadrats or a belt transect.
- Use the mark-release-recapture method to investigate the abundance of a motile species.

Extension task

Using your local newspaper, regional television news or local community websites in your area, identify a scheme in your region designed to conserve a habitat. Find out the purpose of this scheme and the organisations involved in it.

Research the various sources of funds that are available to support conservation projects like the one you have identified.

Draft a letter to one source of funds applying for a specified sum of money to support the aims of your project. Include in your letter a justification for the conservation project and the benefits it will bring to the community. Explain how the money will be used, how it will further the aims of the project and how you will evaluate whether it has been well spent.

Maths skills

In this section you have met the following maths skills:

- Calculating ratios and percentages of the offspring of genetic crosses.
- Understanding and calculating the probability associated with genetic inheritance.
- Using the chi-squared test to test the significance of the difference between observed and expected results of genetic crosses.
- Solving, and changing the subject in, algebraic equations such as the Hardy-Weinberg equation.
- Using a logarithmic scale in relation to quantities that range over several orders of magnitude.
- Using the logarithmic function on a calculator.
- Plotting two variables from experimental data provided.
- Finding arithmetical means

1 A student investigated an area of moorland where succession was occurring. She used quadrats to measure the percentage cover of plant species, bare ground, and surface water every 10 metres along a transect. She also recorded the depth of soil at each quadrat. Her results are shown in the table.

	Percentage cover in each quadrat A to E				
	A	B	C	D	E
Bog moss	55	40	10	–	–
Bell heather	–	–	–	15	10
Sundew	10	5	–	–	–
Ling	–	–	–	15	20
Bilberry	–	–	–	15	25
Heath grass	–	–	30	10	5
Soft rush	–	30	20	5	5
Sheep's fescue	–	–	25	35	30
Bare ground	20	15	10	5	5
Surface water	15	10	5	–	–
Soil depth / cm	3.2	4.7	8.2	11.5	14.8

– Indicates zero percentage cover

(a) Explain how these data suggest that succession has occurred from points **A** to **E** along the transect. *(3 marks)*

(b) The diversity of animal species is higher at **E** than **A**. Explain why. *(2 marks)*

(c) The student used the mark-release-recapture technique to estimate the size of the population of sand lizards on an area of moorland. She collected 17 lizards and marked them before releasing them back into the same area. Later, she collected 20 lizards, 10 of which were marked.

　(i) Give **two** conditions for results from mark-release-recapture investigations to be valid. *(2 marks)*

　(ii) Calculate the number of sand lizards on this area of moorland. Show your working. *(2 marks)*

AQA Jan 2013

2 In a species of snail, shell colour is controlled by a gene with three alleles. The shell may be brown, pink, or yellow. The allele for brown C^B, is dominant to the other two alleles. The allele for pink, C^P, is dominant to the allele for yellow, C^Y.

   (a) Explain what is meant by dominant allele. *(1 mark)*

   (b) Give **all** the genotypes which could result in a brown-shelled snail. *(1 mark)*

   (c) A cross between two pink shelled snails produced only pink-shelled and yellow-shelled snails. Use a genetic diagram to explain why. *(3 marks)*

   (d) The shells of this snail may be unbanded or banded. The absence or presence of bands is controlled by a single gene with two alleles. The allele for unbanded, **B**, is dominant to the allele for banded, **b**.
   A population of snails contained 51% of unbanded snails. Use the Hardy-Weinberg equation to calculate the percentage of this population that you would expect to be heterozygous for this gene. Show your working. *(3 marks)*

   AQA June 2012

3 **(a)** Insect pests of crop plants can be controlled by chemical pesticides or biological agents. Give **two** advantages of using biological agents. *(2 marks)*
   Two-spotted mites are pests of strawberry plants. Ecologists investigated the use of predatory mites to control two-spotted mites. They then recorded the percentage of strawberry leaves occupied by two-spotted mites and by predatory mites over a 16-week period. The results are shown in Figure 1.

▲ **Figure 1**

   (b) Describe how the percentage of leaves occupied by predatory mites changed during the period of this investigation. *(2 marks)*

   (c) The ecologists concluded that in this investigation the control of the two-spotted mite by a biological agent was effective. Explain how the results support this conclusion. *(2 marks)*

   (d) Farmers who grow strawberry plants and read about this investigation might decide **not** to use these predatory mites. Suggest **two** reasons why. *(2 marks)*

   (e) The ecologists repeated the investigation but sprayed chemical pesticide on the strawberry plants after 10 weeks. After 16 weeks no predatory mites were found but the population of two-spotted mites had risen significantly. Suggest an explanation for the rise in the two-spotted mite population. *(2 marks)*

   AQA June 2012

4 In birds, males are XX and females are XY.

(a) Use this information to explain why recessive, sex-linked characteristics are more common in female birds than in male birds. *(1 mark)*

(b) In chickens, a gene on the X chromosome controls the rate of feather production. The allele for slow feather production, F, is dominant to the allele for rapid feather production, f. Figure 3 shows the results produced from crosses carried out by a farmer. *(2 marks)*

male chickens are XX

female chickens are XY

key

□ male – rapid feather production

□ male – slow feather production

○ female – rapid feather production

○ female – slow feather production

▲ **Figure 3**

(i) Explain one piece of evidence from Figure 3 which shows that the allele for rapid feather production is recessive. *(2 marks)*

(ii) Give all the possible genotypes of chicken 5 and chicken 7 from Figure 3. *(2 marks)*

(iii) A cross between two chickens produced four offspring. Two of these were males with rapid feather production and two were females with slow feather production. Give the genotypes of the parents. *(1 mark)*

(c) Feather colour in one species of chicken is controlled by a pair of codominant alleles which are not sex linked. The allele C^B codes for black feathers and the allele C^W codes for white feathers. Heterozygous chickens are blue-feathered.

On a farm, 4% of the chickens were black-feathered. Use the Hardy-Weinberg equation to calculate the percentage of this population that you would expect to be blue-feathered. Show your working. *(3 marks)*

AQA June 2014

5   The graph shows the effects of light intensity on the rate of photosynthesis of three species of tree, **X**, **Y** and **Z**. Each of these species occurs at a different stage in succession.

(a) Species **X** is the first tree to become established in the succession. Use the graph to explain why it is likely to become established earlier in the succession than **Y** or **Z**.   (*3 marks*)

(b) Species **X** may change the environment so that it becomes more suitable for species **Z**. Use the graph to explain why.   (*2 marks*)

AQA Jan 2010

Section 8
The control of gene expression

Introduction

At a cellular level, control of metabolic activities is achieved by regulating which genes of the genome are transcribed and translated, and when this takes place. Although the cells within an organism carry the same genes they translate only part of them. In multicellular organisms, this control of translation enables cells to have specialised functions and to form specific tissues and organs. Cells formed from the zygote are initially able to differentiate into any type of cell – they are totipotent. As these cells become specialised they lose the ability to become a different type of cell. In mature mammals, only a few cells retain the ability to differentiate into other cells. These are called stem cells.

It has long been known that many factors control the expression of genes and, thus, the phenotype of organisms. Some of these factors are external, environmental factors, others are internal factors. What was not generally disputed was the idea that these environmental factors are never inherited by the following generation. Only those processes such as mutations, which caused changes to the nucleotide base sequence in a DNA molecule could be inherited. This view has now been challenged by the discovery that environmental factors can cause heritable changes in gene function without any change to the base sequence of DNA. This so called epigenetic regulation of transcription is being recognised as important.

We are increasingly able to control the expression of genes by altering the epigenome. This allows us to alter an organism's genomes and the proteins they produce (proteomes). Along with our ability to manipulate the transcription and translation of genes, this has opened up many medical and technological applications. The use of DNA technology allows us to clone genes for use in medical techniques such as gene therapy. Other aspects include the use of DNA probes and DNA hybridisation in the diagnosis and treatment of human diseases, as well as the use of genetic fingerprinting for medical, forensic, and breeding purposes.

Working scientifically

In studying this unit there will be opportunities to perform practical exercises and so develop practical skills.

In performing these exercises you will have the chance to develop practical skills such as:

- separating biological compounds using electrophoresis
- using microbiological aseptic techniques.

What you already know

The material in this unit is intended to be self-explanatory, but there is certain information from GCSE that will be useful to your appreciation of this section. This information includes:

- ⬭ Different genes control the development of different characteristics of an organism.
- ⬭ Differences in the characteristics of different individuals of the same kind may be due to differences in:
  - the genes they have inherited (genetic causes)
  - the conditions in which they have developed (environmental causes)
  - a combination of both of the above.
- ⬭ The differences between Darwin's theory of evolution and conflicting theories, such as that of Lamark.
- ⬭ In genetic engineering, genes from the chromosomes of humans and other organisms can be 'cut out' using enzymes and transferred to cells of other organisms.
- ⬭ Each person (apart from identical twins) has unique DNA. This can be used to identify individuals in a process known as DNA fingerprinting.
- ⬭ Embryos can be screened for the alleles that cause genetic disorders.
- ⬭ Genes can also be transferred to the cells of animals, plants or microorganisms at an early stage in their development so that they develop with desired characteristics.
- ⬭ New genes can be transferred to crop plants and crops that have had their genes modified in this way are called genetically modified crops (GM crops). Examples of genetically modified crops include ones that are resistant to insect attack or to herbicides
- ⬭ Genetically modified crops generally show increased yields.
- ⬭ Concerns about GM crops include the effect on populations of wild flowers and insects, and uncertainty about the effects of eating GM crops on human health.
- ⬭ Interpreting information about cloning techniques and genetic engineering techniques.
- ⬭ Making informed judgements about the economic, social and ethical issues concerning cloning and genetic engineering, including genetically modified (GM) crops.

In Topic 9.1, we saw that any change to the quantity or the structure of the DNA of an organism is known as a **mutation** and any change to one or more nucleotide bases, or any rearrangement of the bases, in DNA is known as a **gene mutation**. These gene mutations might arise during the replication of DNA.

We also learnt that any changes to one or more bases in the DNA triplets could result in a change in the amino acid sequence of the polypeptide. Let us now consider gene mutations in more detail by looking at the different types.

Substitution of bases

The type of gene mutation in which a nucleotide in a section of a DNA molecule is replaced by another nucleotide that has a different base is known as a substitution. Depending on which new base is substituted for the original base, there are three possible consequences:

- The formation of one of the three stop **codons** that mark the end of a polypeptide chain. As a result the production of the polypeptide coded for by the section of DNA would be stopped prematurely. The final protein would almost certainly be significantly different and the protein could not perform its normal function.

- The formation of a codon for a different amino acid, meaning that the structure of the polypeptide produced would differ in a single amino acid. The protein of which this polypeptide is a part may differ in shape and not function properly. For example, if it is an enzyme, its active site may no longer fit the substrate and it will not catalyse the reaction. An example of this form of substitution mutation causes a condition called sickle cell anaemia.

- The formation of a different codon but one that produces a codon for the same amino acid as before. This is because the genetic code is degenerate and so most amino acids have more than one codon. The mutation therefore has no effect on the polypeptide produced and so the mutation will have no effect.

Deletion of bases

We saw in Topic 9.1 that the loss of a nucleotide base from a DNA sequence is called a deletion. Minor though the loss of a single base might seem, the impact on the phenotype can be enormous. The one deleted base creates what is known as a **frame shift** because the reading frame that contains each three letters of the code has been shifted to the left by one letter. The gene is now read in the wrong three-base groups and the coded information is altered. Most triplets will then be different, as will the amino acids they code for. The polypetides will be different and lead to the production of a non-functional protein that could considerably alter the phenotype. One deleted base at the very start of a sequence could alter every triplet in the sequence. A deleted base near the end of the sequence is likely to have a smaller impact but can still have consequences (see Hint). An example of the effect of a deletion mutation is shown in Figure 1.

DNA sequence	T G C	A G C	T A C	C
mRNA sequence	A C G	U C G	A U G	G
amino acid sequence	threonine	serine	methionine	

deletion of guanine at position 2

	T	C A G	C T A	C C
	A	G U C	G A U	G G
		serine	arginine	tryptophan

▲ **Figure 1** *Effects of the deletion of a DNA base on the amino acid sequence in the final polypeptide*

Other types of gene mutation

There are a number of other ways in which the base sequence of DNA may be changed. These include:

- **Addition of bases** – an extra base becomes inserted in the sequence. This usually has a similar effect to a base deletion in that there is usually a frame shift and the whole sequence of triplets becomes altered. The frame shift is to the right not to the left as it is when a base is deleted. If three extra bases are added, or any multiple of three bases, there will not be a frame shift. The resulting polypeptide will be different from the one produced from a non-mutant gene, but not to the same extent as if there was a frame shift.

- **Duplication of bases** – one or more bases are repeated. This produces a frame shift to the right.

- **Inversion of bases** – a group of bases become separated from the DNA sequence and rejoin at the same position but in the inverse order (back to front). The base sequence of this portion is therefore reversed and effects the amino acid sequence that results.

- **Translocation of bases** – a group of bases become separated from the DNA sequence on one chromosome and become inserted into the DNA sequence of a different chromosome. Translocations often have significant effects on gene expression leading to an abnormal phenotype. These effects include the development of certain forms of cancer and also reduced fertility.

Causes of mutations

Gene mutations can arise spontaneously during DNA replication. Spontaneous mutations are permanent changes in DNA that occur without any outside influence. Despite being random occurrences, mutations occur with a predictable frequency. The natural mutation rate varies from species to species, but is typically around one or two mutations per 100 000 genes per generation. This basic mutation rate can be increased by outside factors known as **mutagenic agents** or mutagens. These include the following:

- **High energy ionising radiation**, for example, α and β particles as well as short wavelength radiation such as X-rays and ultra violet light. These forms of radiation can disrupt the structure of DNA.

- **Chemicals** such as nitrogen dioxide may directly alter the structure of DNA or interfere with transcription. Benzopyrene, a consitituent of tobacco smoke, is a powerful mutagen that inactivates a tumour-suppressor gene TP53 leading to a cancer. We will learn more about this in Topic 20.5.

▲ **Figure 2** *This albino hedgehog is the result of a mutation that prevents the production of the pigment melanin*

Mutations have both costs and benefits. On the one hand they produce the genetic diversity necessary for natural selection and speciation (see Topics 18.4 and 18.5). On the other hand they are almost always harmful and produce an organism that is less well suited to its environment. Additionally, mutations that occur in body cells rather than in gametes leading to disruption of normal cellular activities, such as cell division, for example, cancer.

Summary questions

1 A translocation mutation is, in effect, a combination of two other different types of gene mutation. Deduce which two types of mutation these are and explain your answer.

2 A section of DNA has the following sequence – AGT TCT GAT CGC TG. State the type of mutation that has taken place in each of the following variants of this DNA.

   a AGT TCT GAT CCT G

   b AGT TCT TAG CGC TG

   c AGT TCT GAG CGC TG

   d AGT TCT GAT CGT CTG

3 Explain why the effects of a single additional base in a sequence of DNA bases may have:

   a a considerable effect on the polypeptide produced.

   b little effect on the polypeptide produced.

4 A mutation causes three bases in the DNA of a gene to become duplicated. Explain how the effects of this mutation might differ if the duplicated bases are consecutive rather than in three separate locations on the DNA molecule.

5 Suggest **two** reasons why the addition of a single base into a DNA sequence may not alter the amino acid sequence in the resultant polypeptide.

Mutagenic agents

Mutations can be induced by external influences called mutagenic agents. These cause damage in a number of ways.

- **Certain chemicals can remove groups from nucleotide bases.** Nitrous acid can remove an $-NH_2$ group from cytosine in DNA, changing it into uracil.

> **1** Suggest what the result of this change might be on the codons on a mRNA molecule that is transcribed from a section of DNA with the triplets GCA CTC ATC.

- **Other chemicals can add groups to nucleotides.** Benzopyrene is a chemical found in tobacco smoke. It adds a large group to guanine that makes it unable to pair with cytosine. When DNA polymerase reaches the affected guanine it inserts any of the other bases.

> **2** What type of mutation is caused by benzopyrene?

- **Ionising radiation,** such as X-rays, can produce highly reactive agents, called free radicals, in cells. These free radicals can alter the shape of bases in DNA so that DNA polymerase can no longer act on them

> **3** Explain why DNA polymerase cannot act on DNA that has been damaged by X-rays.
> **4** State one genetic effect of DNA polymerase being unable to act on DNA.

Ultraviolet radiation from the Sun or tanning lamps affects thymine in DNA, causing it to form bonds with the nucleotides on either side of it. This seriously disrupts DNA replication.

Scientific research and experimentation has enabled us to identify potentially dangerous mutagenic agents. The effects of such agents are complex and the amount of harm they cause is often a matter of debate. Commercial organisations, such as the tobacco industry, manufacturers of sunbeds, and producers and retailers of sun-block lotions all have an interest in the research that is undertaken. They are more likely to fund research that may benefit their business than research that may harm it. It is therefore important that the results of any research are subjected to the scrutiny of other scientists from a wide variety of backgrounds, views, interests and organisations, in a process known as peer review.

This is usually achieved by publishing research findings in reputable scientific journals that have an extensive global readership. The conclusions and claims made by researchers and their sponsors can then be debated and the scientific community at large can test the claims by further experimentation. These claims then become accepted, modified or rejected, depending on the outcome of this further research.

Armed with all this scientific information, decision-makers such as governments and heads of business can take appropriate action that benefits society. Governments, for example, can introduce legislation that controls cigarette sales and smoking, and the use of sunbeds and sets a minimum age at which cigarettes or tanning treatments can be bought. The decisions are often not clear-cut however. X-rays, for example, can be harmful on one hand but are an invaluable diagnostic tool, with countless health benefits, on the other.

> **5** Leaders in business and government have to make decisions about the use of scientific discoveries. Who else, apart from research scientists, might influence the advice that these leaders give to the public on the use of sunbeds?

▲ **Figure 3** *Ultraviolet radiation from sunbeds has the potential to disrupt DNA replication*

Specification reference: 3.8.2.1

Learning objectives

→ State what totipotent cells are.

→ Explain how cells lose their totipotency and become specialised.

→ Describe cell differentiation and cell specialisation.

→ Describe the origins and types of stem cells.

→ Explain how pluripotent stem cells can be used to treat human disorders.

In multicellular organisms, cells are specialised to perform specific functions. The process by which each cell develops into a specialised structure suited to the role that it will carry out is known as **cell differentiation**. Let us investigate the process in more detail.

Cell differentiation and specialisation

Single-celled organisms perform all essential life functions inside the boundaries of a single cell. Although they perform all functions adequately, they cannot be totally efficient at all of them, because each function requires a different type of cellular structure, enzymes, and other proteins. One activity may be best carried out by a long, thin cell, while another might suit a spherically shaped cell. No one cell can provide the best conditions for all functions. The cells of multicellular organisms are each adapted in different ways to perform a particular role. In early development, an organism is made up of a tiny ball of cells, all of which are identical. As it matures, each cell takes on its own individual characteristics that adapt it to the function that it will perform when it is mature.

All the cells in an organism, such as a human, are derived by mitotic divisions of the fertilised egg (zygote). It follows that they all contain exactly the same **genes**. Every cell is therefore capable of making everything that the body can produce. A cell in the lining of the small intestine has the gene coding for **insulin** just as a β cell of the pancreas has the gene coding for maltase. So why do the cells of the small intestine produce maltase rather than insulin and β cells of the pancreas produce insulin rather than maltase? The answer is that, although all cells contain all genes, only certain genes are expressed (switched on) in any one cell at any one time.

Some genes are permanently expressed (switched on) in all cells. For example, the genes that code for essential chemicals, such as the enzymes involved in respiration, are expressed in all cells. Other genes permanently switched on in all cells include those coding for enzymes and other proteins involved in essential processes like transcription, translation, membrane synthesis, ribosomes and tRNA synthesis. Other genes are permanently not expressed (switched off), for example, the gene for insulin in cells lining the small intestine. Further genes are switched on and off as and when they are needed. In this chapter we shall look at how the expression of genes is controlled.

Differentiated cells differ from each other, often visibly so. This is mainly because they each produce different proteins. The proteins that a cell produces are coded for by the genes it possesses or, more accurately, by the genes that are expressed (switched on).

Totipotency

An organism develops from a single fertilised egg. A fertilised egg clearly has the ability to give rise to all types of cells. Cells such as fertilised eggs, which can mature into any body cell, are known as **totipotent cells.** The early cells that are derived from the fertilised egg are also totipotent. These later differentiate and become specialised for a particular function. For example, **mesophyll** cells become specialised for photosynthesis and muscle cells become specialised for contraction. This is because, during the process of cell specialisation, only some of the genes are expressed. This means that only part of the DNA of a cell is translated into proteins. The cell therefore only makes those proteins that it requires to carry out its specialised function. These proteins include those required for essential processes like respiration and membrane synthesis.. Although it is still capable of making all the other proteins, these are not needed and so it would be wasteful to produce them. In order to conserve energy and resources, a variety of stimuli (controlling factors) ensure the genes for these other proteins are not expressed. The ways in which genes are prevented from expressing themselves include:

- preventing transcription and so preventing the production of mRNA
- preventing translation.

Stem cells

If specialised cells still retain all the genes of the organism, can they still develop into any other cell? The answer, depends – there are no hard and fast rules. **Xylem vessels**, which transport water in plants, and red blood cells, which carry oxygen in animals, are so specialised that they lose their nuclei once they are mature. As the nucleus contains the genes, then clearly these cells cannot develop into other cells. In fact, specialisation is irreversible in most animal cells. Once cells have matured and specialised they can no longer develop into other cells. In mature mammals, only a few cells retain the ability to differentiate into other cells. These are called **stem cells**.

Stem cells are undifferentiated dividing cells that occur in adult animal tissues and need to be constantly replaced. They therefore have the ability to divide to form an identical copy of themselves in a process called self-renewal.

Stem cells originate from various sources in mammals:

- **Embryonic stem cells** come from embryos in the early stages of development. They can differentiate into any type of cell in the initial stages of development.
- **Umbilical cord blood stem cells** are derived from umbilical cord blood and are similar to adult stem cells.
- **Placental stem cells** are found in the placenta and develop into specific types of cells.
- **Adult stem cells**, despite their name, are found in the body tissues of the fetus through to the adult. They are specific to a particular tissue or organ within which they produce the cells to maintain and repair tissues throughout an organism's life.

Types of stem cells

There are number of different stem cells which are classified according to their ability to differentiate.

- **Totipotent stem cells** are found in the early embryo and can differentiate into any type of cell. Since all body cells are formed from a zygote, it follows that the zygote is totipotent. As the zygote

Hint

Differentiation results from differential gene expression.

▲ **Figure 1** *SEM of a three-day-old human embryo at the 16-cell stage on the tip of a pin. These cells are totipotent*

divides and matures, its cells develop into slightly more specialised cells called pluripotent stem cells.

- **Pluripotent stem cells** are found in embryos and can differentiate into almost any type of cell. Examples of pluripotent stem cells are embryonic stem cells and fetal stem cells.

- **Multipotent stem cells** are found in adults and can differentiate into a limited number of specialised cells. They usually develop into cells of a particular type, for example, stem cells in the bone marrow can produce any type of blood cell. Examples of multipotent cells are adult stem cells and umbilical cord blood stem cells.

- **Unipotent stem cells** can only differentiate into a single type of cell. They are derived from multipotent stem cells and are made in adult tissue.

Induced pluripotent stem cells

Induced pluripotent stem cells (iPS cells) are a type of pluripotent cell that is produced from unipotent stem cells. The unipotent cell may be almost any body cell. These body cells are then genetically altered in a laboratory to make them acquire the characteristics of embryonic stem cells which, as we have seen, are a type of pluripotent cell. To make the unipotent cell acquire the new characters involves inducing genes and transcriptional factors (see Topic 20.3) within the cell to express themselves. In other words to turn on genes that were otherwise turned off. The fact that these genes are capable of being reactivated shows that adult cells retain the same genetic information that was present in the embryo.

The iPS cells are very similar to embryonic stem cells in form and function. However, although they express some of the same genes that are usually expressed in embryonic stem cells, they are not exact duplicates of them. One feature of particular interest is that they are capable of self-renewal. This means that they can potentially divide indefinitely to provide a limitless supply. As such they could replace embryonic stem cells in medical research and treatment and so overcome many of the ethical issues surrounding the use of embryos in stem cell research.

Pluripotent cells in treating human disorders

There are many possible uses of pluripotent cells. The cells can be used to regrow tissues that have been damaged in some way, either by accident (e.g., skin grafts for serious burn damage) or as a result of disease (e.g., neuro-degenerative diseases, such as Parkinson's disease). Table 1 lists some of the potential uses of human cells produced from stem cells.

Summary questions

1 Explain what is meant by totipotent cells.

2 Distinguish between totipotent, pluripotent, multipotent and unipotent cells.

3 All cells possess the same genes and yet a skin cell can produce the protein keratin but not the protein myosin, while a muscle cell can produce myosin but not keratin. Explain why.

4 Suggest a reason why skin cells retain an ability to divide by being unipotent when the cells of some other organs do not.

▼ **Table 1** *Potential uses of human cells produced from stem cells*

Type of cell	Disease that could be treated
Heart muscle cells	Heart damage, for example, as a result of a heart attack
Skeletal muscle cells	Muscular dystrophy
β cells of the pancreas	Type 1 diabetes
Nerve cells	Parkinson's disease, multiple sclerosis, strokes, Alzheimer's disease, paralysis due to spinal injury
Blood cells	Leukaemia, inherited blood diseases
Skin cells	Burns and wounds
Bone cells	Osteoporosis
Cartilage cells	Osteoarthritis
Retina cells of the eye	Macular degeneration

Human embryonic stem cells and the treatment of disease

Although there are a number of types of stem cell in the human body, it is the first few cells from the division of the fertilised egg that have the greatest potential to treat human diseases. As they come from the early stages of an embryo, they are called human embryonic stem cells. These cells can be grown *in vitro* and then induced to develop into a wide range of different human tissues. The process is illustrated in Figure 2.

At present, embryonic stem cell research is only allowed in the UK under licenced and specified conditions. These conditions include its use as a means of increasing knowledge about embryo development and serious diseases, including their treatment. Embryos used in this type of research are obtained from *in vitro* fertilisation. The process nevertheless presents a number of ethical issues.

One issue surrounds the argument as to whether a human embryo less than 14 days old should be afforded the same respect as a fetus or an adult person. Some people feel that using embryos in this way undermines our respect for human life and could progress to the use of fetuses, and even newborn babies, for research or the treatment of disease. They feel that it is a further move towards reproductive cloning and, even if this remains illegal in the UK, the information gained could be used to clone humans elsewhere. Others disagree, arguing that an embryo at such an early stage of development is just a ball of identical, undifferentiated cells, bearing no resemblance to a human being. They feel that the laws prohibiting cloning, in the UK and elsewhere, provide sufficient protection.

Supporters of human embryonic stem cell research contend that it is wrong to allow human suffering to continue when there is a possibility of alleviating it. They further argue that, since embryos are produced for other purposes, for example, fertility treatments, it makes no sense to destroy spare embryos that could be used in research. Opponents of embryonic stem cell research contend that it is wrong to use humans, including human embryos, as a means to an end, even if that end is the to alleviate human suffering.

However, human embryos are not the only source of stem cells. For example, they can be obtained from the bone marrow of adult humans. As long as a person gives consent, this source of stem cells raises no real ethical issues. At present, these cells have far more restricted medical applications but scientists hope, in time, to be able to make them behave more like embryonic stem cells.

1 Write *two* accounts, each of around 200 words, evaluating the case for and against the continued use of embryos for stem cell research.

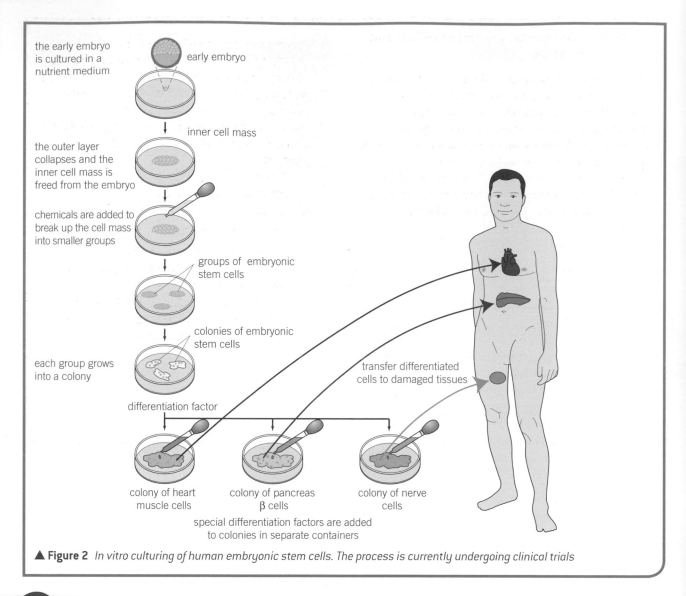

the early embryo is cultured in a nutrient medium

early embryo

inner cell mass

the outer layer collapses and the inner cell mass is freed from the embryo

chemicals are added to break up the cell mass into smaller groups

groups of embryonic stem cells

colonies of embryonic stem cells

each group grows into a colony

differentiation factor

transfer differentiated cells to damaged tissues

colony of heart muscle cells

colony of pancreas β cells

colony of nerve cells

special differentiation factors are added to colonies in separate containers

▲ **Figure 2** *In vitro culturing of human embryonic stem cells. The process is currently undergoing clinical trials*

Growth of plant tissue cultures 🧪

Mature plants have many totipotent cells. Under the right conditions, many plant cells can develop into any other cell. For example, if we take a cell from the root of a carrot, place it in a suitable nutrient medium and give it certain chemical stimuli at the right time, we can develop a complete new carrot plant.

There are many factors that influence the growth of plant tissue cultures from totipotent cells. One group of factors consists of plant growth factors, which are chemicals involved in the growth and development of plant tissues. Plant growth factors have a number of features:

- They have a wide range of effects on plant tissues.
- The effects on a particular tissue depend upon the concentration of the growth factor.

▲ **Figure 3** *Plants growing from tissue cultures in a Petri dish*

- The same concentration affects different tissues in different ways.

- The effect of one growth factor can be modified by the presence of another.

An experiment was carried out to investigate the effects on the development of a plant tissue culture of three growth factors: cytokinin, IAA, and 2,4-D. Samples of totipotent plant cells were grown on a basic growth medium in a series of test tubes. Each test tube contained a mixture of the three growth factors in different concentrations. After two weeks, the tubes were observed to see the effects of the growth factor mixtures on shoot and root growth. The results are shown in Table 2.

▼ **Table 2** *Effect of growth factors on shoot and root development*

Tube no.	Relative concentration of growth factors			Shoot development	Root development
	Cytokinin	IAA	2,4-D		
1	None	Low	None	Moderate	Little
2	Low	High	None	Extensive	Little
3	High	Low	None	Little	Moderate
4	None	High	High	Extensive	Extensive
5	None	None	None	Very little	Very little

1 Name the process by which the totipotent cells of the plant tissue culture change in appearance and develop into shoot or root cells.
2 State the general term used to describe growing living cultures like plants in a laboratory.
3 Plant tissues grown in culture often originate from a single initial cell and are therefore genetically identical.
   a State the name given to this group of genetically identical cells.
   b Name the process by which these genetically identical cells formed.
4 From Table 2, state which *two* growth factors together produced the greatest development of both shoots and roots.
5 Describe one piece of evidence from Table 2 that supports the view that the effects of one growth factor can be modified by another.

Learning objectives

→ Explain how oestrogen affects gene transcription.

→ State what small interfering RNA is.

→ Explain how small interfering RNA affects gene expression.

Specification reference: 3.8.8.2

Hint

Only target cells have the oestrogen receptor and so only these cells respond to the stimulus of oestrogen.

In Topic 20.2, we saw how cell specialisation is the result of the selective expression of certain genes out of the full complement found in every cell. Let us now investigate some ways in which cells control which genes are expressed.

The effect of oestrogen on gene transcription

In Topic 16.3 we learned how hormones, such as **adrenaline**, operate by using a second messenger. Here we will examine another mechanism, which is used by steroid hormones such as oestrogen. Before looking at how oestrogen operates, let us consider the general principles involved in controlling the expression of a gene by controlling transcription.

- For **transcription** to begin the gene is switched on by specific molecules that move from the cytoplasm into the nucleus. These molecules are called **transcriptional factors**.

- Each transcriptional factor has a site that binds to a specific base sequence of the DNA in the nucleus.

- When it binds, it causes this region of DNA to begin the process of transcription.

- Messenger RNA (mRNA) is produced and the information it carries is then translated into a polypeptide.

- When a gene is not being expressed (i.e., is switched off), the site on the transcriptional factor that binds to DNA is not active.

- As the site on the transcriptional factor binding to DNA is inactive it cannot cause transcription and polypeptide synthesis.

Hormones like oestrogen can switch on a gene and thus start transcription by combining with a receptor site on the transcriptional factor. This activates the DNA binding site by causing it to change shape. The process is illustrated in Figure 1 and operates as follows:

- Oestrogen is a lipid-soluble molecule and therefore diffuses easily through the **phospholipid** portion of cell-surface membranes (Figure 1, stage 1).

- Once inside the cytoplasm of a cell, oestrogen binds with a site on a receptor molecule of the transcriptional factor. The shape of this site and the shape of the oestrogen molecule complement one another (Figure 1, stage 2).

- By binding with the site, the oestrogen changes the shape of the DNA binding site on the transcriptional factor, which can now bind to DNA (it is activated) (Figure 1, stage 3).

- The transcriptional factor can now enter the nucleus through a nuclear pore and bind to specific base sequences on DNA (Figure 1, stage 4).

- The combination of the transcriptional factor with DNA stimulates transcription of the gene that makes up the portion of DNA (Figure 1, stage 4).

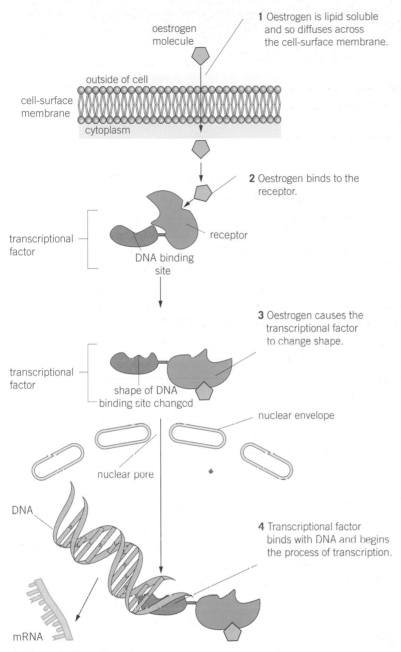

1 Oestrogen is lipid soluble and so diffuses across the cell-surface membrane.

oestrogen molecule

outside of cell
cell-surface membrane
cytoplasm

2 Oestrogen binds to the receptor.

transcriptional factor

receptor

DNA binding site

3 Oestrogen causes the transcriptional factor to change shape.

transcriptional factor

shape of DNA binding site changed

nuclear envelope

nuclear pore

DNA

4 Transcriptional factor binds with DNA and begins the process of transcription.

mRNA

▲ **Figure 1** *The effect of oestrogen on gene transcription*

Synoptic link

The attachment of oestrogen to a receptor causes changes in the shape of the receptor in the same way as the attachment of a non-competitive inhibitor to an enzyme molecule changes its shape and also its active site. This involves the same basic mechanism, which is described in Topic 1.9, Enzyme inhibition, which would be worthwhile revising.

Summary questions

1 What is the role of a transcriptional factor?
2 Describe how oestrogen stimulates the expression of a gene.

Gene expression in haemoglobin

A haemoglobin molecule is made up of four polypeptide chains each known as a globulin. In adult humans two of the polypeptides in a haemoglobin molecule are alpha-globulin and two are beta-globulin. In other words, 50% of the total globulin in all haemoglobin is alpha and 50% is beta. In a human fetus, however, the haemoglobin is different, with much of the beta-globulin being replaced by a third type, gamma-globulin. Fetal haemoglobin has a greater affinity for oxygen than adult haemoglobin. The changes in the production of the three types of globulin during early human development are shown in Figure 2.

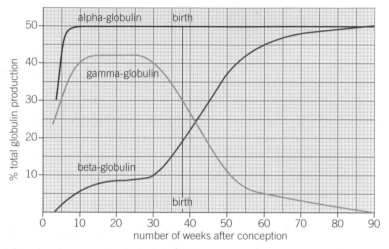

▲ **Figure 2** *Percentage total globulin production during early human development*

Humans have genes that code for the production of all three types of globulin. The production of the different haemoglobins depends upon which gene is expressed. The expression of these genes changes at different times during development.

1 Suggest an advantage of fetal haemoglobin having a greater affinity for oxygen than adult haemoglobin.

2 At birth, what percentage of the total globulin production is of each globulin type?

3 Describe the changes in gene expression that occur at 30 weeks.

4 Outline *two* possible explanations for the change in the expression of the gene for gamma-globulin after 30 weeks.

5 Sickle cell disease is the result of a mutant form of haemoglobin. In Saudia Arabia and India, some individuals have high levels of fetal haemoglobin in their blood, even as adults. Where these individuals have sickle cell disease, their symptoms are much reduced. Suggest how controlling the expression of the genes for globulin might provide a therapy for sickle cell disease.

Ever since James Watson and Francis Crick proposed the double-helix model in 1953, it has been taken as fact that DNA possesses the instructions for making all parts of an organism. In recent years, however, we have come to realise that DNA is only part of the story of heredity. It is accepted that while genes determine the features of an organism, the environment can influence the expression of these genes (Topic 18.2). However, the changes they cause to the **phenotype** were thought not to be inherited by the offspring. We now believe that environmental factors can cause heritable changes in gene function without changing the base sequence of DNA. This process is known as **epigenetics**.

Epigenetics

Epigenetics is a relatively new scientific field that provides explanations as to how environmental influences such as diet, stress, toxins, etc. can subtly alter the genetic inheritance of an organism's offspring. It is helping to explain, and maybe cure, illnesses ranging from autism to cancer. It is even causing scientists to look again at previously discredited theories of evolution that suggested characteristics acquired during an organism's life could be passed on to future generations (Lamarckism).

The epigenome

We learned in Topic 8.2 that DNA is wrapped around proteins called histones. We now know that both the DNA and histones are covered in chemicals, sometimes called tags. These chemical tags form a second layer known as the **epigenome**. The epigenome determines the shape of the DNA-histone complex. For example it keeps genes that are inactive in a tightly packed arrangement and therefore ensures that they cannot be read (it keeps them switched off). This is known as epigenetic silencing. By contrast, it unwraps active genes so that the DNA is exposed and can easily be transcribed (switches them on).

We know that the DNA code is fixed. The epigenome, however, is flexible. This is because its chemical tags respond to environmental changes. Factors like diet and stress can cause the chemical tags to adjust the wrapping and unwrapping of the DNA and so switch genes on and off.

The epigenome of a cell is the accumulation of the signals it has received during its lifetime and it therefore acts like a cellular memory. In early development, the signals come from within the cells of the fetus and the nutrition provided by the mother is important in shaping the epigenome at this stage. After birth, and throughout life, environmental factors affect the epigenome, although signals from within the body, for example, hormones, also influence it. These factors cause the epigenome to activate or inhibit specific sets of genes.

Learning objectives

→ State what is meant by epigenetics.

→ Describe the nature of the epigenome.

→ Explain the effect of epigenetic factors on DNA and histones.

→ Explain the effects of decreased acetylation of histones.

→ Explain the effects of increased methylation of DNA.

Specification reference 3.8.2.2

▲ **Figure 1** *What we eat not only affects us, it may affect our children too*

The environmental signal stimulates proteins to carry its message inside the cell from where it is passed by a series of other proteins into the nucleus. Here the message passes to a specific protein which can be attached to a specific sequence of bases on the DNA. Once attached the protein has two possible effects. It can change:

- acetylation of histones leading to the activation or inhibition a gene
- methylation of DNA by attracting enzymes that can add or remove methyl groups.

Before we look in a little more detail at how each process works it will be helpful to look more closely at the DNA–histone complex.

The DNA–histone complex (chromatin)

Where the association of histones with DNA is weak, the DNA–histone complex is less condensed (loosely packed). In this condition the DNA is accessible by transcription factors, which can initiate production of mRNA, that is, can switch the gene on.

Where this association is stronger, the DNA–histone complex is more condensed (tightly packed). In this condition the DNA is not accessible by transcription factors, which therefore cannot initiate production of mRNA, that is, the gene is switched off.

Condensation of the DNA–histone complex therefore inhibits transcription. It can be brought about by decreased acetylation of the histones or by methylation of DNA. Let us turn our attention to these two processes and how they inhibit transcription.

Decreased acetylation of associated histones

Acetylation is the process whereby an acetyl group is transferred to a molecule. In this case the group donating the acetyl group is acetylcoenzyme A which you may remember from the link reaction in respiration (Topic 12.2). Deacetylation is the reverse reaction where an acetyl group is removed from a molecule.

Decreased acetylation increases the positive charges on histones and therefore increases their attraction to the phosphate groups of DNA. The association between DNA and histones is stronger and the DNA is not accessible to transcription factors. These transcription factors cannot initiate mRNA production from DNA. In other words, the gene is switched off.

Increased methylation of DNA

Methylation is the addition of a methyl group (CH_3) to a molecule. In this case the methyl group is added to the cytosine bases of DNA. Methylation normally inhibits the transcription of genes in two ways:

- preventing the binding of transcriptional factors to the DNA
- attracting proteins that condense the DNA–histone complex (by inducing deacetylation of the histones) making the DNA inaccessible to transcription factors.

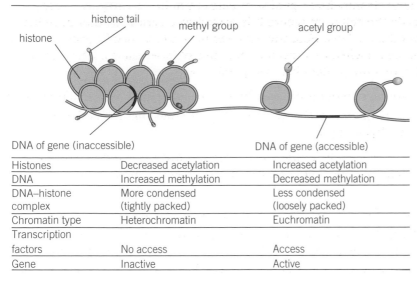

Histones	Decreased acetylation	Increased acetylation
DNA	Increased methylation	Decreased methylation
DNA–histone complex	More condensed (tightly packed)	Less condensed (loosely packed)
Chromatin type	Heterochromatin	Euchromatin
Transcription factors	No access	Access
Gene	Inactive	Active

▲ **Figure 2** *Effects of epigenetic factors such as methyl and acetyl groups on the DNA–histone complex*

Epigenetics and inheritance

Unexpected though it might be, there is now little doubt that epigenetic inheritance takes place. Experiments on rats have shown that female offspring who received good care when young, respond better to stress in later life and themselves nurture their offspring better. Female offspring receiving low-quality care, nurture their offspring less well. Good maternal behaviour in rats transmits epigenetic information onto their offspring's DNA without passing through an egg or sperm.

In humans, when a mother has a condition known as gestational diabetes, the fetus is exposed to high concentrations of glucose. These high glucose concentrations cause epigenetic changes in the daughter's DNA, increasing the likelihood that she will develop gestational diabetes herself.

It is thought that in sperm and eggs during the earliest stages of development a specialised cellular mechanism searches the genome and erases its epigenetic tags in order to return the cells to a genetic 'clean slate'. However, a few epigenetic tags escape this process and pass unchanged from parent to offspring.

Epigenetics and disease

Epigenetic changes are part of normal development and health but they can also be responsible for certain diseases. Altering any of the epigenetic processes can cause abnormal activation or silencing of genes. Such alterations have been associated with a number of diseases including cancer. In some cases the activation of a normally inactive gene can cause cancer, in other cases it is the inactivation of a normally active gene that gives rise to the disease.

In 1983, researchers found that diseased tissue taken from patients with colorectal cancer had less DNA methylation than normal tissue from the same patients. As we saw earlier, increased DNA methylation normally inhibits transcription (switches off genes). This means that these patients with less DNA methylation would have higher than normal gene activity – more genes were turned on.

It is known that there are specific sections of DNA (ones near regions called promoter regions) that have no methylation in normal cells. However, in cancer cells these regions become highly methylated causing genes that should be active to switch off. This abnormality happens early in the development of cancer.

We have seen that epigenetic changes do not alter the sequence of bases in DNA. They can, however, increase the incidence of mutations. Some active genes normally help repair DNA and so prevent cancers. In people with various types of inherited cancer, it is found that increased methylation of these genes has led to these protective genes being switched off. As a result, damaged base sequences in DNA are not repaired and so can lead to cancer.

Treating diseases with epigenetic therapy

As we have seen, many diseases, such as cancer, are triggered by of epigenetic changes that cause certain genes to be activated or silenced. It is therefore logical to try to use epigenetic treatments to counteract these changes. These treatments use drugs to inhibit certain enzymes involved in either histone acetylation or DNA methylation. For example, drugs that inhibit enzymes that cause DNA methylation can reactivate genes that have been silenced. Epigenetic therapy must be specifically targeted on cancer cells. If the drugs were to affect normal cells they could activate gene transcription and make them cancerous, so causing the very disorder they were designed to cure.

Another use of epigenetics in disease treatment has been the development of diagnostic tests that help to detect the early stages of diseases such as cancer, brain disorders and arthritis. These tests can identify the level of DNA methylation and histone acetylation at an early stage of disease. This allows those with these diseases to seek early treatment and so have a better chance of cure.

The effect of RNA interference on gene expression

In **eukaryotes** and some **prokaryotes** the translation of mRNA produced by a gene can be inhibited by breaking mRNA down before its coded information can be translated into a polypeptide. One type of small RNA molecule that may be involved is small interfering RNA (siRNA). The mechanism involving small double-stranded sections of siRNA operates as follows.

- An enzyme cuts large double-stranded molecules of RNA into smaller sections called small interfering RNA (siRNA) (Figure 3, stage 1).
- One of the two siRNA strands combines with an enzyme (Figure 3, stage 2).

> ### Hint
> Double-stranded RNA can be made by in vitro transcription of a DNA template using the polymerase chain reaction. (Topic 21.3)

- The siRNA molecule guides the enzyme to a messenger RNA molecule by pairing up its bases with the complementary ones on a section of the mRNA molecule (Figure 3, stage 3).

- Once in position, the enzyme cuts the mRNA into smaller sections (Figure 3, stage 4).

- The mRNA is no longer capable of being translated into a polypeptide.

- This means that the gene has not been expressed, that is, it has been blocked.

Summary questions

1 Explain what is meant by epigenetics.

2 Name two mechanisms by which changes in the environment can inhibit transcription.

3 One of the two strands of siRNA combines with an enzyme and guides it to an mRNA molecule which it then cuts. Explain why the mRNA is unlikely to be cut if the other siRNA strand combines with the enzyme.

4 Suggest how siRNA could be used to:

  a identify the role of genes in a biological pathway

  b to prevent certain diseases.

5 The enzyme histone deacetylase (HDAC) removes acetyl groups from histones. Suggest what the effect of this enzyme would be on:

  a the arrangement of chromatin (DNA–histone complex)

  b transcription.

▲ **Figure 3** *The effect of siRNA on gene expression*

 ## Nature versus nurture

A small Californian plant, *Potentilla glandulosa*, has a number of genetic forms, each adapted to growing at different altitudes. Experiments were carried out as follows:

- plants of *Potentilla* were collected from three altitudes – high, medium, and low

- one plant from each location was split into three cuttings, each of which therefore had an identical genotype

- one of the cuttings from each location was grown at each altitude (high, medium, and low).

		Where the plants originally came from		
		High altitude	Medium altitude	Low altitude
Where the plants were grown	High altitude	small plant with many leaves	tiny plant with few leaves	plant died
	Medium altitude	large, bushy plant with many leaves	very large, bushy plant with many leaves	small plant with few leaves
	Low altitude	small plant with many leaves	small plant with many leaves	medium-sized, bushy plant with many leaves

The plants in each column had the same genotype

The plants in each row were grown under the same environmental conditions

▲ **Figure 4** *Effect of environment on phenotype – growing genetically identical* Potentilla glandulosa *at different altitudes*

The results are illustrated in Figure 4.

1 Deduce whether genetic or environmental factors have the greatest influence on the phenotype of *Potentilla glandulosa* as illustrated in Figure 4.
2 Justify your answer to question 1 using evidence from Figure 4.
3 Suggest why the differences in phenotype between the three genetically identical plants from low altitude are greater than the differences between the three genetically identical plants from high altitude.

Prader-Willi syndrome

Epigenetic inheritance is thought to be involved in a rare genetic disease called Prader-Willi syndrome. It is the result of seven genes on chromosome 15 being deleted.

In most people, only one copy of the genes (usually from the father) is expressed while the other copy of the genes (from the mother) is silenced through epigenetic inheritance. This means that most people have one working and one epigenetically-silenced set of these genes. However, if a mutation on chromosome 15 of the father deletes the relevant seven genes, any offspring produced will have one set of non-working genes and one set of epigenetically-silenced genes. These individuals will inherit Prader-Willi syndrome.

1 Explain why most offspring do not develop Prader-Willi syndrome despite inheriting epigenetically-silenced genes from their mother.
2 People with Prader-Willi syndrome are often infertile. Suggest how a deletion mutation to chromosome 15 inherited from the father might result in infertility in a person with Prader-Willi syndrome.

The word cancer (Latin for crab), was first used by Hippocrates 2400 years ago. He saw a similarity between the swollen veins radiating from a breast tumour and the legs of a crab. Cancer is a group of diseases caused by damage to the genes that regulate **mitosis** and the cell cycle (Topic 3.8). This leads to unrestrained growth of cells. As a consequence, a group of abnormal cells, called a tumour, develops and constantly expands in size. Cancer is a common and destructive disease. That being said, cancer is to some extent avoidable and, if diagnosed early enough, successfully treatable.

Types of tumour

Not all tumours are cancerous. Those that are cancerous are called **malignant** while those that are non-cancerous are called **benign**. The main characteristics of benign and malignant tumours are compared in Table 1.

▼ **Table 1** *A comparison of benign and malignant tumours*

Benign tumours	Malignant tumours
Can grow to a large size	Can also grow to a large size
Grow very slowly	Grow rapidly
The cell nucleus has a relatively normal appearance	The cell nucleus is often larger and appears darker due to an abundance of DNA
Cells are often well differentiated (specialised)	Cells become de-differentiated (unspecialised)
Cells produce adhesion molecules that make them stick together and so they remain within the tissue from which they arise = primary tumours	Cells do not produce adhesion molecules and so they tend to spread to other regions of the body, a process called **metastasis**, forming secondary tumours
Tumours are surrounded by a capsule of dense tissue and so remain as a compact structure	Tumours are not surrounded by a capsule and so can grow finger-like projections into the surrounding tissue
Much less likely to be life-threatening but can disrupt functioning of a vital organ	More likely to be life-threatening, as abnormal tumour tissue replaces normal tissue
Tend to have localised effects on the body	Often have systemic (whole body) effects such as weight loss and fatigue
Can usually be removed by surgery alone	Removal usually involves radiotherapy and/or chemotherapy as well as surgery
Rarely reoccur after treatment	More frequently reoccur after treatment

Learning objectives

→ Distinguish between benign and malignant tumours.

→ Explain the role of oncogenes and tumour suppressor genes in the development of tumours.

→ Explain the effects of abnormal methylation of tumour suppressor genes and oncogenes.

→ Explain how increased oestrogen levels can cause breast cancer.

Specification reference 3.8.2.3

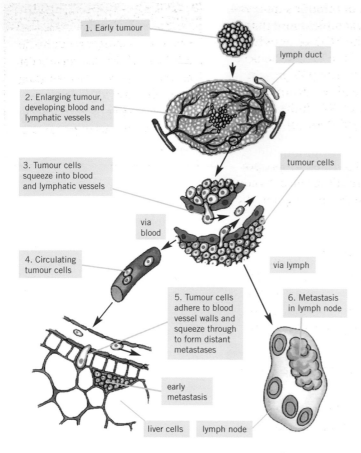

▲ **Figure 1** *A primary tumour and its development and spread into a secondary tumour*

The development of secondary tumours from a primary tumour is illustrated in Figure 1.

Cancer and the genetic control of cell division

DNA analysis of tumours has shown that, in general, cancer cells are derived from a single mutant cell. The initial mutation causes uncontrolled mitosis in this cell. Later, a further mutation in one of the descendant cells leads to other changes that cause subsequent cells to be different from normal in growth and appearance. The two main types of genes that play a role in cancer are tumour suppressor genes and oncogenes.

Oncogenes

Most oncogenes are mutations of proto-oncogenes. As we discovered in Topic 3.8, proto-oncogenes stimulate a cell to divide when growth factors attach to a protein receptor on its cell-surface membrane. This then activates genes that cause DNA to replicate and the cell to divide. If a proto-oncogene mutates into an oncogene it can become permanently activated (switched on) for two reasons:

- The receptor protein on the cell-surface membrane can be permanently activated, so that cell division is switched on even in the absence of growth factors.
- The oncogene may code for a growth factor that is then produced in excessive amounts, again stimulating excessive cell division.

The result is that cells divide too rapidly and out of control, and a tumour or cancer, develops. A few cancers are caused by inherited mutations of proto-oncogenes that cause the oncogene to be activated but most cancer-causing mutations involving oncogenes are acquired, not inherited.

Tumour suppressor genes

Tumour suppressor genes slow down cell division, repair mistakes in DNA, and 'tell' cells when to die – a process called apoptosis (programmed cell death). They therefore have the opposite role from proto-oncogenes. As its name suggests, a normal tumour suppressor gene maintains normal rates of cell division and so prevents the formation of tumours. If a tumour suppressor gene becomes mutated it is inactivated (switched off). As a result, it stops inhibiting cell division and cells can grow out of control. The mutated cells that are formed are usually structurally and functionally different from normal cells. While most of these die, those that survive can make clones of themselves and form tumours. There are a number of forms of tumour suppressor genes including *TP53*, *BRCA1* and *BRCA2*.

Synoptic link

You may recall that we dealt with the roles of proto-oncogenes and tumour suppressor genes in Topic 3.8, The cell cycle. A review of the information there would be useful background to this topic.

Hint

A proto-oncogene's role in cell division is like the accelerator pedal in a car – it controls how fast it goes. When it mutates into an oncogene, it is as if the accelerator pedal has got stuck down and cell division proceeds at a very rapid and uncontrolled rate.

Some cancers are caused by inherited mutations of tumour suppressor genes but most are acquired, not inherited. For example, more than half of human cancers display abnormalities of the *TP53* gene (which codes for the p53 protein). Acquired mutations of the *TP53* gene occur in many cancers, including lung and breast cancer. The p53 protein is involved in the process of apoptosis (programmed cell death). This process is activated when a cell is unable to repair DNA. If the gene for p53 is not functioning correctly, cells with damaged DNA continue to divide leading to cancer.

An important difference between oncogenes and tumour suppressor genes is that while oncogenes cause cancer as a result of the activation of proto-oncogenes, tumour suppressor genes cause cancer when they are inactivated.

Abnormal methylation of tumour suppressor genes

In Topic 20.4 we learnt about the significance of methylation of DNA. It is now known that abnormal DNA methylation is common in the development of a variety of tumours. The most common abnormality is hypermethylation (increased methylation). The process by which hypermethylation may lead to cancer is as follows:

- Hypermethylation occurs in a specific region (promoter region) of tumour suppressor genes.
- This leads to the tumour suppressor gene being inactivated.
- As a result, transcription of the promoter regions of tumour suppressor genes is inhibited.
- The tumour suppressor gene is therefore silenced (switched off).
- As the tumour suppressor gene normally slows the rate of cell division, its inactivation leads to increased cell division and the formation of a tumour.

Abnormal methylation of this type is thought to occur in a tumour suppressor gene known as BRCA1 and leads to the development of breast cancer.

Another form of abnormal methylation is hypomethylation (reduced methylation). This has been found to occur in oncogenes where it leads to their activation and hence the formation of tumours.

Oestrogen concentrations and breast cancer

Oestrogens play a central role in regulating the menstrual cycle in women. It is known that after the menopause, a woman's risk of developing breast cancer increases. This is thought to be due to increased oestrogen concentrations. At first this seemed paradoxical because the production of oestrogens from the ovaries diminishes after the menopause. However, the fat cells of the breasts tend to produce more oestrogens after the menopause. These locally produced oestrogens appear to trigger breast cancer in postmenopausal women. Once a tumour has developed, it further increases oestrogen concentration which therefore leads to increased development of the tumour. It also appears that white blood cells that are drawn to the tumour increase oestrogen production. This leads to even greater development of the tumour.

Hint

A tumour suppressor gene's role in cell division is like the brake pedal on a car – it prevents it from going too quickly. When it mutates, it is as if the brake pedal doesn't work and cell division takes place more rapidly.

How then can oestrogen cause a tumour to develop? We saw in Topic 20.3 the mechanism by which oestrogen effectively activates a gene by releasing an inhibitor molecule that prevents transcription. If the gene that oestrogen acts on is one that controls cell division and growth, then it will be activated and its continued division could produce a tumour. It is known, for example, that oestrogen causes proto-oncogenes of cells in breast tissue to develop into oncogenes. This leads to the development of a tumour (breast cancer).

Summary questions

1 Describe a process by which oestrogen might cause breast cancer in post-menopausal women.

2 Explain why the activation of a proto-oncogene can cause the development of a tumour while it requires deactivation of a tumour suppressor gene to do so.

3 Suggest two reasons why the surgical removal of a benign tumour is usually sufficient treatment to prevent the tumour growing again.

4 Suggest why the surgical removal of a malignant tumour requires follow-up treatments such as chemotherapy and radiotherapy.

5 The enzyme histone deacetylase (HDAC) removes acetyl groups from histones. Phenylbutyric acid is an inhibitor of the enzyme HDAC. Suggest how phenylbutyric acid might be used to treat cancer. Explain your answer.

Risk factors and cancer

Cancer is not a single disease and, likewise, does not have a single cause. Some causal factors are beyond our individual control, for example age and genetic factors. Others are lifestyle factors and therefore within our power to change.

We can do nothing about our genes or our age but our lifestyle can expose us to environmental and **carcinogenic** factors that put us at risk of contracting cancer. It is thought that about half the people who are diagnosed with cancer in the UK could have avoided getting the disease if they had changed their lifestyle. The specific lifestyle factors that contribute to cancer include:

- **smoking**. Not only smokers are in danger, those who passively breathe tobacco smoke also have an increased risk of getting cancer.

- **diet**. What we eat and drink affects our risk of contracting cancer. There is strong evidence that a low-fat, high-fibre diet, rich in fruit and vegetables, reduces the risk.

- **obesity**. Being overweight increases the risk of cancer.

- **physical activity**. People who take regular exercise are at lower risk from some cancers than those who take little or no exercise.

- **sunlight**. The more someone is exposed to sunlight or light from sunbeds, the greater is the risk of skin cancer.

Smoking and cancer

Tobacco was first introduced to Britain in the 16th century. Initially, only men smoked, but women took up smoking in the 1920s. By 1945, the equivalent of 12 cigarettes a day for every British male was being smoked. At the time the public regarded smoking as a harmless pleasure. Doctors, however, were alarmed by a phenomenal increase in deaths from lung cancer. At a 1947 conference, a number of scientists suggested tobacco smoke as a possible cause of the increase.

Epidemiologists collect data on diseases and then look for correlations between these diseases and various factors in the lives of those who have them. The world's longest-running survey of smoking began in the UK in 1951. This survey, and others elsewhere in the world, has revealed a number of statistically significant correlations about smokers.

- A regular smoker is three times more likely to die prematurely than a non-smoker.

- The more cigarettes smoked per day, the earlier, on average, a smoker dies.

- Smokers who give up the habit improve their life expectancy compared with those who continue to smoke.

- Long-term smokers are more likely to die early as a result of smoking.

- The incidence of pulmonary disease increases with the number of cigarettes smoked.

- Smokers make up 98% of emphysema sufferers.

Data like those in Figure 2 were used to help establish a correlation between disease and smoking.

▲ **Figure 2** *Annual number of cases of lung cancer per 100 000 men in the USA correlated to daily consumption of cigarettes*

Epidemiological statistics show correlations between lung cancer and smoking. These include:

- A man smoking 25 cigarettes a day is 25 times more likely to die of lung cancer than a non-smoker.

- The longer a person smokes, the greater the risk of developing lung cancer. Smoking 20 cigarettes a day for 40 years increases the risk of lung cancer eight times more than smoking 40 cigarettes a day for 20 years.

- When a person stops smoking, the risk of developing lung cancer decreases and approaches that of a non-smoker after around 10–15 years (depending on age and amount of tobacco consumed).

- The death rate from lung cancer is 18 times greater in a smoker than in a non-smoker.

Cigarette manufacturers and some smokers argued that these epidemiological correlations were coincidental.

Experimental evidence linking smoking to disease

Scientists carried out experiments in the 1960s in which dogs were made to inhale cigarette smoke. The smoke was either inhaled directly or first passed through a filter tip. Those dogs that inhaled the filtered smoke remained generally healthy. Those inhaling unfiltered smoke developed pulmonary disease and early signs of lung cancer. Scientists then carried out a further series of experiments that allowed them to formulate a new hypothesis from each result, which they could then test experimentally.

- Machines were used to simulate the action of smoking and to collect the harmful constituents that accumulated in the filters.

- These were then analysed chemically and each constituent was tested in the laboratory for its ability to damage epithelial cells and mutate the genes they contain. This was done by adding tar to the skin of mice or to cells that had been grown in culture.

- As a result of such tests it was shown that the tar found in cigarette smoke contained **carcinogens**.

- The constituent chemicals of the tar were each tested and one, benzopyrene (BP), was shown to mutate DNA.

- The scientists still had to demonstrate precisely *how* it caused cancer. They carried out experiments which showed that BP is absorbed by epithelial cells and converted to a derivative. This then binds with a **gene** and mutates it.

- Another experiment showed that this **mutation** led to uncontrolled cell division of epithelial cells and hence the growth of a **tumour**.

- Even this was not proof. In further experiments, scientists showed that the mutations of the gene in a cancer cell occurred at three specific points on the DNA. When the derivative of BP from tobacco smoke was used to mutate the gene, it caused changes to the DNA at precisely the same points.

> 5 Identify the key evidence that smoking is a cause of lung cancer?

The evidence was now conclusive. Smoking tobacco could cause lung cancer. This is not to say that it always does, but simply that there is an increased risk – it is about probabilities not certainties.

These experiments convinced the public of the health risks of smoking and led to reduced use of tobacco in the UK. This changed view in turn persuaded the government to take measures designed to reduce smoking. These included – progressively raising taxes on tobacco, banning tobacco advertising, placing health warnings on tobacco products, banning smoking in work and public places, including bars, pubs and clubs.

> 6 'My father smoked 30 cigarettes a day and lived to be 95.' This type of argument is sometimes used to suggest that smoking is not harmful. Explain why scientists do not accept this reasoning.

▲ **Figure 3** *Smoking these 20 cigarettes would, on average, reduce your life expectancy by $3\frac{1}{2}$ hours*

Cancer – the 'two hit' hypothesis

We have learnt that tumours can develop as a result of a mutation of proto-oncogenes that causes cells to divide more rapidly than normal. Tumours can also develop by a mutation of tumour suppressor genes that prevents them from inhibiting cell division.

It only takes a single mutated allele to activate proto-oncogenes but it takes a mutation of both alleles to inactivate tumour suppressor genes (two-hits). As natural mutation rates are slow, it takes a considerable time for both tumour suppressor alleles to mutate. This explains why the risk of many cancers increases as one gets older. It is thought that some people are born with one mutated allele. These people are at greater risk of cancer as they need only one further mutation, rather than two, to develop the disease. This explains why certain cancers carry an inherited increased risk.

1 Explain why a doctor may enquire about a patient's family medical history before deciding on using X-ray analysis for a condition other than cancer.

2 Suggest a reason why a single mutant allele of a proto-oncogene can cause cancer, but it requires two mutant alleles of the tumour suppressor gene to do so.

3 One experimental treatment for cancer involves introducing tumour suppressor genes into rapidly dividing cells in order to arrest tumour growth. Explain how this treatment might work.

4 Another experimental treatment is the development of an antibiotic drug that will destroy certain protein receptors on membranes of cancer cells. Explain how this treatment might be effective.

Projects to determine the entire DNA nucleotide base sequence of a wide range of organisms, including humans, have taken place over the past few decades. The idea has been to map the DNA base sequences that make up the genes of the organism and then to map these genes on the individual chromosomes of that organism. In this way a complete map of all the genetic material in an organism (the **genome**) is obtained.

Sequencing genomes

When you consider that the human genome consists of over 3 billion base pairs organised into around 20 000 genes, sequencing every one of those bases is a mammoth task and yet it took just 13 years to complete. This would have been impossible without the use of bioinformatics. Bioinformatics is the science of collecting and analysing complex biological data such as genetic codes. It uses computers to read, store, and organise biological data at a much faster rate than previously. It also utilises algorithms (mathematical formulae) to analyse and interpret biological data.

DNA sequencing

Determining the complete DNA base sequence of an organism uses the technique of whole-genome shotgun (WGS) sequencing. This involves researchers cutting the DNA into many small, easily sequenced sections and then using computer algorithms to align overlapping segments to assemble the entire genome. Sequencing methods such as these are continuously updated which, along with the increased automation of the processes involved, have led to extremely rapid sequencing of whole genomes.

The medical advances that have been made as a result of sequencing the human genome are many. For example, over 1.4 million single nucleotide polymorphisms (SNPs) have been found in the human genome. SNPs are single base variations in the genome that are associated with disease and other disorders. Figure 1 shows some of the diseases that have been mapped on the human X chromosome. Medical screening of individuals has allowed quick identification of potential medical problems and for early intervention to treat them. As we saw in Topic 10.4, sequencing the DNA of different organisms has also made it possible to establish the evolutionary links between species.

The proteome

Of greater practical importance to humans is not the genes themselves, but the nature of the proteins these genes code for. These proteins are known as the **proteome**. A general definition of the proteome is all the proteins produced by the genome. However, as a protein is only produced when a gene is switched on, and genes are not switched on all the time, a more specific definition is all the proteins produced

chrondrodysplasia

retinoschisis

adrenal hypoplasia

ornithine enzyme deficiency

retinitis pigmentosa 2

merkes syndrome

androgen insensitivity

cleft palate
spastic paraplegia
deafness with stapes fixation

alport syndrome

PRPS-related gout

lowe syndrome

HPRT-related gout

haemophilia B
haemophilia A
colour blindness
diabetes insipidus

▲ **Figure 1** *Some of the 60 diseases that have been mapped on the human X chromosome . This is shown for illustration only and the chromosome map does not have to be learnt*

in a given type of cell (cellular proteome) or organism (complete proteome), at a given time, under specified conditions. There are differences in the ease with which we can determine the genomes and proteomes of simple and complex organisms.

Determining the genome and proteome of simpler organisms

The first bacterium to have its genome fully sequenced was *Haemophilus influenza* in 1995. *H. influenza* contains 1700 genes comprising 1.8 million bases. The genomes of thousands of prokaryotic and single-celled eukaryotic organisms are currently being sequenced as part of the Human Microbiome Project. It is hoped that the information gained will help cure disease and provide knowledge of genes that can be usefully exploited. For example, ones from organisms that can withstand extreme or toxic environmental conditions and so have potential uses in cleaning up pollutants or in manufacturing biofuels.

Determining the proteome of prokaryotic organisms like bacteria is relatively easy because:

- the vast majority of prokaryotes have just one, circular piece of DNA that is not associated with histones
- there are none of the non-coding portions of DNA which are typical of eukaryotic cells.

Knowledge of the proteome of organisms like bacteria has a number of applications. Of particular interest is the identification of those proteins that act as antigens on the surfaces of human pathogens. These antigens can be used in vaccines against diseases caused by these pathogens. In the case of vaccines, the antigens can be manufactured and then administered to people in appropriate doses. In response to the antigen, memory cells are produced which trigger a secondary response when the antigen is encountered on a second occasion (Topic 5.4).

One example is sequencing of the DNA of *Plasmodium falciparum* which causes malaria. All 5300 genes on *Plasmodium's* 14 chromosomes have been sequenced giving us an insight into its metabolism and knowledge of the proteins it produces. All this will be invaluable in helping us to develop the elusive vaccine against this globally important disease.

Determining the genome and proteome of complex organisms

The success in mapping the human genome in 2003 is a testimony to what can be achieved in mapping DNA sequences of complex organisms. There are around 20 000 genes in the human genome although this number is constantly being revised down as our techniques for identifying genes improves. The problem in complex organisms is translating knowledge of the genome into the proteome. This is because the genome of complex organisms contains many non-coding genes as well as others that have a role in regulating other genes. In humans, it is thought that as few as 1.5% of genes may code for proteins. There is a human proteome project currently underway to identify all the proteins produced by humans. There is also the question of whose DNA is used for mapping. All individuals, except identical twins, have different base sequences on their DNA. The DNA mapped will differ, if only slightly, from everyone else's DNA.

Summary questions

1 Distinguish between a genome and a proteome.

2 Explain why determining the proteome of simple organisms like bacteria is easier than determining the proteome of complex ones like humans.

3 Explain how knowledge of the proteome of a pathogen might help to control the disease it causes.

1 SCID is a severe inherited disease. People who are affected have no immunity. Doctors carried out a trial using gene therapy to treat children with SCID. The doctors who carried out the trial obtained stem cells from each child's umbilical cord.

(a) Give two characteristic features of stem cells. *(2 marks)*

The doctors mixed the stem cells with viruses. The viruses had been genetically modified to contain alleles of a gene producing full immunity. The doctors then injected this mixture into the child's bone marrow. The viruses that the doctors used had RNA as their genetic material. When these viruses infect cells, they pass their RNA and two viral enzymes into the host cells.

(b) One of the viral enzymes makes a DNA copy of the virus RNA. Name this enzyme. *(1 mark)*

The other viral enzyme is called integrase. Integrase inserts the DNA copy anywhere in the DNA of the host cell. It may even insert the DNA copy in one of the host cell's genes.

(c) (i) The insertion of the DNA copy in one of the host cell's genes may cause the cell to make a non-functional protein. Explain how. *(2 marks)*

(ii) Some of the children in the trial developed cancer. How might the insertion of the DNA have caused cancer? *(2 marks)*

(d) Five out of the 20 children in the trial developed cancer. Although the cancer was treated successfully, the doctors decided to stop the trial in its early stages. They then reviewed the situation and decided to continue. Do you agree with their decision to continue? Explain your answer. *(2 marks)*

AQA June 2010

2 Figure 1 shows part of a gene that is being transcribed.

promoter region of gene

enzyme **X**

RNA

▲ **Figure 1**

(a) Name enzyme **X**. *(1 mark)*

(b) (i) Oestrogen is a hormone that affects transcription. It forms a complex with a receptor in the cytoplasm of target cells. Explain how an activated oestrogen receptor affects the target cell. *(2 marks)*

(ii) Oestrogen only affects target cells. Explain why oestrogen does not affect other cells in the body. *(1 mark)*

(c) Some breast tumours are stimulated to grow by oestrogen. Tamoxifen is used to treat these breast tumours. In the liver, tamoxifen is converted into an active substance called endoxifen. **Figure 2** shows a molecule of oestrogen and a molecule of endoxifen.

Oestrogen

Endoxifen

▲ **Figure 2**

Use **Figure 2** to suggest how endoxifen reduces the growth rate of these breast tumours. (*2 marks*)

AQA June 2010

3 (a) Explain how the methylation of tumour suppressor genes can lead to cancer. (*3 marks*)
        Scientists investigated a possible relationship between the percentage of fat
        in the diet and the death rate from breast cancer in women from 10 countries.
        Their data are shown in **Table 3**.

▼ Table 3

Percentage of fat in diet of population	Death rate of women from breast cancer per 100 000 women
9.5	1.5
15.0	7.0
20.0	12.0
25.0	9.0
32.0	15.0
35.0	8.0
35.0	20.0
40.5	18.0
43.0	24.0
45.0	26.0

(b) Describe how you would plot a suitable graph of these data. Explain your choice of type of graph. (*3 marks*)

(c) Use the data to calculate the correlation coefficient, *r*, to test for the strength of any correlation between fat percentage in the dict and the death rate from breast cancer. (*4 marks*)

AQA Specimen 2014 (apart from 3 (c))

$$r = \frac{\sum(x-\bar{x}) \times (y-\bar{y})}{\sqrt{\sum(x-\bar{x})^2 \times \sum(y-\bar{y})^2}}$$

NB you will only be required to carry out this form of calculation as part of your practical work.

Learning objectives

→ Explain how complementary DNA is made using reverse transcriptase.

→ Explain how restriction endonucleases are used to cut DNA into fragments.

Specification reference: 3.8.4.1

Perhaps the most significant scientific advance in recent years has been the development of recombinant DNA technology that allows genes to be manipulated, altered and transferred from organism to organism – even to transform DNA itself. These techniques have enabled us to understand better how organisms work and to design new industrial processes and medical applications.

A number of human diseases result from individuals being unable to produce for themselves various metabolic chemicals. Many of these chemicals are proteins, such as **insulin**. They are therefore the product of a specific length of DNA, that is, the product of a gene. Treatment of such deficiencies previously involved extracting the chemical from a human or animal donor and introducing it into the patient. This presents problems such as rejection by the immune system and risk of infection. The cost is also considerable.

It follows that there are advantages in producing large quantities of 'pure' proteins from other sources. As a result, techniques have been developed to isolate genes, **clone** them and transfer them into microorganisms. These microorganisms are then grown to provide a 'factory' for the continuous production of a desired protein. The DNA of two different organisms that has been combined in this way is called **recombinant DNA**. The resulting organism is known as a **transgenic** or **genetically modified organism** (**GMO**).

How then is it possible that the DNA of one organism is not only accepted by a different species but also functions normally when it is transferred? The answer lies in the fact that the genetic code is the same in all organisms. In other words it is universal and can be used by all living organisms. This explains why the coded information on the transferred DNA can be interpreted, but what about the making of proteins? Well this too is universal in that the mechanisms of transcription and translation are essentially the same in all living organisms. As a result, transferred DNA can be transcribed and translated within the cells of the recipient (transgenic) organism and the proteins it codes for can be manufactured in the same way as they would be within the donor organism. This is all indirect evidence for evolution.

The process of making a protein using the DNA technology of gene transfer and cloning involves a number of stages:

1 **isolation** of the DNA fragments that have the gene for the desired protein
2 **insertion** of the DNA fragment into a vector
3 **transformation**, that is, the transfer of DNA into suitable host cells
4 **identification** of the host cells that have successfully taken up the gene by use of **gene markers**
5 **growth/cloning** of the population of host cells.

Let us consider each stage in detail.

Before a gene can be transplanted, it must be identified and isolated from the rest of the DNA. Given that the required gene may consist

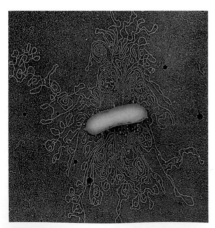

▲ **Figure 1** *An* Escherichia coli *bacterial cell that has been treated so that its DNA is ejected*

of a sequence of a few hundred bases amongst the many millions in human DNA, this is no small feat. There are several methods of producing DNA fragments:

- conversion of mRNA to cDNA using reverse transcriptase
- using restriction endonucleases to cut fragments containing the desired gene from DNA
- creating the gene in a gene machine, usually based on a known protein structure.

Using reverse transcriptase

Retroviruses are a group of viruses, of which the best known is human immunodeficiency virus (HIV) (Topic 5.7). The coded genetic information of retroviruses is in the form of RNA. However, in a host cell they are able to synthesise DNA from their RNA using an enzyme called reverse transcriptase. It is so-named because it catalyses the production of DNA from RNA, which is the reverse of the more usual transcription of RNA from DNA. The process of using reverse transcriptase to isolate a gene is illustrated in Figure 2 and described in Figure 2.

β-cells from islets of Langerhans in the human pancreas. As these cells are specialised to produce insulin, they make a lot of mRNA that codes for insulin.

mRNA coding for insulin from β-cells

mRNA coding for insulin

mRNA acts as a template on which a single-stranded complementary copy of DNA (cDNA) is formed using reverse transcriptase

single-stranded (cDNA) is isolated by hydrolysis of the mRNA with an enzyme

DNA polymerase

double-stranded DNA is formed on the template of the cDNA using DNA polymerase

copy of human insulin gene

▲ **Figure 2** *The use of reverse transcriptase to isolate the gene that codes for insulin*

Synoptic link

DNA polymerase acts in the same way when forming the second DNA strand during DNA replication, as described in Topic 2.2 , DNA replication.

- A cell that readily produces the protein is selected (e.g., the β-cells of the islets of Langerhans from the pancreas are used to produce insulin).
- These cells have large quantities of the relevant mRNA, which is therefore more easily extracted.
- Reverse transcriptase is then used to make DNA from RNA. This DNA is known as **complementary DNA (cDNA)** because it is made up of the **nucleotides** that are complementary to the mRNA.
- To make the other strand of DNA, the enzyme DNA polymerase is used to build up the complementary nucleotides on the cDNA template. This double strand of DNA is the required gene

Using restriction endonucleases

All organisms use defensive measures against pathogens. Bacteria are frequently infected by viruses that inject their DNA into them in order to take over the cell. Some bacteria defend themselves by producing enzymes that cut up the viral DNA. These enzymes are called restriction endonucleases.

There are many types of restriction endonucleases. Each one cuts a DNA double strand at a specific sequence of bases called a recognition sequence. Sometimes, this cut occurs between two opposite base pairs. This leaves two straight edges known as blunt ends. For example, one restriction endonuclease cuts in the middle of the base recognition sequence GTTAAC (Figure 3).

Other restriction endonucleases cut DNA in a staggered fashion. This leaves an uneven cut in which each strand of the DNA has exposed, unpaired bases. An example is a restriction endonuclease that recognises a six-base pair (or six bp) AAGCTT, as shown in Figure 3. In this figure, look at the sequence of unpaired bases that remain. If you read both the four unpaired bases at each end from left to right, the two sequences are opposites of one another, that is, they are a **palindrome**. The recognition sequence is therefore referred to as a

> ### Hint
>
> Each restriction endonuclease recognises and cuts DNA at a specific sequence of bases. These sequences occur in the DNA of all species of organisms – but not in the same places!

a HpaI restriction endonuclease has a recognition site GTTAAC, which produces a straight cut and therefore blunt ends

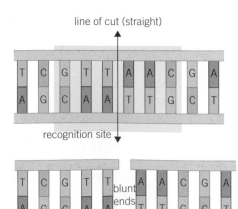

b HindIII restriction endonuclease has the recognition site AAGCTT, which produces a staggered cut and therefore sticky ends

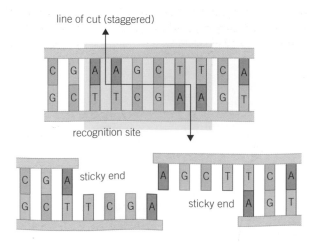

▲ **Figure 3** *Action of restriction endonucleases*

six bp palindromic sequence. This feature is typical of the way restriction endonucleases cut DNA to leave sticky ends. We shall look at the importance of these sticky ends in Topic 21.2.

The 'gene machine'

It is now possible to manufacture genes in a laboratory in the following manner:

- The desired sequence of nucleotide bases of a gene is determined from the desired protein that we wish to produce. The amino acid sequence of this protein is determined. From this, the mRNA codons are looked up and the complementary DNA triplets are worked out.

- The desired sequence of nucleotide bases for the gene is fed into a computer.

- The sequence is checked for biosafety and biosecurity to ensure it meets international standards as well as various ethical requirements.

- The computer designs a series of small, overlapping single strands of nucleotides, called oligonucleotides, which can be assembled into the desired gene.

- In an automated process, each of the oligonucleotides is assembled by adding one nucleotide at a time in the required sequence.

- The oligonucleotides are then joined together to make a gene. This gene doesn't have introns or other non-coding DNA. The gene is replicated using the polymerase chain reaction (Topic 21.3).

- The polymerase chain reaction also constructs the complementary strand of nucleotides to make the required double stranded gene. It then multiples this gene many times to give numerous copies.

- Using sticky ends (Topic 21.2) the gene can then be inserted into a bacterial plasmid. This acts as a vector for the gene allowing it to be stored, cloned or transferred to other organism in the future.

- The genes are checked using standard sequencing techniques (Topic 20.6) and those with errors are rejected.

The advantages of this process are that any sequence of nucleotides can be produced, in a very short time (as little as 10 days) and with great accuracy. A further advantage is that these artificial genes are also free of introns, and other 'non-coding' DNA, so can be transcribed and translated by prokaryotic cells. The process is shown in Figure 4.

Summary questions

In the following passage replace each number with the most appropriate word or words.

Where the DNA of two different organisms is combined, the product is known as (1) DNA. One method of producing DNA fragments is to make DNA from RNA using an enzyme called (2). This enzyme initially forms a single strand of DNA called (3) DNA. To form the other strand requires an enzyme called (4). Another method of producing DNA fragments is to use enzymes called (5), which cut up DNA. Some of these leave fragments with two straight edges, called (6) ends. Others leave ends with uneven edges, called (7) ends. If the sequence of bases on one of these uneven ends is GAATTC, then the sequence on the other end, if read in the same direction, will be (8).

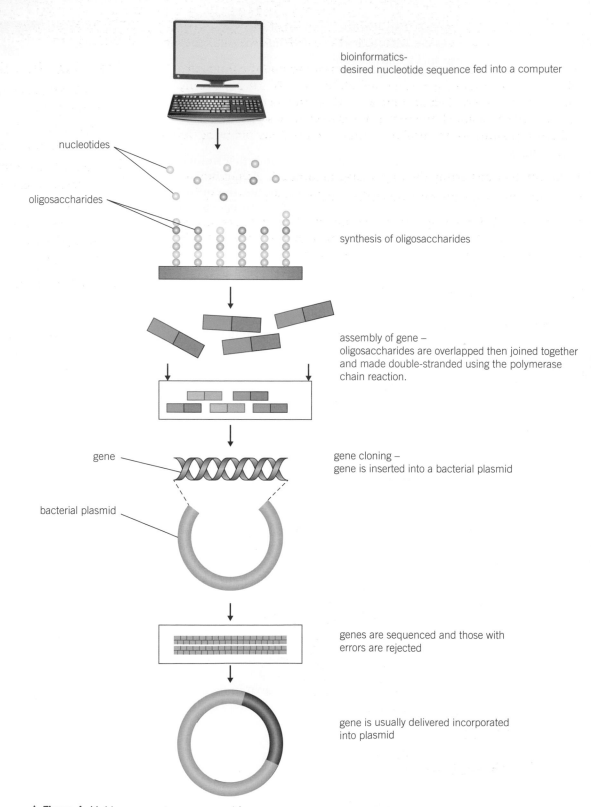

▲ **Figure 4** *Making a gene in a gene machine*

The following labels appear in the figure:

bioinformatics-
desired nucleotide sequence fed into a computer

nucleotides

oligosaccharides

synthesis of oligosaccharides

assembly of gene –
oligosaccharides are overlapped then joined together
and made double-stranded using the polymerase
chain reaction.

gene

gene cloning –
gene is inserted into a bacterial plasmid

bacterial plasmid

genes are sequenced and those with
errors are rejected

gene is usually delivered incorporated
into plasmid

Having cut DNA into fragments, it is necessary to find the fragment which has the required gene amongst all the rest. This is done using a DNA probe as described in Topic 21.4. Once the fragment with the gene has been obtained, the next stage is to clone it so that there is a sufficient quantity for medical or commercial use. This can be achieved in two ways:

- *in vivo*, by transferring the fragments to a host cell using a vector
- *in vitro*, using the polymerase chain reaction (see Topic 21.3).

Before we consider how genes can be cloned within living organisms (*in vivo* cloning), let us look at the importance of the sticky ends left when DNA is cut by **restriction endonucleases**.

Importance of sticky ends

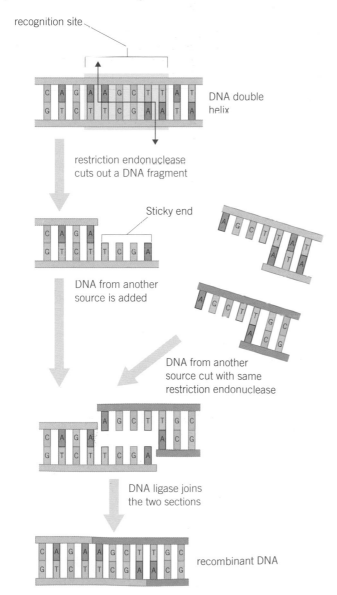

recognition site

DNA double helix

restriction endonuclease cuts out a DNA fragment

Sticky end

DNA from another source is added

DNA from another source cut with same restriction endonuclease

DNA ligase joins the two sections

recombinant DNA

▲ **Figure 1** *The use of sticky ends to combine DNA from different sources*

Learning objectives

→ Explain the importance of sticky ends.

→ Explain how a DNA fragment can be inserted into a vector.

→ Explain how the DNA of the vector is introduced into host cells.

→ Describe the nature of gene markers and explain how they work.

Specification reference: 3.8.4.1

The sequences of DNA that are cut by restriction endonucleases are called recognition sites. If the recognition site is cut in a staggered fashion, the cut ends of the DNA double strand are left with a single strand which is a few **nucleotide** bases long. The nucleotides on the single strand at one side of the cut are obviously complementary to those at the other side because they were previously paired together.

If the same restriction endonuclease is used to cut DNA, then all the fragments produced will have ends that are complementary to one another. This means that the single-stranded end of any one fragment can be joined (stuck) to the single-stranded end of any other fragment. In other words, their ends are sticky. Once the complementary bases of two sticky ends have paired up, an enzyme called **DNA ligase** is used to bind the phosphate-sugar framework of the two sections of DNA and so unite them as one.

Sticky ends have considerable importance because, provided the same restriction endonuclease is used, we can combine the DNA of one organism with that of any other organism (see Figure 1).

Preparing the DNA fragment for insertion

The preparation of the DNA fragment involves the addition of extra lengths of DNA. For the transcription of any gene to take place, the enzyme that synthesises mRNA (RNA polymerase) must attach to the DNA near a gene. The binding site for RNA polymerase is a region of DNA, known as a **promoter**. The nucleotide bases of the promoter attach both RNA polymerase and transcription factors (Topic 20.3) and so begin the process of transcription. If we want our DNA fragment to transcribe mRNA in order to make a protein, it is essential that we attach to it the necessary promoter region to start the process.

In the same way as a region of DNA binds RNA polymerase and begins transcription of a gene, another region releases RNA polymerase and ends transcription. This region of DNA is called a **terminator**. Again we need to add a terminator region to the other end of our DNA fragment to stop transcription at the appropriate point.

Insertion of DNA fragment into a vector

Once an appropriate fragment of DNA has been cut from the rest of the DNA and the promoter and terminator regions added, the next task is to join it into a carrying unit, known as a **vector**. This vector is used to transport the DNA into the host cell. There are different types of vector but the most commonly used is the **plasmid**. Plasmids are circular lengths of DNA, found in bacteria, which are separate from the main bacterial DNA. Plasmids almost always contain genes for antibiotic resistance, and restriction endonucleases are used at one of these antibiotic-resistance genes to break the plasmid loop.

The restriction endonuclease used is the same as the one that cut out the DNA fragment. This ensures that the sticky ends of the opened-up plasmid are complementary to the sticky ends of the DNA

▲ **Figure 2** *Coloured TEM of genetically engineered DNA plasmids from the bacterium* Escherichia coli. *The plasmids (yellow) have had different gene sequences (various colours) inserted into them*

fragment. When the DNA fragments are mixed with the opened-up plasmids, they may become incorporated into them. Where they are incorporated, the join is made permanent using the enzyme DNA ligase. These plasmids now have recombinant DNA. These events are summarised in Figure 4.

Introduction of DNA into host cells

Once the DNA has been incorporated into at least some of the plasmids, they must then be reintroduced into bacterial cells. This process is called **transformation** and involves the plasmids and bacterial cells being mixed together in a medium containing calcium ions. The calcium ions, and changes in temperature, make the bacterial membrane permeable, allowing the plasmids to pass through the cell-surface membrane into the cytoplasm. However, not all the bacterial cells will possess the DNA fragments with the desired gene for the desired protein. Some reasons for this are:

* Only a few bacterial cells (as few as 1%) take up the plasmids when the two are mixed together.
* Some plasmids will have closed up again without incorporating the DNA fragment.
* Sometimes the DNA fragment ends join together to form its own plasmid.

The first task is to identify which bacterial cells have taken up the plasmid. One way to do this is to use the fact that bacteria have evolved mechanisms for resisting the effects of antibiotics, typically by producing an enzyme that breaks down the antibiotic before it can destroy the bacterium. The genes for the production of these enzymes are found in the plasmids.

Some plasmids carry genes for resistance to more than one antibiotic. One example is the R-plasmid, which carries genes for resistance to two antibiotics, ampicillin and tetracycline.

The task of finding out which bacterial cells have taken up the plasmids entails using the gene for antibiotic resistance, which is unaffected by the introduction of the new gene. In Figure 4, this is the gene for resistance to ampicillin. The process works as follows:

* All the bacterial cells are grown on a medium that contains the antibiotic ampicillin.
* Bacterial cells that have taken up the plasmids will have acquired the gene for ampicillin resistance.
* These bacterial cells are able to break down the ampicillin and therefore survive.
* The bacterial cells that have not taken up the plasmids will not be resistant to ampicillin and therefore die.

This is an effective method of showing which of the bacterial cells have taken up the plasmids. However, some cells will have taken up the plasmids and then closed up without incorporating the new gene, and these will also have survived. The next task is to identify the cells without the new gene and eliminate them. This is achieved using marker genes. Gene transfer and cloning are summarised in Figure 3.

Synoptic link

Antibiotic resistance is covered in Topic 9.4, Types of selection.

Marker genes

There are a number of different ways of using marker genes to identify whether a gene has been taken up by bacterial cells. They all involve using a second, separate gene on the plasmid. This second gene is easily identifiable for one reason or another. For example:

- It may be resistant to an antibiotic.
- It may make a fluorescent protein that is easily seen.
- It may produce an enzyme whose action can be identified.

Antibiotic-resistance marker genes

The use of **antibiotic-resistance** genes as markers is a rather old technology and has been superseded by other methods. However, it is an interesting example of how science works, particularly of the way in which scientists use knowledge and understanding to solve new problems, use appropriate methodology and carry out relevant experiments.

Isolation

Production of DNA fragments that have the required gene using reverse transcriptase or restriction endonucleases.

Insertion

Insertion of DNA fragment into a vector, e.g., plasmid, using DNA ligase

Transformation

Introduction of DNA fragment into suitable host cell.

Identification

Identification of host cells that have taken up the DNA using gene markers.

Growth/cloning

Culturing of host cells containing the DNA to produce the protein on a large scale.

▲ **Figure 3** *Outline summary of gene transfer and cloning*

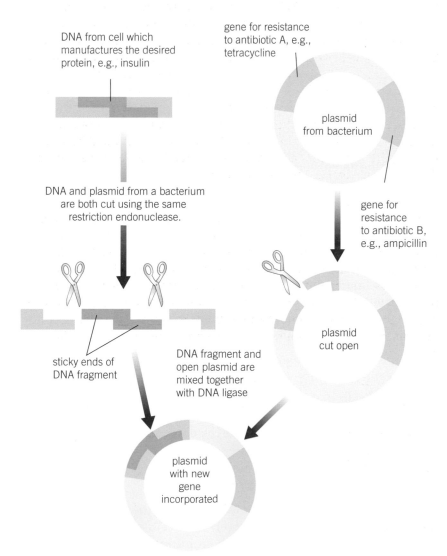

DNA from cell which manufactures the desired protein, e.g., insulin

gene for resistance to antibiotic A, e.g., tetracycline

plasmid from bacterium

DNA and plasmid from a bacterium are both cut using the same restriction endonuclease.

gene for resistance to antibiotic B, e.g., ampicillin

sticky ends of DNA fragment

DNA fragment and open plasmid are mixed together with DNA ligase

plasmid cut open

plasmid with new gene incorporated

▲ **Figure 4** *Inserting a gene into a plasmid vector*

To identify those cells with plasmids that have taken up the new gene we use a technique called **replica plating**. This process uses the other antibiotic-resistance gene in the plasmid: the gene that was cut in order to incorporate the required gene. In Figure 4 this is the gene for resistance to tetracycline. As this gene has been cut, it will no longer produce the enzyme that breaks down tetracycline. In other words, the bacteria that have taken up the required gene will no longer be resistant to tetracycline. We can therefore identify these bacteria by growing them on a culture that contains tetracycline.

The problem is that treatment with tetracycline will destroy the very cells that contain the required gene. However by using a technique called replica plating it is possible to identify living colonies of bacteria containing the required gene.

Fluorescent markers

A more recent and more rapid method is the transfer of a gene from a jellyfish (Figure 5) into the plasmid. The gene in question produces a green fluorescent protein (GFP). The gene to be cloned is transplanted into the centre of the GFP gene. Any bacterial cell that has taken up the plasmid with the gene that is to be cloned will not be able to produce GFP. Bacterial cells that have not taken up the gene will continue to produce GFP and to fluoresce. Unlike the cells that have not taken up the gene, these cells that have taken it up will not fluoresce. As the bacterial cells with the desired gene are not killed, there is no need for replica plating. Results can be obtained by simply viewing the cells under a microscope and retaining those that do not fluoresce. This makes the process more rapid.

Enzyme markers

Another gene marker is the gene that produces the enzyme lactase. Lactase will turn a particular colourless substrate blue. Again, the required gene is transplanted into the gene that makes lactase. If a plasmid with the required gene is present in a bacterial cell, the colonies grown from it will not produce lactase. Therefore, when these bacterial cells are grown on the colourless substrate they will be unable to change its colour. Where the gene has not been taken up by the bacteria, they will not turn the substrate blue. These bacteria can be discounted.

▲ Figure 7

▲ **Figure 5** *The gene in this jellyfish that produces a green fluorescent protein can be transplanted into other organisms and used as a fluorescent marker*

Hint

Interestingly, the gene for the green fluorescent protein has itself been genetically modified by the same techniques it is used to support. As a result, varieties have been engineered that fluoresce more brightly and in a number of different colours.

Summary questions

1 Explain the role of a vector during *in vivo* gene cloning.

2 Explain why gene markers are necessary during *in vivo* gene cloning.

3 Give *one* advantage of using fluorescent gene markers rather than antibiotic gene markers. Explain your answer.

4 Figure 7 shows the results of an experiment using antibiotic-resistance gene markers to find which bacterial cells have taken up a gene X. The circles within each plate represent a colony of growing bacteria. Deduce which colonies on the original plate:

   a did not take up any plasmids with gene X

   b contained plasmids possessing gene X.

   Explain your answers.

Learning objectives

→ Describe the polymerase chain reaction.

→ Explain how the polymerase chain reaction is carried out.

→ Summarise the advantages of *in vitro* and *in vivo* cloning.

Specification reference: 3.8.4.1

Study tip

Remember DNA polymerase causes nucleotides to join together as a strand, not complementary base pairing.

Hint

The polymerase chain reaction is not the same as semi-conservative replication of DNA in cells.

Study tip

Make sure you can describe the polymerase chain reaction, particularly the importance of the temperature changes.

▲ **Figure 1** *This is a thermocycler, a machine that carries out the polymerase chain reaction (PCR)*

After looking at *in vivo* cloning in Topic 21.2, let us now consider *in vitro* cloning using the polymerase chain reaction.

Polymerase chain reaction

The polymerase chain reaction (PCR) is a method of copying fragments of DNA. The process is automated, making it both rapid and efficient. The process requires the following:

- **the DNA fragment** to be copied
- **DNA polymerase** – an enzyme capable of joining together tens of thousands of **nucleotides** in a matter of minutes. One such enzyme, taq polymerase, is obtained from bacteria in hot springs and is therefore tolerant to heat (thermostable) and does not denature during the high temperatures used as part of the process
- **primers** – short sequences of nucleotides that have a set of bases complementary to those at one end of each of the two DNA fragments
- **nucleotides** – which contain each of the four bases found in DNA
- **thermocycler** – a computer-controlled machine that varies temperatures precisely over a period of time (Figure 1).

The polymerase chain reaction is illustrated in Figure 2 and is carried out in three stages:

1 **separation of the DNA strand**. The DNA fragments, primers and DNA polymerase are placed in a vessel in the thermocycler. The temperature is increased to 95 °C, causing the two strands of the DNA fragments to separate due to the breaking of the hydrogen bonds between the two DNA strands.

2 **addition (annealing) of the primers**. The mixture is cooled to 55 °C, causing the primers to join (anneal) to their complementary bases at the end of the DNA fragment. The primers provide the starting sequences for DNA polymerase to begin DNA copying because DNA polymerase can only attach nucleotides to the end of an existing chain. Primers also prevent the two separate strands from simply rejoining.

3 **synthesis of DNA**. The temperature is increased to 72 °C. This is the optimum temperature for the DNA polymerase to add complementary nucleotides along each of the separated DNA strands. It begins at the primer on both strands and adds the nucleotides in sequence until it reaches the end of the chain.

Because both separated strands are copied simultaneously there are now two copies of the original fragment. Once the two DNA strands are completed, the process is repeated by subjecting them to the temperature cycle again, resulting in four strands. The whole temperature cycle takes around two minutes. Over a million copies of the DNA can be made in only 25 temperature cycles and 100 billion copies can be manufactured in just a few hours. The polymerase chain reaction has revolutionised many aspects of science and medicine. Even the tiniest sample of DNA from a single hair or a speck of blood can now be multiplied to allow forensic examination and accurate cross-matching. You will learn more about this in Topic 21.5.

Advantages of *in vitro* and *in vivo* gene cloning

The advantages of *in vitro* gene cloning are:

- **It is extremely rapid.** Within a matter of hours a 100 billion copies of a gene can be made. This is particularly valuable where only a minute amount of DNA is available, for example, at the scene of a crime. This can quickly be increased using the polymerase chain reaction and so there is no loss of valuable time before forensic analysis and matching can take place. A complicating factor is that PCR will also increase massively any other contaminating DNA found at the scene. *In vivo* cloning would take many days or weeks to produce the same quantity of DNA.

- **It does not require living cells.** All that is required is a base sequence of DNA that needs amplification. No complex culturing techniques, requiring time and effort, are needed.

The advantages of *in vivo* gene cloning are:

- **It is particularly useful where we wish to introduce a gene into another organism**. As it involves the use of **vectors**, once we have introduced the gene into a **plasmid**, this plasmid can be used to deliver the gene into another organism, such as a human being (i.e., it can transform other organisms). This is done in a technique called gene therapy.

- **It involves almost no risk of contamination**. This is because a gene that has been cut by the same **restriction endonuclease** can match the sticky ends of the opened-up plasmid. Contaminant DNA will therefore not be taken up by the plasmid. *In vitro* cloning requires a very pure sample because any contaminant DNA will also be multiplied and could lead to a false result.

- **It is very accurate.** The DNA copied has few, if any, errors. At one time, about 20% of the DNA cloned *in vitro* by the PCR was copied inaccurately, but modern techniques have improved the accuracy of the process considerably. However, any errors in copying DNA or any contaminants in the sample will also be copied in subsequent cycles. This problem hardly ever arises with *in vivo* cloning because, although mutations can arise, these are very rare.

- **It cuts out specific genes**. It is therefore a very precise procedure as the culturing of transformed bacteria produces many copies of a specific gene and not just copies of the whole DNA sample.

- **It produces transformed bacteria that can be used to produce large quantities of gene products**. The transformed bacteria can produce proteins for commercial or medical use (e.g., hormones such as insulin).

▲ **Figure 2** *The polymerase chain reaction showing a single cycle*

Summary questions

1 In the polymerase chain reaction (PCR), primers are used. Describe what these are.

2 Explain the role of these primers.

3 Explain why two different primers are required.

4 State what type of bond is broken when DNA strands are separated in the PCR.

5 It is important in the PCR that the fragments of DNA used are not contaminated with any other biological material. Suggest a reason why.

Evaluation of DNA technology

Genetic engineering undoubtedly brings many benefits to mankind, but it is not without its risks. It is therefore important to evaluate the ethical, moral and social issues associated with its use.

The benefits of recombinant DNA technology

- Microorganisms can be modified to produce a range of substances, for example, antibiotics, hormones and enzymes, that are used to treat diseases and disorders.

- Microorganisms can be used to control pollution, for example, to break up and digest oil slicks or destroy harmful gases released from factories. Care needs to be taken to ensure that such bacteria do not destroy oil in places where it is required, for example, car engines. To do this, a suicide gene can be incorporated that causes the bacteria to destroy themselves once the oil slick has been digested.

- Genetically modified plants can be transformed to produce a specific substance in a particular organ of the plant. These organs can then be harvested and the desired substance extracted. If a drug is involved, the process is called plant pharming. One promising application of this technique is in combating disease. This involves the production of plants that manufacture antibodies to pathogens and the toxins they produce. Alternatively the plants can be modified to manufacture **antigens** which, when injected into humans, induce natural **antibody** production.

- Genetically modified crops can be engineered to have financial and environmental advantages. These include making plants more tolerant to environmental extremes, for example, able to survive drought, cold, heat, salt, or polluted soils, etc. This permits crops to be grown commercially in places where they do not grow at present. Globally, each year, an area of land equal to half the United Kingdom becomes unfit for normal crops because of increases in soil salt concentrations. Growing of genetically modified plants, such as salt-tolerant tomatoes, could bring this land back into productivity. In a world where millions lack a basic nutritious diet, and with a predicted 90 million more mouths to feed by 2025, can we ethically oppose the use of such plant crops?

- Genetically modified crops can help prevent certain diseases. A type of rice, called golden rice, can have a gene for vitamin A production added. Can we justify not developing more vitamin A-enriched crops when 250 million children worldwide are at risk from vitamin A deficiency leading to 500 000 cases of irreversible blindness each year?

- Genetically modified animals are able to produce expensive drugs, antibiotics, hormones and enzymes relatively cheaply.

- Replacing defective genes (gene therapy) might be used to cure certain genetic disorders, such as cystic fibrosis and severe combined immunodeficiency (SCID).

- Genetic fingerprinting can be used in forensic science. Details are given in Topic 21.5.

Hint

Ethics is a narrower concept than morals. Ethics are a set of standards that are followed by a particular group of individuals and are designed to regulate their behaviour. They determine what is acceptable in pursuing the aims of the group.

Hint

Social issues relate to human society and its organisation. They concern the mutual relationships of human beings, their interdependence and their cooperation for the benefit of all.

Hint

Evaluating always involves looking at the positives and negatives, that is, the benefits and risks, of a particular issue.

The risks of recombinant DNA technology
Against the benefits of genetic engineering, must be weighed the risks – both real and potential.

- It is impossible to predict with complete accuracy what the ecological consequences will be of releasing genetically engineered organisms into the environment. The delicate balance that exists in any habitat may be irreversibly damaged by the introduction of organisms with engineered genes. There is often no going back once an organism is released although 'suicide genes' can be inserted or the organism engineered so it can only survive when a supplement is added.

- A recombinant gene may pass from the organism it was placed in, to a completely different one. We know, for example, that viruses can transfer genes from one organism to another. What if a virus were to transfer genes for herbicide resistance and vigorous growth from a crop plant to a weed that competed with the crop plant? What if the same gene were transferred in pollen to other plants? How would we then be able to control this weed?

- Any manipulation of the DNA of a cell will have consequences for the metabolic pathways within that cell. We cannot be sure until after the event what unforeseen by-products of the change might be produced. Could these lead to metabolic malfunctions, cause cancer, or create a new form of disease?

- Genetically modified bacteria often have antibiotic resistance marker genes that have been added. These bacteria might spread antibiotic resistance to harmful bacteria.

- All genes mutate. What then, might be the consequences of our engineered gene mutating? Could it turn the organism into a **pathogen** which we have no means of controlling?

- What will be the long-term consequences of introducing new gene combinations? We cannot be certain of the effects on the future evolution of organisms. Will the artificial selection of 'desired' genes reduce the genetic variety that is so essential to evolution?

- What might be the financial consequences of developing plants and animals to grow in new regions? Developing bananas which grow in Britain could have disastrous consequences for the Caribbean economies that rely heavily on this crop for their income.

- How far can we take the technique of replacing defective genes? It may be acceptable to replace a defective gene to cure cystic fibrosis, but is it equally acceptable to introduce genes for intelligence, more muscular bodies, cosmetic improvements, or different facial features?

- Will knowledge of, and ability to change, human genes lead to eugenics, whereby selection of genes leads to a means of selecting one race rather than another?

- What will be the consequences of the ability to manipulate genes getting into the wrong hands? Will unscrupulous individuals, groups or governments use this power to achieve political goals, control opposition or gain ultimate power?

- Is the financial cost of recombinant DNA technology justified, or would the money be better used fighting hunger and poverty, that are the cause of much human misery. Will sophisticated treatments, with their more high-profile images, be put before the everyday treatment of rheumatoid arthritis or haemorrhoids? Will such treatments only be within the financial reach of the better-off?

- Genetic fingerprinting (Topic 21.5), with its ability to identify an individual's DNA accurately, is a highly reliable forensic tool. How easy would it be for someone to exchange a DNA sample maliciously, leading to wrongful conviction?

- Is it immoral to tamper with genes at all? Should we let nature take its own course in its own time?

- How do we deal with the issues surrounding the **human genome project**? Is it right that an individual or company can patent, and therefore effectively own, a gene?

It is inevitable that we remain inquisitive about the world in which we live, and that we will seek to try to improve the conditions around us. Genetic research is bound to continue, but the challenge will be to develop the safeguards and ethical guidelines that will allow recombinant DNA technology to be used in a safe and effective manner.

1 Take any *three* aspects of recombinant DNA technology that are beneficial to humans (as listed above) and present a reasoned argument in each case for the continued use of that technology.
2 Using the same three aspects, present a reasoned argument that an environmentalist or anti-globalisation activist might make against the continued use of that technology.

+ Treatment of severe combined immunodeficiency using gene therapy

Severe combined immunodeficiency (SCID) is a rare inherited disorder. People with this condition do not show a cell-mediated immune response (Topic 5.3), nor are they able to produce antibodies (Topic 5.5). The disorder arises when individuals inherit a defect in the gene that codes for the enzyme adenosine deaminase (ADA). This enzyme destroys toxins that would otherwise kill white blood cells. Survival has depended upon patients being given bone marrow transplants and/or injections of ADA. There have been recent attempts to treat the disorder using a technique called gene therapy as follows:

- The normal ADA gene is isolated from healthy human tissue.

- The ADA gene is inserted into a retrovirus.

- The retroviruses are grown with host cells in the laboratory to increase their number and hence the number of copies of the ADA gene.

- The retroviruses are mixed with the patient's T cells into which they inject a copy of the normal ADA gene.

- The T cells are reintroduced into the patient's blood to provide the coded information needed to make ADA.

The effectiveness of this treatment is limited to 6–12 months and so it has to be repeated at intervals. A more long-term treatment involves introducing the normal gene into bone marrow stem cells rather than T cells.

1 Outline three ways in which the normal ADA gene might be isolated from human tissue.
2 Explain what a retrovirus is.
3 Suggest the likely sequence of events whereby the defective gene could lead to the death of a person with SCID.
4 Suggest why the effectiveness of the treatment only lasts 6–12 months.
5 Suggest why introducing the gene into bone marrow stem cells rather than T cells is a more long-term treatment for SCID.

Many human diseases have a genetic origin. These are often the result of a **gene mutation**.

Recombinant DNA technology has enabled us to diagnose and treat many of these genetic disorders. In doing so, it is often necessary to know exactly where a particular DNA sequence (gene) is located. To achieve this we use labelled DNA probes and DNA hybridisation.

DNA probes

A DNA probe is a short, single-stranded length of DNA that has some sort of label attached that makes it easily identifiable. The two most commonly used probes are:

- **radioactively labelled probes**, which are made up of **nucleotides** with the **isotope** ^{32}P. The probe is identified using an X-ray film that is exposed by radioactivity.

- **fluorescently labelled probes**, which emit light (fluoresce) under certain conditions, for instance when the probe has bound to the target DNA sequence.

DNA probes are used to identify particular alleles of genes in the following way:

- A DNA probe is made that has base sequences that are complementary to part of the base sequence of the DNA that makes up the allele of the gene that we want to find.

- The double-stranded DNA that is being tested is treated to separate its two strands.

- The separated DNA strands are mixed with the probe, which binds to the complementary base sequence on one of the strands. This is known as **DNA hybridisation** (see below).

- The site at which the probe binds can be identified by the radioactivity or fluorescence that the probe emits.

Before we can make a specific probe we need to know the base sequence in the particular allele that we are trying to locate. A number of different methods are used to sequence the exact order of bases in a length of DNA.

DNA hybridisation

DNA hybridisation takes place when a section of DNA or RNA is combined with a single-stranded section of DNA which has complementary bases. Before hybridisation can take place, the two strands of the DNA molecule must be separated. This is achieved by heating DNA until its double strand separates into its two complementary single strands (denaturation). When cooled, the complementary bases on each strand recombine (anneal) with each other to reform the original double strand. Given sufficient time,

Learning objectives

→ Describe what DNA probes are and explain how they work.

→ Explain how DNA hybridisation is used to locate specific alleles of genes.

→ Describe the use of labelled DNA probes to screen for heritable conditions or health risks.

→ Consider the use of genetic screening in genetic counselling.

Specification reference: 3.8.4.2

Synoptic link

The material in this topic brings together information from many other topics including DNA replication (Topic 2.2), The genetic code (Topic 8.1), Gene mutations (Topics 9.1 and 20.1), Studying inheritance (Topic 17.1), DNA sequencing (Topic 20.6), and the Polymerase chain reaction (Topic 21.3).

all strands in a mixture of DNA will pair up with their partners. If, however, other complementary sections of DNA are present in the mixture as the DNA cools, these are just as likely to anneal with one of the separated DNA strands as the two strands are with one another.

Locating specific alleles of genes

Using DNA probes and DNA hybridisation, it is possible to locate a specific allele of a gene. For example, we may wish to determine whether someone possesses a mutant allele that causes a particular genetic disorder. The process is as follows:

- We must first determine the sequence of nucleotide bases of the mutant allele we are trying to locate. This can be achieved using DNA sequencing techniques. However, we now have extensive genetic libraries that store the base sequences of most genetic diseases and so we can simply refer to these to obtain the sequence.

- A fragment of DNA is produced that has a sequence of bases that are complementary to the mutant allele we are trying to locate.

- Multiple copies of our DNA probe are formed using the polymerase chain reaction.

- A DNA probe is made by attaching a marker, for example a fluorescent dye, to the DNA fragment.

- DNA from the person suspected of having the mutant allele we want to locate is heated to separate its two strands.

- The separated strands are cooled in a mixture containing many of our DNA probes.

- If the DNA contains the mutant allele, one of our probes is likely to bind to it because the probe has base sequences that are exactly complementary to those on the mutant allele.

- The DNA is washed clean of any unattached probes.

- The remaining hybridised DNA will now be fluorescently labelled with the dye attached to the probe.

- The dye is detected by shining light onto the fragments causing the dye to fluoresce which can be seen using a special microscope.

The process is summarised in Figure 1.

Genetic screening

Many genetic disorders, such as sickle-cell anaemia, are the result of **gene mutations**. Gene mutations may arise if one or more **nucleotide** bases in DNA are changed in any one of a variety of ways. If the mutation results in a **dominant allele**, all individuals will have the genetic disorder. If the allele is **recessive**, it will only be apparent in those individuals that have two recessive alleles, that is, who are **homozygous** recessive. Individuals that are **heterozygous** will not display symptoms of the disease but will carry one copy of the mutant allele. These individuals are known as carriers. They have the capacity to pass the disease to their offspring if the other parent is also heterozygous or homozygous recessive.

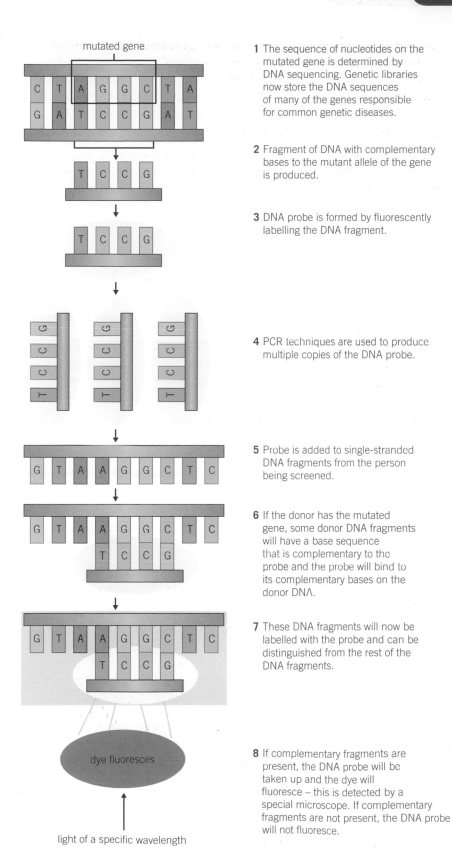

1 The sequence of nucleotides on the mutated gene is determined by DNA sequencing. Genetic libraries now store the DNA sequences of many of the genes responsible for common genetic diseases.

2 Fragment of DNA with complementary bases to the mutant allele of the gene is produced.

3 DNA probe is formed by fluorescently labelling the DNA fragment.

4 PCR techniques are used to produce multiple copies of the DNA probe.

5 Probe is added to single-stranded DNA fragments from the person being screened.

6 If the donor has the mutated gene, some donor DNA fragments will have a base sequence that is complementary to the probe and the probe will bind to its complementary bases on the donor DNA.

7 These DNA fragments will now be labelled with the probe and can be distinguished from the rest of the DNA fragments.

8 If complementary fragments are present, the DNA probe will be taken up and the dye will fluoresce – this is detected by a special microscope. If complementary fragments are not present, the DNA probe will not fluoresce.

▲ **Figure 1** *Summary of the process to locate a specific allele of a gene*

> **Hint**
>
> The donor's DNA must be made single-stranded to allow the use of DNA probes and hybridisation because both depend on the formation of complementary base sequences.

It is important to screen individuals who may carry a mutant allele. Such individuals often have a family history of a disease. Screening can determine the probabilities of a couple having offspring with a genetic disorder. As a result, potential parents who are at risk can obtain advice from a genetic counsellor about the implications of having children, based on their family history and the results of genetic screening.

It is possible to fix hundreds of different DNA probes in an array (pattern) on a glass slide. By adding a sample of DNA to the array, any complementary DNA sequences in the donor DNA will bind to one or more probes. In this way it is possible to test simultaneously for many different genetic disorders by detecting fluorescence that occurs where binding has taken place.

Another area where genetic screening can be valuable is in the detection of oncogenes, which are responsible for cancer. Cancers may develop as a result of mutations that prevent the **tumour suppressor genes** inhibiting cell division. Mutations of both alleles must be present to inactivate the tumour suppressor genes and to initiate the development of a tumour. Some people inherit one mutated tumour suppressor gene. These individuals are at greater risk of developing cancer.

If a mutated gene is detected by genetic screening, individuals who are at greater risk of cancer can then make informed decisions about their lifestyle and future treatment. They can choose to give up smoking, lose weight, eat more healthily and avoid **mutagens** as far as possible. They can also check themselves more regularly for early signs of cancer, which can lead to an early diagnosis and a better chance of successful treatment. They may choose to undergo some form of surgery or other treatment.

Personalised medicine

One of the advantages of genetic screening is personalised medicine. It allows doctors to provide advice and health care based on an individual's genotype. Some people's genes can mean that a particular drug may be either more or less effective in treating a condition. By genetically screening patients, doctors and pharmacists can determine, more exactly, the dose of a drug which will produce the desired outcome. This can save money that would otherwise be wasted on overprescribing the drug. In some cases it avoids medications that could cause harm or avoids raising false hopes.

One example is the prescribing of painkillers. To function effectively many pain medications need a specific enzyme to activate them. About half the population have genes that alter the function of this enzyme. Screening for the presence of these genes allows the dosage to be adjusted to compensate for the ways in which the genes affect an individual's metabolism of the painkiller. This ensures that their use is both safe and effective.

Another example involves vitamin E. It has been shown that among people who have diabetes, vitamin E reduces the risk of cardiovascular disease for those with certain genotypes, but it can increase the risk

▲ **Figure 2** *Amniotic fluid is being taken from this pregnant woman. The fluid can be used to screen the unborn baby for genetic disorders*

for those with a different genotype. It is clearly advantageous to screen a person who has diabetes before advising on whether or not to take vitamin E supplements.

Genetic screening goes hand in hand with genetic counselling. The expert advice provided by a counsellor helps individuals to understand the results and implications of the screening and so make appropriate decisions.

Genetic counselling

Genetic counselling is like a special form of social work, where advice and information are given that enable people to make personal decisions about themselves or their offspring. One important aspect of genetic counselling is to research the family history of an inherited disease and to advise parents on the likelihood of it arising in their children.

Consider a mother who has a family history of sickle-cell anaemia. If the mother herself is unaffected but carries the gene for sickle-cell anaemia, she must be heterozygous for the condition. Suppose that she wishes to have children with a man who has no family history of sickle-cell anaemia. In this case, it can be assumed that the man does not carry the allele for the disease, and therefore none of the children will develop the disease, although they may be carriers. On the other hand, if the man has a family history of the disease, it is possible that he too carries the allele. In this case, the genetic counsellor can make the couple aware that there is a one in four chance of their children being affected and a two in four chance that their children will be carriers.

A counsellor can also inform the couple of the effects of sickle-cell anaemia and its emotional, psychological, medical, social and economic consequences. On the basis of this advice the couple can then choose whether or not to have children. Counselling can also make them aware of any further medical tests that might give a more accurate prediction of whether their children will have the condition, for example IVF with screening of embryos.

Genetic counselling is closely linked to genetic screening and the screening results provide the genetic counsellor with a basis for informed discussion. For example, in cases of cancer, screening can help to detect:

- oncogene mutations, which can determine the type of cancer that the patient has and hence the most effective drug or radiotherapy to use
- gene changes that predict which patients are more likely to benefit from certain treatments and have the best chance of survival. For example, the drug herceptin is most effective at treating certain types of breast cancer
- a single cancer cell among millions of normal cells, thus identifying patients at risk of relapse from certain forms of leukaemia.

This information can help a counsellor to discuss with the patient the best course of treatment and their prospects of survival.

Synoptic link

To remind yourself about how oncogenes and tumour suppressor genes work, re-read Topic 20.5, Gene expression and cancer.

▲ Figure 3 *Genetic counsellor talking with potential parents. Genetic counselling is used to advise on the risk of passing on genetic disorders and can be given to couples in certain risk groups who are intending to have children*

Summary questions

1 Explain what a DNA probe is.

2 Outline the process of genetic screening for a disease.

3 Genetic screening shows that a person has one mutant allele of the tumour suppressor gene.

  a Describe the role of the tumour suppressor gene.

  b Suggest how the person might use the information revealed by genetic screening.

Learning objectives

→ Describe what genetic fingerprinting is.

→ Explain the technique of gel electrophoresis.

→ Explain how genetic fingerprinting is carried out.

→ Explain how the results of genetic fingerprinting are interpreted.

→ Consider the uses of genetic fingerprinting.

Specification reference: 3.8.4.3

Synoptic link

Not all DNA sequences carry obvious genetic information. You will have already come across this idea with the processing of DNA described in Topic 8.4, Protein synthesis – transcription and splicing.

Genetic fingerprinting is a diagnostic tool used widely in forensic science, plant and animal breeding, and medical diagnosis. It is based on the fact that the DNA of every individual, except identical twins, is unique.

Genetic fingerprinting

This technique relies on the fact that the genome of most eukaryotic organisms contains many repetitive, non-coding bases of DNA. Indeed, 95% of human DNA is currently not known to code for any characteristic but may yet be found to be functional. DNA bases which are non-coding are known as **variable number tandem repeats (VNTRs)**. For every individual the number and length of VNTRs has a unique pattern. They are different in all individuals except identical twins, and the probability of two individuals having identical sequences of these VNTRs is extremely small. However, the more closely related two individuals are, the more similar the VNTRs will be.

Gel electrophoresis

Gel electrophoresis is used to separate DNA fragments according to their size. The DNA fragments are placed on to an agar gel and a voltage is applied across it (Figure 1). The resistance of the gel means that the larger the fragments, the more slowly they move. Therefore, over a fixed period, the smaller fragments move further than the larger ones. In this way DNA fragments of different lengths are separated. If the DNA fragments are labelled, for example with radioactive DNA probes (Topic 21.4), their final positions in the gel can be determined by placing a sheet of X-ray film over the agar gel for several hours. The radioactivity from each DNA fragment exposes the film and shows where the fragment is situated on the gel. Only DNA fragments up to around 500 bases long can be sequenced in this way. Larger genes and whole genomes must therefore be cut into smaller fragments by **restriction endonucleases**.

▲ **Figure 1** *Apparatus for carrying out electrophoresis*

▲ **Figure 2** *An electrophoretic gel being loaded with DNA samples*

The making of a genetic fingerprint consists of five main stages: extraction, digestion, separation, hybridisation and development. The complete process of genetic fingerprinting is summarised in Figure 3.

Extraction

Even the tiniest sample of animal tissue, such as a drop of blood or a hair root, is enough to give a genetic fingerprint. Whatever the sample, the first stage is to extract the DNA by separating it from the rest of the cell. As the amount of DNA is usually small, its quantity can be increased by using the polymerase chain reaction (Topic 21.3).

Digestion

The DNA is then cut into fragments, using the same restriction endonucleases (Topic 21.1). The endonucleases are chosen for their ability to cut close to, but not within, the target DNA.

Separation

The fragments of DNA are next separated according to size by gel electrophoresis under the influence of an electrical voltage. The gel is then immersed in alkali in order to separate the double strands into single strands.

Hybridisation

Radioactive (or fluorescent) DNA probes are now used to bind with VNTRs (Topic 21.4). The probes have base sequences which are complementary to the base sequences of the VNTRs, and bind to them under specific conditions, such as temperature and pH. The process is carried out with different probes, which bind to different target DNA sequences

Development

Finally, an X-ray film is put over the nylon membrane. The film is exposed by the radiation from the radioactive probes. (If using fluorescent probes, the positions are located visually.) Because these points correspond to the position of the DNA fragments as separated during electrophoresis, a series of bars is revealed. The pattern of the bands (Figure 5) is unique to every individual except identical twins.

❶ Extraction DNA is extracted from the sample

❷ Digestion Restriction endonucleases cut the DNA into fragments.

❸ Separation Fragments are separated using gel electrophoresis.
gel plate
large fragments
small fragments
Direction of movement

❹ Separation (cont.) DNA fragments are transferred from the gel to nylon membrane.
nylon membrane
gel plate

❺ Hybridisation DNA probes are added to label the fragments. These radioactive probes attach to specific fragments.
DNA probes

❻ Development Membrane with radioactively labelled DNA fragments is placed onto an X-ray film.
nylon sheet with radioactively labelled DNA strands
X-ray film

❼ Development (cont.) Development of the X-ray film reveals dark bands where the radioactive DNA probes have attached.

▲ **Figure 3** *Summary of genetic fingerprinting technique*

Hint

Remember that, in gel electrophoresis, the smallest DNA fragments travel the furthest.

Hint

In theory, the inheritance of VNTRs does not have any influence on the phenotype of an organism.

▲ **Figure 4** *DNA fingerprints of a child and each parent. Note that each band on the child's fingerprint corresponds to a band on one or other parent's fingerprint . As the child only inherits half the VNTR's of each parent, there are inevitably some bands in each parental fingerprint that do not match to bands in the child's fingerprint*

Interpreting the results

DNA fingerprints from two samples, for example, from blood found at the scene of a crime and from a suspect, are visually checked. If there appears to be a match, the pattern of bars of each fingerprint is passed through an automated scanning machine, which calculates the length of the DNA fragments from the bands. It does this using data obtained by measuring the distances travelled during electrophoresis by known lengths of DNA. Finally, the odds are calculated of someone else having an identical fingerprint. The closer the match between the two patterns, the greater the probability that the two sets of DNA have come from the same person.

Uses of DNA fingerprinting

DNA fingerprinting has a variety of uses:

Genetic relationships and variability

DNA fingerprinting can be used to help resolve questions of paternity. Individuals inherit half their genetic material from their mother and half from their father. Therefore each band on a DNA fingerprint of an individual should have a corresponding band in one of the parents parents' DNA fingerprint (Figure 4). This can be used to establish whether someone is the genetic father of a child. Genetic fingerprinting is also useful in determining genetic variability within a population. The more closely two individuals are related the closer the resemblance of their genetic fingerprints. A population whose members have very similar genetic fingerprints has little genetic diversity. A population whose members have a greater variety of genetic fingerprints has greater genetic diversity.

Forensic science

DNA is often left at the scene of a crime, for example, blood at the scene of a violent crime, semen at the scene of a rape and hair at the scene of a robbery. Genetic fingerprinting can establish whether a person is likely to have been present at the crime scene, although this does not prove they actually carried out the crime. Even if there is a close match between a suspect's DNA and the DNA found at the crime scene, it does not follow that the suspect carried out the crime. Other possible explanations need to be investigated. For example:

- The DNA may have been left on some other, innocent occasion.
- The DNA may belong to a very close relative.
- The DNA sample may have been contaminated after the crime, either by the suspect's DNA or by chemicals that affected the action of the restriction endonucleases used in preparing the fingerprint.

Finally, the probability that someone else's DNA might match that of the suspect has to be calculated. This calculation is based on the assumption that the DNA which produces the banding patterns is randomly distributed in the community. This may not always be the case, for example, it may not apply where religious or ethnic groups tend to have partners from within their own small community.

Medical diagnosis

Genetic fingerprints can help in diagnosing diseases such as Huntington's disease. This is a genetic disorder of the nervous system. It results from a three-base sequence (AGC) at one end of a gene on chromosome 4 being repeated over and over again – a sort of genetic stutter. People with fewer than 30 repeats are unlikely to get the disease, while those with more than 38 repeats are almost certain to do so. If they have over 50 repeats, the onset of the disease will occur earlier than average.

A sample of DNA from a person with the allele for Huntington's disease can be cut with restriction endonucleases and a DNA fingerprint prepared. This can then be matched with fingerprints of people with various forms of the disease and those without the disease. In this way, the probability of developing the symptoms, and when, can be determined.

Genetic fingerprints are also used to identify the nature of a microbial infection by comparing the fingerprint of the microbe found in patients with that of known pathogens.

Plant and animal breeding

Genetic fingerprinting can be used to prevent undesirable inbreeding during breeding programmes on farms or in zoos. It can also identify plants or animals that have a particular allele of a desirable gene. Individuals with this allele can be selected for breeding in order to increase the probability of their offspring having the characteristic that it produces. Another application is the determination of paternity in animals and thus establishing the pedigree (family tree) of an individual.

▲ **Figure 5** *The bands in these DNA fingerprints are marked M for mother, C for child and F for father*

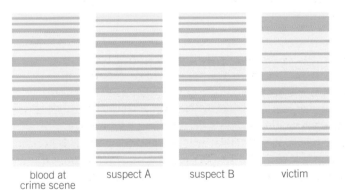

| blood at crime scene | suspect A | suspect B | victim |

▲ **Figure 6**

Summary questions

1 Explain why it is often necessary to use the polymerase chain reaction when producing a genetic fingerprint.

2 Figure 6 shows the genetic fingerprints of four DNA samples collected following a crime.

  a Which of the two suspects do you think was present at the scene of the crime? Give a reason for your answer.

  b Suggest why a genetic fingerprint of a DNA sample from the victim was made.

3 Suggest how chemicals that affect the action of restriction endonucleases can alter the genetic fingerprint of a DNA sample.

4 Suggest how the genetic fingerprint of someone with the allele for Huntington's disease might differ from that of someone who does not have the allele.

5 Explain how genetic fingerprinting can be used to ensure that inbreeding is avoided.

Locating DNA fragments

A section of DNA was cut into fragments and these fragments were separated by electrophoresis. Table 1 shows the number of base pairs in each fragment. The position of the fragments after gel electrophoresis is shown in Figure 7.

1 Name an enzyme that could have been used to cut the DNA.
2 Using the letters (A–F) in the boxes in Table 1, indicate which of the fragments (1–6) in Figure 7 are located in each box. Explain your answer.
3 The enzyme used to cut the DNA does so at a particular sequence of nucleotide bases. How many times does this base sequence occur in the original section of DNA?

▼ Table 1

Fragment	Number of base pairs (kilobases)
A	8.02
B	5.43
C	4.78
D	11.31
E	2.46
F	6.12

▲ Figure 7

 ## Gel electrophoresis and DNA sequencing

Gel electrophoresis is an integral part of the Sanger method for sequencing DNA. You may remember that the Sanger method uses modified nucleotides, called terminators, which cannot attach to the next base in the sequence when they are being joined together. They therefore end the synthesis of a DNA strand. Four different terminator nucleotides are used, each with one of the four bases adenine, thymine, guanine, or cytosine. As explained in Topic 20.6, depending on exactly where the terminator nucleotide binds to the DNA, its synthesis may be terminated after only a few nucleotides or after a long fragment of DNA has been synthesised. As a result, DNA fragments of varying lengths are produced. Each fragment will end in one of the four bases, adenine, thymine, guanine, or cytosine. These fragments can be identified because the primer attached to the other end of the DNA section is labelled radioactively.

The results of one such experiment in DNA sequencing is shown in Figure 8. This used a DNA fragment that was just eight nucleotides long. The results are read from the bottom up because the shortest fragments move the furthest distance. The smallest fragment (labelled 1) is just one nucleotide long and is therefore nearest the bottom. This fragment has a terminator nucleotide with the base adenine. The second fragment (labelled 2) is two bases long and has a terminator nucleotide with the base guanine, and so on. In this way the whole sequence of bases on the terminator nucleotides

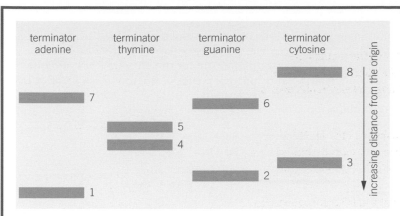

▲ **Figure 8** *Results of a DNA sequencing experiment*

was found to be AGCTTGAC and this is the sequence on one of the strands of the newly formed DNA.

The results of sequencing another fragment of DNA in the same manner are shown in Figure 9.

1 √x̄ using Figure 9 calculate how many adenine bases were present in the fragment of DNA.
2 Determine which nucleotide ends the shortest fragment.
3 Deduce the nucleotide sequence of the longest DNA fragment that has been produced.

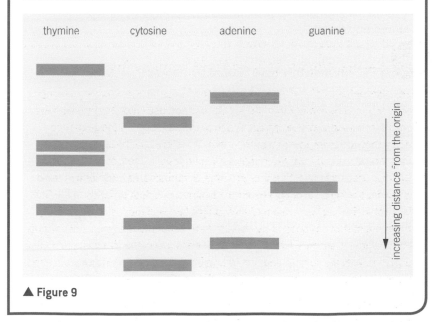

▲ Figure 9

1 Scientists used restriction mapping to investigate some aspects of the base sequence of an unknown piece of DNA. This piece of DNA was 3 000 base pairs (bp) long.
The scientists took plasmids that had one restriction site for the enzyme *Kpn*1 and one restriction site for the enzyme *Bam*H1. They inserted copies of the unknown piece of DNA into the plasmids. This produced recombinant plasmids.
The diagram shows a recombinant plasmid.

(a) When the scientists digested one of the recombinant plasmids with Kpn1, they obtained two fragments. One fragment was measured as 1 000 bp. The other fragment was described as "very large".
   (i) What does this show about the base sequence of the unknown piece of DNA?   (*2 marks*)
   (ii) One of the fragments that the scientists obtained was described as "very large". What is represented by this very large fragment?   (*1 mark*)
(b) When the scientists digested another of the recombinant plasmids with *Bam*H1, they obtained three fragments.
How many *Bam*H1 restriction sites are there in the unknown piece of DNA?   (*1 mark*)
(c) (i) Scientists can separate fragments of DNA using electrophoresis. Suggest how they can use electrophoresis to estimate the number of base pairs in the separated fragments.   (*2 marks*)
   (ii) Scientists need to take precautions when they carry out restriction mapping. They need to make sure that the enzyme they have used has completely digested the DNA. One check they may carry out is to add the sizes of the fragments together. How could scientists use this information to show that the DNA has **not** been completely digested? Explain your answer.   (*2 marks*)
                                                                                              AQA June 2011

2 Silkworms secrete silk fibres, which are harvested and used to manufacture silk fabric.
Scientists have produced genetically modified (GM) silkworms that contain a gene from a spider.
The GM silkworms secrete fibres made of spider web protein (spider silk), which is stronger than normal silk fibre protein.
The method the scientists used is shown in **Figure 5**.
(a) Suggest why the plasmids were injected into the eggs of silkworms, rather than into the silkworms.   (*2 marks*)
(b) Suggest why the scientists used a marker gene and why they used the EGFP gene.   (2 marks)
The scientists ensured the spider gene was expressed only in cells within the silk glands.
(c) What would the scientists have inserted into the plasmid along with the spider gene to ensure

▲ **Figure 5**

that the spider gene was only expressed in the silk glands
of the silkworms? *(1 mark)*

(d) Suggest two reasons why it was important that the spider gene was
expressed only in the silk glands of the silkworms. *(2 marks)*

AQA Specimen 2014

3 Haemophilia is a genetic condition in which blood fails to clot. Factor
IX is a protein used to treat haemophilia. Sheep can be genetically
engineered to produce Factor IX in the milk produced by their
mammary glands. The diagram shows the stages involved in this process.

(a) Name the type of enzyme that is used to cut the gene for Factor IX
from human DNA (Stage **1**). *(1 mark)*

(b) (i) The jellyfish gene attached to the human Factor IX gene
(Stage **2**) codes for a protein that glows green under fluorescent
light. Explain the purpose of attaching this gene. *(2 marks)*

(ii) The promoter DNA from sheep (Stage **3**) causes transcription
of genes coding for proteins found in sheep milk. Suggest the
advantage of using this promoter DNA. *(2 marks)*

(c) Many attempts to produce transgenic animals have failed. Very few
live births result from the many embryos that are implanted.

(i) Suggest **one** reason why very few live births result from the
many embryos that are implanted. *(2 marks)*

(ii) It is important that scientists still report the results from failed
attempts to produce transgenic animals. Explain why. *(2 marks)*

AQA June 2012

5 In gel electrophoresis DNA fragments separate according to their lengths
measured in base pairs. In a gel, standard marker fragments of known
length migrated as follows:

fragment length (bp)	distance migrated (mm)
10000	6.1
8000	9.5
6000	15.5
4000	21.8
2000	36.0
1000	50.0
500	61.2

(a) Plot a calibration graph of $\log_{10}$ fragment length against distance migrated *(3 marks)*

(b) Use the graph to estimate the length of a fragment which migrates 27.0 mm
in the same gel *(2 marks)*

Stage 1 Gene for Factor IX is cut from human DNA

Stage 2 Jellyfish gene is attached to the human gene

Stage 3 Promoter DNA from sheep is attached

Stage 4 Copies of this DNA are inserted into the nuclei of body cells from sheep

Stage 5 Each nucleus is transplanted into a sheep egg cell from which the original nucleus has been removed

Stage 6 The egg cells divide to form an embryo. Each embryo is implanted into the uterus of a different sheep

Section 8 Summary

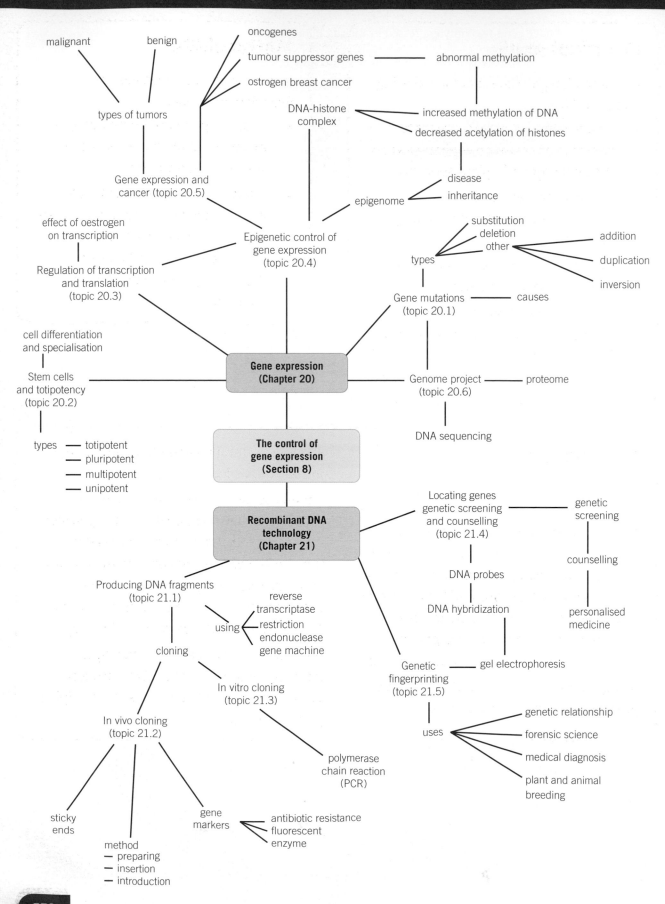

malignant

benign

oncogenes

tumour suppressor genes ——— abnormal methylation

ostrogen breast cancer

types of tumors

DNA-histone complex ——— increased methylation of DNA

decreased acetylation of histones

Gene expression and cancer (topic 20.5)

disease

inheritance

epigenome

effect of oestrogen on transcription

Epigenetic control of gene expression (topic 20.4)

substitution
deletion
other ——— addition

duplication

inversion

types

Regulation of transcription and translation (topic 20.3)

Gene mutations (topic 20.1) ——— causes

cell differentiation and specialisation

Stem cells and totipotency (topic 20.2)

Gene expression (Chapter 20)

Genome project (topic 20.6) ——— proteome

DNA sequencing

types ——— totipotent
——— pluripotent
——— multipotent
——— unipotent

The control of gene expression (Section 8)

Recombinant DNA technology (Chapter 21)

Locating genes genetic screening and counselling (topic 21.4)

genetic screening

counselling

DNA probes

personalised medicine

DNA hybridization

Producing DNA fragments (topic 21.1)

reverse transcriptase

restriction endonuclease

gene machine

using

cloning

In vitro cloning (topic 21.3)

Genetic fingerprinting (topic 21.5) ——— gel electrophoresis

In vivo cloning (topic 21.2)

polymerase chain reaction (PCR)

uses

genetic relationship

forensic science

medical diagnosis

plant and animal breeding

sticky ends

gene markers

antibiotic resistance
fluorescent
enzyme

method
— preparing
— insertion
— introduction

Practical skills

In this section you have met the following practical skills:

- How to carry out an experiment to show the effect of the environment on the phenotype of a plant species.
- How to obtain and evaluate experimental data linking smoking to disease.
- How to carry out *in vivo* and *in vitro* cloning of DNA fragments.
- How to use gene markers to identify whether a gene has been taken up by a bacterial cell.
- How to carry out gel electrophoresis.

Maths skills

In this section you have met the following maths skills:

- Interpret graphical information, for example, about the expression of genes in haemoglobin.
- Interpret and understand tabular information, for example, the results of an experiment to investigate the effects of growth factors on the development of plant tissue cultures.

Extension tasks

Recombinant DNA technology has allowed us to map the genomes of humans and other organisms. This has provided us with vast quantities of information. With information comes power and opportunity; the power to make informed decisions and the opportunity to change what we do. Knowledge of an individual's genetic make-up is perhaps the greatest invasion of an individual's privacy.

Using the internet, find out what legislation governs who is allowed access to an individual's genome and what rules control the use of this information.

Genetic screening allows parents to choose whether to have children that might possess a mutant gene. In time this could lead to the removal of these genes from the human genome. This, along with other human activities that you have covered during your A-level course, reduces genetic diversity.

Does mankind have a responsibility to maintain genetic diversity? Write a short passage in answer to this question justifying your viewpoint.

Section 8 Practice questions

1   Human Immunodeficiency virus (HIV) particles have a specific protein on their surface. This protein binds to a receptor on the plasma membrane of a human cell and allows HIV to enter. This HIV protein is found on the surface of human cells after they have become infected with HIV.

   Scientists made siRNA to inhibit expression of a specific HIV gene inside a human cell. They attached this siRNA to a carrier molecule. The flow chart shows what happens when this carrier molecule reaches a human cell infected with HIV.

Carrier with siRNA attached binds to HIV protein
on plasma membrane of infected cell

↓

Allows carrier with siRNA to enter human cell

↓

siRNA binds to mRNA transcribed
from HIV gene

↓

mRNA from HIV gene not translated

   (a) When siRNA binds to mRNA, name the complementary base pairs holding the siRNA and mRNA together. One of the bases is named for you:
   .... with ......adenine.........
   .... with ....   *(1 mark)*
   (b) This siRNA would **only** affect gene expression in cells infected with HIV. Suggest **two** reasons why.   *(4 marks)*
   (c) The carrier molecule on its own may be able to prevent the infection of cells by HIV. Explain how.   *(2 marks)*

   AQA June 2013

2   Scientists wanted to measure how much mRNA was transcribed from allele **A** of a gene in a sample of cells. This gene exists in two forms, **A** and **a**.
   The scientist isolated mRNA from the cells. They added an enzyme to mRNA to produce cDNA.
   (a) Name the type of enzyme used to produce the cDNA.   *(1 mark)*
   The scientists used the polymerase chain reaction (PCR) to produce copies of the cDNA. They added a DNA probe for allele **A** to the cDNA copies. This DNA probe had a dye attached to it. This dye glows with a green light **only** when the DNA probe is attached to its target cDNA.
   (b) Explain why this DNA probe will only detect allele **A**.   *(2 marks)*

(c) The scientists used this method with cells from two people, H and G. One person was homozygous, **AA**, and the other was heterozygous, **Aa**. The scientists used the PCR and the DNA probe specific for allele **A** on the cDNA from both people.

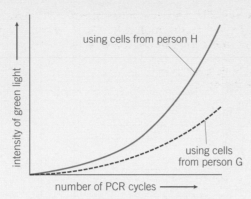

▲ **Figure 1** *shows the scientists' results*

(i)   Explain the curve for person H.   (*3 marks*)
(ii) Which person, H or G, was heterozygous, **Aa**? Explain your answer.   (*2 marks*)

AQA June 2014

3 (a) Describe how a gene can be isolated from human DNA.   (*2 marks*)
   (b) Describe how an isolated gene can be replicated by the polymerase chain reaction (PCR).   (*4 marks*)
   (c) (i)   Describe how a harmless virus, genetically engineered to contain a CFTR gene, can be used to insert the gene into a cystic fibrosis sufferer.   (*2 marks*)
       (ii) A virus used in gene therapy has RNA as its genetic material and has an enzyme called reverse transcriptase. Inside a human cell, reverse transcriptase uses viral RNA to make viral DNA. Explain why the enzyme is called *reverse transcriptase*.   (*1 mark*)

AQA Jan 2004

4 (a) Plasmids are often used as vectors in genetic engineering.
       (i)   What is the role of a vector?   (*1 mark*)
       (ii)   Describe the role of restriction endonucleases in the formation of plasmids that contain donor DNA.   (*2 marks*)
       (iii) Describe the role of DNA ligase in the production of plasmids containing donor DNA.   (*1 mark*)
   (b) There are many different restriction endonucleases. Each type cuts the DNA of a plasmid at a specific base sequence called a restriction site. Figure 2 shows the position of four restriction sites, **J**, **K**, **L**, and **M**, for four different enzymes on a single plasmid. The distances between these sites is measured in kilobases of DNA.

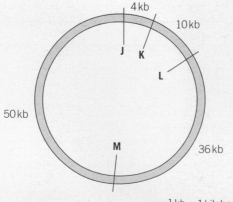

1 kb = 1 kilobase

▲ **Figure 2**

561

The plasmid was cut using only two restriction endonucleases. The resulting fragments were separated by gel electrophoresis. The positions of the fragments are shown in Figure 3.

▲ **Figure 3**

    (i)   Which of the restriction sites were cut?   (*1 mark*)

    (ii) Explain your answer.   (*1 mark*)

AQA June 2006

5 (a) Gene mutations occur naturally.

         Give **one** factor that increases the rate of gene mutations.   (*1 mark*)

 (b) Table 1 shows the DNA base sequences that code for three amino acids.

▼ **Table 1**

DNA base sequence(s) coding for amino acids	Amino acid
CCA	
CCG	
CCT	Glycine
CCC	
TAC	Methionine
TAA	
TAG	Isoleucine
TAT	

Some substitution mutations would affect the sequence of amino acids in a polypeptide, and others would not.

Using only the information in the table, explain why. (3 marks)

AQA Jan 2007

6 Huntington's disease is a genetic condition that leads to a loss in brain function. The gene involved contains a section of DNA with many repeats of the base sequence CAG. The number of these repeats determines whether or not an allele of this gene will cause Huntington's disease.
- An allele with 40 or more CAG repeats will cause Huntington's disease.
- An allele with 36 – 39 CAG repeats may cause Huntington's disease.
- An allele with fewer than 36 CAG repeats will not cause Huntington's disease.
  The graph shows the age at which a sample of patients with Huntington's disease first developed symptoms and the number of CAG repeats in the allele causing Huntington's disease in each patient.

(a) (i) People can be tested to see whether they have an allele for this gene with more than 36 CAG repeats. Some doctors suggest that the results can be used to predict the age at which someone will develop Huntington's disease. Use information in the graph to evaluate this suggestion. *(3 marks)*

(ii) Huntington's disease is always fatal. Despite this, the allele is passed on in human populations. Use information in the graph to suggest why. *(2 marks)*

(b) Scientists took DNA samples from three people, **J**, **K** and **L**. They used the polymerase chain reaction (PCR) to produce many copies of the piece of DNA containing the CAG repeats obtained from each person. They separated the DNA fragments by gel electrophoresis. A radioactively labelled probe was then used to detect the fragments. The diagram shows the appearance of part of the gel after an X-ray was taken. The bands show the DNA fragments that contain the CAG repeats.

(i) Only one of these people tested positive for Huntington's disease. Which person was this? Explain your answer. *(2 marks)*

(ii) The diagram only shows part of the gel. Suggest how the scientists found the number of CAG repeats in the bands shown on the gel. *(1 mark)*

(iii) Two bands are usually seen for each person tested. Suggest why only one band was seen for Person **L**. *(1 mark)*

AQA June 2012

Biology students are often less comfortable with the application of mathematics compared with students such as physicists, for whom complex maths is a more obvious everyday tool. Nevertheless, it is important to realise that biology does require competent maths skills in many areas. It is important to practise these skills so you are familiar with them as part of your routine study of the subject.

Confidence with mental arithmetic is very helpful, but among the most important skills is that of taking care and checking calculations. We may not be required to understand the detailed theory of the maths we use, but we do need to be able to apply the skills accurately, whether simply calculating percentages or means, or substituting numbers into complex-looking algebraic equations, such as in statistical tests.

This chapter is designed to help with some of the regularly encountered mathematical problems in biology.

Working with the correct units

In biology it is very important to be secure in the use of correct units. These must always be written clearly in calculations.

Base units

The units we use are from the Système Internationale – the SI units. In biology we most commonly use the SI base units:

* metre (m) for length, height, distance
* kilogram (kg) for mass
* second (s) for time
* mole (mol) for the amount of a substance.

You should develop good habits right from the start, being careful to use the correct abbreviation for each unit used. For example, seconds should be abbreviated to s, not 'sec' or 'S'.

Derived units

Biologists also use SI derived units, such as:

* square metres (m^2) for area
* cubic metre (m^3) for volume
* cubic centimetre (cm^3), also written as millilitre (ml), for volume
* degree Celsius (°C) for temperature
* mole per litre ($mol\ dm^{-3}$) is usually used for concentration of a substance in solutions (although the official SI derived unit is moles per cubic metre)
* joule (J) for energy
* pascal (Pa) for pressure
* volt (V) for electrical potential.

Maths link $\sqrt{x}$

MS 0.1

Non-SI units

Although examination boards use SI units, you may also encounter non-SI units elsewhere, for example:

- litre (cubic decimetre) (l, L, dm^3) for volume;
- Minute (min) for time;
- hour (h) for time;
- svedberg (S) (for sedimentation rate), used for ribosome particle size.

Unit prefixes

To accommodate the huge range of dimensions in our measurements, they may be further modified using appropriate prefixes. For example, one thousandth of a second is a millisecond (ms). This is illustrated in the Table 1.

▼ **Table 1**

Division	Factor	Prefix	Length		Mass		Volume		Time	
one thousand millionth	10^{-9}	nano	nanometre	nm	nanogram	ng	nanolitre	nl	nanosecond	ns
one millionth	10^{-6}	micro	micrometre	μm	microgram	μg	microlitre	μl	microsecond	μs
one thousandth	10^{-3}	milli	millimetre	mm	milligram	mg	millilitre	ml/cm^3	millisecond	ms
one hundredth	10^{-2}	centi	centimetre	cm						
whole unit			metre	m	gram	g	litre	$l/L/dm^3$	second	s
one thousand times	10^3	kilo	kilometre	km	kilogram	kg				

Converting between units

You may need to convert between units in order to be able to scale and express numbers in sensible forms. For example, rather than refer to the width of a cell in metres you would use micrometres (μm). This allows your measurements to be understood within the relevant scale of the observation.

Divide by 1000 for each step to convert in this direction

nano-	micro-	milli-	whole unit	kilo-
e.g. nm	e.g. μm	e.g. mm	e.g. m	e.g. km

 Multiply by 1000 for each step to convert in this direction

▲ **Figure 1**

Examples:

Convert 1 m to mm: $1 \times 1000 = 1000\,mm$

Convert 1 m to μm: $1 \times 1000 = 1000\,mm$, then $1000 \times 1000 = 1\,000\,000\,μm$

Convert 1 l to cm^3: $1 \times 1000 = 1000\,cm^3$

Convert 20 000 μm to mm: $20\,000 \div 1000 = 20\,mm$

Converting between square or cube units requires a bit more care.

One $m^2 = 1000 \times 1000 = 1\,000\,000\,mm^2$, so your conversion factor becomes $\times$ or $\div\,1\,000\,000$.

One m^3 is $1000 \times 1000 \times 1000 = 1\,000\,000\,000\,mm^3$, so your conversion factor now becomes $\times$ or $\div\,1\,000\,000\,000$.

Examples:

Convert $20\,m^2$ to km^2: $20 \div 1\,000\,000 = 0.000\,02\,km^2$

Convert $1\,m^2$ to mm^2: $1 \times 1\,000\,000 = 1\,000\,000\,mm^2$

Convert $5\,000\,000\,mm^3$ to m^3: $5\,000\,000 \div 1\,000\,000\,000 = 0.005\,m^3$

Convert $0.000\,000\,07\,m^3$ to mm^3: $0.000\,000\,07 \times 1\,000\,000\,000 = 70\,mm^3$

Decimals and standard form

When you are using numbers that are very small, such as dimensions of molecules and organelles, it is useful to use **standard form** to express them more easily. Standard form is also commonly called **scientific notation**.

Standard form is essentially expressing numbers in powers of ten. For example, 10 raised to the power 10 means 10×10, i.e. 100. This may be written down as 10×10^1 or 1×10^2. To get to 1000 you use $10 \times 10 \times 10$, which would be written as 1×10^3.

An easy way to look at this is to imagine the decimal point moving one place per power of ten. For example, to write down $58\,900\,000\,000$ as standard form, you would follow the steps below.

Step 1: write down the smallest number between 1 and 10 that can be derived from the number to be converted. In this case it would be 5.89.

Step 2: write the number of times the decimal place will have to shift to expand this to the original number as powers of ten. On paper this can be done by hopping the decimal over each number like this:

▲ Figure 2

until the end of the number is reached. In this example, that requires 10 shifts, so the standard form should be written as 5.89×10^{10}.

Going the other way, for example expressing $0.000\,007\,8$ as standard form, write the number in terms of the number of places the decimal place would have to hop forward to make the smallest number between 1 and 10, so to get to 7.8 you would have to hop over six times, so this number is written as 7.8×10^{-6}.

Significant figures

There are some simple rules to use when working out significant figures.

Rule 1: All non-zero digits are significant.

For example, 78 has two significant figures, 9.543 has four significant figures and 340 has two significant figures.

Maths link √x̄

MS 0.2

Maths link √x̄

MS 1.1

Rule 2: Intermediate zeros are significant.

For example, 706 has three significant figures and 5.90076 has six significant figures.

Rule 3: Any leading zeroes are not significant.

For example, 0.00567 has three significant figures (5,6 and 7; ignore the leading zeroes)

Rule 4: Zeroes at the ends of numbers containing decimal places are significant.

For example, 45.60 has four significant figures and 330.00 has five significant figures.

Significant figures and rounding

Table 2 shows the effect of rounding numbers to decimal places compared with significant figures. Remember that in rounding, when the next number is 5 or more round up, while if it is 4 or less don't round up. For example, 4.35 rounds to 4.4 and 4.34 rounds to 4.3.

Table 2 shows examples of rounding the number 23.33600 to decimal places and to significant figures.

▼ **Table 2**

Measurements expressed by rounding to decimal places	Number of decimal places	Measurements expressed by rounding to significant figures	Number of significant figures	Measured to the nearest
23.33600	5	23.336	5	100 thousandth
23.3360	4	23.34	4	Ten thousandth
23.336	3	23.3	3	Thousandth
23.34	2	23	2	Hundredth
23.3	1	20	1	Tenth
23	0	—	—	Whole number

Significant figures and standard form

In standard form only the significant figures are written as digits, for example 5.600×10^3 has four significant figures. If this were written as a straight number it would be 5600. But according to the rules above, 5600 only has two significant figures – what does this mean?

In a given number, the significant figures are defined as the ones that contribute to its precision. Writing the number as 5600 implies precision only to the nearest whole hundred. The zeroes in the number could mean that it has simply been rounded, e.g. 5600.44 or even 5633. But if this number were actually more precise, for example it had been measured with equipment genuinely sensitive to the nearest hundredth part (2 decimal places) then 5600.00 is actually very precise and the two zeros have significance because they tell us that the measurement is *exactly* 5600 with no tenths or hundredths at all. So using standard form allows this precision to remain clearly as part of the stated number, because all significant figures are written.

Maths link $\sqrt{x}$

MS 1.2 and 1.6

Averages

An average value is actually a measure of central tendency. The most familiar measurement is the arithmetic mean (mean for short), but median or modal values are sometimes more appropriate to the data.

The arithmetic mean

Usually referred to simply as the mean, this is a measure of central tendency that takes into account the number of times each measurement occurs together with the range of the measurements. When repeated measurements are averaged, the mean will approach the true value, which will lie somewhere in the middle of the observed range, more accurately. This is why it is important to repeat experimental measurements, especially in biology where the natural unpredictability of living systems leads to inevitable fluctuations.

The mean is determined by adding together all the observed values and then dividing by the number of measurements made.

For a range of values of x, the mean $\bar{x} = \dfrac{\sum x}{n}$.

$\bar{x}$ is the mean value.

$\sum x$ is the sum of all values of x.

n is the number of values of x.

For example, five mice were weighed, giving masses of 6.2 g, 7.7 g, 6.7 g, 7.1 g and 6.3 g.

The mean mass is $(6.2 + 7.7 + 6.7 + 7.1 + 6.3) \div 5 = 6.8$ g

Be careful with your decimal places when calculating mean values. Your mean should normally have the same level of precision as the original measurements and therefore the same number of decimal places, otherwise you may be implying that the averaged measurements are more precise than they really are. For example, masses in whole grams would not average to a mass with one or more decimal places. Similarly averaging the numbers of whole objects should result in a whole number; if counts of bubbles in a pondweed experiment were averaged to a decimal place it implies you counted a fraction of one bubble, which is impossible!

The median

The median value in a set of data is calculated by placing the values in numerical order then finding the middle value in the range.

For example, the data set 12, 15, 10, 17, 9, 13, 13, 19, 10, 11 rearranges as 9, 10, 10, 11, 12, 13, 13, 15, 17, 19.

The middle of this range is 12.5.

The median value is very useful when data sets have a few values (outliers) at the extremes, which if included in an arithmetic mean could skew the data. It also allows comparison of data sets with similar means but a clear lack of overlap, skewed data and when there are too few measurements to calculate a reliable mean value.

For example, in the data set 1, 3, 3, 11, 12, 12, 12, 13, 14, 15, the median value is 12, a sensible looking mid point, but the mean would be 9.6, skewed to the left by the numbers at the lower extreme.

The mode

The modal value is the most frequent value in a set of data. It is very useful when interpreting data that is qualitative or in situations where the distribution has more than one peak (bimodal).

For example, in the data set 9, 10, 11, 11, 12, 13, 13, 13, 14, 17, 18, 19, the modal value is 13.

In biology, caution should be used because the sets of data are usually small and can introduce confusion. For example, in the data set 9, 10, 11, 11, 12, 13, 13, 14, 17, 18 there are apparently two modal values, 11 and 13, while in the set 11, 12, 13, 14, 17 there is no most frequent number and the mode is effectively every number and therefore of no value at all.

The modal value is not used very often, but it can be usefully applied when data is collected in categories, for example, numbers of moths attracted to lights of different colours.

Percentages

A percentage is simply expressing a fraction as a decimal. It is important to be confident with calculating percentages, which although straightforward are commonly calculated incorrectly.

Maths link √x̄

MS 0.3

Percentages as proportions and fractions

For example, two shapes of primrose flowers exist depending on stigma length; 'pin eyed' and 'thrum eyed'. In a survey of two areas of grassland, one area had 323 pin and 467 thrum (total 790 plants), the other had 667 pin and 321 thrum (total 988 plants). The percentage of pin eyed plants in each area is calculated as follows:

Area 1: $fraction = \frac{323}{790}$ which gives $decimal$ 0.41, which multiplied by 100 gives $percentage$ 41%.

The percentage of pin eyed flowers in Area 2 is $\frac{667}{988} \times 100 = 67.5\%$.

Percentages as chance

In genetics the likelihood of different offspring phenotypes should always be expressed as a percentage. For example, in a simple genetic cross between two heterozygous parents carrying the cystic fibrosis allele, one out of every four possible children could potentially be affected by the disorder. The chance of a cystic fibrosis child from these parents is therefore $\frac{1}{4} \times 100 = 25\%$.

Percentage change

This often comes up in osmosis experiments where samples (usually of potato tissue) gain and lose mass in different bathing solutions.

For example, a sample weighed 18.50 g at the start and at the end it weighed 11.72 g.

The actual loss in mass = 18.50 − 11.72 g = 6.78 g

The percentage change $= \frac{\text{mass change}}{\text{starting mass}} \times 100 = \frac{6.78}{18.50} \times 100 = -36.7\%$

Note the use of the minus sign to indicate that this is a loss.

Maths link $\sqrt{x}$

MS 2.2, 2.4 and 2.3

Equations

Substituting into equations

There are several equations (mathematical formulae) that you will need to be able to use in advanced level biology. You do not need to learn the theoretical maths from which they are derived, but you do need to be able to put known numerical values in the right place (this is *substituting into the equation*) and then calculate the result of the equation by performing the different steps in the right order (*this is solving the equation*).

An example that you will encounter during ecology studies is called the Simpson's Index of Diversity, which has the formula $= \dfrac{N(N-1)}{\sum n(n-1)}$.

Each symbol (*term*) in the equation has a specific meaning. In this example:

N means the total number of all individual organisms in a survey.

n means the total number of each different species.

Σ means 'the total of' and requires you to add together all the indicated values.

Brackets indicate sub-calculations that must be done, for example $N-1$ means the total of all species found -1.

The figures in brackets need to be multiplied by the figures outside them, e.g. $N(N-1)$ means $N \times (N-1)$.

An example of the data to use could be counts of the plant species found in a certain area. To make life easy, use a table like Table 3.

▼ Table 3

Plant species	Number of plants of each species found (n)	($n-1$)	$n(n-1)$
A	22	$22 - 1 = 21$	$22 \times 21 = 462$
B	30	$30 - 1 = 29$	$30 \times 29 = 870$
C	25	$25 - 1 = 24$	$25 \times 24 = 600$
D	23	$23 - 1 = 22$	$23 \times 22 = 506$
Totals of all plants $= N$	$N = 100$ $N - 1 = 99$		$\sum n(n-1) = 2438$

The brackets in equations always need to be solved first.

Begin by finding $n - 1$ for each plant (see column 3 in Table 3) and $N - 1$ (at the bottom of column 2).

Next work out $n(n - 1)$ (column 4 in Table 3).

Now find $\sum n(n - 1)$ by totalling the figures in column 4.

Substituting the known values into the equation works like this:

$D = \dfrac{N(N-1)}{\sum n(n-1)}$ becomes $D = \dfrac{100(99)}{2438}$ which calculates to $D = \dfrac{9900}{2438}$,

which gives the result D = 4.1.

Rearranging equations

The individual parts or *terms* in equations are all related, but sometimes you might know all the values of the terms except one. The equation can be re-written so that the unknown term can be calculated. This is called rearranging or *changing the subject of* an equation. A very useful example of this arises during the study of microscopy and magnification.

The different terms are magnification, size of image and actual size of the object being observed. The equation that relates them together is:

$$magnification = \frac{size\ of\ image}{size\ of\ real\ object}.$$

Maths link $\sqrt{x}$

MS 1.8

You can use the equation to calculate magnification factors quite simply. For example, if you had a photograph of your pet dog, the magnification of the image would be the height of the image of that dog divided by its real height.

Be very careful to use the same units for each measurement! If the dog is 9 cm tall in the photograph and the real dog is 0.4 m tall you would have to convert the units before starting. 0.4 m is 40 cm, so the sum would be $9 \div 40 = 0.23$. You picture's magnification is ×0.23.

Suppose you only had the photo and the magnification. How would you find out how big the real object was? You may need to do this type of calculation on photomicrographs of cells or parts of cells.

For example, a photograph shows a mitochondrion which is 41 mm long in the picture and is taken at magnification ×34 000. How long is the original mitochondrion?

To find out, rearrange the equation. You might use an equation triangle to help.

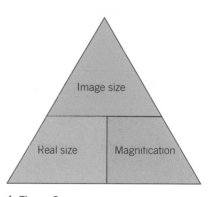

▲ Figure 3

On Figure 3 the horizontal line means divide and the vertical line means multiply.

So $magnification = \dfrac{size\ of\ image}{size\ of\ real\ object}$

rearranges as $size\ of\ image = magnification \times real\ size$

and $real\ size = \dfrac{size\ of\ image}{magnification}.$

You need to find the real size of the mitochondrion, so your sum will be:

$$real\ size = \frac{41}{34000} = 0.0012\ mm.$$

At this point you need to check that your units are sensible. A mitochondrion is so small that the appropriate unit of measurement is a micrometre (μm). The question may even ask you to use this unit. Earlier in this chapter you saw that 1 μm is 1/1000th of a mm, so to convert you need to multiply by 1000. The real mitochondrion is $0.0012 \times 1000 = 1.2\ μm$ long.

Gathering data and making measurements

Estimating results

When measuring and recording data it is useful to be able to make an estimate of the number you should be getting. This will allow you to judge whether the results you actually record seem believable. This

Maths link $\sqrt{x}$

MS 0.4

is especially important when using a calculator, because it is easy to mis-type an entry and get a wrong answer. An estimate is really a sensible guess. It is a good idea to practise this skill, for example when collecting data from practical work in class.

Uncertainties in measurements

When making measurements, even using good quality instruments such as rulers and thermometers, there is a certain level of doubt in the precision of the measurement obtained. This is the *uncertainty of measurement*. The uncertainty can be stated, in which case a margin of error is identified. For example, measurements made using a good mm scale ruler a measurement may be reported as +/- 0.5mm, which is the maximum error likely when using the ruler carefully.

Percentage error is a way of using the maximum error to calculate the possible total error in a given measurement. Some types of instrument have maximum errors written on them, for example a balance may state +/- 0.01g. Other devices such as rulers and thermometers may rely on common sense, e.g. +/- 0.5mm or +/- 0.5 °C when recorded by eye. To find percentage error use the formula:

$$\% \ error = \frac{maximum \ error}{measured \ value \ recorded} \times 100$$

For example, with a ruler the maximum likely error is usually 0.5mm. If an object is measured at 6mm with the ruler, the percentage error = $\frac{0.5}{6} \times 100 = 8.3\%$.

A larger object will have a smaller % error because the +/- 0.5 mm is a lesser part of the total recorded, e.g. an object measured at 87mm has a % error = $\frac{0.5}{87} \times 100 = 0.6\%$.

Working with graphs and charts

Choosing the right type of graph or chart

During your course you will most commonly use line graphs, bar charts and histograms. You need to be able to choose the right one to suit the data and also to be able to draw graphs accurately.

The first part of your decision depends on the type of independent variable that you have measured. When you have used an independent variable that has specific values on a continuous scale, such as temperature, you should use a line graph, e.g. oxygen volume consumed by woodlice in a respirometer at a variety of temperatures. Alternatively your data might be in discrete categories, for example the number of left-handed or right-handed people. For this data a bar chart should be used with a space between each bar. When your categoric data is in groups that can be arranged on a continuous scale, e.g. height categories of plants such as 0 to 1 cm, >1 to 2 cm, >2 to 3 cm and so on, a histogram should be chosen, in which the bars are not separated by gaps.

Plotting the graph or chart

The rules when plotting the graph are:

- Ensure that the graph occupies the majority of the space available (this means more than half the space).
- Mark axes using a ruler and divide them clearly and equidistantly (i.e. 10, 20, 30, 40 not 10, 15, 20, 30, 45.

Maths link $\sqrt{x}$

MS 1.11

Maths link $\sqrt{x}$

MS 3.1, 3.2, 3.4, 3.5, and 1.3

Study tip

Decisions about data also have a large bearing on the choice of statistics and statistical tests you might need to use.

- Ensure that the dependent variable that you measured is on the *y* axis and the independent variable that you varied is on the *x* axis.

- Ensure that both axes have full titles and units clearly labelled, e.g. pH of solution, not just 'pH'; mean height/m, not just 'height'.

- Plot the points accurately using a sharp pencil and '×' marks so the exact position of the point is obvious and is not obscured when you plot a trend line.

- Draw a neat best-fit line, either a smooth curve or a ruled line. It does not have to pass through all the points. Alternatively use a point to point ruled line, which is often used in biology where observed patterns do not necessarily follow mathematically predictable trends!

- Confine your line to the range of the points. Never extrapolate the line beyond the range within which you measured. Extrapolation is conjecture! A common mistake is to try and force the plotted line to go through the origin.

- Distinguish separate plotted trend lines using a key.

- Add a clear concise title.

- Where data ranges fall a long way from zero, a broken axis will save space. For example, if the first value on the *y* axis is 36 it may be sensible to start the axis from 34 rather than zero. This will avoid leaving large areas of your graph blank.

You will be expected to follow these conventions. If you do, then questions that involve drawing a graph become easy.

Adding range bars and error bars to your plotted points

The position of the point on a graph is always subject to uncertainty. It may be a mean value, which will depend on the values averaged or whether you include or exclude any possible anomalies. A way of indicating the level of certainty in the positioning of your points is to use a range bar or error bar. These are ways of pictorially indicating the possible range of positions of the point and reflect the spread or variability in the original measurements that were averaged. The more spread the measurements, the less certain the position of the mean when plotted.

Table 4 shows some example data from an experiment on gas production by a photosynthesising plant at different temperatures, with which the different styles can be demonstrated.

> **Maths link** $\sqrt{x}$
>
> MS 1.10

Range bars

A range bar is the simplest way of showing the showing the spread in the data. Look for the maximum and minimum values in each set of repeats; they are picked out in bold in the table. After plotting the point on your graph mark the positions of the maximum and minimum values above and below the point using a small bar. Join the two extremes with a neat ruler line running vertically through the plotted point (Figure 4).

▼ Table 4

Temperature (°C)	Time taken to collect 10 cm³ of gas (s)								
	1	2	3	4	5	mean	s	mean +s	mean −s
15	87	95	102	121	117	104	14.4	118.4	89.6
20	67	78	61	90	86	76	12.3	88.3	63.7
25	57	59	48	66	51	56	7.0	63	49
30	47	45	39	42	21	39	10.4	49.4	28.6
35	118	123	145	136	132	131	10.7	141.7	120.3

▲ **Figure 4** *Range bars*

▲ **Figure 5** *Error bars*

Maths link $\sqrt{\bar{x}}$

MS 3.5 and 3.6

Notice that the range bars are not always symmetrical above and below the plotted points. The tops and bottoms just show the largest or smallest values among the measurements made.

Error bars

To plot error bars you use the standard deviation (a calculated measure of the spread of the data), to indicate your ranges. This is better than using range bars because it reduces the effect of any extreme values in the dataset. In the table the values of standard deviation are shown in the column headed *s*.

Plot the bars by marking the top and lower limits exactly plus and minus one standard deviation above and below the point. These values are also included in the table. The result is shown in Figure 5.

Notice that the error bars are symmetrical above and below the points, which are now indicated with a range of ± one standard deviation. The length of the bars now indicates not the maximum/minimum values but the mathematical spread in the data. The more the data spread out around the mean, the longer the error bar becomes and the less certain you are that the mean is really accurate.

Calculating rates from graphs

When data have been plotted on a line graph relating measured values on the y axis to time on the x axis, it is possible to calculate a rate of change for the y variable. There are two common graph forms that you will encounter, shown in Figure 6.

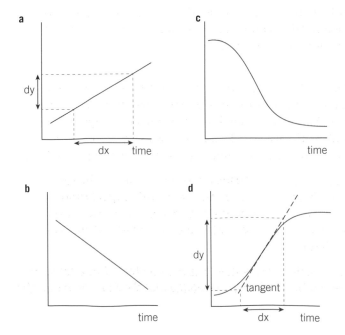

a might represent oxygen production by a photosynthesising plant

b might represent oxygen consumption by a respiring organism

c might represent pH change during a lipid digestion experiment

d might represent growth of a bacterial population in a fermenter

The rate of change is simply the gradient of the graph.

The formula that is used is $\dfrac{change\ in\ y}{change\ in\ x}$ or $\dfrac{dy}{dx}$

With a straight line graph follow these steps, which are marked on Figure 6a

- select any two points on the plotted line
- use a ruler to mark construction lines from the two points to the x and y axes
- measure the difference between the two points on the y axis, this is dy
- measure the time difference between the points on the x axis, this is dx
- substitute the values into the equation and remember to quote suitable units for the result, e.g. cm^3 O_2 per minute.

With a curved line the procedure is the same except that you need to start by marking a tangent against the curve, usually at its steepest point to find the maximum rate of change. This takes a bit of practise and is usually done by eye, although it is possible to calculate a position mathematically.

Once the tangent is drawn, select two points on it and proceed using the same steps that were applied to the straight line. Figure 6d has been marked with a tangent as an example.

Scatter diagrams

A scatter diagram is a method of plotting two variables in order to try and identify a correlation between them. The dependent variable is plotted on the y axis and the independent along the x axis. Once the points are plotted a trend line can be added to show a possible relationship. An example might be plotting incidence of lung cancer against number of cigarettes smoked per day. Once such a plot has been made the relationship can be tested using a statistical test, for example the correlation coefficient, r.

Maths link $\sqrt{x}$

MS 1.7

Probability

When data appears to show a pattern, it is possible to determine whether the pattern is simply due to chance or whether it has an underlying cause. For example, 36 throws of die should give near enough six of each possible number. Throwing 7 4's and 5 6's is likely to be a fluke, but if you threw 23 6's then this is definitely against the rules of probability!

Maths link $\sqrt{x}$

MS 1.4

Probability is assessed using a statistical test, for example chi-squared to test how closely observed measurements fit with expectation, such as in genetic cross results, or student t, which compares the means of sets of data to assess whether they differ, e.g. leaf width of sun versus shade grown ivy plants.

Such tests produce a calculated value that may be found in a table of probability. In these tables, it is the *probability that the data observed differ*

Maths link $\sqrt{x}$

MS 0.3, 1.5, 1.8, 1.9, 1.10, 2.2, 2.3, 2.4 and 4.1.

Maths link $\sqrt{x}$

MS 1.9

by chance alone that is being found. In biology we accept any probability greater than 5% as likely to be just chance or fluke, but probabilities of 5% or below show us that the data do differ significantly and there must be a cause influencing the outcome.

Using statistical tests to calculate probability

Chi-squared (χ^2) test

This test is used to compare the pattern in data collected or observed with the pattern that would be expected by chance. For example, if a die was thrown 36 times, each number should be expected 6 times if only chance determined the outcome. The further the actual numbers thrown deviate from this, the more likely the dice is loaded - a factor other than chance is influencing the outcome.

The test is commonly used to check the results of genetic crosses.

For example, a cross between two heterozygous tall plants has the expected outcome of 3 tall:1 short plant (short is the recessive allele). In the real results, 69 plants were tall and 28 short. We need to test the null hypothesis that there is no significant difference between the observed and expected results.

The formula for the chi-squared (χ^2) test is $\chi^2 = \sum \dfrac{(O-E)^2}{E}$. It is easiest to lay out the calculation in a table.

	Observed (O)	Expected (E)	(O–E)	(O–E)^2	$\dfrac{(O-E)^2}{E}$
Tall pea plants	69	72.75	3.75	14.06	0.19
Dwarf pea plants	28	24.25	3.75	14.06	0.58
				Total :	$\chi^2 = 0.77$

The probability value can now be found by using the table of chi-squared distribution (Table 6).

First, work out the degrees of freedom, which is the number of categories minus 1. Here there are two categories (tall or short), so 1 degree of freedom.

The critical value of χ^2 for one degree of freedom at the p = 5% level is 3.84. The calculated value of 0.77 is less than 3.84 and gives a p value of between 0.1 and 0.5 (10% and 50%) of the data collected being just chance. In this case we should accept the null hypothesis; there is no evidence of a significant difference between the observed and expected numbers.

Student *t* test

This test is used to judge the significance of any difference between the means of sets of data that are collected from two groups. There must be enough data that a reliable mean can be calculated and it should be normally distributed. The number in each sample does not need to be the same, but will ideally be more than 15 samples in each set.

For example, limpet diameters were measured at two sites to find out whether there was any difference due to aspect. One site faced east and the other west. 28 animals were measured at each site. The null hypothesis was that there would be no difference between the sites.

Maths link $\sqrt{x}$

MS 1.9

	Site 1 (east)	Site 2 (west)
n	28	28
Mean limpet diameter / mm	35.64	37.36
Variance (s^2) (= the square of standard deviation) See Table 5.	77.17	74.4

First, calculate the value of t, using this formula:

$$t = \frac{\overline{x}_1 - \overline{x}_2}{\sqrt{\dfrac{s_1^2}{n_1} + \dfrac{s_2^2}{n_2}}}$$

Where s_1^2 and s_2^2 are the variances at each site, n_1 and n_2 are the numbers sampled at each site and $\overline{x}_1$ and $\overline{x}_2$ are the means for each site.

So substituting in the values from the table

$$t = \frac{35.64 - 37.36}{\sqrt{\dfrac{77.17}{28} + \dfrac{74.4}{28}}} = \frac{-1.72}{2.33} \text{ (ignore the } - \text{ sign)} = 0.74$$

Next, calculate the degrees of freedom. For an unpaired test this is $(n_1 + n_2) - 2 = 54$

The t value can be looked up in Table 8.

In this case the probability of the difference between the means being due to chance alone is more than 10%, so chance must have caused the difference. The null hypothesis is accepted and there is no significant difference between the two sites.

Correlation coefficient, r (Pearson's product moment correlation coefficient)

When sampling collects data from two variables it is possible to use a calculation to determine whether the two variables correlate in any way. For example, does rock pool algal diversity increase as pool surface area increases? The variables compared need to be plotted on scatter graphs which will indicate possible relationships that can then be tested.

The formula for correlation coefficient is $r = \dfrac{\Sigma(x - \overline{x}) \times (y - \overline{y})}{\sqrt{\Sigma(x - \overline{x})^2 \times \Sigma(y - \overline{y})^2}}$

In the equation

x = the values of the first variable

$\overline{x}$ = the mean of the values of the first variable

y = the values of the second variable

$\overline{y}$ = the mean of the values of the second variable

Σ = the sum of

For example, data were collected to investigate the possible negative correlation of wrinkling on horse chestnut seeds (conkers) with seed mass. The calculation is most conveniently laid out in a table to make it easy to follow the steps.

Maths link $\sqrt{x}$

MS 1.9

Seed mass (g) $= x$	Number of wrinkles per seed $= y$	$x - \bar{x}$	$y - \bar{y}$	$(x - \bar{x}) \times (y - \bar{y})$	$(x - \bar{x})^2$	$(y - \bar{y})^2$
12	1	5	−14	−70	25	196
10	3	3	−12	−36	9	144
8	8	1	−7	−7	1	49
6	15	−1	0	0	1	0
4	27	−3	12	−36	9	144
2	36	−5	21	−105	25	441
$\Sigma x = 42$	$\Sigma y = 90$			$\Sigma = -254$	$\Sigma = 70$	$\Sigma = 974$
$\bar{x} = 42 \div 6 = 7$	$\bar{y} = 90 \div 6 = 15$					

Now substitute the values from the table into the equation to find r

$$r = \frac{\Sigma(x - \bar{x}) \times (y - \bar{y})}{\sqrt{\Sigma(x - \bar{x})^2 \times \Sigma(y - \bar{y})^2}}$$

$$r = \frac{-254}{\sqrt{70 \times 974}} = \frac{-254}{\sqrt{68180}} = \frac{-254}{261} = -0.97$$

This suggests a negative correlation. The strength of this correlation may be found by looking up the calculated value of r in the table of r values (Table 7).

In this test the number of degrees of freedom = the number of values for the two variables $(n) - 2$ ($df = n - 2$). In this example $df = 12 - 2 = 10$.

At 10 degrees of freedom a value of 0.97 gives a p value of <0.001 or <0.1% of the observed data being due only to chance. Thus the correlation is 99.9% certain.

▼ **Table 5** *Formulae commonly used in biology*

Circumference of a circle	$\pi \times d$	d = diameter
Surface area of a cube or cuboid	$2(ab) + 2(ac) + 2(bc)$	a, b and c are side lengths
Surface area of a sphere	$4\pi r^2$	r is the radius
Surface area of a cylinder	$2\pi r^2 + (\pi d \times h)$	h is the length or height of a cylinder
Volume of a cube or cuboid	$a \times b \times c$	a, b and c are side lengths
Volume of a sphere	$\frac{4}{3}\pi r^3$	r is the radius
Volume of a cylinder	$\pi r^2 h$	r is the radius
magnification	$magnification = \frac{image\ size}{real\ size}$	Rearrange to find the other quantities
pH	$pH = -log_{10}[H^+]$	$[H^+]$ is the concentration of the hydrogen ion in moles per litre
Pulmonary ventilation rate	$PVR = tidal\ volume \times breathing\ rate$	
Cardiac output	$CO = stroke\ volume \times heart\ rate$	

Species diversity index	$D = \dfrac{N(N-1)}{\sum n(n-1)}$	N is the grand total of all species sampled n is the number of each individual species sampled
Lincoln index	$N = \dfrac{S_1 \times S_2}{R}$	S_1 = total captured and marked, S_2 = total number recaptured, R = number of marked animals recaptured
Efficiency of energy transfer	$\dfrac{energy\ transferred}{energy\ intake} \times 100\%$	
Net Primary Production	$NPP = GPP - R$	GPP is gross primary production and R is energy loss in respiration.
Net production by consumers, N	$N = I - (F + R)$	I = potential energy in ingested food, F = energy lost in faeces and urine, R = energy lost in respiration.
Standard deviation, s Used to assess spread or dispersion in a set of data	$s = \sqrt{\dfrac{\sum x^2 - \dfrac{(\sum x)^2}{n}}{n-1}}$	x refers to the values of the measurements taken n = the number of measurements $\sum$ means "the sum of"
An alternative sum for standard deviation	$s = \sqrt{\dfrac{\sum(x - \bar{x})^2}{n-1}}$	$\bar{x}$ is the mean value of the values of x
Chi-squared (χ^2) test, used to compare agreement between sample and expectation	$x^2 = \sum \dfrac{(O - E)^2}{E}$	O are the values you actually measure E are the values you expected to see
Student t test, used to compare the means of two sets of data	$t = \dfrac{\bar{x}_1 - \bar{x}_2}{\sqrt{\dfrac{s_1^2}{n_1} + \dfrac{s_2^2}{n_2}}}$	s^2 is the variance of a set of data Subscript denotes the data being compared, e.g. n_1 is the number of values of the first set s_2^2 is the variance of the second set $\bar{x}$ is the mean value of the values of x
Variance s^2, which is also a measure of dispersion in a set of data	$s^2 = \dfrac{\sum(x - \bar{x})^2}{n-1}$	
Correlation coefficient, r	$r = \dfrac{\sum(x-\bar{x}) \times (y-\bar{y})}{\sqrt{\sum(x-\bar{x})^2 \times \sum(y-\bar{y})^2}}$	use this formula to test correlation between two variables. Values will lie between 1 (perfect positive correlation) and 1 (perfect negative correlation).
Hardy-Weinberg formula	$p^2 + 2pq + q^2 = 1$	P = frequency of dominant allele, q = frequency of recessive allele

Statistical test tables

Table 6 *Table of values of chi-squared*

df	0.99	0.95	0.90	0.50	0.10	0.05	0.01	0.001	df
1	0.0016	0.0039	0.016	0.46	2.71	3.84	6.63	10.83	1
2	0.02	0.10	0.21	1.39	4.60	5.99	9.21	13.82	2
3	0.12	0.35	0.58	2.37	6.25	7.81	11.34	16.27	3
4	0.30	0.71	1.06	3.36	7.78	9.49	13.28	18.46	4
5	0.55	1.14	1.61	4.35	9.24	11.07	15.09	20.52	5
6	0.87	1.64	2.20	5.35	10.64	12.59	16.81	22.46	6
7	1.24	2.17	2.83	6.35	12.02	14.07	18.48	24.32	7
8	1.65	2.73	3.49	7.34	13.36	15.51	20.09	26.12	8
9	2.09	3.32	4.17	8.34	14.68	16.92	21.67	27.88	9
10	2.56	3.94	4.86	9.34	15.99	18.31	23.21	29.59	10
11	3.05	4.58	5.58	10.34	17.28	19.68	24.72	31.26	11
12	3.57	5.23	6.30	11.34	18.55	21.03	26.22	32.91	12
13	4.11	5.89	7.04	12.34	19.81	22.36	27.69	34.53	13
14	4.66	6.57	7.79	13.34	21.06	23.68	29.14	36.12	14
15	5.23	7.26	8.55	14.34	22.31	25.00	30.58	37.70	15

Table 7 *Values of r, correlation coefficient*

df	0.1	0.05	0.01	0.001
1	0.9877	0.9969	0.9999	1.0000
2	0.9000	0.9500	0.9900	0.9990
3	0.8054	0.8783	0.9587	0.9912
4	0.7293	0.8114	0.9172	0.9741
5	0.6694	0.7545	0.8745	0.9507
6	0.6215	0.7067	0.8343	0.9249
7	0.5822	0.6664	0.7977	0.8982
8	0.5494	0.6319	0.7646	0.8721
9	0.5214	0.6021	0.7348	0.8471
10	0.4973	0.5760	0.7079	0.8233
11	0.4762	0.5529	0.6835	0.8010
12	0.4575	0.5324	0.6614	0.7800
13	0.4409	0.5139	0.6411	0.7603
14	0.4259	0.4973	0.6226	0.7420
15	0.4124	0.4821	0.6055	0.7246
16	0.4000	0.4863	0.5897	0.7084
17	0.3887	0.4555	0.5751	0.6932
18	0.3783	0.4438	0.5614	0.6787
19	0.3687	0.4329	0.5487	0.6652
20	0.3598	0.4227	0.5368	0.6524

Table 8 *Values of* t

Degree of freedom (df)	0.10	0.05	0.01	0.001
1	6.31	12.71	63.66	636.60
2	2.92	4.30	9.92	31.60
3	2.35	3.18	5.84	12.92
4	2.13	2.78	4.60	8.61
5	2.02	2.57	4.03	6.87
6	1.94	2.45	3.71	5.96
7	1.89	2.36	3.50	5.41
8	1.86	2.31	3.36	5.04
9	1.83	2.26	3.25	4.78
10	1.81	2.23	3.17	4.59
12	1.78	2.18	3.05	4.32
14	1.76	2.15	2.98	4.14
16	1.75	2.12	2.92	4.02
18	1.73	2.10	2.88	3.92
20	1.72	2.09	2.85	3.85
α	1.64	1.96	2.58	3.29

Chapter 23 Practical skills

Practical skills are at the heart of Biology, a good foundation in practical skills will help you take your skills to a higher level. Biology is a dynamic subject in which our understanding constantly changes, largely as a result of developments in practical research. In the A level specification there is a separate practical endorsement which requires you to carry out 12 practicals across the two years of the A level course. The practical endorsement is not graded as part of your A level qualification. It is assessed by your teachers as a pass or a fail, a pass will be reported separately on your A level certificate. You will also be required to apply your understanding and knowledge of practicals to the written exams, practical-based questions account for 15% of the total assessment – the majority of these questions are in paper 3.

By undertaking the set practical activities in this course, it will not only develop your manipulative skills with specific apparatus and techniques but will also help you to gain a deeper understanding into the processes of scientific investigations. Skills such as researching, planning, implementing by making and processing measurements safely, analysing, and evaluating results will be reinforced and enhanced.

It is advantageous for you to answer practical questions when you have completed the practical – any questions on practical skills will have been written with the expectation that you will have carried out the practical activities. Having undertaken the practical, this helps with the teaching and learning of concepts in the specification. A richer practical experience will be gained if you do more practicals than the following twelve set practical activities in Table 1. Table 1 shows the 12 practicals which will be assessed in exams. For the practical endorsement the 12 practicals can consist of the required practicals or teacher devised practicals. For each activity, Table 1 references the relevant topic(s) in this book, the first six you will have already covered in your AS year of study.

▼ **Table 1** *A level required practical activities*

	Practical	Topic
1	Investigation into the effect of a named variable on the rate of an enzyme-controlled reaction	1.8 Factors affecting enzyme action
2	Preparation of prepared squashes of cells from plant root tips; set-up and use of an optical microscope to identify the stages of mitosis in these stained squashes and calculation of a mitotic index	3.1 Methods of studying cells 3.7 Mitosis
3	Production of a dilution series of a solute to produce a calibration curve with which to identify the water potential of plant tissue	4.3 Osmosis
4	Investigation into the effect of a named variable on the permeability of cell-surface membranes	4.1 Structure of the cell-surface membrane
5	Dissection of animal or plant gas exchange systems, a mass transport system or of an organ within such a system	6.2 Gas exchange in insects 6.3 Gas exchange in fish 6.4 Gas exchange in the leaf of a plant 6.6 Mammalian lungs
6	Use of aseptic technique to investigate the effect of antimicrobial substances on microbial growth	9 Genetic diversity and adaptation
7	Use of chromatography to investigate the pigments isolated from leaves of different plants, e.g., leaves from shade-tolerant and shade intolerant plants or leaves of different colours	3.5.1 Photosynthesis
8	Investigation into the effect of a named factor on the rate of dehydrogenase activity in extracts of chloroplasts	3.5.1 Photosynthesis

▼ **Table 1** *continued*

	Practical	Topic
9	Investigation into the effect of a named variable on the rate of respiration of cultures of single-celled organisms	3.5.2 Respiration
10	Investigation into the effect of an environmental variable on the movement of an animal using either a choice chamber or a maze	3.7.4 Populations in ecosystems
11	Production of a dilution series of a glucose solution and use of colorimetric techniques to produce a calibration curve with which to identify the concentration of glucose in an unknown 'urine' sample	3.6.4.2 Control of blood glucose
12	Investigation into the effect of a named environmental factor on the distribution of a given species	3.7.2 Populations

Practical questions

The following questions are designed to give you some practice at this practical style of question. If you haven't completed the practical yet, just think of similar practicals you have done or when you have used the apparatus and this will help you.

Practical 1 – The effect of pH on catalases

A celery extract was liquidised and prepared by the technician as a source of the enzyme catalase. A burette had been filled up to the 50 cm³ mark with hydrogen peroxide. 10 cm³ of celery extract was added and the height of the upper level of the frothing liquid was recorded. The class was asked to repeat the procedure adding the following to the H_2O_2:

- Add 2 drops HCl / 2 drops distilled water
- Add 4 drops HCl
- Add 2 drops NaOH / 2 drops distilled water
- Add 4 drops NaOH

The pH of each solution was tested before starting the experiment.

1 (a) Sketch a graph of your expected results. Remember to label your axes.
  (b) List all variables that need to be controlled and how you would control them.

Describe how you could change the method to make it:

2 (a) more reliable
  (b) more valid

Practical 2 – The mitotic index

A student performed a root tip squash on tissue from a garlic root tip using acetic orcein stain. She counted the number of cells she could see in one of the stages of mitosis. In total there were 150 cells and 12 cells were undergoing mitosis.

1 (a) Calculate the mitotic index for these cells. Show your working. *(2 marks)*
  (b) Another student didn't follow the exact procedure and as a result did not see any cells undergoing mitosis.
    Suggest two reasons why she did not see any cells in any stages of mitosis. *(2 marks)*

In a further investigation to see the effect of cells environment on cell division, cells were taken from varying distances from the root tip and the number of cells undergoing mitosis was counted. To make the results quantitative, the student calculated the mitotic index for each sample and plotted it on graph 1.

Graph 1

2 (a) Using graph 1, describe the results. *(2 marks)*
  (b) Suggest one reason why this relationship exists? *(1 mark)*

Practical 3 – Water potential in plant tissue

A practical was carried out to estimate the water potential in plant tissue. Six solid cylinders of potato were prepared each with identical dimensions. The students were given 100 cm³ of a stock 0.5 mol dm⁻³ solution sucrose. They were instructed to make up a series of six different dilutions. The mass (g) of the potato cylinder was measured before being submerged in the solution for 1 hour and then measured again after 1 hour.

Concentration of sucrose $(mol\,dm^{-3})$	Mass before submerging in solution (g)	Mass after submerging in solution (g)	Percentage change in mass of potato tissue (%)
0.0	4.5	5.0	
0.1	3.9	4.3	
0.2	4.3	4.5	
0.3	4.1	4.2	
0.4	4.4	3.7	
0.5	4.4	3.6	

1 Construct a table to show how to make up 20 cm³ of each of the six dilutions required.

2 (a) Calculate the % change in mass for all results in the results table. Show your working.
   (b) Why is it important to calculate a % change of mass in this experiment?

3 (a) Plot a graph of sucrose concentration against % change in mass.
   (b) Use this graph to find the concentration of sucrose.
   (c) Describe how you can estimate the water potential with this value.

Practical 4 - Effect of temperature on beetroot cell-surface membranes

Core samples of beetroot were washed and put into tubes containing distilled water. Each tube was left in a different temperature for 20 minutes. The distilled water in the tube became coloured and was transferred to a colorimeter and a reading was taken.

Graph 2

1 (a) Explain why the distilled water in the tubes becomes coloured?
   (b) Using the graph, describe the effect of changing temperature on the permeability of the cell-surface membrane.
   (c) Explain how the structure of the cell-surface membrane changes at temperatures above 50°C?

Practical 5 – Dissection

A live locust was being examined in class. Using a magnifying glass, tiny holes on each side of the segments were visible.

1 (a) Name these small holes?
   (b) Further observation showed these holes to be opening and closing. What benefit does this give the locust?
   (c) Some other insects have hairs around these holes.
        What environmental condition does this help them survive in and how does it aid their survival?

The teacher dissected a locust and exposed the inside of the body cavity. The teacher located the tracheal tubes and mounted a small sample of tissue onto a slide. The tracheal tubes were seen to be highly branched.

2 (a) Give one advantage of the tubes being highly branched.
   (b) What substances cause the ring thickening of the tracheae?
   (c) The tracheae branch into tubes of a much smaller diameter with little thickening. Suggest a reason for this structure.

Practical 6 – Aseptic technique

The technician set up a petri dish with nutrient agar jelly using equipment that had been in an autoclave. He inoculated it with non-pathogenic bacteria and left it to incubate for 48 hours.

1 (a) Why did he use equipment that had been in an autoclave?
   (b) Describe the steps he took to transfer the bacteria from the bottle to the petri dish. Only make reference to steps concerning keeping conditions aseptic.
   (c) Explain why he disinfected the work surface and washed his hands after the experiment was finished?

Practical 7

A student used paper chromatography to separate the pigments from a sample of petals. The procedure followed by the student is shown in the diagram below.

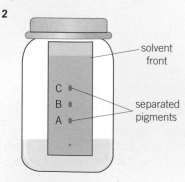

(a) State one precaution which the student should have taken in this investigation.

(1 mark)

(b) Outline the chemical principle illustrated by this technique.

(2 marks)

The table shows the distance moved by the solvent and pigments.

Substance	Distance moved (mm)
Solvent front	93
Pigment A	18
Pigment B	35
Pigment C	36

(c) (i) Define the term Rf value. (2 marks)

(ii) Calculate the Rf values for pigments B and C. Show your working. (2 marks)

(iii) Suggest one way by which greater separation of pigments B and C could have been achieved. (1 mark)

Practical 8

In photosynthesis the light-dependent reactions produce a reducing agent. This normally reduces NADP, but in this experiment the electrons arc accepted by the blue dye DCPIP. Reduced DCPIP is colourless. Spinach leaves were blended and placed in a cold isolation medium then filtered. The filtrate was then centrifuged until there was a small pellet of chloroplasts. The pellet was then suspended in solution. The pellet now distributed in the solution was divided between 5 test tubes and set up as below then DCPIP was added. Tubes 1, 2, 4 and 5 were placed near a light source, tube 3 was placed in darkness. The time taken for the DCPIP to decolourise in each tube was measured.

Tube	Leaf extract (cm³)	Supernatant (cm³)	Isolation medium (cm³)	Distilled water (cm³)	DCPIP solution (cm³)
1	0.5	–	–	–	5
2	–	–	0.5	–	5
3	0.5	–	–	–	5
4	0.5	–	–	5	–
5	–	0.5	–	–	5

Tube 3 (incubated in the dark) gave a reading of 4.9 absorption units after 10 minutes. Tube 2 (DCPIP with no leaf extract) was 6.4 absorption units

1 Describe and explain the changes observed in the five tubes. Compare the results and make some concluding comments about what they show. (5 marks)

2 The rate of photosynthesis in intact leaves can be limited by several factors including light, temperature and carbon dioxide. Which of these factors will have little effect on the reducing capacity of the leaf extract? (2 marks)

3 Describe how you might extend this practical to investigate the effect of light intensity on the light-dependent reactions of photosynthesis. (2 marks)

Practical 9

An investigation was carried out to determine the effect of temperature on the rate of cellular respiration in yeast. Five experimental groups, each containing five fermentation tubes, were set up. The fermentation tubes all contained the same quantities of water, glucose, and yeast. Each group of five tubes was placed in a water bath at a different temperature. After 30 minutes, the volume of gas produced (D) in each fermentation tube was measured in millilitres. The mean for each group was calculated. A sample setup and the data collected are shown below.

Mean volume of gas produced (D) after 30 minutes at various temperatures

Group	Temperature (°C)	D (cm³)
1	5	0
2	20	5
3	40	12
4	60	6
5	80	3

(a) Using the information in the data table, construct a line graph to show the relationship between temperature and the volume of gas produced. *(3 marks)*

(b) Deduce from the graph the temperature at which the maximum rate of cellular respiration in yeast occurred. *(1 mark)*

(c) Compared to the other tubes at the end of 30 minutes, state which of the following the tubes in group 3 contained:

(1) smallest volume of CO_2

(2) smallest quantity of glucose

(3) smallest quantity of ethanol

(4) same quantities of glucose, ethanol, and CO_2 *(2 marks)*

Practical 10

The diagram shows a choice chamber that can be used to investigate woodlice behaviour.

(a) Name two environmental factors that might attract woodlice to live underneath dead leaves. *(2 marks)*

(b) Design an experiment to show how you would use the choice chamber to investigate one of the factors you have named. *(3 marks)*

(c) Predict the results of your experiment *(1 mark)*

(d) Suggest one advantage to the woodlice of behaving in the way you predicted. *(1 mark)*

Practical 11

A student wanted to identify the concentration of glucose in an unknown sample. They used the following method

- A range of known glucose solutions were used starting with 20g dm^{-3}
- Each solution was heated with Benedict's solution. Once there was no more colour change the liquid was then cooled and filtered.
- The absorbance of the liquid was measured with a colorimeter.
- The student's results are shown below

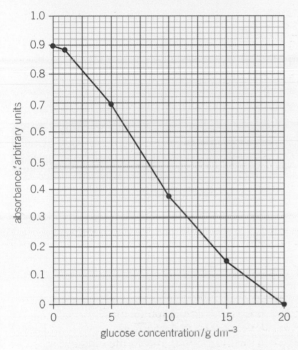

(a) State two precautions that the student should have taken during the procedure to ensure that the results give a valid comparison between the different glucose solutions *(2 marks)*

(b) The student tested the unknown sample and the absorbance reading obtained was 0.60 arbitrary units. Use the graph to determine the concentration of the sample. *(1 mark)*

(c) The procedure used does **not** test for non-reducing sugars such as sucrose. How could the student alter the procedure to determine the concentration of non-reducing sugar in the sample? *(2 marks)*

Practical 12

A group of students decided to investigate the relationship between light intensity and the distribution of herbs in a deciduous woodland. Light intensity was measured at randomly chosen sampling points within the woodland. At each location a sampling quadrat of diameter 1.0 m^2 was set up and the number of three herb species within the quadrat was recorded. The outline of the woodland and the sampling pattern used are shown in the diagram below.

Key: ● = sampling point

(a) Outline a suitable procedure for randomly selecting sampling points in this investigation. *(2 marks)*

(b) State one advantage and one disadvantage of random sampling. *(2 marks)*

(c) A section of one student's records are shown in the table below.

Sample point	Number of herbs			Mean light intensity (% of maximum incident)
	Species A	Species B	Species C	
1	27	18	4	70
2	25	22	0	85
3	18	18	3	65
4	19	14	7	60
5	29	24	6	90
6	10	12	20	30
7	38	26	4	80
8	0	2	20	15
9	39	29	4	80
10	0	0	16	5

(i) Plot a graph showing the relationship between mean light intensity and the number of herb species A and C. *(5 marks)*

(ii) State the relationship between average light intensity and the number of herb species C. *(1 mark)*

(iii) Suggest one possible explanation for the distribution of herb species C. *(1 mark)*

(d) Suggest two adaptations which may be seen in the leaves of plants growing in low light intensities. *(2 marks)*

1 **(a)** Gas exchange in fish takes place in gills. Explain how two features of gills allow efficient gas exchange. *(2 marks)*

(b) A zoologist investigated the relationship between body mass and rate of oxygen uptake in four species of mammal. The results are shown in the graph.

(i) The scale for plotting body mass is a logarithmic scale. Explain why a logarithmic scale was used to plot body mass. *(1 mark)*

(ii) Describe the relationship between body mass and oxygen uptake. *(1 mark)*

(iii) The zoologist measured oxygen uptake per gram of body mass. Explain why he measured oxygen uptake per gram of body mass. *(2 marks)*

(iv) Heat from respiration helps mammals to maintain a constant body temperature. Use this information to explain the relationship between body mass and oxygen uptake shown in the graph. *(3 marks)*

AQA, Jan 2010

2 **(a)** Describe how DNA is replicated. *(6 marks)*

(b) The graph shows information about the movement of chromatids in a cell that has just started metaphase of mitosis.

key

- - - - - = distance between chromatids

────── = distance between each chromatid and the pole to which it is moving

(i) What was the duration of metaphase in this cell? *(1 mark)*

(ii) Use line **X** to calculate the duration of anaphase in this cell. *(1 mark)*

(iii) Complete line **Y** on the graph. *(2 marks)*

(c) A doctor investigated the number of cells in different stages of the cell cycle in two tissue samples, **C** and **D**. One tissue sample was taken from a cancerous tumour. The other was taken from non-cancerous tissue. The table shows his results.

Stage of the cell cycle	Percentage of cells in each stage of the cell cycle	
	Tissue sample C	Tissue sample D
Interphase	82	45
Prophase	4	16
Metaphase	5	18
Anaphase	5	12
Telophase	4	9

(i) In tissue sample **C**, one cell cycle took 24 hours. Use the data in the table to calculate the time in which these cells were in interphase during one cell cycle. Show your working. *(2 marks)*

(ii) Explain how the doctor could have recognised which cells were in interphase when looking at the tissue samples. *(1 mark)*

(iii) Which tissue sample, **C** or **D**, was taken from a cancerous tumour? Use information in the table to explain your answer. *(2 marks)*

AQA Jan 2013

3 Taxol is a drug used to treat cancer. Research scientists investigated the effect of injecting taxol on the growth of tumours in mice. Some of the results are shown in **Figure 3**.

Number of days of treatment	Mean volume of tumour / mm^3	
	Control group	Group injected with taxol in saline
1	1	1
10	7	2
20	21	11
30	43	20
40	114	48
50	372	87

▲ Figure 3

(a) Suggest how the scientists should have treated the control group. *(2 marks)*

(b) Suggest and explain **two** factors which should be considered when deciding the number of mice to be used in this investigation. *(2 marks)*

(c) The scientists measured the volume of the tumours. Explain the advantage of using volume rather than length to measure the growth of tumours. *(1 mark)*

(d) The scientists concluded that taxol was effective in reducing the growth rate of the tumours over the 50 days of treatment. Use suitable calculations to support this conclusion. *(2 marks)*

(e) In cells, taxol disrupts spindle activity. Use this information to explain the results in the group that has been treated with taxol. *(3 marks)*

(f) The research scientists then investigated the effect of a drug called OGF on the growth of tumours in mice. OGF and taxol were injected into different mice as separate treatments or as a combined treatment. **Figure 4** and **Figure 5** show the results from this second investigation.

▲ Figure 4

Treatment	Mean volume of tumour following 70 days treatment / mm³ (± standard deviation)
OGF	322 (± 28.3)
Taxol	207 (± 22.5)
OGF and taxol	190 (± 25.7)
Control	488 (± 32.4)

▲ Figure 5

(i) What information does standard deviation give about the volume of the tumours in this investigation? *(1 mark)*

(ii) Use **Figure 4** and **Figure 5** to evaluate the effectiveness of the two drugs when they are used separately and as a combined treatment. *(4 marks)*

AQA Jan 2010

4 Read the following passage.

Gluten is a protein found in wheat. When gluten is digested in the small intestine, the products include peptides. Peptides are short chains of amino acids. These peptides cannot be absorbed by facilitated diffusion and leave the gut in faeces. Some people have coeliac disease. The epithelial cells of people with coeliac disease do not absorb the products of digestion very well. In these people, some of the   5
peptides from gluten can pass between the epithelial cells lining the small intestine and enter the intestine wall. Here, the peptides cause an immune response that leads to the destruction of microvilli on the epithelial cells.

Scientists have identified a drug which might help people with coeliac disease. It reduces the movement of peptides between epithelial cells. They have   10
carried out trials of the drug with patients with coeliac disease.

Use the information in the passage and your own knowledge to answer the following questions.

(a) Name the type of chemical reaction which produces amino acids from proteins. *(1 mark)*

(b) The peptides released when gluten is digested cannot be absorbed by facilitated diffusion (lines 2 – 3). Suggest why. *(3 marks)*

(c) The epithelial cells of people with coeliac disease do not absorb the products of digestion very well (lines 4 – 5). Explain why. *(3 marks)*

(d) Explain why the peptides cause an immune response (lines 7 – 8). *(1 mark)*

(e) Scientists have carried out trials of a drug to treat coeliac disease (lines 10 – 11). Suggest **two** factors that should be considered before the drug can be used on patients with the disease. *(2 marks)*

AQA June 2012

5   (a)   Haemoglobin contains iron. One type of anaemia is caused by a lack of iron. This type of anaemia can be treated by taking tablets containing iron. A number of patients were given a daily dose of 120 mg of iron. **Figure 8** shows the effect of this treatment on the increase in the concentration of haemoglobin in their red blood cells.

▲ Figure 8

(i)   Give one difference in the response of adults and children to this treatment.   (*1 mark*)

(ii) You could use the graph to predict the effect of this treatment on the increase in haemoglobin content of an adult after 40 days. Explain how.   (*2 marks*)

(iii) Haemoglobin has a quaternary structure. Explain what is meant by a quaternary structure.   (*1 mark*)

(b)   (i)   Pernicious anaemia is another type of anaemia. One method of identifying pernicious anaemia is to measure the diameter of the red blood cells in a sample of blood that has been diluted with an isotonic salt solution. Explain why an isotonic salt solution is used to dilute the blood sample.   (*3 marks*)

(ii) A technician compared the red blood cells in two blood samples of equal volume.

One sample was from a patient with pernicious anaemia, the other was from a patient who did not have pernicious anaemia. **Figure 9** shows some of the results she obtained.

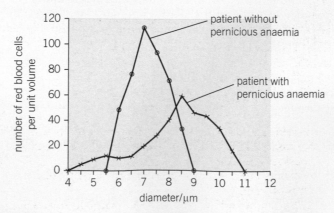

▲ Figure 9

Describe **two** differences between the blood samples.   (*2 marks*)

(c)   Scientists' analysis of blood proteins has indicated a lack of genetic diversity in populations of some organisms. Describe the processes that lead to a reduction in the genetic diversity of populations of organisms.   (*6 marks*)

AQA June 2010

6 Students investigated the effect of different concentrations of sodium chloride solution on discs cut from an apple. They weighed each disc and then put one disc into each of a range of sodium chloride solutions of different concentrations. They left the discs in the solutions for 24 hours and then weighed them again. Their results are shown in the table.

Concentration of sodium chloride solution / mol dm^{-3}	Mass of disc at start / g	Mass of disc at end / g	Ratio of mass at start to mass at end
0.00	16.1	17.2	0.94
0.15	19.1	20.2	0.95
0.30	24.3	23.2	1.05
0.45	20.2	18.7	1.08
0.60	23.7	21.9	
0.75	14.9	13.7	1.09

(a) **(i)** Calculate the ratio of the mass at the start to the mass at the end for the disc placed in the 0.60 mol dm^{-3} sodium chloride solution. *(1 mark)*

(ii) The students gave their results as a ratio. What is the advantage of giving the results as a ratio? *(2 marks)*

(iii) The students were advised that they could improve the reliability of their results by taking additional readings at the same concentrations of sodium chloride. Explain how. *(2 marks)*

(b) **(i)** The students used a graph of their results to find the sodium chloride solution with the same water potential as the apple tissue. Describe how they did this. *(2 marks)*

(ii) The students were advised that they could improve their graph by taking additional readings. Explain how. *(2 marks)*

AQA Jan 2010

7 The diagram shows the structure of the cell-surface membrane of a cell.

(a) Name **A** and **B**. *(2 marks)*

(b) **(i)** **C** is a protein with a carbohydrate attached to it. This carbohydrate is formed by joining monosaccharides together. Name the type of reaction that joins monosaccharides together. *(1 mark)*

(ii) Some cells lining the bronchi of the lungs secrete large amounts of mucus. Mucus contains protein.
Name **one** organelle that you would expect to find in large numbers in a mucus-secreting cell and describe its role in the production of mucus. *(2 marks)*

AQA June 2013

8 Students investigated the effect of removing leaves from a plant shoot on the rate of water uptake. Each student set up a potometer with a shoot that had eight leaves. All the shoots came from the same plant. The potometer they used is shown in the diagram.

(a) Describe how the students would have returned the air bubble to the start of the capillary tube in this investigation. *(1 mark)*

(b) Give **two** precautions the students should have taken when setting up the potometer to obtain reliable measurements of water uptake by the plant shoot. *(2 marks)*

(c) A potometer measures the rate of water uptake rather than the rate of transpiration. Give two reasons why the potometer does not truly measure the rate of transpiration. *(2 marks)*

(d) The students' results are shown in the table.

Number of leaves removed from the plant shoot	Mean rate of water uptake / cm^3 per minute
0	0.10
2	0.08
4	0.04
6	0.02
8	0.01

Explain the relationship between the number of leaves removed from the plant shoot and the mean rate of water uptake. *(3 marks)*

AQA Jan 2013

9 Read the following passage.
Some foods contain substances called flavenoids. Flavenoids lower blood cholesterol concentration and reduce the risk of developing coronary heart disease.

Some types of dark chocolate have a high concentration of flavenoids. One group of scientists investigated the effect of eating dark chocolate on the risk of developing coronary heart disease. 5

The scientists randomly divided healthy volunteers into two groups. Every day one group was given dark chocolate containing flavenoids to eat. The other group acted as a control.

The scientists measured the diameter of the lumen of the main artery in the arms of the volunteers every week. At the end of a month, the diameter of the lumen of the main artery in the arm of the volunteers who had eaten dark chocolate containing flavenoids had increased. 10

Use information from the passage and your own knowledge to answer the questions.

(a) High blood cholesterol concentration is a risk factor associated with coronary heart disease.

Give **two** other risk factors associated with coronary heart disease. *(2 marks)*

(b) (i) The scientists used healthy volunteers in this investigation (line 7). Why was it important that the volunteers were healthy? *(1 mark)*

(ii) The scientists randomly divided the volunteers into two groups (line 7). Explain why they divided them randomly. *(1 mark)*

(c) (i) Describe how the control group should have been treated. *(2 marks)*

(ii) Why was it important to have a control group in this investigation? *(1 mark)*

(d) Suggest why an increase in the diameter of the lumen of the main artery in the arm (lines 11–12) is associated with a reduced risk of coronary heart disease. *(3 marks)*

AQA June 2010

10 (a) What is intraspecific variation? *(1 mark)*

(b) Schizophrenia is a mental illness. Doctors investigated the relative effects of genetic and environmental factors on the development of schizophrenia. They used sets of identical twins and non-identical twins in their investigation. At least one twin in each set had developed schizophrenia.

- Identical twins are genetically identical.
- Non-identical twins are not genetically identical.
- The members of each twin pair were raised together.

The table shows the percentage of cases where both twins had developed schizophrenia.

Type of twin	Percentage of cases where both twins had developed schizophrenia
Identical	50
Non-identical	15

(i) Explain why both types of twin were used in this investigation. *(2 marks)*

(ii) What do these data suggest about the relative effects of genetic and environmental factors on the development of schizophrenia? *(1 mark)*

(iii) Suggest two factors that the scientists should have taken into account when selecting the twins to be used in this study. *(2 marks)*

AQA Jan 2013

11 (a) Students measured the rate of transpiration of a plant growing in a pot under different environmental conditions. Their results are shown in the table.

Conditions		Transpiration rate/gh^{-1}
A Still air	15 °C	1.2
B Still air	15 °C	1.7
C Still air	25 °C	2.3

During transpiration, water diffuses from cells to the air surrounding a leaf.

(i) Suggest an explanation for the difference in transpiration rate between conditions **A** and **B**. *(2 marks)*

(ii) Suggest an explanation for the difference in transpiration rate between conditions **A** and **C**. *(2 marks)*

(b) Scientists investigated the rate of water movement through the xylem of a twig from a tree over 24 hours. The graph shows their results. It also shows the light intensity for the same period of time.

(i) Describe the relationship between the rate of water movement through the xylem and the light intensity. *(1 mark)*

(ii) Explain the change in the rate of water movement through the xylem between 06.00 and 12.00 hours. *(2 marks)*

(iii) The scientists also measured the diameter of the trunk of the tree on which the twig had been growing. The diameter was less at 12.00 than it was at 03.00 hours. Explain why the diameter was less at 12.00 hours. *(2 marks)*

AQA Jan 2011

12 Imatinib is a drug used to treat a type of cancer that affects white blood cells. Scientists investigated the rate of uptake of imatinib by white blood cells. They measured the rate of uptake at 4 °C and at 37 °C.

Their results are shown in the table.

Concentration of imatinib outside cells / μmol dm^{-3}	Mean rate of uptake of imatinib into cells / μg per million cells per hour	
	4 °C	**37 °C**
0.5	4.0	10.5
1.0	10.7	32.5
5.0	40.4	420.5
10.0	51.9	794.6
50.0	249.9	3156.1
100.0	606.9	3173.0

(a) The scientists measured the rate of uptake of imatinib in μg per million cells per hour. Explain the advantage of using this unit of rate in this investigation. *(2 marks)*

(b) Calculate the percentage increase in the mean rate of uptake of imatinib when the temperature is increased from 4 °C to 37 °C at a concentration of imatinib outside the cells of 1.0 μmol dm^{-3}.
Give your answer to one decimal place. *(2 marks)*

(c) Imatinib is taken up by blood cells by active transport.

(i) Explain how the data for the two different temperatures support this statement. *(2 marks)*

(ii) Explain how the data for concentrations of imatinib outside the blood cells at 50 and 100 μmol dm^{-3} at 37 °C support the statement that imatinib is taken up by active transport. *(2 marks)*

AQA June 2013

1 Metastatic melanoma (MM) is a type of skin cancer. It is caused by a faulty receptor protein in cell-surface membranes. There have been no very effective treatments for this cancer.

Dacarbazine is a drug that has been used to treat MM because it appears to increase survival time for some people with MM.

Doctors investigated the use of a new drug, called ipilimumab, to treat MM. They compared the median survival time (ST) for two groups of patients treated for MM:

- a control group of patients who had been treated with dacarbazine
- a group of patients who had been treated with dacarbazine and ipilimumab.

The ST is how long a patient lives after diagnosis.

The doctors also recorded the percentage of patients showing a significant reduction in tumours with each treatment.

The total number of patients in the investigation was 502.

Table 2 shows the doctors' results.

▼ **Table 2**

Treatment	Median survival time (ST) / months	Percentage of patients showing significant reduction in tumours
Dacarbazine	9.1	10.3
Dacarbazine and ipilimumab	11.2	15.2

1 (a) The doctors compared median survival times for patients in each group. How would you find the median survival time for a group of patients?   (*2 marks*)

(b) In many trials of new drugs, a control group of patients is given a placebo that does not contain any drug. The control group in this investigation had been treated with dacarbazine. Suggest why they had not been given a placebo.   (*1 mark*)

(c) A journalist who read this investigation concluded that ipilimumab improved the treatment of MM. Do the data in **Table 2** support this conclusion? Give reasons for your answer.   (*4 marks*)

(d) MM is caused by a faulty receptor protein in cell-surface membranes. Cells in MM tumours can be destroyed by the immune system. Suggest why they can be destroyed by the immune system.   (*3 marks*)

AQA Specimen 2014

2 (a) Insect pests of crop plants can be controlled by chemical pesticides or biological agents. Give two advantages of using biological agents.   (*2 marks*)

Two-spotted mites are pests of strawberry plants. Ecologists investigated the use of predatory mites to control two-spotted mites. They released predatory mites on strawberry plants infested with two-spotted mites. They then recorded the percentage of strawberry leaves occupied by two-spotted mites and by predatory mites over a 16-week period. The results are shown on the graph.

(b) Describe how the percentage of leaves occupied by predatory mites changed during the period of this investigation. *(2 marks)*

(c) The ecologists concluded that in this investigation control of the two-spotted mite by a biological agent was effective. Explain how the results support this conclusion. *(2 marks)*

(d) Farmers who grow strawberry plants and read about this investigation might decide **not** to use these predatory mites. Suggest **two** reasons why.

(e) The ecologists repeated the investigation but sprayed chemical pesticide on the strawberry plants after 10 weeks. After 16 weeks no predatory mites were found but the population of two-spotted mites had risen significantly. Suggest an explanation for the rise in the two-spotted mite population. *(2 marks)*

AQA June 2012

3 Mountains are harsh environments. The higher up the mountain, the lower the temperature becomes. The diagram shows a forest growing on the side of a mountain. The upper boundary of the forest is called the tree line. Trees do not grow above the tree line.

(a) (i) The position of the tree line is determined by abiotic factors. What is meant by an abiotic factor? *(1 mark)*

(ii) Other than temperature, suggest **one** abiotic factor that is likely to affect the position of the tree line on the mountain. *(1 mark)*

(b) Scientists measured the concentration of carbon dioxide in the air in one part of the forest. They took measurements at different times of day and at two different heights above the ground. Their results are shown in the bar chart.

Use your knowledge of photosynthesis and respiration to explain the data in the bar chart. *(4 marks)*

(c) The population of trees in the forest evolved adaptations to the mountain environment. Use your knowledge of selection to explain how. *(3 marks)*

AQA Jan 2012

4 Plant physiologists attempted to produce papaya plants using tissue culture. They investigated the effects of different concentrations of two plant growth factors on small pieces of the stem tip from a papaya plant. Their results are shown in the table.

Concentration of auxin / μmol dm^{-3}	Concentration of cytokinin / μmol dm^{-3}		
	5	25	50
0	No effect	No effect	Leaves produced
1	No effect	Leaves produced	Leaves produced
5	Leaves produced	No effect	Leaves and some plantlets produced
10	Callus produced	Leaves and some plantlets produced	Plantlets produced
15	Callus produced	Callus and some leaves produced	Callus and some leaves produced

Callus is a mass of undifferentiated plant cells. Plantlets are small plants.

(a) Explain the evidence from the table that cells from the stem tip are totipotent. *(2 marks)*

(b) Calculate the ratio of cytokinin : auxin that you would recommend to grow papaya plants by this method. *(2 marks)*

(c) (i) Papaya plants reproduce sexually by means of seeds. Papaya plants grown from seeds are very variable in their yield. Explain why. *(2 marks)*

(ii) Explain the advantage of growing papaya plants from tissue culture rather than from seeds. *(1 mark)*

AQA June 2011

5 A Sri Lankan scientist investigated the effect of human disturbance on the organisms living on a rocky seashore. He chose three areas for the study. These areas had different amounts of human disturbance. The scientist measured human disturbance by walking from one end of the beach to the other. He recorded the number of people he encountered. Figure 1 shows his results.

	Site R	Site G	Site U
Mean number of people encountered per hour (± standard deviation)	2.2 (± 2.1)	17.6 (± 9.6)	34.6 (± 11.6)

▲ Figure 1

(a) (i) What conclusions can you draw about the number of people visiting Site **R** compared with the number of people visiting the other two sites? Give evidence from **Figure 1** to support your answer. *(2 marks)*

(ii) The scientist believed that there was a significant difference between the numbers of people visiting site R and the other sites. Select and carry out a suitable statistical test to compare the sites. *(3 marks)*

(iii) Comment on the scientist's conclusion about visitor numbers. *(2 marks)*

(b) The scientist used quadrats to find the number of species at each of the three sites. He carried out a preliminary investigation and recorded the total number of species in an increasing number of quadrats. **Figure 2** shows the results.

▲ **Figure 2**

(i) Use **Figure 2** to explain why 10 would not be an appropriate number of quadrats to use. (*1 mark*)

(ii) Use **Figure 2** to explain why 25 would not be an appropriate number of quadrats to use. (*1 mark*)

(c) The scientist measured the dry biomass of seaweeds at each of sites **R**, **G** and **U**. He collected all the organisms of a particular species in a quadrat and incubated them in an oven at a temperature of 80°C.

The scientist incubated the seaweeds at 80°C. Suggest why incubating them at a higher temperature would **not** produce valid results. (*1 mark*)

As well as measuring the dry biomass of the seaweeds, the scientist measured the dry mass of the animals present. He also measured the abundance of each species. **Figure 3** shows the data he collected.

	Site R	Site G	Site U
Mean number of people per hour	2.2	17.6	34.6
Mean number of species of seaweed per quadrat	4.2	2.1	1.3
Ratio of dry biomass of animals to dry biomass of seaweeds	0.15	0.06	0.03
Ratio of dry biomass of animals to abundance of animals	0.20	0.10	0.09
Ratio of dry biomass of seaweeds to abundance of seaweeds	0.79	1.57	3.24

▲ **Figure 3**

(d) The ratio of the dry biomass of animals to the dry biomass of seaweeds is always a lot less than one. Explain why. (*2 marks*)

(e) (i) Conservation officers were working on the beaches used in this investigation. They noticed that there were fewer larger seaweeds on beaches used by a large number of people than on beaches visited by only a few people. Explain how the data in **Figure 3** support this. (*2 marks*)

(ii) What conclusions can you draw from the data in **Figure 3** about the effect of human disturbance on the animals living on the seashore? Explain your answer. (*4 marks*)

AQA June 2010 (apart from 5a (ii) and (iii))

6 Shrews are small mammals. Three species of shrew live in mainland Britain. The table shows the body mass of ten shrews in three different shrew species.

	Body mass (g)							
Common shrew	10.2	11.6	7.9	12.7	13.5	6.5	8.7	9.7
Pygmy shrew	5.9	5.6	4.9	4.1	4.9	5.9	4.6	4.8
Water shrew	14.2	13.7	12	13.9	12.5	12.3	13.1	12.8

a (i) Calculate mean body mass values for the three species. *(2 marks)*
   (ii) Calculate the standard deviation (see page 579) for each of the
        three means. *(3 marks)*
   (iii) Comment on the variation in the three populations *(2 marks)*

A team of biologists investigated a method of estimating the abundance of shrews. They used plastic tubes, called hair tubes. Some of the hairs from a shrew that enters one of these tubes stick to glue in the tube. These hairs can be used to identify the species of shrew. The diagram shows a set of these hair tubes.

part of entrance blocked with tape

hair tubes

card with faunagoo underneath. Hairs from shrews stick to the faunagoo

(b) (i) Faunagoo is a glue that remains sticky after wetting and drying. Explain
        the advantage of using Faunagoo in these hair tubes. *(1 mark)*
   (ii) The diagram shows that the biologists partly blocked the entrances to the
        tubes with tape. Suggest why they partly blocked the entrances. *(1 mark)*

(c) The biologists needed to find a way of distinguishing between the hairs of the three species of shrew. They collected hairs from shrews of each species. For each species, they selected hairs at random and made different measurements.

Explain why the biologists selected the hairs at random. *(1 mark)*

(d) Repeatable measurements are measurements of the same feature that are very similar. In this investigation, each measurement was made by two observers. This helped the team to check the repeatability of these measurements.
   (i) Explain why it was important to check the repeatability of the measurements.
        (2 marks)
   (ii) You could use a scatter diagram to check the repeatability of measurements
        made by two observers. Describe how. *(2 marks)*

(e) The biologists used hair tubes to find the abundance of shrews along the edges of some fields. They also used traps that caught shrews without harming them. They selected areas where all three species of shrew were present.
   • They put sets of hair tubes at 5 m intervals along the edges of the fields. They inspected the tubes one week later and recorded the number of sets of tubes that contained shrew hairs. They called this the hair tube index.
   • At each site where they used hair tubes, they set traps immediately after using the hair tubes. They recorded the number of different shrews caught in these traps.
   (i) The research team found the hair tube index. Explain why they could
        not use the hair tubes to find the total number of shrews present. *(1 mark)*
   (ii) The research team set the traps immediately after using the hair tubes.
        Explain why setting the traps immediately after using the hair tubes

would make comparisons between the two methods more reliable. (*2 marks*)
The graphs are types of scatter diagram called bubble plots. They show hair tube index plotted against the number of shrews caught in traps. The area of the bubble is proportional to the number of records plotted.

(f) Explain why a statistical test was necessary in analysing the results for the common shrew. Use the terms chance and probability in your answer. (*2 marks*)

(g) **(i)** The biologists concluded that hair tubes were a reliable way of measuring the abundance of common shrews. Give evidence from the graph to support this conclusion. (*1 mark*)

  (ii) Use information in this question to evaluate the use of hair tubes as a way of measuring the abundance of pygmy shrews. (*2 marks*)

AQA Jan 2010 (apart from 6 a (i)–(iii))

7 **(a)** Increased intensity of exercise leads to an increased heart rate. Explain how. (*3 marks*)

(b) Scientists investigated the effect of taking omega-3 fatty acids in fish oil on heart rate during exercise. They recruited two large groups of volunteers, **A** and **B**. For each group, they measured the mean heart rates at different intensities of exercise. The volunteers were then given capsules to take for 8 weeks.
  • Group **A** was given capsules containing omega-3 fatty acids in fish oil.
  • Group **B** was given capsules containing olive oil.
After 8 weeks, they repeated the measurements of mean heart rates at different intensities of exercise. The graph shows their results. The bars represent the standard deviations.

  (i) Group B was given capsules containing olive oil. Explain why. (*1 mark*)

  (ii) The scientists concluded that omega-3 fatty acids lower the heart rate during exercise. Explain how the information in the graph supports this conclusion. (*3 marks*)

AQA June 2012

Question A

Sickle-cell anaemia is a disease caused by a gene mutation in the gene which codes for haemoglobin.

In the DNA molecule that produces one of the amino acid chains in haemoglobin, the normal DNA triplet on the template strand is changed from CTC to CAC. As a result, the mRNA produced has a different code.

1 (a) Identify the type of gene mutation that causes sickle cell anaemia.
   (b) Deduce the
        (i)   normal mRNA codon produced from the DNA.
        (ii) mRNA codon produced as a result of the mutation.

The changed mRNA codes for the amino acid valine rather than for glutamic acid. This produces a molecule of haemoglobin (called haemoglobin S) that has a 'sticky patch'. At low oxygen concentrations haemoglobin S molecules tend to adhere to one another by their sticky patches causing them to form long fibres within the red blood cells. These fibres distort the red blood cells, making them inflexible and sickle (crescent) shaped. These sickle cells are unable to carry oxygen and may block small capillaries.

2 Suggest
   (a) how a change in a single amino acid might lead to the change in protein structure described.
   (b) why sufferers of sickle-cell anaemia easily become tired.

Sickle-cell anaemia disables and kills individuals and so the gene causing it has been eliminated from most populations. However, the gene is relatively common among black populations of African origin. This is because the malarial parasite, *Plasmodium*, is unable to exist in sickled red blood cells.

3 Suggest a process that might have eliminated the mutant gene from most populations.

Sickle cell anaemia is the result of a gene that has two codominant alleles, Hb^A (normal) and Hb^S (sickled).

4 What is meant by the term 'codominant'?

The three possible genotype combinations of these two codominant alleles and their corresponding phenotypes are as follows:

- homozygous for haemoglobin S ($Hb^S Hb^S$). Individuals suffer from sickle-cell anaemia and are considerably disadvantaged if they do not receive medical attention. They rarely live long enough to pass their genes on to the next generation.
- homozygous for haemoglobin A ($Hb^A Hb^A$). Individuals lead normal healthy lives, but are susceptible to malaria in areas of the world where the disease is endemic.
- heterozygous for haemoglobin ($Hb^A Hb^S$). Individuals are said to have sickle-cell trait, but are not badly affected. Sufferers may become tired more easily but, in general, the condition is symptomless. They do, however, have resistance to malaria.

5 (a) In parts of the world where malaria is prevalent, the heterozygous state ($Hb^A Hb^S$) is selected for at the expense of both homozygous states. Consider the information above and suggest why this is the case.
   (b) Name the type of selection taking place. Explain your answer.

6 (a) If two heterozygous individuals produce offspring, calculate the chance of any one of them having sickle-cell anaemia.
   (b) Genetic screening for sickle-cell anaemia can be carried out. Explain why the advice given by a genetic counsellor to individuals with the same genotypes might differ depending on where they live.

7 In a population of 175 individuals the frequency of the Hb^A allele is 0.6 and the frequency of the Hb^S allele is 0.4.

(a) Calculate the frequencies of the Hb^A and Hb^S alleles that would be expected in the next generation if the individuals mated randomly.

(b) Using the Hardy–Weinberg equation, calculate the number of individuals with each phenotype. Show your working.

Question B

Figure 1 shows the energy flow through a freshwater ecosystem.

1 (a) (i) State which organisms in this food chain are the primary consumers.

(ii) Calculate the energy lost in respiration and waste products by the freshwater snails. Show your working.

(iii) Calculate the net primary production of the leeches. Show your working.

(iv) Calculate the percentage efficiency of energy transfer from leeches to sticklebacks. Show your working and give your answer to three significant figures.

The sun's energy for food chains like the one shown in Figure 1 is converted to chemical energy by the process of photosynthesis. Photosynthesis has two distinct stages, the light independent stage and the light dependent stage. Figure 2 is a simplified sequence of the light dependent stage.

The units shown are in kilojoules per metre squared per year (kJ m^{-2}year^{-1})

▲ **Figure 1**

2 (a) Name the substances P, Q, R and S.

(b) State the number of carbon atoms in ribulose bisphosphate.

The products of photosynthesis are transported from the leaves to the regions of the plant using or storing them through the tissue shown in Figure 3.

▲ **Figure 2**

▶ **Figure 3**

3 **(a)** **(i)** Name the tissue shown in Figure 3.
(ii) Figure 3 shows two types of cells that are typical of this tissue. Name the two types of cell.
(b) **(i)** Name two organic substances that are transported in this tissue.
(ii) Outline four pieces of evidence that support the view that organic substances are transported in the phloem.
(c) Explain the advantages of studying cells such as these with an electron microscope rather than a light microscope.

Question C

Figure 1 shows a nephron from a mammalian kidney.

1 **(a)** Name the parts labelled A – H
(b) **(i)** Name a substance found in the blood plasma in the afferent arteriole but **not** present in the structure labelled H.
(ii) Explain why this substance is absent from structure H.
(c) **(i)** Name a substance found in the structure labelled B but absent from the fluid in the structure labelled E.
(ii) Explain why this substance is absent from structure E.
2 Name the two blood vessels through which blood passes on its journey from the heart to the afferent arteriole.
Figure 2 represents a cell from the wall of structure H.
3 **(a)** Calculate the magnification of the cell in Figure 2. Show your working.
(b) **(i)** Name the structures labelled X.
(ii) Describe the function of the structures labelled X.
(c) **(i)** Name the structures labelled Y.
(ii) Explain why these cells have a large number of structure Y.
Goodpasture's disease is a very rare condition in which antibodies attack an antigen that is found in cells of the glomerulus of the kidney. One symptom of the disease is blood in the urine.
4 **(a)** Suggest an explanation for the sufferers of the disease having blood in the urine.
(b) **(i)** Name the type of cell that produces antibodies in the body.
(ii) Antibodies do not directly destroy cells with the antigen they are complementary to, but rather prepare these cells for destruction. Explain the role of antibodies in preparing cells with the antigen for destruction.

▲ Figure 1

25 μm

▲ Figure 2

Question D

Cystic fibrosis is caused by a mutant recessive allele in which three DNA bases, adenine-adenine-adenine, are missing from the cystic fibrosis trans-membrane-conductance regulator (CFTR) gene. The deletion results in a single amino acid being left out of the protein produced by this gene. As a result the protein is unable to perform its role of transporting chloride ions across epithelial membranes. CFTR is a chloride-ion channel protein that transports chloride ions out of epithelial cells, and water naturally follows keeping epithelial membranes moist.

1 Explain how it is possible for parents without cystic fibrosis to have a child who suffers from the disease.
2 Name the process by which water will follow chloride ions across epithelial membranes. In a patient with cystic fibrosis, the epithelial membranes are dry and the mucus they produce remains viscous and sticky. The symptoms this causes include:
- mucus congestion in the lungs leading to a much higher risk of infection, breathing difficulties and less efficient gaseous exchange
- accumulation of thick mucus in the pancreatic ducts, preventing pancreatic enzymes, such as lipases, from reaching the duodenum.

3 **(a)** Explain why mucus congestion in the lungs can lead to a higher risk of infection.

(b) Describe precisely the action of lipases.

Where there is a history of the disease in both families, parents may choose to be genetically screened to see whether they carry the allele. Genetic screening involves trying to determine if the mutant allele of the CFTR gene is present.

4 Outline how the presence of a mutant allele of the CFTR might be detected using DNA probes and DNA hybridisation.

Research is taking place to treat cystic fibrosis using a technique called gene therapy. This involves replacing or supplementing the defective gene with a healthy one. One method is to introduce cloned normal genes into the epithelial cells of the lungs but the treatment needs to be repeated as often as every few days. The long-term aim is to target the stem cells that give rise to lung epithelial tissue.

5 **(a)** State the meaning of the word 'cloned'.

(b) Explain why the treatment of lung epithelial tissue has to be repeated frequently.

(c) Suggest an advantage of delivering the healthy gene to stem cells rather than mature lung epithelial tissue.

Viruses make useful vectors for the transfer of the normal CFTR gene into the epithelial cells. They are grown in epithelial cells in the laboratory along with plasmids that have had the normal CFTR gene inserted. The CFTR gene becomes incorporated into the DNA of the viruses which are isolated and purified before being introduced into the nostrils of the patients.

6 **(a)** Define each of the following terms:

(i) virus

(ii) vector

(iii) plasmid.

(b) From your knowledge of viruses, suggest a reason why they are used to introduce the healthy CFTR gene into lung epithelial cells.

(c) Suggest a possible disadvantage of using viruses in this way.

An alternative method of delivering plasmids containing the healthy CFTR gene into lung epithelial cells is to wrap them in lipid molecules to form a liposome. The liposomes are sprayed into the nostrils of the patient as an aerosol and are drawn down into the lungs during inhalation.

7 **(a)** Describe the process of inhalation in a human in terms of the structures involved and the associated pressure and volume changes.

(b) Suggest a reason why liposomes are able to deliver the CFTR gene into lung epithelial cells.

Answers to questions

Atoms, isotopes, and the formation of ions

1 a hydrogen

   b isotope

   c 100% (it is doubled as neutrons have the same mass as protons)

   d It is unchanged (the atomic number is the number of protons and this remains as one).

2 a hydrogen ion

   b It is not changed (mass number is the number of protons and neutrons)

1.2

1 Carbon atoms readily link to one another to form a chain.

2 polymer

3 Sugar donates electrons that reduce blue copper(II) sulfate to orange copper(I) oxide.

Semi-quantitative nature of Benedict's test

1 B, E, A, D, C

2 Dry the precipitate in each sample and weigh it. The heavier the precipitate the more reducing sugar is present.

3 Once all the copper(II) sulfate has been reduced to copper(I) oxide, further amounts of reducing sugar cannot make a difference.

1.3

1 a glucose + galactose; b glucose + fructose; c α glucose only

2 $C_{12}H_{22}O_{11}$ ($C_6H_{12}O_6 + C_6H_{12}O_6 - H_2O$)

3 Enzymes are denatured at higher temperatures and this prevents them functioning / enzymes lower the activation energy required.

1.4

1 starch

2 glycogen

3 α-glucose, β-glucose, starch, cellulose

4 starch, cellulose, glycogen

5 α-glucose

6 cellulose

7 starch, cellulose, glycogen

8 α-glucose, β-glucose

1.5

1 a triglycerides; b glycerol;
   c polyunsaturated; d two; e hydrophobic

2 triglyceride: 3 fatty acids / no phosphate group / nonpolar; phospholipid: 2 fatty acids / 1 phosphate group / hydrophilic 'head' and hydrophobic 'tail

3 Lipids provide more than twice as much energy as carbohydrate when they are oxidised. If fat is stored, the same amount of energy can be provided for less than half the mass. It is therefore a lighter storage product – a major advantage if the organism is motile.

1.6

1 peptide bond

2 condensation reaction

3 amino group (—NH$_2$), carboxyl group (—COOH), hydrogen atom (—H), R group

Protein shape and function

1 It has three polypeptide chains wound together to form a strong, rope-like structure that has strength in the direction of pull of a tendon.

2 prevents the individual polypeptide chains from sliding past one another and so they gain strength because they act as a single unit

3 The junctions between adjacent collagen molecules are points of weakness. If they all occurred at the same point in a fibre, this would be a major weak point at which the fibre might break.

1.7

1 a substance that alters the rate of a chemical reaction without undergoing permanent change

2 They are not used up in the reaction and so can be used repeatedly.

3 The changed amino acid may no longer bind to the substrate, which will then not be positioned correctly, if at all, in the active site.

4 The changed amino acid may be one that forms hydrogen bonds with other amino acids. If the new amino acid does not form hydrogen bonds the tertiary structure of the enzyme will change, including the active site, so that the substrate may no longer fit.

Lock and key model of enzyme action

1 It more clearly matches current observations such as enzyme activity being changed when molecules bind at sites other than the active site. This suggests

that enzyme molecules change shape when other molecules bind to them.

1.8

1 To function, enzymes must physically collide with their substrate. Lower temperatures decrease the kinetic energy of both enzyme and substrate molecules, which then move around less quickly. They hence collide less often and therefore react less frequently.

2 The heat causes hydrogen and other bonds in the enzyme molecule to break. The tertiary structure of the enzyme molecule changes, as does the active site. The substrate no longer fits the active site.

3 a High temperatures denature the enzymes and so they cannot spoil the food;

  b Vinegar is very acidic and the very low pH will denature the enzymes and so preserve the food

4 pH = 4

Enzyme action

1 a Enzyme X is not very specific as it acts on a number of different proteins.

    Enzyme Y is highly specific as it acts on a single protein.

  b Enzyme X could be used in biological washing powders to digest/remove stains from clothes.

    Enzyme Y could be used in making yoghurt/cheese from milk.

  c Milk is the only food in the diet of young mammals. The enzyme coagulates the milk causing it to remain in the stomach for longer. This gives time for enzymes there to act on it so that it can be broken down into products which can then be absorbed. If it had remained liquid, it would pass through the stomach more quickly and only be partially digested.

  d Enzyme X functions at much higher temperatures than enzyme Y and so must have a much more stable tertiary structure to prevent it becoming denatured. The bonds holding the polypeptide chain in its precise 3-D arrangement that makes up the active site must therefore be less easily broken than those of enzyme Y. It is therefore likely that the bonds of enzyme X are mostly, or entirely, disulfide bonds as these are not easily broken by heat. Enzyme Y is likely to have more of the heat-sensitive ionic and hydrogen bonds.

2 a

Temperature / °C	Time / min for hydrolysis of protein	Rate of reaction / $\frac{1}{time}$
15	5.8	0.17
25	3.4	0.29
35	1.7	0.59
45	0.7	1.43
55	0.6	1.67
65	0.9	1.11
75	7.1	0.14

b

c The optimum temperature is found by dropping a vertical line from the highest point on the curve and reading the temperature where it transects the temperature (x) axis. The value is in the range 50–55 °C.

d Carry out the experiment in exactly the same way but use narrower temperature intervals (e.g. 1 °C) over the range 45–60 °C.

1.9

1 Competitive inhibitors occupy the active site of an enzyme while non-competitive inhibitors attach to the enzyme at a site other than the active site.

2 Increase the substrate concentration. If the degree of inhibition is reduced, it is a competitive inhibitor; if it stays the same, it is a non-competitive inhibitor.

Control of metabolic pathways

1 pH / substrate concentration (not temperature)

2 In a metabolic pathway, the product of one reaction acts as the substrate for the next reaction. By having the enzymes in appropriate sequence there is a greater chance of each enzyme coming into contact with its substrate than if the enzymes are floating freely in the organelle. This is a more efficient means of producing the end product.

3 a it would increase; b it would be unchanged

4 Advantage – the level of the end product does not fluctuate with changes in the level of substrate.

Explanation – Non-competitive inhibition occurs at a site on the enzyme other than the active site. Hence it is not affected by the substrate concentration. Therefore, in non-competitive inhibition, changes in the level of substrate do not affect the inhibition of the enzyme, nor the normal level of the end product.

Competitive inhibition involves competition for active sites. In this case the end product needs to compete with the substrate for the active sites of enzyme A. A change in the level of substrate would therefore affect how many end product molecules combine with the active sites. As a result the degree of inhibition would fluctuate and so would the level of the end product.

2.1

1 pentose(sugar), phosphate group, organic base

2 The bases are linked by hydrogen bonds. The molecular structures could be such that hydrogen bonds do not form between adenine and cytosine and between guanine and thymine.

3 ACCTCTGA

4 30.1%. If 19.9% is guanine then, as guanine always pairs with cytosine, it also makes up 19.9% of the bases in DNA, so together they make up 39.8%. This means the remaining 60.2% of DNA must be adenine and thymine and, as these also pair, each must make up half of this, i.e. 30.1%.

Unravelling the role of DNA

1 Alternative theories can be explored and investigated. As a result, new facts may emerge and so a new theory is put forward or the existing one is modified. In this way, scientific progress can be made.

2 A suggested explanation of something based on some logical scientific reasoning or idea.

3 The harmful bacteria in the sample could be tested to ensure they were dead, e.g. by seeing if they multiply when grown in ideal conditions. Dead bacteria cannot multiply.

4 The probability of the mutation happening once is very small. The probability of the same mutation occurring each time the experiment is repeated is so minute that it can be discounted.

5 Society will probably be affected by new discoveries and so is entitled to say how they can or cannot be used.

A prime location

1 It means 'in life'. In other words the synthesis of DNA by a living organism rather than in a laboratory.

2 Enzymes are very specific. They have active sites that are of a specific shape that fits their substrate. The shape of the 3' end of the molecule with its hydroxyl group fits the active site of DNA polymerase whereas the shape of the 5' end does not.

2.2

1 TACGATGC

2 because half the original DNA is built into the new DNA strand

3 The linking together of the new nucleotides could not take place. While the nucleotides would match up to their complementary nucleotides on the original DNA strand, they would not join together to form a new strand.

Evidence for semi-conservative replication

1 the organic bases (adenine, guanine, cytosine and thymine)

2 Each DNA molecule is made up of one strand containing ^{15}N (the original strand) and one strand containing ^{14}N (the new strand). In other words, replication is semi-conservative.

3

4

5 75%

2.3

1 ATP releases its energy very rapidly. This energy is released in a single step and is transferred directly to the reaction requiring it. It is too unstable to be a long term store.

2 ATP provides a phosphate that can attach to another molecule, making it more reactive and so lowering its activation energy. As enzymes work by lowering activation energy they have less 'work' to do and so function more readily.

3 Any 3 from: building up macromolecules (or named example of macromolecule) / active transport / secretions (formation of lysosomes) / activation of molecules.

2.4

a dipolar

b electrons

c hydrogen bonds

d surface tension

e hydrolysis

f photosynthesis

3.1

1 Magnification is how many times bigger the image is compared to the real object. Resolution is the minimum distance apart that two objects can be in order for them to appear as separate items.

2 200 times

3 10 mm

4 500 nm ($0.5 \mu m$)

5 Keep the plant cells in a cold, buffered solution with the same water potential as the cells. Break up the cells using a mortar and pestle / homogeniser. Filter the homogenate to remove cell debris. Centrifuge the homogenate at 1000 times gravity and remove the supernatant liquid (leaving nuclei behind in the sediment). Then centrifuge the supernatant liquid at 2000–3000 times gravity. The sediment produced will be rich in chloroplasts.

6 **a** $1.5 \mu m$; **b** 19 nm

3.2

1 The EM uses a beam of electrons that has a much smaller wavelength than light.

2 Electrons are absorbed by the molecules in air and, if present, this would prevent the electrons reaching the specimen.

3 **a** plant cell and bacteria; **b** all of them; c plant cell, bacterium and virus

4 The preparation of the specimens may not be good enough. A higher energy electron beam is required and this may destroy the specimen

5 The organelle measures 25 mm (= $25 000 \mu m$) long and represents $5 \mu m$. Magnification is therefore $25 000 \mu m \div 5 \mu m = 5000$ times.

3.3

1	light	2	eyepiece
3	stage	4	calibrated
5	2	6	8
7	20	8	3000

3.4

1 protein synthesis

2 glucose, fructose, galactose

3 **a** mitochondrion; **b** nucleus; **c** Golgi apparatus; **d** lysosome

4 **a** mitochondria, nucleus; **b** Golgi apparatus, lysosomes; **c** rough endoplasmic reticulum / ribosomes, mitochondria / smooth endoplasmic reticulum

3.5

1 a collection of similar cells aggregated together to perform a specific function.

2 An artery is made up of more than one tissue (epithelial muscle connective), whereas a blood capillary is made up of only one tissue (epithelial).

3 **a** organ **b** tissue **c** organ **d** tissue

3.6

1 A = absent; B = present; C = present; D = sometimes; E = sometimes; F = sometimes; G = present; H = present; I = sometimes; J = absent; K = present; L = present; M = absent; N = present

2 $6 \mu m = 6000$ nm; $\dfrac{6000 \, nm}{150 \, nm} = 40$ times

3.7

1 **a** interphase; **b** prophase; **c** spindle; **d** nuclear envelope; **e** nucleolus; **f** metaphase; **g** anaphase;

Importance of mitosis

1 **advantage** – as the genetic make up of the parent has enabled it to survive and reproduce, if the offspring have the same genetic make up, they are also likely to survive and reproduce.

disadvantage – genetic variety is limited – if environmental conditions change the species may not have individuals with the necessary genes to survive in the new conditions. It could fail to adapt and become extinct.

Recognising the stages of mitosis

1 A = telophase – chromosomes in two sets, one at each pole; B = prophase – chromosomes visible but randomly arranged; C = interphase – no chromosomes visible; D = metaphase – chromosomes lined up on equator; E = anaphase – chromatids in two sets, each being drawn towards pole

2 24 minutes. Number of cells in metaphase ÷ total number of cells observed × time for one cycle (in minutes), i.e. $20 \div 1000 \times 1200 = 24$

3 11% chromosomes visible in prophase, metaphase, anaphase and telophase $(73 + 20 + 9 + 8) \div 1000 \times 100$

3.8

1 interphase, nuclear division and cell division

2 a 12 hours and 24 hours;
   b 6–9 hours and 18–21 hours

Cancer and its treatment

1 0.2 million / 200 000

2 50%

3 8.33 times $(0.5 \div 0.06)$

4 More cancer cells are killed because they divide more rapidly than healthy cells and so are more susceptible to the drug.

5 a If the frequency was increased the healthy body cells would not have time to increase their numbers to near normal again between treatments. Their numbers would decline rapidly after a few treatments and possibly kill the patient.

   b The increased dose would kill even more healthy cells each time and again their numbers would decline rapidly after a few treatments and possibly kill the patient.

4.1

1 to control the movement of substances in and out of the cell

2 hydrophobic tail

3 a phospholipid;   **b** protein (carrier or channel)

4 Any 2 from: lipid-soluble / small in size / have no electrical charge (or if it does, the charge should be opposite to that on the protein channels).

4.2

1 Any three from: concentration gradient / area over which diffusion takes place / thickness of exchange surface / temperature.

2 Facilitated diffusion only occurs at channels on the membrane where there are special protein carrier molecules.

3 There is no ATP from respiration used in the process. The only energy used is the in-built (kinetic) energy of the molecules themselves.

4 Only lipid-soluble substances diffuse across the phospholipid bilayer easily. Water-soluble substances like glucose diffuse only very slowly.

5 It could increase its surface area with microvilli and it could have more proteins with pores that span the phospholipid bilayer. (Note: the thickness of

the cell-surface membranes does not vary to any degree).

6 a increases two times / doubles;

   b no change;

   c decreases four times / it is one quarter;

   d increase two times / doubles (The CO_2 concentration is irrelevant).

4.3

1 a membrane that is permeable to water molecules (and a few other small molecules) but not to larger molecules

2 zero

3 C, D, A, B

Osmosis and plant cells

1 Both cells have a lower water potential than pure water and so water enters them by osmosis. The animal cell is surrounded only by a thin cell-surface membrane and so it swells until it bursts. The plant cell is surrounded by a rigid cellulose cell wall. Assuming the cell is turgid, water cannot enter as the cellulose cell wall prevents the cell expanding and hence it bursting.

2 A = turgid, B = incipient plasmolysis, C = plasmolysed, D = turgid

3 solutions A, B and D (all except C)

4.4

1 Similarity – both use carrier proteins in the plasma membrane. Difference – active transport requires energy (ATP) / occurs against a concentration gradient.

2 Active transport requires energy in the form of ATP. Mitochondria supply ATP in cells and therefore they are numerous in cells carrying out active transport.

3 Diffusion, at best, can only reabsorb 50% of the glucose lost from the blood. The other 50% will be lost from the body. Active transport can absorb all the glucose, leaving none to be lost from the body.

4.5

1 by increasing the concentration gradient either side of it / by increasing the surface area / by increasing the density of protein channels (carrier proteins)

2 because glucose molecules and sodium ions move into the cells coupled together

3 a active;   **b** passive;   **c** passive

Oral rehydration therapy

1 Glucose stimulates the uptake of sodium ions from the intestine and provides energy as it is a respiratory substrate.

2 The sodium ions replace those lost from the body and encourage the use of the sodium-glucose transporter proteins to absorb more sodium ions.

3 Boiling the water will kill any diarrhoeal pathogens that would otherwise make the patient's condition worse.

4 Potassium in the banana replaces the potassium ions that have been lost. It also stimulates the appetite and so aids recovery.

5 Banana improves the taste and so makes it easier for children to drink the mixture.

6 Too much glucose might lower the water potential within the intestine to a level below that within the epithelial cells. Water will then pass out of the cells by osmosis, increasing dehydration.

7 Starch is a large insoluble molecule that has no osmotic effects.

8 Partially digest the starch with amylase. The smaller and more soluble molecules that result produce a less viscous drink.

9 Each species has different physical and chemical features and therefore may respond differently to the same drug. What is safe for some other animal may be harmful to a human.

10 It acts like a control experiment. Changes in the patients taking the real drug can be compared with patients taking the placebo to see whether they are due to the drug or to some other factor.

11 There is no risk of any deliberate or unwitting bias by the patients. Those knowing they are on the real drug might wrongly attribute changes in their symptoms to the drug.

5.1

1 A specific mechanism distinguishes between different pathogens but responds more slowly than a non-specific mechanism. A non-specific mechanism treats all pathogens in the same way but responds more rapidly than a non-specific mechanism.

2 The lymphocytes that will finally control the pathogen need to build up their numbers and this takes time.

3 The body responds immediately by 'recognising' the pathogen (and by phagocytosis); the delay is in building up numbers of lymphocytes and therefore controlling the pathogen.

5.2

1 a phagocytosis   b phagosome   c lysozyme
   d lysosome

2 The protective covering of the eye, and especially the tear ducts, are potential entry points for pathogens. The eyes are vulnerable to infection because the coverings are thin to allow light through. Lysozyme will break down the cell walls of any bacterial pathogens and so destroy them before they can cause harm.

5.3

1 An organism or substance, usually a protein, that is recognised as foreign by the immune system and therefore stimulates an immune response.

2 Any 2 from: both are types of white blood cell / have a role in immunity / are produced from stem cells

3 T cells mature in the thymus gland while B cells mature in the bone marrow; T cells are involved in cell-mediated immunity while B cells are involved in humoral immunity.

Bird flu

1 H5N1 infects the lungs, leading to a massive production of T cells. Accumulation of these cells may block the airways / fill the alveoli and cause suffocation.

2 Birds carry H5N1 virus. They can fly vast distances across the world in a very short space of time.

5.4

1 In the primary response, the antigens of the pathogen have to be ingested, processed and presented by B cells. Helper T cells need to link with the B cells that then clone, some of the cells developing into the plasma cells that produce antibodies. These processes occur consecutively and therefore take time. In the secondary response, memory cells are already present and the only processes are cloning and development into plasma cells that produce antibodies. Fewer processes means a quicker response.

2 Examples of differences include:

Cell-mediated immunity	Humoral immunity
Involves T cells	Involves mostly B cells
No antibodies	Antibodies produced
First stage of immune response	Second stage of immune response after cell-mediated stage
Effective through cells	Effective through body fluids

3 rough endoplasmic reticulum – to make and transport the proteins of the antibodies; Golgi apparatus – to sort, process and compile the proteins; mitochondria – to release the energy needed for such massive antibody production

5.5

1 There must be a massive variety of antibodies as each responds to a different antigen, of which there are millions. Only proteins have the diversity of molecular structure to produce millions of different types.

2 An antigen is a molecule that triggers an immune response by lymphocytes while an antibody is the molecule that has a complementary shape to the antigen and is produced in response to it.

3 Any accurate response that includes an argument in favour (e.g. removes the risk of healthy volunteers being harmed / terminally ill patients have most to gain and less to lose) and an argument against (e.g. response of terminally ill might be different from those in the early stages of the disease and results therefore could be unreliable / sample size likely to be smaller / not typical).

Producing monoclonal antibodies

1 Detergents affect the lipid component of membranes causing them to develop 'holes'. When the detergent is washed out, the membranes reform, sometimes in combination with those of other cells that are adjacent.

2 to ensure the B cells and tumour cells repeatedly come into physical contact – essential if they are to fuse

3 because B cells are short-lived and do not divide outside of the body. Tumour cells are long-lived and divide outside the body. Using both of them leads to long-lived B cells that can be grown outside the body.

4 B cells with B cells, and tumour cells with tumour cells

5 because monoclonal antibodies from mouse tissue will be recognised as foreign (non-self) and will be destroyed by human antibodies if not 'humanised'

6 The introduction of antibodies into humans could cause a reaction / disease that could be dangerous. The antigen could stimulate an over-response of the immune system.

5.6

1 Active immunity – individuals are stimulated to produce their own antibodies. Immunity is normally long-lasting.

Passive immunity – antibodies are introduced from outside rather than being produced by the individual. Immunity is normally only short-lived.

2 The influenza virus displays antigen variability. Its antigens change frequently and so antibodies no longer recognize the virus. New vaccines are required to stimulate the antibodies that complement the new antigens.

MMR vaccine

1 the MMR vaccine is given at 12–15 months – the same time as autism symptoms appear

2 It might: present the findings in an incomplete / biased fashion, ignore unfavourable findings, fund only further research that seems likely to produce the evidence that its seeks rather than investigating all possible outcomes, withdraw funding for research that seems likely to produce unfavourable findings.

5.7

1 It possesses RNA and the enzyme reverse transcriptase which can make DNA from RNA – a reaction that is the reverse of that carried out by transcriptase.

2 HIV is a virus – the human immunodeficiency virus – while AIDS (acquired immune deficiency syndrome) describes the condition caused by infection with HIV.

3 People with impaired immune systems, such as those with AIDS, are far less able to protect themselves from TB infections and so are more likely to contract and spread TB to others. Widespread use of condoms helps prevent HIV infection and so can reduce the number of people with impaired immune systems who are consequently more likely to contract TB.

6.1

1 respiratory gases, nutrients, excretory products and heat

2 0.6

3 Any 3 from: surface area / thickness of cell-surface membrane / permeability of cell-surface membrane to the particular substance / concentration gradient of substance between inside and outside of cell / temperature

Significance of surface area to volume ratio in organisms

1 They are very small and so have a very large surface area to volume ratio.

2 The blue whale has a very small surface area to volume ratio and so loses less heat to the water than it would if it were small.

Calculating a surface area to volume ratio

In making the calculation it is important to note that the cylinder sits on the rectangular box. This means that one end does not form part of the external surface. At the same time the equivalent area of the rectangular box also does not form part of the external surface. The area of these two discs is equal to $2 \times \pi r^2$ and must be subtracted from our calculation.

Since the surface area of the cylinder is calculated as $2\pi rh + 2\pi r^2$, we can therefore ignore the $2\pi r^2$ because this is the same as the area that must be subtracted from our calculation. The surface area of the cylinder is therefore taken to be $2\pi rh$ or $2 \times 3.14 \times 2 \times 8 = 100.48\,cm^2$. The surface area of the rectangular box is $2 \times (6 \times 5) + 2 \times (5 \times 12) + 2 \times (6 \times 12) = 324\,cm^2$. The total surface area = $100.48 + 324 = \mathbf{424.48\,cm^2}$.

The volume of the cylinder is calculated using the surface area of the base (πr^2) multiplied by its height (h). This equals $3.14 \times (2 \times 2) \times 8 = 100.48\,cm^3$. The volume of the rectangular box = $12 \times 6 \times 5 = 360\,cm^3$. The total volume is therefore $360 + 100.48 = \mathbf{460.48\,cm^3}$.

The surface area to volume ratio is $424.48 \div 460.48 = \mathbf{0.92}$.

6.2

1 diffusion over the body surface

2 Gas exchange requires a thin permeable surface with a large area. Conserving water requires thick, waterproof surfaces with a small area.

3 because it relies on diffusion to bring oxygen to the respiring tissues. If insects were large it would take too long for oxygen to reach the tissues rapidly enough to supply their needs.

Spiracle movements

1 It falls steadily and then remains at the same level.

2 Cells use up oxygen during respiration and so it diffuses out of the tracheae and into these cells. With the spiracles closed, no oxygen can diffuse in from the outside to replace it. Ultimately, all the oxygen is used up and so the level ceases to fall.

3 the increasing level of carbon dioxide

4 It helps conserve water because the spiracles are not open continuously and therefore water does not diffuse out continuously.

5 It contained more oxygen.

6.3

1 the movement of water and blood in opposite directions across gill lamellae

2 because a steady diffusion gradient is maintained over the whole length of the gill lamellae. Therefore more oxygen diffuses from the water into the blood.

3 Mackerel have more gill lamellae / gill filaments / larger surface area compared to plaice.

4 Less energy is required because the flow does not have to be reversed (important as water is dense and difficult to move).

6.4

1 Any 2 from: no living cell is far from the external air / diffusion takes place in the gas phase / need to avoid excessive water loss / diffuse air through pores in their outer covering (can control the opening and closing of these pores).

2 Any 2 from: insects may create mass air flow – plants never do / insects have a smaller surface area to volume ratio than plants / insects have special structures (tracheae) along which gases can diffuse – plants do not / insects do not interchange gases between respiration and photosynthesis – plants do.

3 Helps to control water loss by evaporation / transpiration.

Exchange of carbon dioxide

1 respiration

2 photosynthesis

3 At this light intensity the volume of carbon dioxide taken in during photosynthesis is exactly the same as the volume of carbon dioxide given out during respiration.

4 increase is $160 - 115 = 45\,cm^3\,h^{-1}$. Percentage increase $= \dfrac{45}{115} \times 100 = 39.13$.

5 With stomata closed, there is little, if any, gas exchange with the environment. While there will still be some interchange of gases produced by respiration and photosynthesis, neither process can continue indefinitely by relying exclusively on gases produced by the other. Some gases must be obtained from the environment. In the absence of this supply, both photosynthesis and respiration will ultimately cease and the plant will die.

6 The rate of respiration (in the dark)

6.5

1 Efficient gas exchange requires a thin, permeable surface with a large area. On land these features can lead to a considerable loss of water by evaporation.

2 waterproof covering to the body / ability to close the openings of the gas-exchange system (stomata and spiracles)

3 Plants photosynthesise and therefore need a large surface area to capture light.

4 a Water evaporating from the leaf is trapped. The region within the rolled up leaf becomes saturated with water vapour. There is no water potential gradient between the inside and outside of the leaf and so water loss is considerably reduced.

b Almost all stomata are on the lower epidermis. This would be exposed to air currents that would reduce the water potential immediately outside the leaf. The water potential gradient would be increased and a lot of water vapour would be lost.

Not only desert plants have problems obtaining water

1 The rain rapidly drains through the sand out of reach of the roots. Sand dunes are usually in windy situations, which reduces water potential and so increases the water potential gradient, leading to increased water loss.

2 The soil solution is very salty, i.e. it has a very low water potential, making it difficult for root hairs to draw water in by osmosis.

3 because in winter the water in the soil is frozen and therefore cannot be absorbed by osmosis

4 Being enzyme-controlled, photosynthesis is influenced by temperature. In cold climates enzymes work slowly and this limits the rate of photosynthesis. Therefore there is a reduced need for light as photosynthesis is taking place only slowly. In warm climates, photosynthesis occurs rapidly and therefore a large leaf area is needed to capture sufficient light.

6.6

1 Any 2 from: humans are large / have a large volume of cells; humans have a high metabolic rate / high body temperature.

2 alveoli, bronchioles, bronchus, trachea, nose

3 The cells produce mucus that traps particles of dirt and bacteria in the air breathed in. The cilia on these cells move this debris up the trachea and into the stomach. The dirt / bacteria could damage / cause infection in the alveoli.

6.7

1 17.14 breaths min^{-1}. Measure the time interval between any two corresponding points on either graph that are at the same phase of the breathing cycle (e.g. two corresponding peaks on the volume graph or two corresponding troughs on the pressure graph). The interval is always 3.5 s. This is the time for one breath. The number of breaths in a minute (60 s) is therefore 60 s ÷ 3.5 s = 17.14.

2 It is essential to first convert all figures to the same units. For example 3000 cm^3 is equal to 3.0 dm^3. From the graph you can calculate that the exhaled volume is 0.48 dm^3 less than the maximum inhaled volume. The exhaled volume is therefore 3.0 − 0.48 = 2.52 dm^3. If working in cm^3, the answer is 2520 cm^3.

3 The muscles of the diaphragm contract, causing it to move downwards. The external intercostals muscles contract, moving the rib cage upwards and outwards. Both actions increase the volume of the lungs. Consequently the pressure in the alveoli of the lungs is reduced.

Pulmonary ventilation

1 17 min^{-1}

6.8

1 **a** The rate of diffusion is more rapid the shorter the distance across which the gases diffuse.

b There is a very large surface area in 600 million alveoli (2 lungs) and this makes diffusion more rapid.

c Diffusion is more rapid the greater the concentration gradient. Pumping of blood through capillaries removes oxygen as it diffuses from the alveoli into the blood. The supply of new carbon dioxide as it diffuses out of the blood into the alveoli helps to maintain a concentration gradient that would otherwise disappear as the concentrations equalised.

d Red blood cells are flattened against the walls of the capillaries to enable them to pass through. This slows them down, increasing the time for gas exchange and reducing the diffusion pathway, thereby increasing the rate of diffusion.

2 four times greater

Correlations and causal relationships

1 Correlation between the incidence of lung cancer in men and the number of cigarettes smoked per day.

2 There is no experimental evidence in the data provided to show that smoking causes cancer. Hence there is no causal link between the two variables.

Risk factors for lung disease

1 Any 4 from: smoking / air pollution / genetic make-up / infections / occupation.

2 Allow figures in the range 50–60%.

3 two times. Around 80% of non-smokers live to age 70 compared to around 40% of people who smoke more than 25 cigarettes a day.

4 In general terms, she will live longer. More specifically she has a 50% chance of living to be 65 years if she carries on smoking but a 50% chance of living to 80 years if she gives up. Her life expectancy could increase by 15 years.

5 The aim of all the following measures is to reduce the number of cigarettes or other tobacco products

smoked. The less tobacco smoked, the lower the incidence of lung disease. Measures include:

Banning smoking in public places/at work/ in shops/restaurants – reduces opportunities for smokers to smoke and reduces lung disease as a result of secondary smoking.

Ban on tobacco advertising – less encouragement for smokers to buy tobacco, and for people, especially the young, to start smoking.

Minimum age to buy tobacco products – to help prevent young people starting to smoke and becoming addicted at an early age,

Plain packaging – to reduce the appeal of tobacco.

Health warnings on packaging – to educate/ persuade/shock people of the dangers so they reduce smoking or stop altogether.

6 Much air space within the lungs is occupied by fibrous tissue. This means that less air, and hence oxygen, is being taken into the lungs at each breath. In addition, the thickened epithelium of the alveoli means that the diffusion pathway is increased and so the diffusion of oxygen into the blood is extremely slow. The loss of elasticity makes ventilating the lungs very difficult. this makes it hard to maintain a diffusion gradient across the exchange surface and the patient becomes breathless in an attempt to compensate by breathing faster.

7 FEV will be lower/less because the expulsion of air when breathing out is due to the lungs springing back in the same way as a deflating balloon. To achieve this the lungs must be elastic. Fibrosis reduces elasticity and makes it difficult to breath out.

Smoking and lung cancer
1 a 1940–1950 b 1970–1980
2 The lines for cigarettes smoked and deaths from lung cancer are a similar shape for both sexes (although separated in time).
3 Any 3 from: an increase in smoking / more air pollution / an overall increase in the population of the UK / any other reasonable answer.
4 Lung cancer develops slowly and so the patient dies many years after it has been first contracted.

6.9
1 the breakdown of molecules by the addition of water to the bonds that hold these molecules together
2 salivary glands; pancreas
3 Villi and microvilli increase surface area to speed up the absorption of soluble molecules. As the food in the stomach has not yet been hydrolysed into

soluble molecules they cannot be absorbed and so villi and microvilli are unnecessary.
4 α-glucose
5 maltase, sucrase, lactase

Lactose intolerance
1 a respiration
  b Carbon dioxide is formed in aerobic respiration, whereas conditions in the colon are anaerobic.
2 Modern storage and distribution methods mean that milk and milk products are readily available. Without these our ancestors rarely consumed milk as adults.
3 Low water potential in the colon causes water to move from epithelial cells into the lumen of the colon creating watery stools.

6.10
1 Endoplasmic reticulum to re-synthesise triglycerides from monoglycerides and fatty acids.

Golgi apparatus to form chylomicrons from triglycerides, cholesterol and lipoproteins.

Mitochondria to provide ATP required for the co-transport of glucose and amino acid molecules.
2 Sodium ions
3 An increase in the number/density of protein channels and carrier proteins.

Absorption of fatty acids
1 When glycerol or phosphate is added on its own, there is no increase in the absorption of fatty acids compared to when neither is present.

When glycerol and phosphate are both added together the relative amount of absorption doubles compared to when neither is present.

When glycerol and phosphate are provided as a compound the relative amount of absorption increases even further (to almost three times as much) compared to when neither is present.
2 There is no absorption when the inhibitor of phosphorylation (iodoacetate) is present.

7.1
1 2 pairs of polypeptides (α and β) link to form a spherical molecule. Each polypeptide has a haem group that contains a ferrous (Fe^{2+}) ion
2 Different base sequences in DNA – different amino acid sequences – different tertiary / quaternary structure and shape – different affinities for oxygen
3 If all oxygen molecules were released, there would be none in reserve to supply tissues when they were more active.

4 Carbon monoxide will gradually occupy all the sites on haemoglobin instead of oxygen. No oxygen will be carried to tissues such as the brain. These will cease to respire and to function, making the person lose consciousness.

7.2

1 a 5.5 kPa; **b** 90%; **c** 72% (97%–25%)

2 a the curve is shifted to the right;

b haemoglobin has become less saturated

3 Exercising muscles release heat, shifting the curve to the right and causing the haemoglobin to release more oxygen to fuel the muscular activity.

4 At this partial pressure of oxygen, lugworm haemoglobin is 90% saturated, more than enough to supply sufficient oxygen to the tissues of a relatively inactive organism. Human haemoglobin, by contrast, is only 10% saturated – insufficient to supply enough oxygen to keep tissues alive.

5 Respiration produces carbon dioxide. This builds up in the burrow when the tide is out. If lugworm haemoglobin exhibited the Bohr effect, it would not be able to absorb oxygen when it was present in only very low concentrations in the burrow.

6 The higher part of the beach is uncovered by the tide for a much longer period of time than the lower part. During this longer period all the oxygen in the burrow would be used up and the lugworm might die before the next tide brings in a new supply of oxygen.

Activity counts

1 shifted to the right because this means that oxygen is more readily released to the tissues and so the haemoglobin supplies more oxygen to enable the muscles to respire rapidly.

2 sigmoid-shaped curves, with plaice to the left of mackerel

Size matters

1 Unloading pressure of human haemoglobin is 5 kPa and of mouse haemoglobin is 9 kPa. Difference = 9 − 5 = 4 kPa.

2 a it unloads more easily

b Oxygen is more easily released from haemoglobin to the tissues. This helps the tissues to respire more and so produce more heat, which helps to maintain the body temperature of the mouse.

c Even at an oxygen partial pressure of 21 kPa, mouse haemoglobin is still loaded to the maximum with oxygen.

3 Sigmoid-shaped curves, from left to right – elephant, human, shrew – because surface area to volume ratio increases in this order.

4 The temperatures in Antarctic waters are very low. This means respiration rates of cold-blooded groups such as fish are also very low. As respiration needs oxygen, their oxygen requirements are also very low. Without haemoglobin, ice fish must rely on water alone to transport their oxygen. The amount of oxygen dissolved in water, while very little, is still adequate to supply their needs.

7.3

1 a pulmonary artery; **b** aorta; **c** renal vein; **d** pulmonary vein; **e** aorta

2 low surface area to volume ratio; a high metabolic rate

3 It increases blood pressure and therefore the rate of blood flow to the tissues.

7.4

1 coronary artery

2 a deoxygenated; **b** deoxygenated; **c** oxygenated

3 pulmonary vein   left atrium   left ventricle
aorta   vena cava   right atrium
right ventricle   pulmonary artery

4 The mixing of oxygenated and deoxygenated blood would result in only partially oxygenated blood reaching the tissues and lungs. This would mean the supply of oxygen to the tissues would be inadequate and there would be a reduced diffusion gradient in the lungs, limiting the rate of oxygen uptake.

Risk factors associated with coronary heart disease

1 Reducing blood pressure – for any given cholesterol level, this will reduce the risk more than giving up smoking, e.g. at $8\,mmol\,dm^{-3}$ the risk falls from 23% to 15% by giving up smoking but falls to 12% by lowering blood pressure.

2 At $5\,mmol\,dm^{-3}$, the risk is 5%. At $8\,mmol\,dm^{-3}$, the risk is 15%. The risk is therefore 15 ÷ 5 = 3 times greater.

3 The man who increases his blood cholesterol level is at greater risk. The risk of the man who starts to smoke increases from 2.5% to 3.5% = +1%. The risk of the man who increases his cholesterol level increases from 2.5% to 5.5% = +3%.

7.5

1 left ventricle

2 a true; **b** true; **c** false; **d** false (it is 0.07)

4 The xylem and phloem could have been separated over the 225mm portion of the control branch and then rejoined but without the wax paper. This is an improvement as it eliminates the physical disruption caused by separating xylem and phloem as the explanation for there being no lateral movement of ^{42}K.

8.1

1 a base sequence of DNA that codes for the amino acid sequence of a polypeptide or functional RNA

2 18

3 A different base might code for a different amino acid. The sequence of amino acids in the polypeptide produced will be different. This change to the primary structure of the protein might result in a different shaped tertiary structure. The enzyme shape will be different and may not fit the substrate. The enzyme–substrate complex cannot be formed and so the enzyme is non-functional.

4 a 5

  b the first and last (5th) / the two coded for by the bases TAC

  c because some amino acids have up to six different codes, while others have just one triplet

Interpreting the genetic code

1 trp – UGG and met – AUG

2 a leu

  b lys

  c asp

3 a try-ala-ile-pro-ser

  b arg-phe-lys-gly-leu

8.2

1 In prokaryotic cells the DNA is smaller, circular and is not associated with proteins (i.e. does not have chromosomes). In a eukaryotic cell it is larger, linear and associated with proteins / histones to form chromosomes.

2 It fixes the DNA into position.

3 it is looped and coiled a number of times

4 a 50 mm (46 chromosomes in every cell);

  b 2.3 m (all diploid cells have same quantity of DNA)

8.3

1 mRNA is larger, has a greater variety of types and is shaped as a long single helix. tRNA is smaller, has fewer types and is clover-leaf in shape.

2 Any 3 from: RNA is smaller than DNA / RNA is usually a single strand and DNA a double helix / the sugar in RNA is ribose while the sugar in DNA is deoxyribose / in RNA the base uracil replaces the base thymine found in DNA

3 A codon is the triplet of bases on messenger RNA that codes for an amino acid. An anticodon is the triplet of bases on a transfer RNA molecule that is complementary to a codon.

Comparison of DNA, mRNA and tRNA

1 a The amount of DNA in a gamete is half that in a body cell.

  b It allows gametes to fuse during sexual reproduction without doubling the total amount of DNA at each generation. In so doing it increases genetic variety by allowing the genetic information of two parents to be combined in the offspring.

2 a DNA needs to be stable to enable it to be passed from generation to generation unchanged and thereby allow offspring to be very similar to their parents. Any change to the DNA is a mutation and is normally harmful.

  b mRNA is produced to help manufacture a protein, e.g. an enzyme. It would be wasteful to produce the protein continuously when it is only needed periodically. mRNA therefore breaks down once it has been used and is produced again only when the protein is next required.

8.4

1 The enzyme RNA polymerase moves along the template DNA strand, causing the bases on this strand to join with the individual complementary nucleotides from the pool that is present in the nucleus. The RNA polymerase adds the nucleotides one at a time, to build a strand of pre-RNA until it reaches a particular sequence of bases on the DNA that it recognises as a 'stop' code.

2 DNA helicase – This acts on a specific region of the DNA molecule to break the hydrogen bonds between the bases, causing the two strands to separate and expose the nucleotide bases in that region.

3 Splicing is necessary because pre-mRNA has nucleotide sequences derived from introns in DNA. These introns are non-functional and, if left on the mRNA, would lead to the production of non-functional polypeptides or no polypeptides at all. Splicing removes these non-functional introns from pre-mRNA.

4 a UACGUUCAGGUC

b 4 amino acids (1 amino acid is coded for by 3 bases so 12 bases code for 4 amino acids)

5 Some of the base pairs in the genes are introns (non-functional DNA). These introns are spliced from pre-mRNA so the resulting mRNA has fewer nucleotides.

8.5

1 ribosome

2 a UAG on tRNA

  b TAG on DNA

3 A tRNA molecule attaches an amino acid at one end and has a sequence of 3 bases, called an anticodon, at the other end. The tRNA molecule is transferred to a ribosome on an mRNA molecule. The anticodon on tRNA pairs with the complementary codon sequence on mRNA. Further tRNA molecules, with amino acids attached, line up along the mRNA in the sequence determined by the mRNA bases. The amino acids are joined by peptide bonds. Therefore the tRNA helps to ensure the correct sequence of amino acids in the polypeptide.

4 One of the codons is a stop codon that indicates the end of polypeptide synthesis. Stop codons do not code for any amino acid so there is one less amino acid than there are codons.

Protein synthesis

1 X = ribosome;   Y = mRNA

2 amino group

3 AUG

4 Val-Thr-Arg-Asp-Ser

5 CAATGGGCT

6 The mutation changes CAG to UAG. UAG is a stop codon that signifies the end of an amino acid sequence at which point the polypeptide is complete and is 'cast off'. The polypeptide chain is therefore shorter than it should be and may not function as normal.

7 a Glutamine has two codes GAG and GAA. The reversal of GAG produces the same codon and so still translates as glutamine and hence the polypeptide that is formed is unchanged. Reversal of GAA changes the codon to AAG which translates to a different amino acid – Lys. As a result, the polypeptide has a different primary structure which may affect bonding within the molecule and so change its tertiary structure also.

  b Enzyme function depends on the substrate becoming loosely attached to an enzyme within its active site.

If the mutation has changed the amino acid from glutamine to Lys., then the primary structure of the polypeptide will be different. Hydrogen and ionic bonds between the amino acids of the polypeptide may not be formed in the same way as before and so its tertiary shape may be changed. This change may alter the shape of the active site of the enzyme of which the polypeptide is a part and so the substrate no longer fits.

Glutamine may have been one of the amino acids in the active site to which the substrate normally attaches. Its replacement by another amino acid may mean that, although the shape of the active site is unchanged, the substrate cannot attach normally.

Cracking the code

1 DNase is an enzyme that breaks down DNA. The DNA of the cell needs to be destroyed because it would produce its own mRNA and so make it be impossible to determine which of the many polypeptides produced was due to the synthetic DNA.

2 The codon UUU – because the very radioactive polypeptide (39 800 counts min^{-1}) was only produced from the mixture containing poly U. This polypeptide must be made up of phenylalanine because this is the only radioactive amino acid present. As the synthetic mRNA contains only the base sequence UUUUUUU, etc., one codon for phenylalanine must be UUU.

3 As a control experiment to show that the radioactivity was due to the labelled phenylalanine rather than some other factor, e.g. background radiation.

4 It may not be possible to say whether the mRNA sequence starts with U and therefore reads UGU GUG UGU, etc. or starts with G and reads GUG UGU GUG. Equally it may not be possible to say whether the polypeptide sequence begins with cysteine or valine. It may therefore be impossible to relate a particular codon to a particular amino acid.

5 As most amino acids have more than one codon, the code is degenerate. However, any one triplet only codes for a single amino acid and so it is not ambiguous.

6 Because they do not code for any amino acid – they are 'stop' codons that mark the end of a polypeptide.

9.1

1 A deletion because the fifth nucleotide (A) has been lost. The sequence prior to and after this is the same. (Note: The last base in the mutant version was previously the 13th in the sequence and therefore not shown in the normal version.)

2   In a deletion, all codons after the deletion are affected. Therefore most amino acids coded for by these codons will be different and the polypeptide will be significantly affected. In a substitution, only a single codon, and therefore a single amino acid, will be affected. The effect on the polypeptide is likely to be less severe.

3   The mutation may result from the substitution of one base in the mRNA with another. Although the codon affected will be different, as the genetic code is degenerate, the changed codon may still code for the same amino acid. The polypeptide will be unchanged and there will be no effect.

4   These errors may be inherited and may therefore have a permanent affect on the whole organism. Errors in transcription usually affect only specific cells, are temporary and are not inherited. They are therefore less damaging.

Hybridisation and polyploidy

1   The numbers of chromosomes in the hybrids do not allow them to form homologous pairs during prophase I of meiosis I therefore they cannot produce gametes. This would arise if the hybrid has an odd, rather than an even, number of chromosomes.

2   They are less likely to blow over in storms and are easier to harvest.

3   They cannot breed to produce fertile offspring.

9.2

1   haploid because 27 is an odd number. Diploid cells have 2 sets of chromosomes and so their total must be an even number.

2   independent segregation of homologous chromosome and recombination by crossing over.

3   roller and blood group A, non-roller and blood group A.

4   Gametes are produced by meiosis. In meiosis, homologous chromosomes pair up. With 63 chromosomes precise pairings are impossible. This prevents meiosis and hence gamete production, making them sterile.

5   1024 (the haploid number of 5 is the same as the number of homologous chromosomes)

9.3

1   a increase; b decrease; c increase

2   Different DNA – different codes for amino acids – different amino acids – different protein shape – different protein function (e.g. non-functional

enzyme) – change in a feature determined by that protein – altered appearance – greater genetic diversity

9.3 Natural Selection in action

1   Against the black background the dark form was less conspicuous than the light natural form. As a result, the light form was eaten by birds, or other predators, more frequently than the dark form. More black-coloured moths than light-coloured moths survived and successfully reproduced. Over many generations, the frequency of the 'advantageous' dark-colour allele increased at the expense of the 'less advantageous' light-colour allele.

9.4

1   Selection is the process by which organisms that are better adapted to their environment survive and breed, while those less well adapted fail to do so.

2

Directional selection	Stabilising selection
Favours / selects phenotypes at one extreme of a population	Favours / selects phenotypes around the mean of a population
Changes the characteristics of a population	Preserves the characteristics of a population
Distribution curve remains the same shape but the mean shifts to the left or right	Distribution curve becomes narrower and higher but the mean does not change

3   Directional selection – because birds to one side of the mean (heavier birds) were being selected for, while those to the other side of the mean (lighter birds) were being selected against. The population's characteristics are being changed, not preserved.

They must be cuckoo!

1   Removing cuckoo eggs means there will be more food for the magpie's own chicks. These chicks have a greater probability of being successfully raised to adulthood.

2   Alleles for this type of behaviour are obviously present in the adult birds. There is a high probability that some of the chicks will inherit these alleles. Removing cuckoo eggs increases the probability of more of these chicks surviving to breed and therefore passing on the alleles for this behaviour to subsequent generations.

3   Displaying this behaviour has previously been of no advantage to magpies and so no selection for this behaviour has taken place. Although cuckoos have now arrived, it will take many generations

for selection to operate and for allele frequencies to change.

4 Directional selection – because the population's characteristics are being changed, not preserved.

5 Magpies that do not remove cuckoo eggs will raise both cuckoo and magpie chicks. There will be less food/space for the magpie chicks compared to those raised in magpie nests and so less chance of them surviving to adulthood and breeding. The chicks of magpies that remove cuckoo eggs will be selected for in preference to the chicks of parents that do not. Over many generations the population will change to have a greater proportion of magpies that remove cuckoo eggs at the expense of those that do not.

10.1

1 They are capable of breeding to produce offspring which are themselves fertile

2 It is based on evolutionary relationships between organisms and their ancestors; it classifies species into groups using shared characteristics derived from their ancestors; it is arranged in a hierarchy in which groups are contained within larger composite groups with no overlap.

3 To ensure that mating only takes place between members of the same species as only they can produce fertile offspring.

4 The courtship display that most closely resembles that of the first species is likely to be the closest relative.

5 1 phylum, 2 class, 3 order, 4 family, 5 *Rana*, 6 species, 7 *temporaria*

6 a lizards

  b birds

  c common ancestor of lizards and snakes

  d Dinosaurs are extinct but all the other groups are still living and so they are shown extending further along the time line – as far as 'present'.

The difficulties of defining species

1 Fossil records are normally incomplete and not all features can be observed (there is no biochemical record) and so comparisons between individuals are hard to make. Fossil records can never reveal whether individuals could successfully mate.

2 Species change and evolve over time, sometimes developing into different species. There is considerable variety within a species. Fossil records are incomplete / non-existent. Current classifications only reflect current scientific knowledge and, as this changes, so does the naming and classifying of organisms.

3 No, it does not. Only fertile female mules are known, so interbreeding (a feature of any species) is impossible. The event is so rare that it can be considered abnormal and it would be wrong to draw conclusions from it. If a mule were a species, it would mean that the parents were the same species – however, donkeys and horses are sufficiently different to be recognised as separate species.

10.2

1 the number of different species and the proportion of each species within a given area / community

2

Species	Numbers in salt marsh	$n(n-1)$
Salicornia maritima	24	24(23) = 552
Halimione portulacoides	20	20(19) = 380
Festuca rubra	7	7(6) = 42
Aster tripolium	3	3(2) = 6
Limonium humile	3	3(2) = 6
Suaeda maritima	1	1(0) = 0
$\Sigma n(n-1)$ 986		

$$D = \frac{58(57)}{986} = \frac{3306}{986} = \text{Answer } 3.35$$

3 It measures both the number of species and the number of individuals. It therefore takes account of species that are only present in small numbers.

Species diversity and ecosystems

1 Greenhouse gases lead to climate change. Communities with a high species diversity index are likely to include at least one species adapted to withstand the change and therefore survive. When the index is low, the community is less likely to include a species adapted to withstand the change and is therefore at greater risk of being damaged.

2 a The community fluctuates in line with environmental change – rising and falling in the same way but a little later in time.

  b Communities with a high species diversity are more stable because they have a greater variety of species and therefore are more likely to have species that are adapted to the changed environment. Those with a low species diversity are less stable because they have fewer species and are less likely to include a species adapted to the change.

10.3

1 The few species possessing desirable qualities are selected for and bred. Other species are excluded,

as far as possible, by culling or the use of pesticides. Many individuals of a few species = low species diversity.

2 Because forests, with their many layers, have many habitats with many different species, i.e. a high species diversity. Grasslands have a single layer, fewer habitats, fewer species and lower species diversity.

3 Ponds provide a habitat for a wide range of aquatic species that are unlikely to find alternative habitats as aquatic habitats are few and far between. Ponds may be a source of food and water for terrestrial species which may also not survive without them. Hedgerow species are likely to have a larger range of alternative habitats as most of the area around will be terrestrial with other sources of food and shelter and so fewer species are likely to be lost.

Human activity and loss of species in the UK

1 500 000 km (350 000 × 100 ÷ 70)

2 Mixed woodlands comprise many species while the commercial conifer plantations that replace them are largely of a single predominant species

3 Any 1 from each: benefit – cheaper grazing / fodder for animals and hence cheaper food / more efficient food production; risk – loss of species diversity / less stable ecosystem / more fertilizers and pesticides needed

4 It provides evidence to inform and support decision making. Data show where the most change has occurred and therefore the habitats most at risk. These can be prioritised and measures taken to conserve them, e.g. by giving them special protection. Funds can be directed towards reverting land to its former use, e.g. by grants to farmers to create hay meadows / convert set-aside to woodland / re-establish hedgerows. Helps decision-makers form appropriate rules / legislation to prevent habitat destruction, e.g. ban on drainage of certain sites / rock removal. Informs decisions on planning applications for planting forests / reclaiming land.

5 Hedges provide more habitats / niches / food sources and therefore more species can survive. Species diversity is therefore increased

Hedge rows!

1 There is more land on which to grow crops / as hedges harbour pests, diseases and weeds, especially in winter, their removal reduces the chances of these affecting crops which therefore produce greater yields / hedges compete with the crop for light, water and nutrients and therefore reduce yields.

2 Hedges are a habitat for a wide range of organisms, including some that are natural predators of crop pests and therefore provide a means of biological control / pollinating insects live in hedges and these are essential for the production of certain crops like fruits / hedges provide wind-breaks and so reduce soil erosion.

10.4

1 mutations

2 species 2 and 3 because their amino acid sequences are identical

Establishing relationships

1 chimpanzee, gorilla, orang-utan, lemur, gibbon

2 the chimpanzee and the gorilla because they both show the same % precipitation (95%)

3 the gibbon because it shows only a 3% difference (85–82) in precipitation between itself and the orang-utan. All the other primates show a greater difference.

4 These data suggest that the gibbon is much more closely related to humans than the lemur. The haemoglobin study suggested the lemur was a closer relative. The chimpanzee is shown to be most closely related to humans in the haemoglobin study, while in the gene bases study it is the gorilla.

5 There are fewer differences between the bases in the gene of a human and that of an orang-utan (29) than there are between the genes of a human and a lemur (48). This suggests that the evolution of humans and lemurs diverged earlier than that of humans and orang-utans, giving more time for the amino acid differences to occur.

6 No, it does not. This study suggests that gorillas (with fewer base differences) are more closely related to humans than chimpanzees. The other studies suggest chimpanzees are more closely related to humans. The position of the orang-utan is the same in all three studies. The position of the lemur is the same as in the immunological study but different from that in the haemoglobin study.

10.5

1 sampling bias; chance variation

2 by using random sampling – effectively using a computer to generate sampling sites

3 In a given set of values, the mean is the sum of a set of values divided by the number of items, the mode is the single value that occurs most often while the median is the central or middle value when the values are arranged in ascending order.

11.1

1 carbon dioxide and water

2 glucose and oxygen

3 a grana / thylakoids

  b stroma

4 a reduced NADP, ATP and oxygen

  b sugars and other organic molecules

11.2

1 on the thylakoid membranes (of the grana in the chloroplast)

2 Water molecules are split to form electrons, protons and oxygen, as a result of light exciting electrons / raising the energy levels of electrons in chlorophyll molecules.

3 a reduction b reduction c oxidation

Chloroplasts and the light-dependent reaction

1 A = (double) membrane of chloroplast / chloroplast envelope

  C = granum D = stroma

2 C (granum)

3 starch

4 Any 2 from: the light-dependent reaction does not produce sufficient ATP for the plants' needs / photosynthesis does not take place in the dark / cells without chlorophyll cannot produce ATP in this way and ATP cannot be transported around the plant.

5 Length X–Y on Figure 4 = 24 mm (= 24 000 µm)

  Actual length X–Y = 2 µm

  Magnification – $\dfrac{24\,000}{2}$ = 12 000 times

11.3

1 It accepts / combines with a molecule of CO_2 (to produce 2 molecules of glycerate-3-phosphate).

2 It is used to reduce (donate hydrogen) glycerate-3-phosphate to triose phosphate.

3 ATP

4 Stroma of the chloroplasts.

5 The Calvin cycle requires ATP and reduced NADP in order to operate. Both are the products of the light-dependent reaction, which needs light. No light means no ATP or reduced NADP are produced and so the Calvin cycle cannot continue once any ATP or reduced NADP already produced have been used up.

Factors affecting photosynthesis

1 Volume of oxygen produced / CO_2 absorbed.

2 Light intensity – because an increase in light intensity produces an increase in photosynthesis over this region of the graph.

3 Raising the CO_2 level to 0.1% – because this increases the rate of photosynthesis more than increasing the temperature to 35 °C.

4 Because light is limiting photosynthesis and so an increase in temperature will not increase the rate of photosynthesis.

5 More CO_2 is available to combine with RuBP to form more GP, then more triose phosphate and ultimately more glucose.

Measuring photosynthesis

1 Because any air escaping from or entering the apparatus will respectively decrease or increase the volume of gas measured, which will give an unreliable result.

2 So that any changes in the rate of photosynthesis can be said to be the result of changes in light intensity and not changes in temperature.

3 To ensure there is sufficient CO_2 and so it does not limit the rate of photosynthesis.

4 To prevent other light falling on the plant as this may fluctuate and will affect the light intensity and hence the rate of photosynthesis, leading to an unreliable result.

5 To prevent photosynthesis and to allow any oxygen produced before the experiment begins, to disperse.

6 Because the volume of oxygen produced will be less than that produced by photosynthesis as some of the oxygen will be used up in cellular respiration / dissolved oxygen (and other gases) may be released from, or absorbed by the water.

Using a lollipop to work out the light-independent reaction

1 To allow the substances into which it becomes incorporated to be identified / to allow the sequence of substances produced to be identified.

2 The radioactive carbon is initially found in glycerate 3-phosphate (5 seconds) and is next found in triose phosphate (10 seconds).

3 The high temperature and / or the methanol denature the enzymes that catalyse reactions.

4 The quantity of GP begins to decrease almost immediately. The rate of decrease becomes less until, after about 4.5 minutes, the quantity of GP becomes constant, but at around a quarter of its original level. The quantity of RuBP rises almost immediately. The rate of increase is steady

at first, but then slows, peaking at 3.5 minutes. The quantity of RuBP then falls until it becomes constant at around 4.5 minutes, but at around double its original level. The quantities of GP and RuBP are the same after 2.5 minutes.

5   RuBP combines with CO_2 to form GP during the light-independent reaction / Calvin cycle of photosynthesis. GP is ultimately used to regenerate RuBP. When the CO_2 level is decreased, there is less to combine with RuBP and so less GP is formed but it is still being used up and so its level falls. There is still some CO_2 and so some GP is made, but much less than originally. With less CO_2 to combine with, the RuBP accumulates because it cannot be converted to GP. Its quantity rises to a new higher level due to the lower level of CO_2.

12.1

1	cytoplasm	6	triose phosphate
2	glucose	7	hydrogen
3	phosphate	8	NAD
4	ATP	9	pyruvate
5	phosphorylated glucose	10	ATP

12.2

1   3   2   acetylcoenzyme A

3   matrix of mitochondria

4   True – a, b, c, d, g, h, l, n
    False – e, f, i, j, k, m, o, p, q, r

Coenzymes in respiration

1   To show that the yeast suspension was responsible for any changes that occurred and the glucose did not change methylene blue nor did methylene blue change by itself.

2   a   Yeast uses glucose as a respiratory substrate producing hydrogen atoms that are taken up by methylene blue causing it to become reduced and changing from blue to colourless.

    b   As in (2a), except that the yeast uses stored carbohydrate as a respiratory substrate that has to be converted to glucose and so the production of hydrogen atoms is slower / reduced.

3   Contents of tube might have remained blue because the enzymes involved in respiration are denatured at 60 °C and so respiration, and hence the reduction of methylene blue, ceases / the enzymes involved in hydrogen transport have been denatured and so the indicator is not reduced by hydrogen.

4   Air contains oxygen, which would re-oxidise methylene blue, turning it blue.

5   This is a single experiment. The same results would need to be obtained on many occasions to increase reliability.

12.3

1   The movement of electrons along the chain is due to oxidation. The energy from the electrons combines inorganic phosphate and ADP to form ATP = phosphorylation.

2   It provides a large surface area of membrane incorporating the coenzymes (NAD / FAD) and electron carriers that transfer the electrons along the chain.

3   Oxygen is the final acceptor of the electrons and hydrogen ions (protons) in the electron transfer chain. Without it the electrons would accumulate along the chain and respiration would cease.

4   water molecule

Sequencing the chain

1   Sequence – C, A, D, B

    Explanation – Electron carriers become reduced by electrons from glycolysis and the Krebs cycle. Enzymes catalyse the transfer of these electrons to the next carrier. If an enzyme is inhibited all molecules prior to that enzyme will not be able to pass on their electrons and so will be reduced and those after it will be oxidised. The first molecule in the chain will be reduced with all inhibitors, the second with 2 out of 3 inhibitors, the third with 1 out of 3 and the last in the chain with none (i.e. it is always oxidised).

12.4

a D	**b** A, C, D	**c** A, D	**d** A, B
e A, D	**f** B, C, D	**g** A	

Investigating where certain respiratory pathways take place in cells

1   Homogenate is spun at slow speed. Heavier particles (e.g. nuclei) form a sediment. Supernatant is removed, transferred to another tube and spun at a greater speed. Next heaviest particle is removed. Process is repeated.

2   Nuclei and ribosomes – because neither CO_2 nor lactate (products of respiration) are formed in any of the samples.

3   a   mitochondria

    b   Krebs cycle produces CO_2 and results show that CO_2 is produced when mitochondria only are incubated with pyruvate.

4   (remaining) cytoplasm (Note: The complete homogenate is not a 'portion' of the homogenate.)

5 Cyanide prevents electrons passing down the transport chain. Reduced NAD therefore accumulates and blocks Krebs cycle where CO_2 is produced. Glycolysis can still occur because the reduced NAD it produces is used to make lactate. Glucose can therefore be converted to lactate, but not into CO_2, in the presence of cyanide.

6 The conversion of glucose to CO_2 involves glycolysis (occurs in cytoplasm) and Krebs cycle (occurs in mitochondria). Only the complete homogenate contains both cytoplasm and mitochondria.

7 ethanol and CO_2

8 xylem vessel – no mitochondria as mature xylem vessels are dead and the cell contents have been lost.

liver cell–many mitochondria as it is metabolically very active and requires much ATP.

epithelial cell of intestine – many mitochondria needed to provide the ATP required for the active transport of glucose, amino acids etc.

myofibril – many mitochondria to provide ATP for contraction of fibre.

9 The absence of mitochondria leaves extra space for haemoglobin and so increases the oxygen carrying capacity of red blood cells. As mitochondria carry out oxidative phosphorylation, they could use up some oxygen leaving less to be carried to the tissues by the red blood cell.

13.1

1 dragonfly nymphs
2 unicellular and filamentous algae
3 sticklebacks
4 the direction of energy flow
5 saprobionts/decomposers

13.2

1 Any 3 from: some of the organism is not eaten; some parts are not digested and so are lost as faeces; some energy is lost as excretory materials; some energy is lost as heat

2 The proportion of energy transferred at each trophic level is small (less than 20%). After four trophic levels there is insufficient energy to support a large enough breeding population.

3 $40000 \div 25 = 1600 \text{ kJ m}^{-2} \text{ year}^{-1}$

Calculating the efficiency of energy transfers

1 a $\frac{1250 \times 100}{6300} = 19.84\%$

b $\frac{50 \times 100}{42000} = 0.12 \ (0.119)\%$

Adding up the totals

1 saprobionts/decomposers
2 insect-eating birds
3 $\frac{42500 \times 100}{1.7 \times 106} = 2.5\%$
4 Any 3 from: most (90%+) solar energy is reflected by clouds, dust or absorbed by the atmosphere / not all light wavelengths are used in photosynthesis / much of the light does not fall on the chloroplast/chlorophyll molecule / factors may limit the rate of photosynthesis or photosynthesis is inefficient / respiration by producers means energy is lost (as heat).
5 $4120 - (1010 + 810) = 2300 \text{ kJ m}^{-2} \text{ year}^{-1}$

Productivity and farming practices

1 A longer dark period means more time is spent resting, less energy is expended, and more energy is converted into body mass.

2 The pesticide might kill beneficial organisms (e.g. ones that prey on organisms that are harmful to the farmed organism).

If the pesticide kills most of the pests then the population of organisms (predators) feeding on it will fall. With no predators controlling it, the pest population will increase again, possibly to a level higher than before. The crop will be even more affected by the pest, leading to lower productivity.

3 As the number of weeds increases, the productivity of wheat decreases. The reduction in productivity is initially large, between 0 and 40 weeds m^{-2}, but lessens as the number of weeds increases, from 40 to 50 m^{-2}, the curve then flattens out.

4 Soya bean because it has an increase in productivity of 50% (1000 to 1500 kg ha^{-1}) while wheat only increases by 33% (4500 to 6000 kg ha^{-1}).

5 No. Cost of herbicide per hectare = £100. Reducing weeds from 40–20 m^{-1} increases wheat productivity from 4500 to 5000 kg ha^{-1} – an increase of 500 kg or half a tonne. Wheat is sold at £150 per tonne, so increased income is £75 per hectare which is £25 per hectare less than the cost of treating with herbicide.

A mighty problem

1 Description – In both experiments the spider mite populations rise slowly during the first 15 days and then very rapidly up to around 50 days.

In experiment 1 the spider mite population remains high up to 150 days but fluctuates (between 400 and 900). (Note: The scale is the square root of the numbers and so the figures on the y-axis need to be squared to give actual numbers.)

In experiment 2 the spider mite population falls over the period 50–150 days until it reaches the starting level.

Explanation – In experiment 1 the population of the spider mite increases until some factor (e.g. food supply) limits its size. It remains fairly constant as an equilibrium is reached with the limiting factor, fluctuating slightly as the factor fluctuates.

In experiment 2 the population of the spider mite increases up to 50 days, by which time the population of the predatory mite has increased considerably. The predatory mites feed on the spider mites, causing their population to drop to a very low level by 150 days.

2   Predatory mites are effective in controlling the population of spider mites as their presence reduces the spider mite population from around 400–900 when the predatory mite is absent to around 4 when the predatory mite is present.

3   The two populations will probably remain small as they remain in balance. They will fluctuate because, as the spider mite population falls, there will be less food for the predatory mite and so, a short time later, its population is also likely to fall. The fall in the predatory mite's population means there will be less predation on the spider mite, whose population is likely to increase, followed in turn by an increase in the predatory mite's population.

13.3

1   nitrogen fixation
2   plants
3   nitrate ions
4   root hairs
5   proteins / amino acids / nucleic acids
6   saprobionts/ decomposers
7   ammonia / ammonium ions
8   nitrifying
9   nitrate
10 denitrifying
11 A absorption
    B feeding and digestion
    C excretion and decomposition
    D erosion
    E excretion

13.4

1   Crops are grown repeatedly and intensively on the same area of land. Mineral ions are taken up by the crops, which are transported and consumed away from the land. The mineral ions they contain are not returned to the same area of land and so the levels in the soil are reduced, which can limit the rate of photosynthesis. Fertilisers need to be applied to replace them if photosynthesis / productivity is to be maintained.

2   100 kg ha^{-1} – although 150 kg ha^{-1} gives a slightly better yield, this is marginal and the cost of using 50% more fertiliser makes it uneconomical.

3   Some other factor is limiting photosynthesis, e.g. light, CO_2, and only the addition of this factor will increase photosynthesis and hence productivity.

4   Natural fertilisers are organic and come from living organisms in the form of dead remains, urine or faeces (manure).

    Artificial fertilisers are inorganic and are mined from rocks and deposits.

Different forms of nitrogen fertilisers

1   manure, bone meal and urea

2   To act as a control to show that any changes in productivity were the result of the nitrogen-containing fertiliser being added.

3   Nitrogen is needed for proteins / amino acids / chlorophyll and DNA and therefore for plant growth. Nitrogen shortage may limit the production of proteins and DNA and hence growth. Its addition increases productivity.

4   Some forms of fertilisers contain more actual nitrogen than others and so different masses are added to ensure that the total nitrogen added was always the same (140 kg ha^{-1}).

5   The data do not support the view. While ammonium nitrate brings about the greatest increase in productivity, ammonium sulphate produces a smaller increase than both urea and bone meal. Therefore the investigation suggests that only some ammonium salts are better.

6   The farmer should spread the manure a few months before the main growing season for the crop.

13.5

1   Eutrophication is the process by which salts build up in bodies of water.

2   The concentration of algae near the surface becomes so dense that no light penetrates to deeper levels. No light means no photosynthesis and hence no carbohydrate for respiration and so plants at lower levels die.

3   Dead plants are used as food by saprobionts. With an increased supply of this food, the population of saprobionts increases exponentially. Being aerobic they use up the oxygen in the water leading to the death of the fish, which cannot respire without it.

Troubled waters

1   It has taken 10 days for the fertiliser that has dissolved in the rainwater to leach through the soil and into the lake.

2   In normal circumstances, a low level of nitrate (or other ions) is the limiting factor to algal growth.

The fertilizer leaching into the lake contains nitrate (and other ions) and removes this limit on growth. The algal population grows rapidly, increasing in density.

3   Description – As the density of the algae increases so the clarity of the water decreases, i.e. there is a negative correlation. For the first 20 days the algal density (30 cells cm^{-3}) and water clarity (Secchi = 9 m) remain constant.

From day 20 to day 100 the algal density increases from 30 to 120 cells cm^{-3} while the water clarity decreases from 9 to 1 m (Secchi depth). However, there is an anomaly between day 40 and day 50 when the water clarity suddenly falls from 7 to 4 m.

Explanation – As the density of algae increases, more light is absorbed / reflected by them and so less light penetrates / water clarity is reduced. Between day 40 and day 50 some factor (e.g. water turbulence stirring up sediment) other than algal density is reducing the water clarity.

4   Days 0–10: oxygen level is constant (at 10 ppm) because there is a balance between oxygen produced in photosynthesis of plants and algae, and oxygen used up in respiration of all organisms.

Days 10–25: oxygen level rises (up to around 13 ppm) due to increased photosynthesis by the larger population of algae.

Days 25–100: oxygen level decreases (more rapidly at first and then less so down to around 3 ppm) due to higher density of algae blocking out the light to lower depths and reducing the rate of photosynthesis of plants / algae at these depths. In time, light is blocked out altogether at lower depths → no photosynthesis → plants / algae die → saprobionts decompose them → their population increases → they use up much oxygen in respiration → oxygen levels fall.

14.1

1   (Negative chemo-) taxis – wastes are often removed from an organism because they are harmful. Moving away prevents the waste harming the organism and so increases its chance of survival.

2   (Positive chemo-) taxis – increases the chances of sperm cells fertilising the egg cells of other mosses and so helps to produce more moss plants / future generations. Cross-fertilisation increases genetic variability, making species better able to adapt to future environmental changes.

3   (Negative gravi-) tropism – takes the seedlings above the ground and into the light, where they can photosynthesise.

More photosynthesis means more carbohydrate and so a better chance of survival.

14.2

1   More IAA moves towards the shaded side of shoots than the light side when the light is unidirectional. In response to this uneven distribution of IAA, the cells on the shaded side elongate faster than those on the light side and the shoot bends towards the light. This ensures that the shoot and the leaves attached to it have a greater chance of being well illuminated. As light is essential for photosynthesis, the process by which organic material for respiration is manufactured, the plant has a greater chance of survival.

2   Response ensures that roots grow down into the soil, anchoring the plant firmly and bringing them closer to water (needed for photosynthesis).

3   The fact that IAA is readily absorbed, easily synthesized and is lethal to plants in low concentrations makes it useful as a herbicide. The fact that it more readily kills broad-leaved plants than narrow-leaved ones is an advantage because many agricultural crops are narrow-leaved while the weeds that compete with them are broad-leaved. As a result, application of IAA at appropriate concentrations will kill only the weeds with little, or no, harm to the crop. As IAA is not easily broken down means it will persist in the soil and continue to act as a selective weedkiller for some time. This may prevent a broad-leaved crop being grown on the land for some time after application of IAA. There is also a danger that IAA might accumulate along food chains with possible harm to animals in those chains.

Discovering the role of IAA in tropisms

1   Experiment 1

2   As mica conducts electricity it will not prevent electrical messages passing from the shoot tip but it will prevent chemical messages passing. As there is no response, the message must be chemical and must pass down the shaded side.

3   Displacement of the tip means that the chemical initially only moves down the side of the shoot that is in contact with the tip. This side grows more rapidly, causing bending away from that side.

4   It prevents chemicals / IAA, but not light, passing from one side to the other.

5   Results support the hypothesis that IAA is transported from the lighter side to the shaded side of the shoot.

Experiment 8 shows that the total IAA produced and collected is the same whether the shoot is in the light or the dark. This discounts the theory that light destroys IAA or inhibits its production.

Experiment 9 shows that the amount of IAA produced at either side of the tip is the same. The glass plate prevents any sideways transfer.

Experiment 10 shows that the IAA is transferred from the light to the shaded side of the shoot soon after it is produced because more than twice as much IAA is found on the shaded side of the shoot than on the light side.

14.3

1 brain / spinal cord
2 brain / spinal cord
3 motor
4 sensory
5 involuntary
6 stimulus
7 (temperature) receptor
8 sensory
9 intermediate
10 motor
11 effectors

14.4

1 Stretch-mediated sodium channel – a special type of sodium channel that changes its permeability to sodium when it changes shape / is stretched.

2 pressure on Pacinian corpuscle → corpuscle changes shape → stretches membrane of neurone → widens stretch mediated sodium ion channels → allows sodium ions into neurone → changes potential of (depolarises) membrane → produces generator potential

3 Only rod cells are stimulated by low-intensity (dim) light. Rod cells cannot distinguish between different wavelengths / colours of light, therefore the object is perceived only in a mixture of black and white, i.e. grey.

4 Light reaching Earth from a star is of low intensity. Looking directly at a star, light is focused on to the fovea, where there are only cone cells. Cone cells respond only to high light intensity so they are not stimulated by the low light intensity from the star and it cannot be seen. Looking to one side of the star means that light from the star is focused towards the outer regions of the retina, where there are mostly rod cells. These are stimulated by low light intensity and therefore the star is seen.

14.5

1 Autonomic nervous system – controls the involuntary activities of internal muscles and glands.

2 Sympathetic nervous system stimulates effectors and so speeds up an activity; prepares for stressful situations, e.g. the fight or flight response.

Parasympathetic nervous system inhibits effectors and slows down an activity; controls activities under resting conditions, conserving energy and replenishing the body's reserves.

3 Blood pressure remains high because the parasympathetic system is unable to transmit nerve impulses to the SA node, which decreases heart rate and so lowers blood pressure.

4 a Heart rate remains as it was before taking exercise – after exercise, blood pressure increases and CO_2 concentration of blood rises (causing blood pH to be lowered). The changes are detected by pressure and chemical receptors in the wall of the carotid arteries. As the nerve from here to the medulla oblongata is cut, no nerve impulse can be sent to the centres that control heart rate.

   b Blood CO_2 concentration increases as a result of increased respiration during exercise.

15.1

1 (nerve) impulses / action potentials
2 nucleus
3 rough endoplasmic reticulum
4 dendrites
5 Schwann cells
6 insulation
7 myelin
8 motor
9 sensory
10 intermediate

11 Hormone response is slow, widespread and long-lasting. Nervous response is rapid, localised and short-lived.

Aging in neurones

1 dendrites become longer with age.

2 dendrites are fewer, are much shorter and are less branched.

3 After 10 years (age 60) there will be
$2000 - (\frac{5}{100} \times 2000) = 1900$ neurones remaining.
After a further 10 years (age 70) there will be 1900
$- (\frac{5}{100} \times 1900) = 1805$ neurones remaining.

15.2

1 Active transport of sodium ions out of the axon by sodium–potassium pumps is faster than active transport of potassium ions into the axon. Potassium ions diffuse out of the axon but few, if any, sodium ions diffuse into the axon because the sodium 'gates' are closed. Overall, there are more positive ions outside than inside and therefore the outside is positive relative to the inside.

2 A = closed B = open C = closed D = closed E = closed F = open

Measuring action potentials

1 sodium and potassium ions

2 At resting potential (0.5 ms) there is a positive charge on the outside of the membrane and a negative charge inside, due to the high concentration of sodium ions outside the membrane. The energy of the stimulus causes the sodium voltage-gated channels in the axon membrane to open and therefore sodium ions diffuse in through the channels, along their electrochemical gradient. Being positively charged, they begin a reversal in the potential difference across the membrane. As sodium ions enter, so more sodium ion channels open, causing an even greater influx of sodium ions and an even greater reversal of potential difference: from −70 mv up to +40 mv at 2.0 ms.

3 Two action potentials take place in 10 ms.

Each action potential takes 10 ÷ 2 = 5 ms / action potentials are 5ms apart.

There are 1000 (10^3) ms in 1 second.

Therefore there are 1000 ÷ 5 = 200 action potentials in 1 second (2 × 10^2 ms^{-1})

15.3

1 a node of Ranvier

  b Because the remainder of the axon is covered by a myelin sheath that prevents ions being exchanged / prevents a potential difference being set up.

  c It moves along in a series of jumps from one node of Ranvier to the next.

  d saltatory (conduction)

  e It is faster than in an unmyelinated axon.

2 It remains the same / does not change.

15.4

1 During the refractory period the sodium voltage-gated channels are closed so no sodium ions can move inwards and no action potential is possible. This means there must be an interval between one impulse and the next.

2 All-or-nothing principle – There is a particular level of stimulus that triggers an action potential. At any level above this threshold, a stimulus will trigger an action potential that is the same regardless of the size of the stimulus (the 'all' part). Below the threshold, no action potential is triggered (the 'nothing' part).

3 Mammals have myelinated neurons and so have saltatory conduction. Mammals are endothermic and their constant, usually higher, body temperature increases the rate of diffusion of ions

across the axon membrane and hence the speed of conduction of the action potential.

Different axons different speeds

1 The greater the diameter of an axon the faster the speed of conductance. Comparing the data for the two myelinated axons shows that the 20 μm diameter axon conducts at 120 ms^{-1} while the 10 μm diameter axon conducts at only 50 ms^{-1}. Likewise, the data for the two unmyelinated axons show that the 500 μm diameter axon conducts at 25 ms^{-1} while the 1μm diameter axon conducts at 2 ms^{-1}.

2 In myelinated axons, the myelin acts as an electrical insulator. Action potentials can only form where there is no myelin (at nodes of Ranvier). The action potential therefore jumps from node to node (= saltatory conduction) which makes its conductance faster.

3 Schwann cells

4 The presence of myelin has the greater effect because a myelinated human sensory axon conducts an action potential at twice the speed of the squid giant axon, despite being only 1/50th of its diameter. (Note: Similar comparisons can be made between other types of axon, e.g. squid and human motor axons.)

5 Temperature affects the speed of conductance of action potentials. The higher the temperature, the faster the conductance. The conductance of action potentials in the squid will therefore change as the environmental temperature changes. It will react more slowly at lower temperatures.

6 Area of circle = πr^2. Radius of axon = $\frac{500}{2}$ = 250 μm. Area of axon = $\pi \times 250^2$ = 196 349 μm^2.

Expressed as mm^2 to five significant figures 0.19635 mm^2.

15.5

1 It possesses many mitochondria and large amounts of endoplasmic reticulum.

2 It has receptor molecules for neurotransmitters e.g., acetylcholine, on its membrane.

3 Neurotransmitter is released from vesicles in the presynaptic neurone into the synaptic cleft when an action potential reaches the synaptic knob. The neurotransmitter diffuses across the synapse to receptor molecules on the postsynaptic neurone to which it binds, thereby setting up a new action potential.

4 Only one end can produce neurotransmitter and so this end alone can create a new action potential in the neurone on the opposite side of the synapse. At the other end there is no neurotransmitter that can be released to pass across the synapse and so no new action potential can be set up.

5 a The relatively quiet background noise of traffic produces a low-level frequency of action potentials in the sensory neurones from the ear. The amount of neurotransmitter released into the synapse is insufficient to exceed the threshold in the postsynaptic neurone and to trigger an action potential and so the noise is 'filtered out' / ignored. Louder noises create a higher frequency and the amount of neurotransmitter released is sufficient to trigger an action potential in the postsynaptic neurone and so there is a response. This is an example of temporal summation. (Note: An explanation in terms of spatial summation is also valid: many sound receptors with a range of thresholds → more receptors respond to the louder noise → more neurotransmitter → response.)

b Reacting to low-level stimuli (background traffic noise) that present little danger can overload the (central) nervous system and so organisms may fail to respond to more important stimuli. High-level stimuli (sound of horn) need a response because they are more likely to represent a danger.

6 As the inside of the membrane is more negative than at resting potential, more sodium ions must enter in order to reach the potential difference of an action potential, i.e. it is more difficult for depolarisation to occur. Stimulation is less likely to reach the threshold level needed for a new action potential.

7 a Increase in speed $64 - 40 = 24$ ms^{-1}

Percentage increase $\dfrac{24}{40} \times 100 = 37.5\%$

b Refex arcs allow rapid responses to potentially harmful situations. Information passes across synapses relatively slowly compared to the speed it passes along an axon. The fewer synapses there are, the shorter the overall time taken to respond to a stimulus – an advantage where a rapid response is required.

15.6

1 a sodium ions   **c** ATP

b acetylcholine   **d** calcium ions

2 To recycle the choline and ethanoic acid; to prevent acetylcholine from continuously generating a new action potential in the postsynaptic neurone.

Effects of drugs on synapses

1 They will reduce pain.

2 They act like endorphins by binding to the receptors and therefore preventing action potentials being created in the neurones of the pain pathways.

3 Prozac might prevent the elimination of serotonin from the synaptic cleft (Note: Any biologically accurate answer that results in more serotonin in the synaptic cleft is acceptable.)

4 By increasing the concentration of serotonin in the synaptic cleft, its activity is increased, reducing depression, which is caused by reduced serotonin activity.

5 It will reduce muscle contractions (cause muscles to relax).

6 Valium increases the inhibitory effects of GABA so therefore there are fewer action potentials on the nerve pathways that cause muscles to contract.

7 The molecular structure of Vigabatrin is similar to GABA so it may be a competitive inhibitor (compete) for the active site of the enzyme that breaks down GABA. As less GABA is broken down by the enzyme, more of it is available to inhibit neurone activity. Or Vigabatrin might bind to GABA receptors on the neurone membrane and mimic its action, thereby inhibiting neuronal activity.

15.7

1 Muscles require much energy for contraction. Most of this energy is released during the Krebs cycle and electron transport chain in respiration. Both these take place in mitochondria.

2 A = Z-line B = H-zone C = I-band (isotropic band) D = A-band (anisotropic band).

3 The actin and myosin filaments lie side by side in a myofibril and overlap at the edges where they meet. If cut where they overlap, both filaments can be seen. If cut where they do not overlap, we see one or other filament only.

4 Slow-twitch fibres contract more slowly and provide less powerful contractions over a longer period.

Fast-twitch fibres contract more rapidly and produce powerful contractions but only for a short duration.

5 Slow-twitch fibres have myoglobin to store oxygen, much glycogen to provide a source of metabolic energy, a rich supply of blood vessels to deliver glucose and oxygen, and numerous mitochondria to produce ATP.

Fast-twitch fibres have thicker and more numerous myosin filaments, a high concentration of enzymes involved in anaerobic respiration and a store of phosphocreatine to rapidly generate ATP from ADP in anaerobic conditions.

15.8

1 Myosin is made of two proteins. The fibrous protein is long and thin in shape, which enables it to combine with others to form a long thick filament along which the actin filament can move. The globular protein forms two bulbous structures (the head) at the end of a filament (the tail). This shape allows it to exactly fit recesses in the actin molecule, to which it can become attached. Its shape also means it can be moved at an angle. This allows it to change its angle when attached to actin and so move it along, causing the muscle to contract.

2 Phosphocreatine stores the phosphate that is used to generate ATP from ADP in anaerobic conditions. A sprinter's muscles often work so strenuously that the oxygen supply cannot meet the demand. The supply of ATP from mitochondria during aerobic respiration therefore ceases. Sprinters with the most phosphocreatine have an advantage because ATP can be supplied to their muscles for longer, and so they perform better.

3 A single ATP molecule is enough to move an actin filament a distance of 40 nm.

Total distance moved by actin filament = 0.8 µm (= 800 nm).

Number of ATP molecules required = 800 ÷ 40 = 20.

4 One role of ATP in muscle contraction is to attach to the myosin heads, thereby causing them to detach from the actin filament and making the muscle relax. As no ATP is produced after death, there is none to attach to the myosin, which therefore remains attached to actin, leaving the muscle in a contracted state, i.e. rigor mortis.

16.1

1 Homeostasis is the maintenance of a constant internal environment in organisms.

2 Maintaining a constant temperature is important because enzymes function within a narrow range of temperatures.

Fluctuations from the optimum temperature mean enzymes function less efficiently. If the variation is extreme, the enzyme may be denatured and cease to function altogether. A constant temperature means that reactions occur at a predictable and constant rate.

3 Maintaining a constant blood glucose concentration is important in ensuring a constant water potential. Changes to the water potential of the blood and tissue fluids may cause cells to shrink and expand (even to bursting point), due to water leaving or entering by osmosis. In both instances the cells cannot operate normally. A constant blood glucose concentration also ensures a reliable source of glucose for respiration by cells.

Thermoregulation in ectotherms and endotherms

1 It allows accurate comparisons to be made even though the animals have different body masses. An increase in body size or body mass means there is increased heat generation.

2 a Both increase proportionally up to 25 °C. Above 25 °C, heat generation increases more rapidly (gradient / slope of line increases), whereas evaporative heat loss increases at the same rate (gradient / slope of line remains the same).

  b In a mammal, the relationship is the inverse / opposite, i.e. as evaporative heat loss increases, heat generation decreases.

3 Above 25 °C, the metabolic heat generation in reptiles becomes much more rapid. They therefore generate heat faster than they can lose it. As a result, their body temperature increases and enzymes may be denatured, leading to death. As reptiles have no physiological means of cooling, they must seek shade in order to reduce their body temperature.

4 Sweating or panting increases.

16.2

1 If the information is not fed back once an effector has corrected any deviation and returned the system to the set point, the receptor will continue to stimulate the effector and an over-correction will lead to a deviation in the opposite direction from the original one.

2 It gives a greater degree of homeostatic control.

Positive feedback

1 Positive feedback means the contractions get stronger and more frequent over time leading to the birth of the baby. Negative feedback would mean the contractions became weaker and less frequent until they stopped altogether and the baby would not be born.

Negative feedback in temperature control

1 The blood temperature would be become progressively colder (because positive feedback increases the current process i.e. blood cooling).

2 Cutting the nerves would mean that the thermoreceptors would not be able to communicate the rise in blood temperature to the heat loss centre. The centre would not be able to initiate actions that could lower blood temperature. Blood temperature would rise. If the increase were great enough, enzymes vital to keep the organism alive e.g. respiratory enzymes would be denatured, cease to function and death would result.

3   The heat loss centre might be connected to effectors, other than the skin, which could lower the blood temperature e.g. regions of the brain that control behaviour, and so the individual might be able to reduce blood temperature by resting, sheltering from the sun etc.

4   vena cava, pulmonary artery, pulmonary vein, aorta.

16.3

1 respiration	8 gluconeogenesis
2 brain	9 glycogen
3 osmotic / water potential	10 respiration
4 carbohydrate	11 islets of Langerhans
5 glycogen	12 insulin
6 muscles	13 glucagon
7 amino acids	14 adrenaline

16.4

1   Type I is caused by an inability to produce insulin.

Type II is caused by receptors on body cells losing their responsiveness to insulin.

2   Type I is controlled by the injection of insulin.

Type II is controlled by regulating the intake of carbohydrate in the diet and matching this to the amount of exercise taken.

3   Diabetes is a condition in which insulin is not produced by the pancreas. This leads to fluctuations in the blood glucose level. If the level is below normal, there may be insufficient glucose for the release of energy by cells during respiration. Muscle and brain cells in particular may therefore be less active, leading to tiredness.

4   Match your carbohydrate intake to the amount of exercise that you take. Avoid becoming overweight by not consuming excessive quantities of carbohydrate and by taking regular exercise.

Effects of diabetes on substance concentrations in the blood

1   adrenaline

2   The rise in insulin level is both greater and more rapid in group Y than in group X.

3   Glucose is removed from blood by cells using it during respiration.

4   Glucose concentration rises at first because the glucose that is drunk is absorbed into the blood (glucose line on graph rises). This rise in blood glucose causes insulin to be secreted from cells (B cells) in the pancreas (insulin line rises steeply). Insulin causes increased uptake of glucose into liver and muscle cells, activates enzymes that convert glucose into glycogen and fat, and

increases cellular respiration. The effect of all these actions is to reduce glucose concentration (glucose line falls from 2.5 hours onwards). As the glucose concentration rises after 1 hour, so the glucagon concentration falls. The reduction in glucagon concentration decreases glucose production from other sources (glycogen, amino acids, and glycerol) and so also helps to reduce blood glucose concentration. As the blood glucose concentration falls (after 2.5 hours) so the glucagon concentration increases to help maintain the blood glucose at its optimum concentration.

5   Group X has diabetes and therefore the glucose intake does not stimulate insulin production (insulin concentration shown on the graph is low). The glucose concentration in the blood therefore continues to rise (glucose line rises steeply) as there is no insulin to reduce its concentration. Blood glucose concentration remains high, falling only slightly as it is respired by cells.

6   As it is respired by cells, the glucose concentration will decrease steadily until it falls below the optimum concentration.

16.5

1 renal (Bowman's) capsule	6 epithelial cells
	7 microvilli
2 glomerulus	8 loop of Henle
3 afferent	9 distal
4 podocytes	10 collecting duct
5 proximal	

Control of blood water potential

1   As sweating involves a loss of water from the blood, its water potential will decrease (be lower or more negative).

2   a   osmotic cells (in the hypothalamus)

    b   kidney

3   Being a hormone, it is transported in the blood plasma.

4   Absorption (taking in or consumption or drinking) of water because water has been lost during sweating. As the water potential of the blood returns to normal, the lost water must have been replaced. However, the kidney only excretes less water, it does not replace it. Therefore process X must be the way in which water is replaced.

5   negative feedback

The glomerulus – a unique capillary network

1   The efferent arteriole later divides up into a second capillary bed that surrounds the loop of Henle. These then combine to form the venule.

2 By looking at the structure of its wall. Arterioles have thicker walls with more muscle tissue than venules.

16.6

1 proximal convoluted tubule

2 glomerulus; renal capsule; proximal convoluted tubule; loop of Henle; distal convoluted tubule; collecting duct

3 microvilli to provide a large surface area to reabsorb substances from the filtrate; infoldings at their bases to give a large surface area to transfer reabsorbed substances into blood capillaries; a high density of mitochondria to provide ATP for the active transport.

4 Animals in dry environments would have longer loops of Henle to give a longer counter current multiplier and so more absorption of water by the collecting duct.

16.7

1 hypothalamuss

2 **a** less–because the water drunk causes a rise in water potential of the blood

b more–because intense exercise leads to sweating and the loss of water leading to a fall in water potential of the blood

3 ADH binds to receptors on the cell-surface membrane of the cells lining the collecting duct and activate phosphorylase within the cell. The activation of phosphorylase causes vesicles containing pieces of plasma membrane that have numerous water channels/aquaporins to fuse with the cell-surface membrane. This increases the number of water channels and makes the cell-surface membrane much more permeable to water.

4. 0.05 g dm^{-1}

The significance of glucose in the urine

1 Insulin – lowers blood sugar level

Glucagon – raises blood sugar level

Adrenaline – raises blood sugar level.

2 The graph measures the concentration of substances and not their actual quantities. Because water is progressively removed from filtrate as it passes along the nephron, there is the same amount of urea but in a smaller volume of water. Therefore the **concentration** of urea increases.

3 Glucose because it is totally reabsorbed in proximal convoluted tubule.

4 Sodium ions because they enter the descending limb of loop of Henle initially and so increase in concentration. They are then actively transported out of the ascending limb and so decrease in concentration. The removal of water from the collecting duct causes their concentration to slowly rise again.

5 The reabsorption of water from the collecting ducts depends on there being a large water potential gradient between the fluid in the collecting duct and that in the blood capillaries. The presence of glucose in the fluid in the collecting duct reduces this gradient and leads to more water being lost in urine leading to dehydration.

17.1

1	genotype	**6**	locus
2	mutation	**7**	homozygous
3	phenotype	**8**	heterozygous
4	nucleotides/bases	**9**	recessive
5	polypeptides	**10**	codominant

17.2

1 Let allele for Huntington's disease = H
Let allele for normal condition = h

	father's gametes	
mother's gametes	(H)	(h)
(h)	Hh	hh
(h)	Hh	hh

half (50%) of offspring will have Huntington's disease (Hh).
half (50%) of offspring will be normal (hh).

2   **a**   Let allele for black coat = B

Let allele for red coat = b

parental phenotypes   female with black coat   male with red coat

parental genotypes

BB   bb

gametes   (B) (B)   (b) (b)

offspring:

	male gametes	
female gametes	(b)	(b)
(B)	Bb	Bb
(B)	Bb	Bb

All (100%) offspring will have black coats (Bb).

b   Let allele for black coat = B

Let allele for red coat = b

parental phenotypes   female with black coat   male with black coat

parental genotypes   Bb   Bb

gametes   (B) (b)   (B) (b)

offspring:

	male gametes	
female gametes	(B)	(b)
(B)	BB	Bb
(b)	Bb	bb

3 offspring (75%) with black coat (BB, Bb and Bb).

1 offspring (25%) with red coat (bb).

probability of offspring having red coat = 1 in 4 (25%/0.25)

17.3

1   **a**   homozygous dominant (**GG**)

    b   We cannot be absolutely certain because if the unknown genotype were heterozygous (**Gg**) the gametes produced would contain alleles of two types: either dominant (**G**) or recessive (**g**). It is a matter of chance which of these gametes fuses with those from our recessive parent – all these gametes have a recessive allele (**g**). It is just possible that, in every case, it is the gametes with the dominant allele that fuse and so all the offspring show the dominant character. Provided the sample of offspring is large enough, however, we can be reasonably sure that the unknown genotype is homozygous dominant.

2   **a**   heterozygous (**Gg**)

    b   We can be certain because 7 of the offspring display the recessive character (in our case yellow pods). These plants are homozygous recessive and must have obtained one recessive allele from each parent. Our unknown parental genotype must therefore have a recessive allele and be heterozygous (in our case **Gg**). It is theoretically possible that the plants with yellow pods were due to a mutation but this is most unlikely. The unexpectedly low number of plants with yellow pods is the result of random fusion of the gametes.

    c   50%   **d**   7.29%

17.4

1   Red eyes and normal wings are dominant because these characteristics are expressed in the F_1 generation while pink eyes and vestigial wings are not expressed in the F_1 generation and so these are recessive. Also red eyes and normal wings appear around 3 times more often in the F_2 generation than pink eyes and vestigial wings.

2   **R** for red eyes and **r** for pink eyes, **N** for normal wings, **n** for vestigial wings.

3 Parental cross

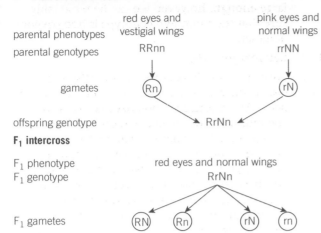

parental phenotypes | red eyes and vestigial wings | pink eyes and normal wings

parental genotypes | RRnn | rrNN

gametes | Rn | rN

offspring genotype | RrNn

F₁ intercross

F_1 phenotype | red eyes and normal wings
F_1 genotype | RrNn

F_1 gametes | RN | Rn | rN | rn

F_2 offspring

		♂ gametes			
		RN	Rn	rN	rn
♀ gametes	RN	RRNN	RRNn	RrNN	RrNn
	Rn	RRNn	RRnn	RrNn	Rrnn
	rN	RrNN	RrNn	rrNN	rrNn
	rn	RrNn	Rrnn	rrNn	rrnn

9 red eyes and normal wings (R-N-)
3 red eyes and vestigial wings (R-nn)
3 pink eyes and normal wings (rrN-)
1 pink eyes and vestigial wings (rrnn)

Better late than never

1 The F_1 ratio is 3:1 is a typical ratio for a monohybrid cross and supports Mendel's law of segregation. The F_2 ratio of 9:7 looks initially unusual but has 16 genotypes and so could be a modified 9:3:3:1 ratio that is the typical ratio for a dihybrid cross and supports Mendel's law of independent assortment.

2 The two different varieties when self-fertilised produce three green offspring for each white one. The green offspring must have allele **A** and allele **B** i.e. must be **A-B-** (where – is either dominant or recessive). The white offspring must lack either allele **A** or allele **B** or both i.e. must have the genotype **aabb**, **aaB-** or **A-bb**. To obtain white offspring the green parent must provide a recessive allele so that a double recessive can occur in the offspring. There are three possible genotypes that fulfil these criteria: **AABb**, **AaBB** and **AaBb**. The latter however, when self-fertilised produces a 9:7 ratio of green to white (see later) and not a 3:1 ratio. This leaves the remaining two. As there are two different varieties that produce this 3:1 ratio

when self-fertilised, these must be the genotypes of the parents. Answer = **AABb** and **AaBB**.

proof for variety 1 (AABb)

parental genotypes | AABb | AABb

gametes | AB | Ab | AB | Ab

F_1 offspring

		♂ gametes	
♀ gametes		AB	Ab
	AB	AABB	AABb
	Ab	AABb	AAbb

3 offspring have alleles A and B and are therefore green.
1 offspring (AAbb) lacks allele B and is therefore white.

proof for variety 2 (AaBB)

parental genotype | AaBB | AaBB

gametes | AB | aB | AB | aB

F_1 offspring

		♂ gametes	
♀ gametes		AB	aB
	AB	AABB	AaBB
	aB	AaBB	aaBB

3 offspring have alleles A and B and are therefore green.
1 offspring (aaBB) lacks allele A and is therefore white.

3

	variety 1	variety 2
parental genotype	AABb	AaBB
gamete	AB Ab	AB aB

F_1 offspring		variety 1 gametes	
variety 2 gametes		AB	Ab
	AB	AABB	AABb
	aB	AaBB	AaBb

All offspring have both allele A and allele B and so are green

4 Examination of the results of cross 2 shows that two of the four genotypes (**AABb** and **AaBB**) were self-fertilised in cross 1 and produced a 3 green to 1 white ratio and so can be discounted. Self-fertilisation of one of the two remaining genotypes, **AABB**, clearly will only produce green offspring. This leaves just the genotype **AaBb**. The results of self-fertilisation of plants with this genotype are as follows:

parental genotype

F_2 offspring

♀ Gametes	♂ Gametes			
	AB	Ab	aB	ab
AB	AABB	AABb	AaBB	AaBb
Ab	AABb	AAbb	AaBb	Aabb
aB	AaBB	AaBb	aaBB	aaBb
ab	AaBb	Aabb	aaBb	aabb

Nine boxes have a genotype with both allele A and allele B and so can produce chlorophyll and the plants are green.

The seven other boxes lack either allele A, allele B or both and so cannot produce chlorophyll and the plants are white.

5 Each gene controls the production of a polypeptide. Dominant alleles of genes A and B are required to form their respective polypeptides. These polypeptides could form part of a single enzyme or two different enzymes in the biochemical pathway that forms chlorophyll. Only plants carrying both dominant alleles will be able to synthesise chlorophyll.

17.5

1 The man is not the father.

Reasons – child has blood group AB and therefore has alleles $I^A I^B$. The mother is blood group A and therefore either $I^A I^O$ or $I^A I^A$. In either case she could have provided the I^A alleles to the child but not the I^B allele. The I^B allele must have come from the real father. The supposed father is blood group O and therefore has alleles $I^O I^O$. He cannot provide an I^B allele and so cannot be the father.

2

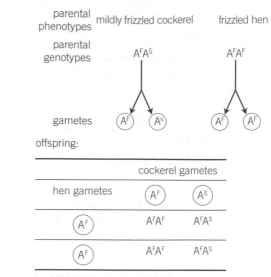

hen gametes	cockerel gametes	
	A^F	A^S
A^F	$A^F A^F$	$A^F A^S$
A^F	$A^F A^F$	$A^F A^S$

half (50%) frizzled fowl ($A^F A^F$)
half (50%) mildly frizzled fowl ($A^F A^S$)

Coat of many colours

1

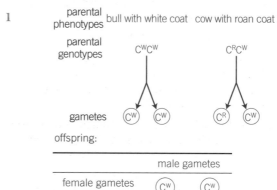

female gametes	male gametes	
	C^W	C^W
C^R	$C^R C^W$	$C^R C^W$
C^W	$C^W C^W$	$C^W W^W$

half (50%) roan coat ($C^R C^W$)
half (50%) white coat ($C^W C^W$)

2 a 100% **b** 50% **c** 50% **d** 50%

3 Kittens develop inside their mother and so are kept warm / at a uniform temperature. As the kitten's coat is light-coloured, tyrosinase must have been denatured / inactivated at this warm temperature. After birth, a kitten is exposed to cooler environmental temperatures and its extremities (ears, face, feet and tail) will be the coolest as they are furthest from the main body, where heat is generated and have a large surface area to volume ratio. Cooler temperature means tyrosine is activated / not denatured. Tyrosinase therefore catalyses the production of dark pigment in these areas.

17.6

1 E = **XX**   F = **XY**

2 A = not colour blind/normal vision B = not colour blind/normal vision D = colour blind

3 G = $X^R X^r$ H = $X^R Y$ I = $X^R X^R$ J = $X^r Y$

4 0% – because sons inherit their X chromosome from their mother and she has only alleles for normal vision (X^R).

5 By mutation (of the **R** allele).

A right royal disease

1 Because the ancestors from whom they are descended (Edward VII and Victoria) did not have, or carry, alleles for haemophilia.

2 a The disease of haemophilia only occurs in males and not females.

   b Parents without the disease are shown to have children with the disease. Alexandra and Tsar Nicholas II do not have the disease but their son Tsarevitch Alexis does. (Note: There are many other examples.)

3 a $X^H X^H$ b $X^h Y$ c $X^H X^h$

4 Anastasia could have either genotype $X^H X^h$ or $X^H X^H$, depending on whether she inherited an X^H or an X^h from her mother Alexandra. Waldemar's genotype must be $X^h Y$. Therefore:

   a Sons would inherit a **Y** from Waldemar and either an X^H or X^h from Anastasia (mother). Therefore the possible genotypes are $X^H Y$ or $X^h Y$.

   b Daughters must inherit X^h from Waldemar (father) and either X^H or X^h from Anastasia (mother). Therefore the possible genotypes are $X^h X^h$ or $X^H X^h$.

17.7

1 In sex-linkage the linked genes are on the same sex chromosome (usually the X chromosome) whereas in autosomal linkage they are on any chromosome other than the sex chromosomes.

2

parental phenotype	short, white fur	long, grey fur
parental genotype	hh gg	Hh Gg
Gametes produced by meiosis	(hg)	(HG) (hg)

	Male gametes	
Female gametes	(HG)	(hg)
(hg)	HhGg	hhgg

1 rabbit with long, grey fur (HhGg)

1 rabbit with short, white fur (hhgg)

Tales of the unexpected

1 a **Y** = allele for yellow flowers

      y = allele for white flowers

      R = allele for red fruit

      r = allele for yellow

2 **YyRr**

3 3 yellow flowers and red fruit

   1 white flowers and yellow fruit.

4 9 yellow flowers and red fruit

   3 yellow flowers and yellow fruit

   3 white flowers and red fruit

   1 white flowers and yellow fruit

5 a yellow flowers and yellow fruit

      white flowers and red fruit

   b crossing over between chromatids giving rise to recombinants

17.8

1 a mouse 1 = albino(white); mouse 2 = agouti

   b **AABb; AaBb; AABb; Aabb; AaBb; aaBb; Aabb; aabb**

   c 4 albino : 3 agouti : 1 black

2 a **AaBb**

   b F_1 phenotype   Purple seeds

      F_1 genotype   AaBb

      F_1 gametes   (AB) (Ab) (aB) (ab)

      F_2 offspring

		male gametes			
		(AB)	(Ab)	(aB)	(ab)
Female gametes	(AB)	AABB	AABb	AaBB	AaBb
	(Ab)	AABb	AAbb	AaBb	Aabb
	(aB)	AaBB	AaBb	aaBB	aaBb
	(ab)	AaBb	Aabb	aaBb	aabb

   c i **AABB; AABb; AaBB; AaBb; AABb; AaBb; AaBB; AaBb; AaBb.**

      ii They all possess at least one dominant allele **A** and one dominant allele B.

   d The production of anthocyanin uses a biochemical pathway that requires two functional enzymes each coded for by the dominant allele of both genes A and B. If either gene is represented by two recessive alleles the enzyme it codes for is non-functional and the pathway cannot be completed. This is an example of epistasis because it affects the other gene in that, even if it is functional and produces its enzyme, its effects cannot be expressed because no pigment can be manufactured.

Against: wrong to use humans, including potential humans as a means to an end; embryos are human, they have human genes, and deserve the same respect and treatment as adult humans; is a 'slippery slope' to the use of older embryos and fetuses for research; could lead to research and development of human cloning and, although banned in UK the information could be used elsewhere; undermines respect for life; adult stem cells are an available alternative and energies should be directed towards developing these.

Growth of plant tissue cultures

1 differentiation

2 'in vitro'

3 **a** clone **b** mitosis

4 IAA and 2,4-D

5 In test tube 1 the low concentration of IAA produces moderate shoot development but when a high concentration of cytokinin is added (test tube 3) the presence of cytokinin influences the effects of the IAA by reducing shoot development to a 'little'.

20.3

1 Transcriptional factors stimulate transcription of a gene.

2 Oestrogen diffuses through the phospholipid portion of a cell-surface membrane into the cytoplasm of a cell, where it binds with a site on a receptor portion of the transcriptional factor. Oestrogen changes the shape of the DNA binding site on the transcription factor so it can now bind with DNA. The transcriptional factor now enters the nucleus through a nuclear pore and binds with DNA, stimulating transcription of the gene that makes up that portion of DNA, i.e. it stimulates gene expression.

Gene expression in haemoglobin

1 It allows the fetus to load its haemoglobin with oxygen from the mother's haemoglobin where the two blood supplies come close to each other (at the placenta).

2 alpha = 50% beta = 20% gamma = 30%

3 The gene for gamma-globulin is expressed less while the gene for beta-globulin is expressed more.

4 Expression of the gene for gamma-globulin is progressively reduced as a result of either preventing transcription, and hence preventing the production of mRNA, or by the breakdown of mRNA before its genetic code can be translated.

5 A possible therapy would be to express (switch on) the gene for gamma-globulin and prevent the expression of (switch off) the gene for beta-globulin. This would result in haemoglobin being of the fetal rather than the adult type.

20.4

1 Epigenetics is the process by which environmental factors can cause heritable changes in gene function without changing the base sequence of DNA.

2 decreased histone acetylation and increased DNA methylation.

3 The other strand would have complementary bases (i.e. GCUA instead of CGAU respectively). It is unlikely that these opposite base pairings would complement a sequence on the mRNA. The siRNA, with enzyme attached, would therefore not bind to the mRNA and so would be unaffected.

4 **a** Some siRNA that blocks a particular gene could be added to cells. By observing the effects (or lack of them) we could determine what the role of the blocked gene is.

  b siRNA could be used to prevent the disease by blocking the gene that causes it.

5 **a** Chromatin would be more condensed (tightly packed).

  b Transcription would cease.

Nature versus nurture

1 Environmental factors

2 If the influence was totally genetic the plants that were genetically identical (those in the same column) would have the same phenotype regardless of where they were grown. The greater the environmental influence the greater the differences between the genetically identical plants. There are major differences so the main influence is environmental.

3 Environmental conditions at high altitude are more extreme than those at low altitude and less suited for photosynthesis (colder, windier, less soil). Plants from high altitude have evolved to survive in these extremes. The conditions at medium and low altitude therefore present few problems and they thrive. Plants that have evolved at low altitude, by contrast, find harsher conditions at medium and high altitude and struggle to grow.

Prader-Willi syndrome

1 Epigenetic tags are usually erased in sperm and eggs or during early fetal development and so do not silence the genes.

2 The genes that are deleted might code for polypeptides/proteins/enzymes/hormones that are essential to the reproductive process. For example, they might code for enzymes/proteins needed to make functioning ovaries/eggs or testes/sperm or for the synthesis of hormones needed for the development of these organs/gametes.

20.5

1 Fat cells of the breasts tend to produce more oestrogens after the menopause. These locally produced oestrogens release an inhibitor molecule that prevents transcription causing proto-oncogenes of breast tissue to develop into oncogenes. These oncogenes increase the rate of cell division leading to the development of a tumour (breast cancer).

2 Proto-oncogenes **increase** the rate of cell division and so their activation produces a mass of cells (tumour) but tumour suppressor genes **decrease** the rate of cell division and so their deactivation produces tumour.

3 Cells of a benign tumour produce adhesion molecules that make them stick together and are surrounded by a capsule of dense tissue. The tumour therefore remains as a compact structure and so surgical removal is likely to remove **all** tumour cells.

4 Malignant tumours spread to other regions of the body and so even though surgery can remove the more obvious larger ones, tiny ones will require other therapies to prevent these re-growing into new tumours.

5 HADC removes acetyl groups from histones, inhibiting transcription and switching off the gene. Some cancers are the result of genes (e.g. tumour suppressor genes) that normally help repair DNA (and so prevent cancers) being switched off. By inhibiting HADC, phenylbutyric acid could prevent the removal of acetyl groups from histones and switch the 'protective' gene back on.

Risk factors and cancer

1 air pollution / inhaled substances (carcinogens), e.g. asbestos at work

2 a positive correlation between the number of cases of lung cancer in men and the number of cigarettes smoked

3 It is unlikely that a coincidence would have occurred many times over.

4 No. While the data clearly point to the likelihood that smoking causes lung cancer, they do not provide experimental evidence that specifically links smoking to lung cancer.

5 The experiment that showed that the derivative of benzopyrene caused changes to DNA at precisely the same three points as the mutations of the gene in a cancer cell.

6 This is a single case. The link between early death and lung cancer is about probabilities not certainties. Statistically it is unlikely, but not impossible, for smokers to live to be very old.

20.6

1 A genome is all the genetic material in an organism. A proteome is all the proteins produced by the genome.

2 Simple organisms generally have just one, circular piece of DNA that is not associated with histones and there is little, if any, non-coding DNA. Complex organisms have considerably more DNA and the majority of this does not code for proteins.

3 It allows identification of those proteins that act as antigens on the surfaces of the pathogens. These antigens can then be used to produce effective vaccines against the disease.

21.1

1 recombinant

2 reverse transcriptase

3 complementary (cDNA)

4 DNA polymerase

5 restriction endonucleases

6 blunt

7 sticky

8 CTTAAG

21.2

1 A vector transfers genes (DNA) from one organism into another.

2 To show which cells (bacteria) have taken up the plasmid (gene).

3 Results can be obtained more easily and more quickly – because, with antibiotic-resistance markers, the bacterial cells with the required gene are killed, so replica plating is necessary to obtain the cells with the gene. With fluorescent gene markers, the bacterial cells are not killed and so there is no need to carry out replica plating.

4 a B, C, D, J, K and L – because those that did not take up the plasmid will not have taken up the gene for ampicillin resistance and so will be the ones that are killed on the ampicillin plate, i.e. the colonies that have disappeared.

  b E, F and I – because those with the plasmid containing gene X will have a non-functional gene for tetracycline resistance and therefore the colonies will have been killed on the tetracycline plate, i.e. the colonies will have disappeared.

21.3

1 Primers are short pieces of DNA that have a set of bases complementary to those at the end of the DNA fragment to be copied.

2 Primers attach to the end of a DNA strand that is to be copied and provide the starting sequences for DNA polymerase to begin DNA cloning. DNA polymerase can only attach nucleotides to the end

of an existing chain. They also prevent the two separate strands from rejoining.

3 Because the sequences at the opposite ends of the two strands of DNA are different.

4 Hydrogen bonds

5 Biological contaminants may contain DNA and this DNA would also be copied.

Evaluation of DNA technology

Arguments should be reasoned, logical and based on sound science, and should use specific examples rather than vague references.

1 Whichever aspects are chosen the beneficial aspects to humans should be clear, e.g. genetically modified crops that can be grown in extreme conditions – greater productivity; more food; less poverty and hunger in some human populations.

2 Arguments must relate directly to the aspects chosen in question 1 and should oppose the use of the technology, for example, genetically modified crops that can be grown in extreme conditions – risk of damaging the ecological balance; risk of the gene passing to other organisms; dangers from unforeseen by-products of the plants' metabolism; dangers from possible mutations of the genes; the economic consequences for developing countries.

Treatment of severe combined immunodeficiency using gene therapy

1 Using restriction endonucleases to cut the gene from the length of DNA that carries it / using reverse transcriptase to make cDNA from mRNA produced by the gene / manufacturing it in a 'gene machine'.

2 A virus with RNA as its nucleic acid and which can make a copy of DNA from this RNA using the enzyme reverse transcriptase.

3 Defective gene – ineffective or no enzyme ADA produced – toxins not destroyed by enzyme ADA – T cells destroyed by toxins – no/little immunity to infection – an infection causes death.

4 T cells only live for 6-12 months and those that replace them do not possess the gene.

5 Bone marrow stem cells divide to produce T cells and so there is a constant supply of the ADA gene and hence the enzyme ADA.

21.4

1 A DNA probe is a short, single-stranded section of DNA that has some label attached that makes it easily identifiable.

2 Determine the order of nucleotides on the mutated gene by DNA sequencing – produce a fragment of DNA that has complementary bases to the mutated portion of the gene – label the fragments to form

a DNA probe – make multiple copies of the DNA probe using PCR techniques – add the probe to DNA fragments from the individual being tested. If the donor has a mutant allele, the probe will bind to the complementary bases on the donor DNA. These fragments will now be labelled and can be distinguished from the rest of the DNA.

3 a Tumour suppressor genes inhibit cell division.

b He/she might change their lifestyle to reduce the risk of cancer, e.g. by giving up smoking, losing weight, eating more healthily and avoiding mutagens as far as possible; checking more regularly for early symptoms of cancer; choosing to undergo treatment.

21.5

1 PCR is used to increase the quantity of DNA because the quantity available, e.g. at a crime scene, is often very small.

2 a Suspect B – because the bands on this suspect's genetic fingerprint match those of the genetic fingerprint of blood found at the crime scene.

b To eliminate the victim as the source of the blood sample found at the scene.

3 The chemicals may inhibit some of the restriction endonucleases, which would then fail to cut some sections of DNA. There would therefore be a greater number of longer DNA fragments than normal and the fingerprint would be different.

4 In a person with the allele for Huntington's disease, some of the DNA fragments will be larger than those in a person without the allele because of the extra repeating units on the gene. These will travel a shorter distance in the electrophoresis gel and so there will be more thick bands nearest the start of the fingerprint (where the initial sample was located).

5 Genetic fingerprints can determine how closely any two individuals are related. The closer the match between their fingerprints, the closer they are related. Therefore, to avoid the problems caused by inbreeding, it is advisable to mate animals whose fingerprints differ the most.

Locating DNA fragments

1 restriction endonuclease

2 A = 2 B = 4 C = 5 D = 1 E = 6 F = 3

Explanation – the shorter the fragments (those with fewer base pairs) the further they travel and the longer the fragments the less distance they travel.

3 5 times – because 5 cuts produce 6 fragments.

Gel electrophoresis and DNA sequencing

1 two   3   CACTGTTCAT

2 cytosine

Practical questions – answers

Practical 1

1 **(a)** Sketch a graph of your expected results. Remember to label your axes.

(b) List all variables that need to be controlled and how you would control them.
Temperature – Keep celery extract and H_2O_2 in a thermostatically controlled water bath at 30°C
Enzyme concentration – use the same source of celery extract which has been mixed evenly
Substrate concentration – use same volume and concentration of H_2O_2

2 **(a)** Repeat each pH at least twice and calculate a mean.
(b) This method is very subjective to decide on the highest point of the froth.
Change method to using a gas syringe to collect the O_2 gas released.
Celery extract may contain varying concentrations of enzyme.
Change method to use a pure source of a specific concentration of enzyme.

Practical 2

1 **(a)** $\dfrac{12}{150} \times 100 = 8\%$

(b) No stain used / not root tip / cells not dividing in this small sample / more than one layer of cells as not squashed firmly enough.

2 **(a)** As distance increases from the root tip, the mitotic index decreases.
Above 1 mm an increase in distance from root tip has little effect on the mitotic index / plateaus.
Correctly quote paired set of data.

(b) Meristem tissue only nearest the tip has the ability to divide and there is less meristem tissue as the distance increases from the tip.
Nearest the tip gets more damage, therefore needs to do more cell division to repair the tissue.

Practical 3

1

Concentration of sucrose (mol dm^{-3})	Volume of distilled water (cm^3)	Volume of 0.5 mol dm^{-3} stock solution sucrose (cm^3)
0.0	20	0
0.1	16	4
0.2	12	8
0.3	8	12
0.4	4	16
0.5	0	20

2 **(a)** Calculate the % change in mass of potato tissue.

Concentration of sucrose (mol dm^{-3})	Mass before submerging in solution (g)	Mass after submerging in solution (g)	Mass change (g)	Percentage change in mass of potato tissue (%)
0.0	4.5	5.0	+0.5	**+11.1**
0.1	3.9	4.3	+0.4	**+10.3**
0.2	4.3	4.5	+0.2	**+4.7**
0.3	4.1	4.2	−0.1	**−2.4**
0.4	4.4	3.7	−0.7	**−15.9**
0.5	4.4	3.6	−0.8	**−18.2**

(b) There are different starting and finishing masses.
The mass change is very small; therefore a % change is easier to compare real differences.

3   (a) Correctly labelled axes with units;
uniform axes;
plots taking up over $\frac{1}{2}$ space of graph;
accurate plots;
smooth line of best fit.
(b) Use this graph to find the concentration of sucrose
(where curve crosses x-axis). Between 0.25 and 0.3 mol dm^{-3}

(c) Use a data resource with listed sucrose concentrations and water potentials to find the water potential for the sucrose solution read off the graph.

Practical 4

1   (a) The red pigment is water soluble and held in the vacuole;
The cell-surface membrane is selectively permeable and some pigment diffuses out.
(b) As temperature increases from 20 to 40°C, there is a small increase in absorbance reflecting a small increase in the permeability of the cell-surface membrane.
Above 50°C there is a steep increase in the permeability of the cell-surface membrane.
(c) The proteins embedded in the cell-surface membrane become denatured.
The structure of the cell-surface membrane has been permanently disrupted so is now fully permeable and most of the pigment diffuses out.

Practical 5

1   (a) Spiracles.
(b) Control water vapour loss by closing spiracles if need to conserve water.
(c) High temperature environment causes more water to evaporate;
hairs trap water <u>vapour</u> and this reduces water potential gradient and therefore water vapour loss.
2   (a) Penetrate deep into muscle tissues;
delivers more air / oxygen to muscles.
(b) Chitin
(c) Smaller diameters are more permeable to gases and get closer to body cells for gaseous exchange by diffusion.

Practical 6

1   (a) To sterilise the equipment/ to kill any microbes on the equipment.
(b) 1   Washing hands / cleaning work surface with disinfectant
2   Flame sterilising the inoculating loop
3   Flaming the neck of the culture tube containing the bacteria
4   Streaking the plate with the inoculating loop **quickly**
5   Only lifting the lid of the petri dish a small amount
(c) To kill any harmful / pathogenic bacteria so they don't harm anyone.

Practical 7

(a) Ensure pigment spot is above solvent/ensure atmosphere in container was saturated with solvent before running.

(b) Solutes/pigments dissolve in solvent; solvent moves up paper; distance moved by solutes/pigments depends on their relative solubility/molecular size.

(c) (i) Relative flow (Rf) is a physical constant; for a specific solute in a specific solvent; it is the distance moved by the solute divided by the distance moved by the solvent (front);
Rf = distance moved by the compound ÷ distance moved by the solvent.

(ii) $B = \dfrac{35}{93} = 0.38$; (0.376) $C = \dfrac{36}{93} = 0.39$; (0.387)

(iii) 2-way chromatography/run with a different solvent

Practical 8

1 Colour change and inferences that can made from the results:
Tube 1 (leaf extract + DCPIP) colour changes until it is the same colour as tube 4 (leaf extract + distilled water).
Tube 2 (isolation medium + DCPIP) no colour change. This shows that the DCPIP does not decolourise when exposed to light.
Tube 3 (leaf extract + DCPIP in the dark) no colour change. It can therefore be inferred that the loss of colour in tube 1 is due to the effect of light on the extract.
Tube 4 (leaf extract + distilled water) no colour change. This shows that the extract does not change colour in the light. It acts as a colour standard for the extract without DCPIP.
Tube 5 (supernatant + DCPIP) no colour change if the supernatant is clear; if it is slightly green there may be some decolouring.
The results should indicate that the light-dependent reactions of photosynthesis are restricted to the chloroplasts that have been extracted.

2 Carbon dioxide will have no effect, because it is not involved in the light-dependent reactions.

3 Students should describe a procedure in which light intensity is varied but temperature is controlled.

Practical 9

1 a

mean volume of gas produced
at various temperatures

b) 40°C

c) the smallest quantity of glucose – answer 2

Practical 10

1 **(a)** Darkness/shelter/humidity/moisture/protection from predators

(b) Cover one side to put into darkness/ make one side more moist – place equal numbers of woodlice in each side – leave for a set time and then compare the numbers in each side and note the preference.

(c) The woodlice would prefer the moist / dark side

(d) Protection from predators / prevention of water loss

Practical 11

(a) Use the same volume or concentration of Benedict's solution, use the same volume of glucose solution, calibrate the colorimeter, boil for the same length of time

(b) 6.5

(c) Boil/heat with Benedict's then test the filtrate. Treat the filtrate with sucrose/ invertase/acid

Practical 12

(a) Set up coordinate grid/use tapes along 2 sides; generate random numbers for co-ordinates; co-ordinate indicates centre of sampling quadrat.

(b) Advantage: unbiased/allows statistical testing; Disadvantage: coverage may be uneven/unrepresentative/large areas may be missed.

(i) Correct axes (light intensity on x-axis) / suitable scale / accurate plotting / joining points with a ruled / straight line / key/curves labelled.

(ii) Number increases as average light intensity decreases/converse.

(iii) Shade tolerant/loving plant/able to photosynthesise efficiently at low light intensities/low compensation point.

(d) Thinner epidermis / large or many chloroplasts / chloroplasts concentrated towards upper surface / high chlorophyll concentration / larger leaves.

Glossary

A

abiotic an ecological factor that makes up part of the non-biological environment of an organism, e.g. temperature, pH, rainfall and humidity. See also *biotic*.

acetylcholine one of a group of chemicals, called *neurotransmitters*, released by *neurones*. It diffuses across the gap (*synapse*) between adjacent neurones and so passes an impulse from one neurone to the next.

action potential change that occurs in the electrical charge across the membrane of an *axon* when it is stimulated and a nerve impulse passes.

actin filamentous protein which is involved in contraction within cells, especially muscle cells. See also *myosin*.

activation energy energy required to bring about a reaction. The activation energy is lowered by the presence of *enzymes*.

active immunity resistance to disease resulting from the activities of an individual's own immune system whereby an *antigen* induces plasma cells to produce *antibodies*.

active site a group of amino acids that makes up the region of an *enzyme* into which the *substrate* fits in order to catalyse a reaction.

active transport movement of a substance from a region where it is in a low concentration to a region where it is in a high concentration. The process requires the expenditure of *metabolic* energy.

adenosine triphosphate (ATP) an activated nucleotide found in all living cells that acts as an energy carrier. The *hydrolysis* of ATP leads to the formation of adenosine diphosphate (ADP) and inorganic phosphate, with the release of energy.

adrenaline a hormone produced by the adrenal glands in times of stress that prepares the body for an emergency.

aerobic connected with the presence of free oxygen. Aerobic respiration requires free oxygen to release energy from glucose. See also *anaerobic*.

allele one of a number of alternative forms of a *gene*. For example, the gene for the shape of pea seeds has two alleles: one for 'round' and one for 'wrinkled'.

allele frequency the number of times an allele occurs within the *gene pool*.

allergen a normally harmless substance that causes the immune system to produce an immune response. See also *allergy*.

allergy the response of the immune system to an *allergen*. Examples include hay fever and *asthma*.

anaerobic connected with the absence of oxygen. Anaerobic respiration releases energy from glucose or other foods without the presence of oxygen. See also *aerobic*.

antibiotic a substance produced by living organisms that can destroy or inhibit the growth of microorganisms.

antibiotic resistance the development in microorganisms of mechanisms that prevent *antibiotics* from killing them.

antibody a protein produced by *lymphocytes* in response to the presence of the appropriate antigen.

anticodon a sequence of three adjacent *nucleotides* on a molecule of transfer RNA that is complementary to a particular *codon* on a messenger RNA molecule.

antidiuretic hormone (ADH) a hormone produced by the *hypothalamus* that passes to the posterior *pituitary gland* from where it is secreted. ADH reduces the volume of water in urine by increasing water reabsorption in the kidneys.

antigen a molecule that triggers an immune response by *lymphocytes*. ATP see adenosine triphosphate.

antioxidant chemical which reduces or prevents *oxidation*. Often used as an additive to prolong the shelf-life of certain foods.

apoplastic pathway route through the cell walls and intercellular spaces of plants by which water and dissolved substances are transported. See also *symplastic pathway*.

artificial selection breeding of organisms by human selection of parents/gametes in order to perpetuate certain characteristics and/or to eliminate others.

asthma a chronic illness in which there is resistance to air flow to the alveoli of the lungs as a result of the airways becoming inflamed due to an allergic response to an *allergen*.

atheroma fatty deposits in the walls of arteries, often associated with high *cholesterol* levels in the blood.

ATP (adenosine triphosphate) *nucleotide* found in all living organisms, which is produced during respiration and is important in the transfer of energy.

autonomic nervous system part of the nervous system, controlling the muscles and glands, that is not under voluntary control.

autosome a chromosome which is not a sex chromosome.

axon a process extending from a *neurone* that conducts *action potentials* away from the cell body.

B

B cell (B lymphocyte) type of white blood cell that is produced and matures within the bone marrow. B lymphocytes produce *antibodies* as part of their role in *immunity*. See also *T cell*.

Benedict's test a simple biochemical reaction to detect the presence of reducing sugars.

biodiversity the range and variety of genes, species and habitats within a particular region.

biomass the total mass of living material, normally measured in a specific area over a given period of time.

biotic an ecological factor that makes up part of the living environment of an organism. Examples include food availability, competition and predation. See also *abiotic*.

biosensor a device that uses biological molecules to measure the level of certain chemicals.

Biuret test a simple biochemical reaction to detect the presence of protein.

body mass index (BMI) a person's body mass in kilograms divided by the square of their height in metres.

C

Calvin cycle a biochemical pathway that forms part of the *light-independent reaction* of photosynthesis, during which carbon dioxide is reduced to form carbohydrate.

cancer a disease, resulting from cells that break away from an original tumour to form secondary *tumours* elsewhere in the body.

carcinogen a chemical, a form of radiation, or other agent that causes *cancer*.

cardiac cycle a continuous series of events which make up a single heart beat.

cardiac muscle type of muscle found only in the heart. It has fewer striations than *skeletal muscle* and can contract continuously throughout life without stimulation by nerve impulses. See also *smooth muscle*.

cardiac output the total volume of blood that the heart can pump each minute. It is calculated as the volume of blood pumped at each beat (*stroke volume*) multiplied by the number of heart beats per minute (heart rate).

carrier molecule (carrier protein) a protein on the surface of a cell that helps to transport molecules and ions across plasma membranes.

Casparian strip a distinctive band of suberin around the endodermal cells of a plant root that prevents water passing into xylem via the cell walls. The water is forced through the living part (*protoplast*) of the endodermal cells.

centrifugation process of separating out particles of different sizes and densities by spinning them at high speed in a centrifuge.

cholinesterase enzyme that breaks down and therefore inactivates the *neurotransmitter*, *acetylcholine*, in the *synapse*.

cholesterol lipid that is an important component of cell-surface membranes. Excess in the blood can lead to *atheroma*.

chromatid one of the two strands of a *chromosome* that are joined together by a single centromere prior to cell division.

chromatin the material that makes up *chromosomes*. It consists of DNA and the protein histone.

chromosome a thread-like structure made of protein and DNA by which hereditary information is physically passed from one generation to the next.

climax community the organisms that make up the final stage of ecological succession.

clone a group of genetically identical cells or organisms formed from a single parent as the result of asexual reproduction or by artificial means.

co-dominance condition in which both *alleles* for one *gene* in a *heterozygous* organism contribute to the *phenotype*.

codon a sequence of three adjacent *nucleotides* in mRNA that codes for one amino acid.

cohesion attraction between molecules of the same type. It is important in the movement of water up a plant.

collagen fibrous protein that is the main constituent of connective tissues such as tendons, cartilage and bone.

community all the living organisms present in an *ecosystem* at a given time.

complementary DNA DNA that is made from messenger RNA in a process that is the reverse of normal transcription.

condensation chemical process in which two molecules combine to form a more complex one with the elimination of a simple substance, usually water. Many biological *polymers*, such as polysaccharides and polypeptides, are formed by condensation. See also *hydrolysis*.

conservation method of maintaining *ecosystems* and the living organisms that occupy them. It requires planning and organisation to make best use of resources while preserving the natural landscape and wildlife.

consumer any organism that obtains energy by 'eating' another. Organisms feeding on plants are known as primary consumers and organisms feeding on primary consumers are known as secondary consumers. See also *producer*.

continuous variation variation in which organisms do not fall into distinct categories but show gradations from one extreme to the other.

coronary arteries arteries that supply blood to the cardiac muscle of the heart.

coronary heart disease (CHD) any condition, for example, *atheroma* and *thrombosis*, affecting the *coronary arteries* that supply heart muscle.

correlation when a change in one variable is reflected by a change in the second variable.

co-transport the transport of one substance coupled with the transport of another substance across a plasma membrane in the same direction through the same protein carrier.

countercurrent system a mechanism by which the efficiency of exchange between two substances is increased by having them flowing in opposite directions.

covalent bond type of chemical bond in which two atoms share a pair of *electrons*, one from each atom.

crossing over the process whereby a *chromatid* breaks during *meiosis* and rejoins to the chromatid of its *homologous chromosome* so that their *alleles* are exchanged.

cuticle exposed non-cellular outer layer of certain animals and the leaves of plants. It is waxy and impermeable to water. It therefore helps to reduce water loss.

D

deciduous term applied to plants that shed all their leaves together at one season.

denaturation permanent changes due to the unravelling of the three-dimensional structure of a protein as a result of factors such as changes in temperature or pH.

dendrite a process, usually branched, extending from the cell body of a *neurone*, which conducts impulses towards the cell body.

denitrifying bacteria bacteria that convert nitrates to nitrogen gas as part of the nitrogen cycle.

depolarisation temporary reversal of charges on the cell-surface membrane of a *neurone* that takes place when a nerve impulse is transmitted.

diabetes a metabolic disorder in which the body is unable to regulate the level of blood glucose. There are two forms of the disorder – Type I and Type II diabetes.

diastole the stage in the *cardiac cycle* when the heart muscle relaxes. See also *systole*.

differentiation the process by which cells become specialised for different functions.

diffusion the movement of molecules or ions from a region where they are in high concentration to one where their concentration is lower.

diploid a term applied to cells in which the nucleus contains two sets of *chromosomes*. See also *haploid*.

dipolar having a pair of equal and opposite electrical charges.

directional selection selection that operates towards one extreme in a range of variation.

discontinuous variation variation shown when the characteristics of organisms fall into distinct categories, e.g. blood groups in humans. See also *continuous variation*.

DNA helicase enzyme that acts on a specific region of the DNA molecule to break the hydrogen bonds between the bases causing the two strands to separate and expose the *nucleotide* bases in that region.

DNA replication the process in which the double helix of a DNA molecule unwinds and each strand acts as a template on which a new strand is constructed.

dominant allele a term applied to an allele that is always expressed in the phenotype of an organism. See also *recessive allele*

E

ecological niche describes how an organism fits into its environment. It describes what a species is like, where it occurs, how it behaves, its interactions with other species and how it responds to its environment.

ecosystem all the living and nonliving components of a particular area.

ectothermic an animal that uses the environment to regulate its body temperature. See also *endotherm*.

effector an organ that responds to stimulation by a nerve impulse resulting in a change or response.

electron negatively charged subatomic particle that orbits the positively charged nucleus of all atoms.

electron carrier molecules a chain of carrier molecules along which *electrons* pass, releasing energy in the form of *ATP* as they do so.

emphysema a disease in which the walls of the alveoli break down, reducing the surface area for gaseous exchange, thereby causing breathlessness in the patient.

endocytosis the inward transport of large molecules through the cell-surface membrane.

endotherm an animal maintaining its body temperature by physiological mechanisms. See also *ectotherm*.

enzyme a protein or RNA that acts as a catalyst and so alters the speed of a biochemical reaction.

epidemiology the study of the spread of disease and the factors that affect this spread.

eukaryotic cell a cell that has a membrane-bound nucleus and *chromosomes*. The cell also possesses a variety of other membranous organelles, such as mitochondria and endoplasmic reticulum. See also *prokaryotic cell*.

exocytosis the outward bulk transport of materials through the cell-surface membrane.

F

facilitated diffusion diffusion involving the presence of protein *carrier molecules* to allow the passive movement of substances across plasma membranes.

G

gamete reproductive (sex) cell that fuses with another gamete during fertilisation.

gel electrophoresis a technique used to separate DNA fragments of different lengths by placing them on agar gel and passing a voltage across them.

gene section of DNA on a *chromosome* coding for one or more polypeptides.

gene pool the total number of *alleles* in a particular population at a specific time.

gene marker a section of DNA that is used to indicate the location of a *gene* or other section of DNA.

gene mutation a change to one or more *nucleotide* bases in DNA resulting in a change in *genotype* which may be inherited.

gene therapy a mechanism by which genetic diseases, e.g. *cystic fibrosis*, may be cured by masking the effect of the defective *gene* by inserting a functional gene.

generator potential *depolarisation* of the membrane of a receptor cell as a result of a stimulus.

genetic engineering see *recombinant DNA technology*.

genetically modified organism (GMO) organism that has had its DNA altered as a result of *recombinant DNA technology*.

genotype the genetic composition of an organism. See also *phenotype*.

glomerulus a cluster of blood capillaries enclosed by the renal (Bowman's capsule) in the kidney.

glucagon a hormone produced by α cells of the *islets of Langerhans* in the pancreas that increases blood glucose levels by initiating the breakdown of glycogen to glucose.

gluconeogenesis the conversion of non-carbohydrate molecules to glucose.

glycogenesis the conversion of glucose to glycogen.

glycogenolysis the conversion of glycogen to glucose.

glycolysis first part of cellular respiration in which glucose is broken down anaerobically in the cytoplasm to two molecules of pyruvate.

glycoprotein substance made up of a carbohydrate molecule and a protein molecule. Parts of cell surface membrane and certain hormones are glycoproteins.

granum (plural grana) a stack of *thylakoids* in a chloroplast that resembles a pile of coins. This is the site of the *light-dependent reaction* of photosynthesis. See also *stroma*.

guard cell one of a pair of cells that surround a *stoma* in plant leaves and control its opening and closing.

H

habitat the place where an organism normally lives and which is characterised by physical conditions and the types of other organisms present.

haemoglobin globular protein in blood that readily combines with oxygen to transport it around the body. It comprises four polypeptide chains around an iron-containing haem group.

haploid term referring to cells that contain only a single copy of each *chromosome*, e.g. the sex cells (*gametes*).

heterozygous condition in which the *alleles* of a particular gene are different.

histones proteins which together with DNA, make up the chromosomes of eukaryotic cells.

homeostasis the maintenance of a more or less constant internal environment.

homologous chromosomes a pair of *chromosomes*, one maternal and one paternal, that have the same gene *loci* and therefore determine the same features.

They are not necessarily identical, however, as individual *alleles* of the same *gene* may vary, e.g. one chromosome may carry the allele for blue eyes, the other the allele for brown eyes. Homologous chromosomes are capable of pairing during *meiosis*.

homozygous condition in which the *alleles* of a particular *gene* are identical.

human genome the totality of the DNA sequences on the *chromosomes* of a single human cell.

human genome project international scientific project to map the entire sequence of all the base pairs of the genes in a single human cell.

hydrogen bond chemical bond formed between the positive charge on a hydrogen atom and the negative charge on another atom of an adjacent molecule, e.g. between the hydrogen atom of one water molecule and the oxygen atom of an adjacent water molecule.

hydrolysis the breaking down of large molecules into smaller ones by the addition of water molecules. See also *condensation*.

hyperthermia a condition that results from the core body temperature rising above normal.

hypothalamus region of the brain adjoining the pituitary gland that acts as the control centre for the *autonomic nervous system* and regulates body temperature and fluid balance.

hypothermia a condition that results from the core body temperature falling below normal.

I

immunity the means by which the body protects itself from infection.

insulin a hormone, produced by the α cells of the *islets of Langerhans* in the pancreas, which decreases blood glucose levels by, amongst other things, increasing the rate of conversion of glucose to glycogen.

intercropping the practice of growing two or more crops in close proximity usually to produce a greater yield on a piece of land.

interspecific competition competition between organisms of different species.

interspecific variation differences between organisms of different species.

intraspecific competition competition between organisms of the same species.

intraspecific variation differences between organisms of the same species.

intrinsic proteins proteins of the cell-surface membrane that completely span the *phospholipid* bilayer from one side to the other.

introns portions of DNA within a gene that do not code for a polypeptide. The introns are removed from pre-messenger RNA after *transcription*.

ion an atom or group of atoms that has lost or gained one or more *electrons*. Ions therefore have either a positive or negative charge.

ion channel a passage across a cell-surface membrane made up of a protein that spans the membrane and opens and closes to allow *ions* to pass in and out of the cell.

islets of Langerhans groups of cells in the pancreas comprising large α cells, which produce the hormone *glucagon*, and small β cells, which produce the hormone *insulin*.

isotope variations of a chemical element that have the same number of protons and *electrons* but different numbers of neutrons. While their chemical properties are similar they differ in mass. One example is carbon which has a relative atomic mass of 12 and an isotope with a relative atomic mass of 14.

in vitro refers to experiments carried out outside the living body, e.g. in test tubes.

in vivo refers to experiments that are carried out within living bodies.

K

Krebs cycle series of aerobic biochemical reactions in the matrix of the mitochondria of most *eukaryotic cells* by which energy is obtained through the oxidation of acetylcoenzyme A produced from the breakdown of glucose.

L

latent heat of vaporisation heat taken in by a liquid in order to transform it into a vapour.

ligament a tough, fibrous connective tissue, rich in collagen, that joins bone to bone. See also *tendon*.

light-dependent reaction stage of photosynthesis in which light energy is required to produce *ATP* and reduced *NADP*. See also *light-independent reaction*.

light-independent reaction stage of photosynthesis which does not require light energy directly but does need the products of the *light-dependent reaction* to reduce carbon dioxide and so form carbohydrate.

limiting factor a variable that limits the rate of a chemical reaction.

link reaction the process linking *glycolysis* with the *Krebs cycle* in which hydrogen and carbon dioxide are removed from pyruvate to form acetylcoenzyme A in the matrix of the mitochondria.

locus the position of a gene on a *chromosome*/DNA molecule.

loop of Henle the portion of the *nephron* that forms a hairpin loop that extends into the medulla of the kidney. It has a role in the reabsorption of water.

lumen the hollow cavity inside a tubular structure such as the gut or a *xylem vessel*.

lymph a slightly milky fluid found in lymph vessels and made up of *tissue fluid*, fats and *lymphocytes*.

lymphocytes types of white blood cell responsible for the immune response. They become activated in the presence of *antigens*. There are two types: *B lymphocytes* and *T lymphocytes*.

M

meiosis the type of nuclear division in which the number of *chromosomes* is halved. See also *mitosis*.

mesophyll tissue found between the two layers of epidermis in a plant leaf comprising an upper layer of *palisade cells* and a lower layer of spongy cells.

metabolism all the chemical processes that take place in living organisms.

microvilli tiny finger-like projections from the cell-surface membrane of some animal cells.

middle lamella layer made up of pectins and other substances found between the walls of adjacent plant cells.

mitosis the type of nuclear division in which the daughter cells have the same number of *chromosomes* as the parent cell. See also *meiosis*.

monoclonal antibody an antibody produced by a single clone of *cells*.

monomer one of many small molecules that combine to form a larger one known as a *polymer*.

mono-unsaturated fatty acid fatty acid that possesses a carbon chain with a single double bond. See also *polyunsaturated fatty acid*.

motor neurone *neurone* that transmits *action potentials* from the central nervous system to an effector, e.g. a muscle or gland.

multiple alleles term used to describe a *gene* that has more than two possible *alleles*.

mutagen any agent that induces a *mutation*.

mutation a sudden change in the amount or the arrangement of the genetic material in the cell.

mutualism a nutritional relationship between two species in which both gain some advantage.

myelin a fatty substance that surrounds *axons* and *dendrites* in certain *neurones*.

myocardial infarction otherwise known as a heart attack, results from the interruption of the blood supply to the heart muscle, causing damage to an area of the heart with consequent disruption to its function.

myosin the thick filamentous protein found in *skeletal muscle*.

N

NAD (nicotinamide adenine dinucleotide) a molecule that carries *electrons* and hydrogen ions during *aerobic* respiration.

NADP (nicotinamide adenine dinucleotide phosphate) a molecule that carries *electrons* produced in the *light-dependent reaction* of photosynthesis.

negative feedback a series of changes, important in homeostasis, that result in a substance being restored to its normal level. See also *positive feedback*.

nephron basic functional unit of the mammalian kidney responsible for the formation of urine.

neurone a nerve cell, comprising a cell body, *axon* and *dendrites*, which is adapted to conduct *action potentials*.

neuromuscular junction a *synapse* that occurs between a neurone and a muscle.

neurotransmitter one of a number of chemicals that are involved in communication between adjacent neurones or between nerve cells and muscles. Two important examples are *acetylcholine* and *noradrenaline*.

niche see *ecological niche*.

nitrifying bacteria microorganisms that convert ammonium compounds to nitrites and nitrates.

nitrogen fixation the incorporation of atmospheric nitrogen gas into organic nitrogen-containing compounds.

normal distribution a bell-shaped curve produced when a certain distribution is plotted on a graph.

node of Ranvier a gap in the *myelin* sheath that surrounds the axon of a neurone.

normal distribution a bell-shaped curve produced when a certain distribution is plotted on a graph.

nucleotides complex chemicals made up of an organic base, a sugar and a phosphate. They are the basic units of which the nucleic acids DNA and RNA are made.

O

oestrus the period in the oestrous cycle immediately after ovulation when the female is most fertile.

oncogenes mutated versions of proto-oncogenes that result in increased cell division leading to the growth of a tumour.

oral rehydration solution (ORS) means of treating dehydration involving giving, by mouth, a balanced solution of salts and glucose that stimulates the gut to reabsorb water.

osmosis the passage of water from a region of high *water potential* to a region where its *water potential* is lower, through a selectively permeable membrane.

oxidation chemical reaction involving the loss of *electrons*.

oxidation-reduction a chemical reaction in which electrons are transferred from one substance to another substance. The substance losing electrons is oxidised and the substance gaining the electrons is reduced.

oxidative phosphorylation the formation of *ATP* in the electron transport system of *aerobic* respiration.

P

palisade cells long, narrow cells, packed with chloroplasts, that are found in the upper region of a leaf and which carry out photosynthesis.

parasite an organism that lives on or in a host organism. The parasite gains a nutritional advantage and the host is harmed in some way.

passive immunity resistance to disease that is acquired from the introduction of *antibodies* from another individual, rather than an individual's own immune system, e.g. across the placenta or in the mother's milk. It is usually short-lived.

pathogen any microorganism that causes disease.

pentose sugar a sugar that possesses five carbon atoms. Two examples are ribose and deoxyribose.

peptide bond the chemical bond formed between two amino acids during *condensation*.

phagocytosis mechanism by which cells engulf particles to form a vesicle or a vacuole.

phenotype the characteristics of an organism, often visible, resulting from both its *genotype* and the effects of the environment.

phloem plant tissue that transports the products of photosynthesis from leaves to the rest of the plant. See also *xylem*.

phospholipid triglyceride in which one of the three fatty acid molecules is replaced by a phosphate molecule. Phospholipids are important in the structure and functioning of plasma membranes.

photolysis splitting of a water molecule by light such as occurs during the *light-dependent reaction* of photosynthesis.

photomicrograph photograph of an image produced by a microscope.

pioneer species a species that can colonise bare rock or ground.

plasmid a small circular piece of DNA found in bacterial cells.

plasmodesmata fine strands of cytoplasm that extend through pores in adjacent plant cell walls and connect the cytoplasm of one cell with another.

plasmolysis the shrinkage of cytoplasm away from the cell wall that occurs as a plant cell loses water by *osmosis*.

polygenes group of genes that are responsible for controlling a characteristic.

polymer large molecule made up of repeating smaller molecules (*monomers*).

polymerase chain reaction (PCR) process of making many copies of a specific sequence of DNA or part of a gene. It is used extensively in *gene technology* and genetic fingerprinting.

polymerases group of enzymes that catalyse the formation of long-chain molecules (*polymers*) from similar basic units (*monomers*).

polyunsaturated fatty acid (PUFA) fatty acid that possesses carbon chains with many double bonds. See also *mono-unsaturated fatty acid*.

population a group of individuals of the same species that occupy the same *habitat* at the same time.

positive feedback process which results in a substance that departs from its normal level becoming further from its norm. See also *negative feedback*.

primary structure of a protein the sequence of amino acids that makes up the polypeptides of a protein.

primary succession the progressive colonisation of bare rock or other barren terrain by living organisms.

producer an organism that synthesises organic molecules from simple inorganic ones such as carbon dioxide and water. Most producers are photosynthetic and form the first trophic level in a food chain. See also *consumer*.

prokaryotic cell a cell of an organism belonging to the kingdom Prokaryotae that is characterised by lacking a nucleus and membrane-bound organelles. Examples include bacteria. See also *eukaryotic cell*.

proton positively charged sub-atomic particle found in the nucleus of an atom. See also *electron*.

protoplast the living portion of a plant cell, that is, the nucleus and cytoplasm along with the organelles it contains.

Q

quaternary structure of a protein a number of polypeptide chains linked together, and sometimes associated with non-protein groups, to form a protein.

R

receptor a cell adapted to detect changes in the environment.

recessive allele the condition in which the effect of an allele is apparent in the *phenotype* of a *diploid* organism **only** in the presence of another identical allele. See also *dominant allele*.

recognition site a nucleotide sequence, usually of 4, 6 or 8 *nucleotides*, that is recognised by a *restriction endonuclease* and to which it attaches.

recombinant DNA technology general term that covers the processes by which genes are manipulated, altered or transferred from organism to organism. Also known as *genetic engineering*.

reduction chemical process involving the gain of *electrons*.

reflex arc the nerve pathway in the body taken by an action potential that leads to a rapid, involuntary response to a stimulus.

refractory period period during which the membrane of the axon of a neurone cannot be *depolarised* and no new *action potential* can be initiated.

renal capsule the cup shaped portion of the start of the nephron that encloses the glomerulus.

repolarisation return to the *resting potential* in the axon of a neurone after an *action potential*.

resting potential the difference in electrical charge maintained across the membrane of the axon of a neurone when not stimulated. See also *action potential*.

restriction endonucleases a group of enzymes that cut DNA molecules at a specific sequence of bases called a recognition sequence.

RNA polymerase enzyme that joins together *nucleotides* to form messenger RNA during *transcription*.

S

saltatory conduction propagation of a nerve impulse along a *myelinated dendron* or *axon* in which the action *potential* jumps from one *node of Ranvier* to another.

saprobiotic microorganism also known as a saprophyte, this is an organism that obtains its food from the dead or decaying remains of other organisms.

sarcomere a section of myofibril between two Z-lines that forms the basic structural unit of *skeletal muscle*.

saturated fatty acid a fatty acid in which there are no double bonds between the carbon atoms.

Schwann cell cell around a neurone whose cell-surface membrane wraps around the *dendron* or *axon* to form the *myelin* sheath.

secondary structure of a protein the way in which the chain of amino acids of the polypeptides of a protein is folded.

secondary succession the recolonisation of an area after an earlier community has been removed or destroyed.

selection process that results in the best-adapted individuals in a *population* surviving to breed and so pass their favourable *alleles* to the next generation.

selection pressure the environmental force altering the frequency of alleles in a *population*.

selective breeding breeding of organisms by human selection of parents/*gametes* in order to perpetuate certain characteristics and/or eliminate others.

semi-conservative replication the means by which DNA makes exact copies of itself by unwinding the double helix so that each chain acts as a template for the next. The new copies therefore possess one original and one new strand of DNA.

sensory neurone a *neurone* that transmits an *action potential* from a sensory receptor to the central nervous system.

serum clear liquid that is left after blood has clotted and the clot has been removed. It is therefore blood plasma without the clotting factors.

sickle-cell anaemia inherited blood disorder in which abnormal haemoglobin leads to red cells becoming sickle-shaped and less able to carry oxygen.

sinoatrial node (SAN) an area of heart muscle in the right atrium that controls and coordinates the contraction of the heart. Also known as the pacemaker.

skeletal muscle the muscle that makes up the bulk of the body and which works under conscious control. Also known as voluntary muscle. See also *smooth muscle*.

smooth muscle also known as involuntary or unstriated muscle, smooth muscle is found in the alimentary canal and the walls of blood vessels. Its contraction is not under conscious control. See also *skeletal muscle*.

sodium–potassium pump protein channels across cell-surface membranes that use ATP to move sodium *ions* out of the cell in exchange for potassium ions that move in.

speciation the evolution of two or more species from existing species.

species a group of similar organisms that can breed together to produce fertile offspring.

species diversity the number of different species and the number of individuals of each species within any one *community*.

stabilising selection selection that tends to eliminate the extremes of the *phenotype* range within a *population*. It arises when environmental conditions are constant.

stem cells undifferentiated dividing cells that occur in embryos and in adult animal tissues that require constant replacement, e.g. bone marrow.

stimulus a detectable alteration in the internal or external environment of an organism that produces some change in that organism.

stoma (plural stomata) a pore, mostly found in the lower epidermis of a leaf, through which gases diffuse in and out of the leaf.

stroke volume the volume of blood pumped at each ventricular contraction of the heart.

stroma matrix of a chloroplast where the *light-independent reaction* of photosynthesis takes place.

substrate a substance that is acted on or used by another substance or process. In microbiology, the nutrient medium used to grow microorganisms.

substrate-level phosphorylation the formation of ATP by the direct transfer of a phosphate group from a reactive intermediate to ADP.

supernatant liquid the liquid portion of a mixture left at the top of the tube when suspended particles have been separated out at the bottom during *centrifugation*.

symplastic pathway route through the cytoplasm and *plasmodesmata* of plant cells by which water and dissolved substances are transported. See also *apoplastic pathway*.

synapse a junction between *neurones* in which they do not touch but have a narrow gap, the synaptic cleft, across which a *neurotransmitter* can pass.

systole the stage in the *cardiac cycle* in which the heart muscle contracts. It occurs in two stages: atrial systole when the atria contract and ventricular systole when the ventricles contact. See also *diastole*.

T

tendon tough, flexible, but inelastic, connective tissue that joins muscle to bone. See also *ligament*.

tertiary structure of a protein the folding of a whole polypeptide chain in a precise way, as determined by the amino acids of which it is composed.

threshold level/value the minimum intensity that a stimulus must reach in order to trigger an *action potential* in a *neurone*.

thrombosis formation of a blood clot within a blood vessel that may lead to a blockage.

thylakoid series of flattened membranous sacs in a chloroplast that contain chlorophyll and the associated molecules needed for the *light-dependent* reaction of photosynthesis.

tidal volume the volume of air breathed in and out during a single breath when at rest.

tissue a group of similar cells organised into a structural unit that serves a particular function.

tissue fluid fluid that surrounds the cells of the body. Its composition is similar to that of blood plasma except that it lacks proteins. It supplies nutrients to the cells and removes waste products.

T cell (T lymphocyte) type of white blood cell that is produced in the bone marrow but matures in the thymus gland. T lymphocytes coordinate the immune response and kill infected cells. See also *B cell*.

transcription formation of messenger RNA molecules from the DNA that makes up a particular *gene*. It is the first stage of protein synthesis.

transducer cells cells that convert a non-electrical signal, such as light or sound, into an electrical (nervous) signal and vice versa.

transduction the process by which one form of energy is converted into another. In microbiology, the natural process by which genetic material is transferred between one host cell and another by a virus.

transpiration evaporation of water from a plant.

triglyceride an individual lipid molecule made up of a glycerol molecule and three fatty acids.

trophic level the position of an organism in a food chain.

tumour a swelling in an organism that is made up of cells that continue to divide in an abnormal way.

tumour suppressor gene a gene that maintains normal rates of cell division and so prevents the development of tumours.

turgid a plant cell that contains the maximum volume of water it can. Additional entry of water is prevented by the cell wall stopping further expansion of the cell.

U

ultrafiltration filtration assisted by blood pressure, e.g. in the formation of *tissue fluid*.

unsaturated fatty acid a fatty acid in which there are one or more double bonds between the carbon atoms.

V

vaccination the introduction of a vaccine containing appropriate disease *antigens* into the body, by injection or mouth, in order to induce artificial *immunity*.

vasoconstriction narrowing of the internal diameter of blood vessels. See also *vasodilation*.

vasodilation widening of the internal diameter of blood vessels. See also *vasoconstriction*.

vector a carrier. The term may refer to something such as a *plasmid*, which carries DNA into a cell, or to an organism that carries a *parasite* to its host.

voltage-gated channel protein channel across a cell-surface membrane that opens and closes according to changes in the electrical potential across the membrane.

W

water potential the pressure created by water molecules. It is the measure of the extent to which a solution gives out water. The greater the number of water molecules present, the higher (less negative) the water potential. Pure water has a water potential of zero.

X

xerophyte a plant adapted to living in dry conditions.

xylem vessels dead, hollow, elongated tubes, with lignified side walls and no end walls, that transport water in most plants.

Index

Acknowledgements

The authors wish to thank Graham Read for his invaluable help with the manuscript, Mitch Fitton for her meticulous editing, James Penny for the mathematical aspects, Louise Garcia for her work on the practice questions, Ellena Bale and Simon Ditchfield with their help on the practical elements and, of course, Alison Schrecker and Amy Johnson from OUP for their support, hard work and encouragement.

cover: Blend images Photography/Veer.com; **p30:** Darin Burks/Shutterstock; **p09:** Martin Shields/Alamy; **p25:** J.C. Revy, ISM/Science Photo Library; **p62:** Biophoto Associates/Science Photo Library; **p95:** Steve Gschmeissner/Science Photo Library; **p97(L):** Deepblue-photographer/Shutterstock; **p97(R):** Andy Crawford/Getty Images; **p117:** Oksana Kuzmina/Shutterstock; p118: Saturn Stills/Science Photo Library; **p155(L):** Manfred Kage/Science Photo Library; **p161:** Francis Leroy, Biocosmos/Science Photo Library; **p168:** Claudia Paulussen/Shutterstock; pg 184: Science Photo Library; **p202:** J.C. Revy, ISM/Science Photo Library; **p221:** Bon Appetit/Alamy; **p222(TL):** WILDLIFE GmbH/Alamy; pg 222(BL): Blickwinkel/Alamy; **p222(R):** Tim Gainey/Alamy; **p238(T):** Claude Nuridsany & Marie Perennou/Science Photo Library; **p238(M):** Chris2766/Shutterstock; **p238(B):** Michal Ninger/Shutterstock; **p246(C):** AC Rider/Shutterstock; **p247:** Oticki/Shutterstock; **p81:** Image Point Fr/Shutterstock; **p141:** Pawel Kazmierczak/Shutterstock; **p154:** Istetiana/Shutterstock; **p241:** Nancy Tripp Photography/Shutterstock; **p230:** Michael W. Tweedie/Science Photo Library; **p02-03** Background: Vitstudio/Shutterstock; **p53** Background: Vitstudio/Shutterstock; **p56-57** Background: Fusebulb/Shutterstock; **p125** Background: Fusebulb/Shutterstock; **p128-129** Background: Triff/Shutterstock; **p111:** Alfred Pasieka/Science Photo Library; **p200-201** Background: Aquapix/Shutterstock; **p261** Background: Aquapix/Shutterstock; **p239:** Duncan Usher/Alamy; **p209:** Alfred Pasieka/Science Photo Library; **p234:** David Hosking/Alamy; **p13:** Power and Syred/Science Photo Library; **p48:** Hermann Eisenbeiss/Science Photo Library; **p49:** Rido/Shutterstock; **p60:** Chris Priest/Science Photo Library; **p61:** Mauro Fermariello/Science Photo Library; **p63(T):** David Mccarthy/Science Photo Library; **p63(B):** Susumu Nishinaga/Science Photo Library; **p66:** Dr Gopal Murti/Science Photo Library; **p68:** CNRI/Science Photo Library; **p69:** Dr.Jeremy Burgess/Science Photo Library; **p70(T):** Don Fawcett/Science Photo Library; **p70(B):** Science Photo Library; **p76:** Dr Gopal Murti/Science Photo Library; **p79(A):** Pr. G Gimenez-Martin/Science Photo Library; **p79(B):** Pr. G Gimenez-Martin/Science Photo Library; **p79(C):** Pr. G Gimenez-Martin/Science Photo Library; **p79(D):** Pr. G Gimenez-Martin/Science Photo Library; **p82:** DR GOPAL MURTI/SCIENCE PHOTO LIBRARY; p91(L): Prof. P. Motta/Dept. Of Anatomy/University "La Sapienza", Rome/Science Photo Library; **p91(R):** J.C. Revy, ISM/Science Photo Library; **p102:** Lowell Georgia/Science Photo Library; **p103:** K.R. Porter/Science Photo Library; **p104(T):** Eye of Science/Science Photo Library; **p104(B):** Science Photo Library; **p108:** Eye of Science/Science Photo Library; **p110:** Dr Gopal Murti/Science Photo Library; **p114:** Dr Jeremy Burgess/Science Photo Library; **p187:** Ed Reschke/Getty Images; **p113(B):** Dr Rob Stepney/Science Photo Library; **p113(T):** Author; **p115:** Volker Steger/Science Photo Library; **p120:** Dr. Hans Gelderblom, Visuals Unlimited / Science Photo Library; **p134:** Microfield Scientific Ltd/Science Photo Library; **p137:** Dr Jeremy Burgess/Science Photo Library; **p139(T):** Author; **p139(B):** Author; **p140:** Author; **p143(T):** CNRI/Science Photo Library; **p143(B):** Proff. Motta, Correr & Nottola/ University "La Sapienza", Rome/Science Photo Library; **p146:** Manfred Kage/Science Photo Library; **p155(R):** Eye of Science/Science Photo Library; **p165(L):** William J. Howes/FLPA/Alamy; **p165(R):** Author; **p166(T):** Ron Steiner/Alamy; **p166(B):** Natural Visions/Alamy; **p175:** Proff. P. Motta/G. Macchiarelli/University "La Sapienza", Rome/Science Photo Library; **p179:** CNRI/Science Photo Library; **p180(T):** Steve Gschmeissner/Science Photo Library; **p180(B):** Susumu Nishinaga/Science Photo Library; **p183:** Power and Syred/Science Photo Library; **p189:** J.C. Revy, Ism/Science Photo Library; **p206:** Biophoto Associates/Science Photo Library; **p229(A):** Author; **p229(B):** Author; **p229(C):** Author; **p229(D):** Author; **p231:** David Levenson/Alamy; **p232:** WRPublishing/Alamy; **p242:** Mark J. Barrett/Alamy; **p243(T):** Photodisc; **p243(B):** SIMON FRASER/SCIENCE PHOTO LIBRARY; **p245:** Art Kowalsky/Alamy; **p246(A):** Author; **p246(B):** Author; **p248(L):** Worldwide Picture Library/Alamy; **p248(M):** Leslie J Borg/Science Photo Library; **p248(R):** Photodisc; **p249:** James King-Holmes/Science Photo Library; **p253(T):** George Ranalli/Science Photo Library; **p253(C):** Tony Wood/Science Photo Library; **p253(B):** Bob Gibbons/Science Photo Library; **p207:** CNRI/Science Photo Library; **p197:** Levente Gyori/Shutterstock.

p266-267: Maks Narodenko/Shutterstock; **p269:** Dr Jeremy Burgess/Science Photo Library; **p270:** Biophoto Associates/Science Photo Library; **p273:** Dr Kenneth R Miller/Science Photo Library; **p278:** Martyn F Chillmaid/Science Photo Library; **p289:** ISM/Science Photo Library; **p293:** Eye of Science/Science Photo Library; **p294:** David Levenson/Alamy; **p299:** Suzanne L & Joseph T Collins/Science Photo Library; **p302:** Stuart Wilson/Science Photo Library; **p305:** Martin Dohrn/Science Photo Library; **p308:** Hugh Spencer/Science Photo Library; **p309:** Chris Gomersall/Alamy; **p311:** Nigel Cattlin/Alamy; **p313(T):** Vladimir Sazonov/Shutterstock; **p313(B):** Djgis/Shutterstock; **p314:** Geogphotos/Alamy; **p319:** Maks Narodenko/Shutterstock; **p324-325:** Martynowi.Cz/Shutterstock; **p326:** Stefan Sollfors/